D0820387

Modification of Polymers

Charles E. Carraher, Jr., EDITOR
Wright State University

Minoru Tsuda, EDITOR
Chiba University

Based on a symposium sponsored by
the Division of Organic Coatings
and Plastics at the
ACS/CSJ Chemical Congress,
Honolulu, Hawaii,
April 2–6, 1979.

ACS SYMPOSIUM SERIES **121**

AMERICAN CHEMICAL SOCIETY
WASHINGTON, D. C. 1980

Library of Congress CIP Data

Modification of polymers.

(ACS symposium series; 121 ISSN 0097-6156)

"Based on a symposum sponsored by the Division of Organic Coatings and Plastics [Chemistry] at the ACS/CSJ Chemical Congress, Honolulu, Hawaii, April 2–6, 1979."

Includes bibliographies and index.

1. Polymers and polymerization—Congresses.
I. Carraher, Charles E. II. Tsuda, Minoru, 1935– . III. American Chemical Society. Division of Organic Coatings and Plastics Chemistry. IV. ACS/CSJ Chemical Congress, Honolulu, 1979. V. Series: American Chemical Society. ACS symposium series; 121.

QD380.M6 547.8'4 79-28259
ISBN 0-8412-0540-X ASCMC8 121 1–500 1980

PRINTED IN THE UNITED STATES OF AMERICA

ACS Symposium Series

FOREWORD

The ACS Symposium Series was founded in 1974 to provide a medium for publishing symposia quickly in book form. The format of the Series parallels that of the continuing Advances in Chemistry Series except that in order to save time the papers are not typeset but are reproduced as they are submitted by the authors in camera-ready form. Papers are reviewed under the supervision of the Editors with the assistance of the Series Advisory Board and are selected to maintain the integrity of the symposia; however, verbatim reproductions of previously published papers are not accepted. Both reviews and reports of research are acceptable since symposia may embrace both types of presentation.

CONTENTS

RADIATION INTERACTIONS

NATURAL POLYMERS

PREFACE

Some intimate that macromolecular chemistry has become a mature science and that it is no longer at the frontier of scientific endeavor. In actuality, if we use the analogy of a human body, macromolecular science has only developed its skeleton, composed largely of homopolymers such as polyethylene, polyesters, and polyamides. The body is just beginning to develop.

The investigation of macromolecules has just begun to unfold its potential in our lives. Polymer modification is a major frontier introducing needed subtle or gross changes that allow biocompatability, enhanced thermal stability, increased solvent stability, etc. to the modified polymer.

Polymer modification is a quite broad and rapidly expanding area of science. The enclosed chapters are meant only to present glimpses of many of the most important areas. The contributions were selected from over 100 possible papers. The contributors include eminent scientists from many countries giving the book the necessary international flavor.

The book, divided into four sections, begins with a brief chapter describing some present problems in need of research and future trends related to polymer modification. The volume is not exhaustive but chapters were selected to illustrate specific aspects of more general areas of polymer modification.

The first section, Chemical Reactions on Polymers, deals with aspects of chemical reactions occurring on polymers—aspects relating to polymer size, shape, and composition are described in detail. One of the timely fields of applications comprises the use of modified polymers as catalysts (such as the immobilization of centers for homogeneous catalysis). This topic is considered in detail in Chapters 2, 3, 8, 9, and 11 and dealt with to a lesser extent in other chapters. The use of models and neighboring group effect(s) is described in detail. The modification of polymers for chemical and physical change is also described in detail in Chapters 2 (polystyrene); 4 (polyvinyl chloride); 5 (polyacrylic acid, polyvinyl alcohol, polyethyleneimine, and polyacrylamide); 6 (polyimides); 7 (polyvinyl alcohol); 8 (polystyrene sulfonate and polyvinylphosphonate); 10 (polyacrylamide); and 12 (organotin carboxylates).

The second section, Radiation Interactions, contains nine chapters ranging from preliminary aspects of radiation-induced polymer modifications to industrial applications, and includes topics related to IR, UV, plasma, visible light, and corona radiations. Various aspects related to mechanisms of radiation interactions with the polymers are discussed in Chapters 13-16, 18, 20, and 21. When a polymer is to be used as a film, plate, fiber, or molded material, the surface properties are often as important as the bulk properties. Aspects of surface modification are described in Chapter 15. Deep UV lithography, expected to be a near-future technique for the production of microelectronic devices, is discussed in Chapter 18 as related to poly(methylmethacrylate).

A detailed description of the Rigilon plate, a photopolymer relief printing plate for making matrix master plates, is given in Chapter 17. Rigilon plates are now used extensively in the printing (mainly newspapers) industry in Japan and Europe. They have features close to those of metal plates but also possess a number of advantages; they are solid, water-"washoutable" plates having good matrix and heat resistance, and good reproducibility, and require a short plate-making time.

Recently the biodegradability of polymers has become very important to the chemical industry and society in general. Chapter 19 describes studies on the biodegradability of polyamides by a number of bacteria.

The use or modeling related to natural polymers as building materials and chemical reagents is described in the third section on Natural Polymers. Chapter 22 describes modeling systems that mimic the high stereoselectivity of certain biological systems for the development of stereoselective catalysts. Modeling for conformational effects and shapes of nucleic acids is described in Chapter 23. The intimate, three dimensional modification of cellulose derived from cotton, chitin, amylose, dextran, and amylopectin, leading to building materials with good biological resistance, is described in Chapters 24 and 25. Chapter 26 describes the surface modification of cellulose and determination of surface density of hydroxyl groups.

Formation of industrially usable interpenetrating polymer networks derived from caster oil is described in Chapter 27. Products can vary from soft and flexible to hard and tough.

The final section contains chapters related to the modification of polymer properties with changes in polymer structure, such as the use of different phosphorus-containing polyesters (as copolymers and polymer blends) as flame retardants (Chapter 28). Chapter 29 deals with the electronic cooperativity of modified polymers derived from poly(vinylbenzyl chloride). Property–structural aspects of carborane–silonane polymers are described in Chapter 30, while new synthetic techniques leading to siloxane-modified poly(arylene carbonates) are covered in Chapter 31. The final chapter reports on the compatability of clay with polyolefins, with the clay acting as a property-enhancing reinforcing agent as well as an extender.

We thank the authors for their valuable contributions and the Division of Organic Coatings and Plastics for its support of the symposium. Scientific reviewers' cooperations are also acknowledged.

Finally we thank Setsuko Oikawa and Joan Comstock for their cooperation in the organization of the symposium and the editorial work on this book.

Wright State University
Dayton, Ohio 45435

CHARLES E. CARRAHER, JR.

Chiba University
Chiba 280, Japan

MINORU TSUDA

January 2, 1980

1

Introduction: Polymer Modification—Some Problems and Possibilities—Areas in Need of Research

CHARLES E. CARRAHER, JR.
Department of Chemistry, Wright State University, Dayton, OH 45435

MINORU TSUDA
Laboratory of Physical Chemistry, Chiba University, 1-33, Yayoi, Chiba (280) Japan

The modification of polymers has been practiced since the dawn of mankind with the working of animal hides and natural fibers. In spite of this ancient beginning significant advances are occurring almost daily. The "black art" of polymer modification is increasingly yielding to scientific investigation and as new insights become available, new applications are found for this information - the advances in knowledge and applications coupled.

The modification of polymers is interdisciplinary in nature cutting across traditional boundaries of chemistry, biochemistry, medicine, physics, biology and materials science and engineering. Because of this interdisciplinary nature, persons involved with polymer modification should be broadly trained to permit the best application of revealed information.

Polymer modifications are intended to impute different, typically desired properties to the new modified material-properties such as enhanced thermal stability; multiphase physical responses; biological resistance, compatibility or degradability; impact response; flexibility; rigidity; etc.

Today modifications can be roughly grouped into two categories - a. physical modifications including entanglement and entrapment and radiation induced changes and b. chemical modifications where chemical reactions on the polymer are emphasized. This distinction is often unclear at best.

Following is a brief summary of only some of the areas in need of study in the broad area of polymer modification.

As signaled within this book, modification through exposure to radiation, (thermal, light and particle) continues to be at the forefront of many areas of polymer modification. A major problem involves use of industrial radiation curing of

0-8412-0540-X/80/47-121-001$05.00/0

coatings surfaces because of the present practical limitation of depth of cure penetration. This problem was cited in the 1978 Workshop on Organic Coatings held at Kent State University.

The problem is common to the application of all industrial coatings. Potential solutions are numerous including a. repeatable coatings application (negative features include time, adhesion of the separate coats, and increased energy requirements and equipment housing and complexity); b. increased energy of radiation (currently largely ruled out due to energy, safety and cost considerations); c. formulation of polymer mixtures that can be "set" with radiation, but which continue to cure on standing by a slower mechanism; and d. addition of species which can transfer "captured" radiation to greater depths. Regarding the latter, polydyes have been synthesized using Group IVB Cp_2MCl_2 compounds condensed with dyes such as xanthene and sulfonphthalein dyes. The Group IVB Cp_2MCl_2 compounds are known "ultraviolet sinks". By proper coupling of the metal, dye and radiation it is possible impact material impregnated with a polydye with the polydye accepting and reemitting the radiation permitting greater depth of penetration by the effects of radiation.

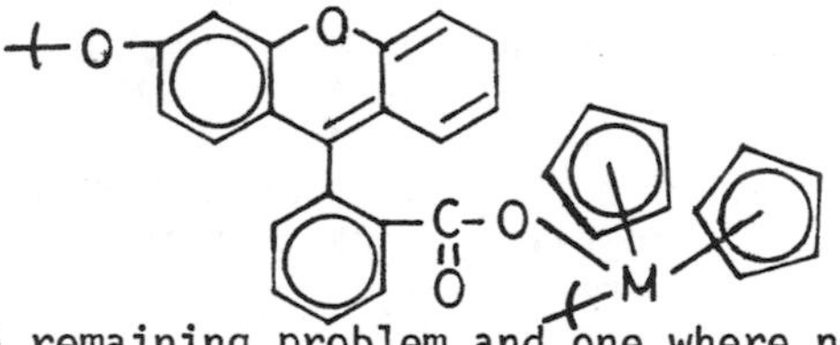

A remaining problem and one where no real widespread solution has even been (experimentally) proposed is the adequate description of molecular weight of crosslinked materials and the innerrelationship(s) of amount and type of crosslinking, polymer molecular weight and physical and chemical characteristics. Related to this is the need to better control extent and location (i.e. random, homogeneous, etc.) of modifications on polymers. Some of the good NMR work concerning identification of sequence with copolymers can be utilized in the description of many graph and block copolymers. Mass spectrophotometry utilizing laser excitation of modified polymers may enable a better description of the actual framework of many crosslinked modified materials since laser excitation allows the examination of both small and large (to greater than 1000 amu) fragments.

The construction of a powerful, continuously variable wavelength laser is approaching reality. Such a laser could be of great use in tailoring polymer modifications through activation of only selected sites for reaction. A number of groups are currently conducting selected reactions utilizing laser energy so the needed technology is becoming available.

While much of the current and near past research has emphasized modification of synthetic polymers, increasing efforts will undoubtedly focus on the modification of regenerable polymers and the blending of natural polymers and natural polymers with synthetic polymers through block, graft, etc. approaches.

The need for replacements of objects currently derived from nonregenerable materials (most plastics, rubbers, elastomers, metals) with objects derived from regenerable materials is critical and must be continually emphasized in our research efforts. It is the editor's opinion that this is one of the future areas of science which offers the greatest lasting benefits to society.

This book presents several chapters relating recent advances in the modification of regenerable materials and many other recent books and symposia have portions devoted to this topic. Some of the work related to total modifications may be in time extended to the modification of currently wasted materials such as leaves, sea weed, flower and weed stocks, corn stocks, grass, etc. all of which typically contain high degrees of cellulosic material which when suitably solubilized should rapidly permit suitable modifications to ensue. As an interesting side note, paper mill workers have been able to increase the amount of usable "cellulosic" material through electron bombardment or other suitable radiation treatment of the "raw" ground wood presumably through initiation of crosslinking reactions between the cellulosic portions and other materials such as the lignin. It should be possible to routinely graft onto raw ground wood giving materials which can directly be pressed to give a product superior in thermal stability, hardness, etc. to "simple" pressed board, with the grafted portions containing units to enhance color, flame retardance, adhesion, etc.

The advent of computer chips and laser signal and controlling devices will permit more complex modifications to be carried out on an industrial scale.

Another area in need of work is the on-site grafting, attachment of polymeric materials on biological sites such as particularly badly broken bones where the leg is surgically opened and a polymeric material chemically attached after suitable bone activation with the polymeric material degrading after its use period is up. This area is mentioned only to reinforce the notion that interdisciplinary team efforts and polymer chemists with broad training are needed to make the best use of applications of polymer modifications.

The area of delivery of biologically active materials also will involve in great part polymer modifications. For instance, Gebelein describes the ideal polymer - for good drug delivery as being composed of three parts - one to give the overall polymer the desired solubility, the second part containing the drug to be delivered and the third part containing chemical units which will direct the overall material only to the site where the drug is to be delivered. It may be possible to combine several of these aspects by a judicious choice of polymeric units but presently more fruitful approaches include grafting of desired components together forming the needed overall polymeric properties. As a side comment, relatively little work has been done with the generation of "directing groups" and this is an area where much work is needed if the advantages of polymeric drugs are to be recognized.

In summary, much has been done and much remains to be done in the area of polymer modification. Significant problems await solution.

RECEIVED July 12, 1979.

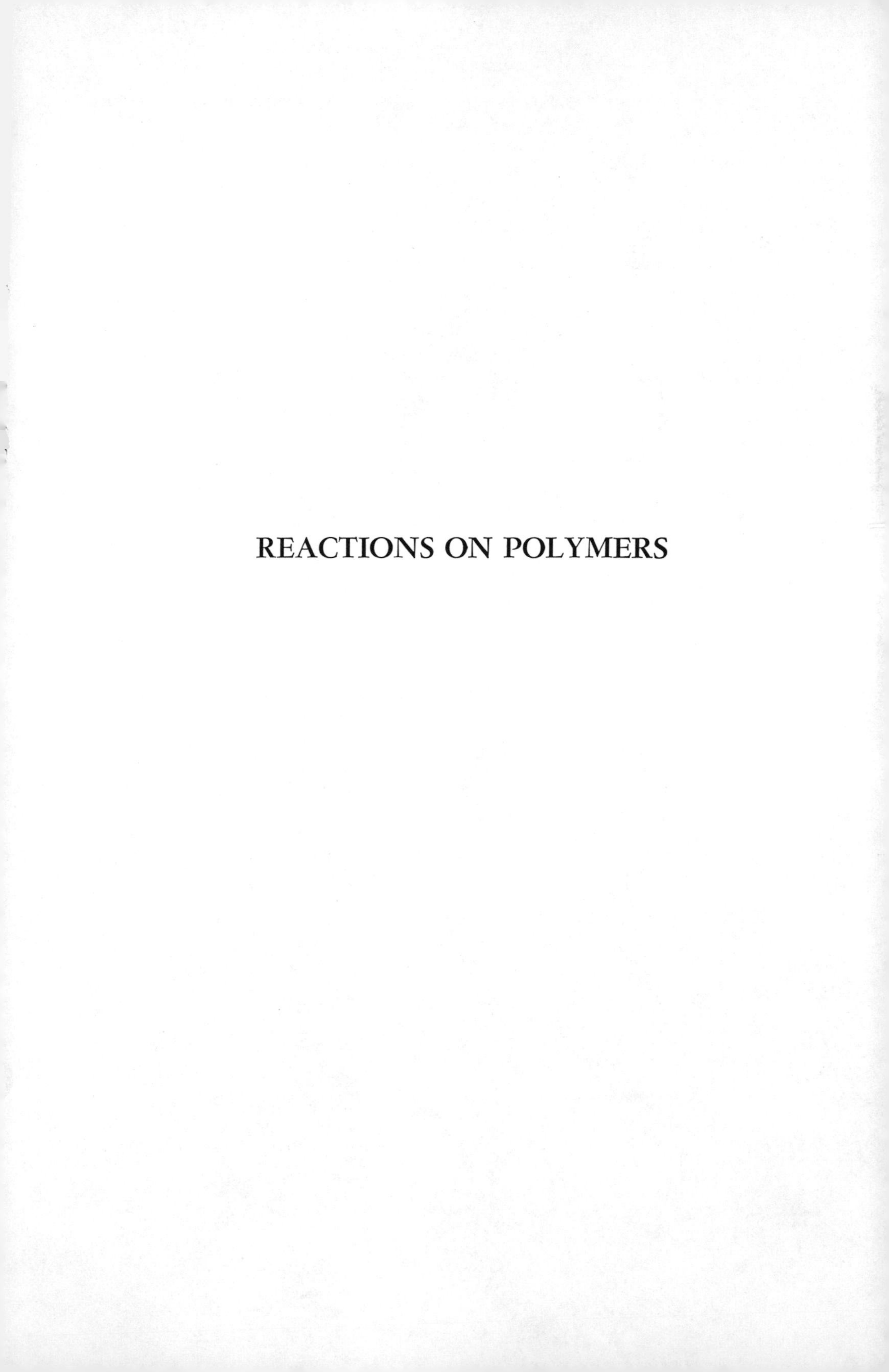

REACTIONS ON POLYMERS

2

Aminated Polystyrene–Copper Complexes as Oxidation Catalysts: The Effect of the Degree of Substitution on Catalytic Activity

G. CHALLA, A. J. SCHOUTEN, G. TEN BRINKE, and H. C. MEINDERS

State University of Groningen, Laboratory of Polymer Chemistry, Nijenborgh 16, 9747 AG Groningen, Netherlands

Modification of polymers is a topic in polymer science, because new highly valued or improved applications often require sophisticated chemical structures along the polymer chains. One of such timely domains of interest comprises the development of modified polymers as catalysts for chemical processes. Of course, we do not have in mind catalysts, wherein polymers function as inert supports for the active centers and nomore. In fact, our aim is to develop polymeric catalysts, which combine advantages of the other type of catalysts, viz.

(i) the specificity of homogeneous catalysts.
(ii) the separability and high stability of heterogeneous catalysts.
(iii) the high activity and selectivity of enzymes.

In other words, we try to mimic enzymes by attaching centers for homogeneous catalysis to polymer chains; we want to learn from nature how to conduct chemical processes in a cleaner, more selective and milder way. In this respect it is of great importance that we can adapt, just like in enzymes, the micro-environment of the catalytic centers by modification of neighbouring polymer chain segments.

From the above it will be clear that the polymer chain carrying catalytic centers has to play an active role during each catalytic cycle. Therefore, we prefer to speak of macromolecular catalysis rather than of polymer catalysis, the more so, as we omitted crosslinked carriers from our studies in order to prevent that diffusion of reactants and products would become rate-determining. Consequently, the practical combination of simple separability and really macromolecular catalysis should be realized by

0-8412-0540-X/80/47-121-007$05.00/0

attaching whole catalytically active macromolecules to inert nonporous or macroporous supports. The basic research for such developments will still imply the study of loose, modified macromolecules as microphases containing catalytic centers and surrounded by solvent without such centers. The concept of an isolated reactive macromolecule was introduced by Morawetz (1) and further developped by him and others, e.g. Overberger (2) and Kunitake (3) for polymeric catalysis of esterhydrolysis, Ise (4) for catalysis of ionic reactions by polyelectrolytes and Tsuchida (5) for catalysis by coordination complexes of transition metals with polymeric ligands. In addition, Kabanov (6) tried to optimize polymer catalysis in a practical way by applying block or graft-polymers which partly associate yielding gel-like structures with catalytic domains which remain quite accessible.

We were interested in the behaviour of polymeric catalysts in order to confirm that typical polymer effects may occur. Oxidative coupling of 2,6-disubstituted phenols, as developped by Hay (7), was chosen as a model reaction and the catalytic activities of coordination complexes of copper with several polymeric tertiary amines were compared with the activities of their low molecular weight analogs. The overall reaction scheme is presented in scheme 1.

Scheme 1

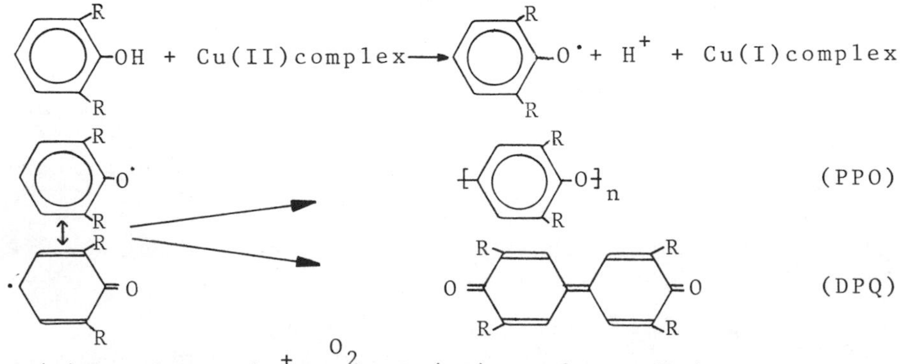

R, R — OH + Cu(II)complex ⟶ R, R — O• + H^+ + Cu(I)complex

R, R — O• ↕ R, R, •, =O ⟶ [R, R — O]$_n$ (PPO)

⟶ O= R, R, R, R =O (DPQ)

$$\text{Cu(I)complex} + 2H^+ \xrightarrow{O_2} \text{Cu(II)complex} + H_2O$$

A similar oxidation by electron transfer from phenolate anion to Cu(II) is also an important step in the phenol oxidation by copper-containing enzymes like laccase and tyrosinase (8,9,10). Many important products like lignins, tannins, pigments, antibiotics and alkaloids are produced through this step. Instead of the biopolymeric ligands in the enzymes we introduced synthetic polydentates for copper complexation like those listed in scheme 2.

Scheme 2

	Polymeric ligands	Low molecular weight analogs	
(I)	$[CH_2-CH(C_6H_5)]_{1-\alpha}$ co $[CH_2-CH(C_6H_4CH_2NMe_2)]_{\alpha}$	$C_6H_5-CH_2-NMe_2$	(DMBA)
(II)	$[CH_2-CH(C_6H_5)]_{1-\alpha}$ co $[CH_2-CH(C_5H_4N)]_{\alpha}$	C_5H_5N	(Pyr.)
(III)	$[CH_2-CH(C_6H_5)]_{1-\alpha}$ co $[CH_2-CH(N\text{-imidazolyl})]_{\alpha}$	N-Me imidazole	(NMIm)

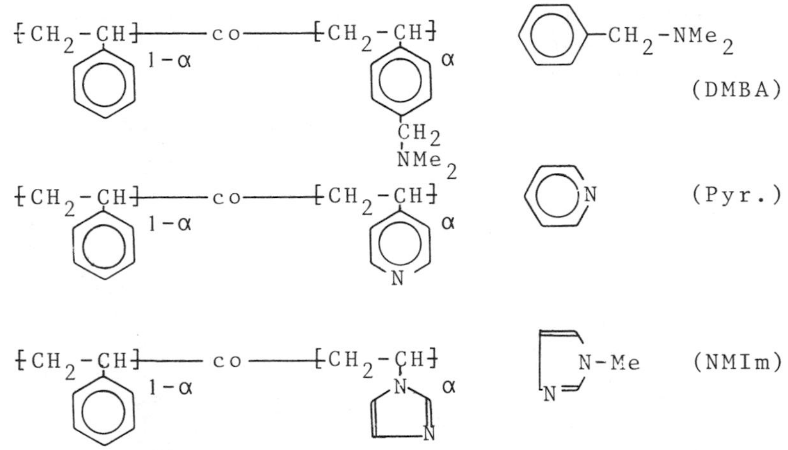

The dimethylaminomethylated polystyrene (I) was prepared by chloromethylation of atactic polystyrene according to Galeazzi (11) using methylal and thionylchloride instead of an excess of the dangerous chlorodimethylether:

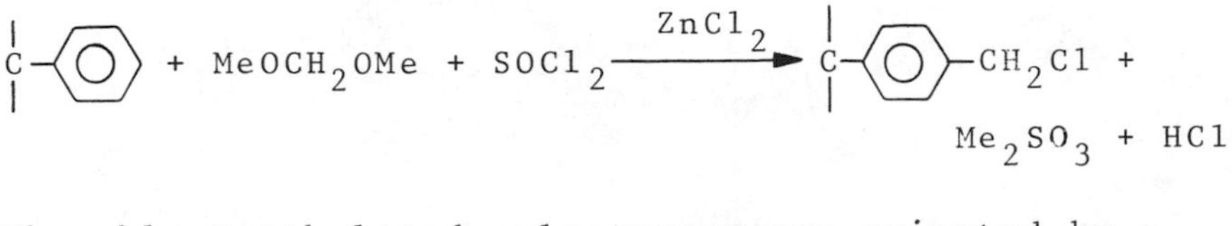

$$-C-C_6H_5 + MeOCH_2OMe + SOCl_2 \xrightarrow{ZnCl_2} -C-C_6H_4-CH_2Cl + Me_2SO_3 + HCl$$

The chloromethylated polystyrene was aminated by a large excess of dimethylamine during 1 week at 20°C in dioxane:

$$\text{-C-}\,C_6H_4CH_2Cl + 2\,HNMe_2 \longrightarrow \text{-C-}\,C_6H_4CH_2NMe_2 + Me_2NH_2^+Cl^-$$

Copolymers of styrene with 4-vinylpyridine (II) and N-vinylimizadole (III) were obtained by copolymerization for one day at 60°C in 25 wt% comonomer solutions in toluene, using AIBN as initiator. In all cases the degrees of substitution, α, of the functionalized polymers with ligand groups were derived from the nitrogen contents found by elemental analyses.

In the following sections we shall discuss: (i) the structure and behaviour of the various copper complexes with the ligands listed in scheme 2; (ii) the activities of the polymeric catalysts in comparison with the low molecular weight analogs; (iii) the effect of the degree of substitution, α, on the activities of the polymeric catalysts.

Structure and Behaviour of the Copper Complexes

The basic study was performed on copper complexes with N,N,N',N'-tetramethylethane-1,2-diamine (TMED), which were known to be very effective oxidative coupling catalysts (7,12). From our first kinetic studies it appeared that binuclear copper complexes are the active species as in some copper-containing enzymes. By applying the very strongly chelating TMED we were able to isolate crystals of the catalyst and to determine its structure by X-ray diffraction (13). Figure 1 shows this structure for the TMED complex of basic copper chloride Cu(OH)Cl prepared from CuCl by oxidation in moist pyridine.

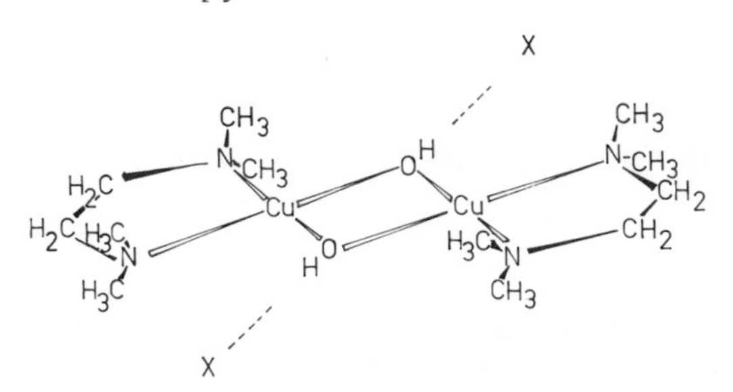

Figure 1. The structure of [TMED · Cu(OH)₂Cu · TMED]²⁺ · 2Cl⁻ as determined by x-ray diffraction

The same complex could be obtained starting from $CuCl_2$ and subsequent substitution of both bridged chlorides by adding hydroxyl ions. Scheme 3 describes the formation and interconversion of both binuclear copper complexes.

Scheme 3

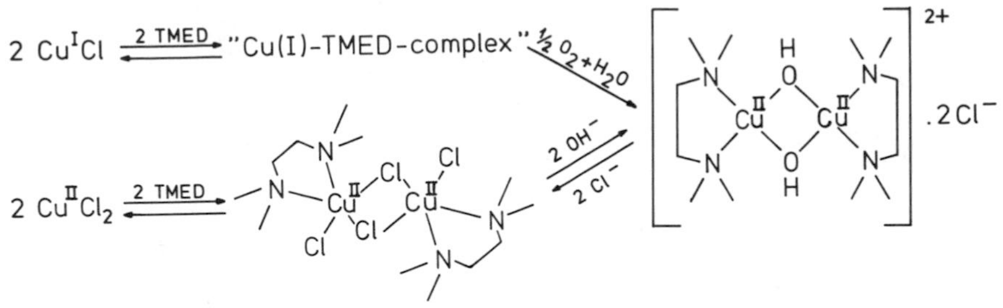

$2\,Cu^{I}Cl \rightleftharpoons$ (2 TMED) "Cu(I)-TMED-complex" $\xrightarrow{\frac{1}{2}O_2 + H_2O}$

$2\,Cu^{II}Cl_2 \rightleftharpoons$ (2 TMED)

2 OH^- / 2 Cl^-

$2+$. $2Cl^-$

The bridged binuclear structure could be corroborated by several techniques: (i) infrared spectroscopy gave absorption bands for bridged OH and Cu-O vibrations; (ii) elemental analyses give the calculated contents for hydroxo-bridged complex; (iii) ESR measurements did not produce signals of mononuclear Cu(II); (iv) magnetic susceptibility increased with temperature, an antiferromagnetic behaviour; (v) during oxidative coupling O_2 is reduced to H_2O like for binuclear copper enzymes, whereas H_2O_2 is usually produced by mononuclear copper complexes.

Titration of $CuCl_2$ with ligand produced an increased near-i.r. absorption at 880 nm as shown in figure 2. It is clear that the polymeric ligand (I) is more effective than its low molecular weight analog DMBA. It gives the maximum absorbance exactly at the theoretical ratio N/Cu = 2 (14), whereas a large excess of DMBA is needed to achieve coordination of each Cu(II) with two ligands. This is a good demonstration of the so-called polychelate effect within the seperate macromolecular coils. When the titration with polymeric ligand was stopped half-way, e.g. at N/Cu = 1, ESR signals revealed that part of the $CuCl_2$ was still unchanged and the other part formed directly the ESR inactive binuclear complexes with N/Cu = 2. In fact, this situation appeared to yield the highest reaction rate for the chloro-bridged catalyst because free $CuCl_2$ could liberate protons which are needed for the

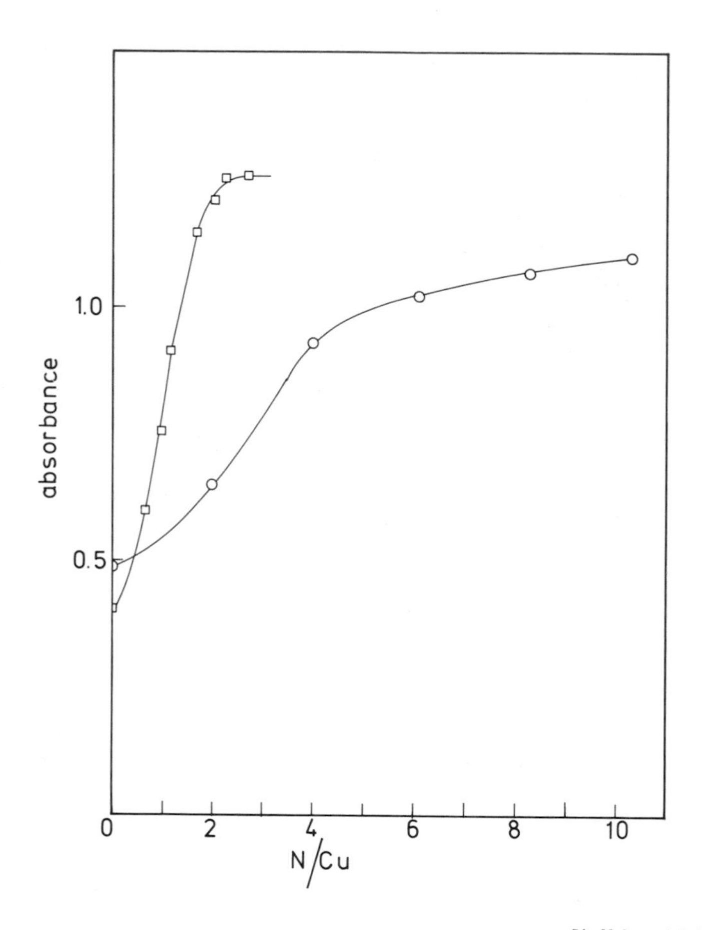

Die Makromolekulare Chemie

Figure 2. Titration of a copper(II)chloride solution with DMBA (○) and polymer ligand (I) (□). $[CuCl_2]_o$ = 4.46mM; solvent: 1,2-dichlorobenzene/methanol (13:2, v/v); room temperature. The curves are not corrected for dilution (14).

reoxidation of Cu(I) as shown in scheme 1 (14).

When we titrated from the other side by adding $CuCl_2$ to a dilute solution of polymeric ligand (I), another phenomenon could be detected, viz. a decrease in reduced viscosity of the polymer solution (14). This points to contraction of the separate coils of the polymeric ligand due to intramolecular crosslinking via binuclear copper complexes. We prevented gel formation due to intermolecular crosslinking via complex formation by applying only low polymer concentrations.

Finally, we report the effect of the bridged ligand on the specificity of the catalysts. It could be shown that the chloro-bridged catalyst generally promotes C-C coupling to DPQ, whereas the hydroxo-bridged catalyst is somewhat specific for C-O coupling to polymer PPO (see scheme 1). This tendency is clearly demonstrated for TMED complex in Figure 3, wherein both the fraction DPQ formation and the catalytic activity are plotted against the ratio NaOH/Cu; for NaOH/Cu = 1 all chloro-bridges are substituted by hydroxo-bridges.

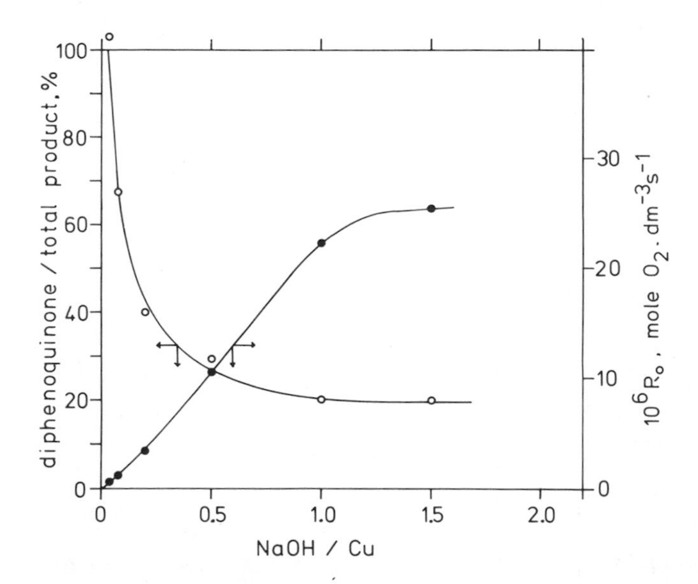

Figure 3. Effect of mineral base (NaOH) on the catalytic activity and specificity of the copper(II)–TMED complex. $[CuCl_2]_o = 3.33mM$; $[DMP]_o = 0.06M$; temp 25°C; solvent: 1,2-dichlorobenzene/methanol (9:1, v/v). The fraction DPQ was determined spectroscopically at 420 nm after 35% conversion.

In case of the complexes with polymeric ligands II and III C-O coupling could be further promoted by changing the solvent and increasing the ratio ligand/copper (13,15). Both factors seem to force the substrates to enter the catalytic complex as phenolate anions by substitution of OH^- at the bridge positions and formation of water. This mechanism leading to polymer formation is quite different from that for C-C coupling which probably involves substrate coordination to Cu at the free z-position, since the phenols cannot substitute the strongly coordinating chloride bridges. This situation is met for complexes with the polymeric ligand (I) and its analog DMBA, because they do not form stable hydroxo-bridged complexes. Both mechanisms are presented in Figure 4.

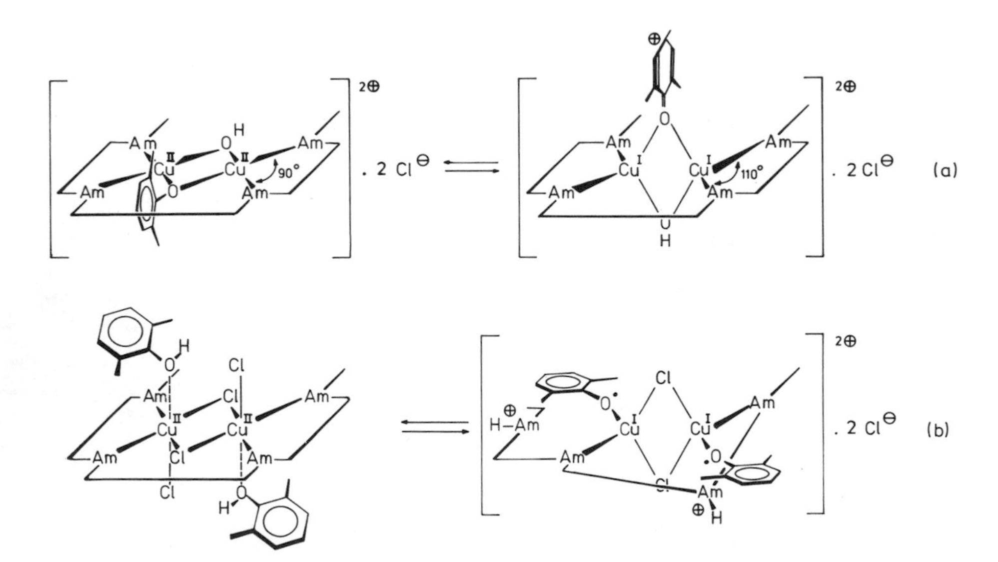

Figure 4. Schematic of electron transfer processes for 2,6-disubstituted phenol. The ligand groups are indicated as Am and the intermediate polymer chain segments as straight lines. (a) Hydroxo-bridged catalyst (b) chloro-bridged catalyst.

Activities of Polymeric Catalysts and Analogs

We always applied the polymeric ligands in concentrations below those for homogeneous segmental distribution. In other words we dealt with separate polymer coils containing the active centers, which were described as "cooperative microphases" by Williams (16). The crowding of the ligand groups in the interior of a polymer coil leads to an enhanced local concentration and to stronger steric interaction. The latter might stabilize binuclear complexes, which are the real catalysts. The effect of the enhanced local concentration was already indicated in the previous section when dealing with the higher coordinating efficiency of the polymeric ligand (I) as compared to an equivalent amount of the analog DMBA. So, both effects maintain an enlarged local concentration of active catalytic centers and cause the rate of oxidative coupling with polymeric catalysts to be higher than with equivalent amounts of low molecular weight analogs, especially for low ligand/copper ratios. This rate enhancement is clearly demonstrated in Figure 5 for polydentates (I) vs. DMBA (17), and was also found for polydentate (II) vs. pyridine (18).

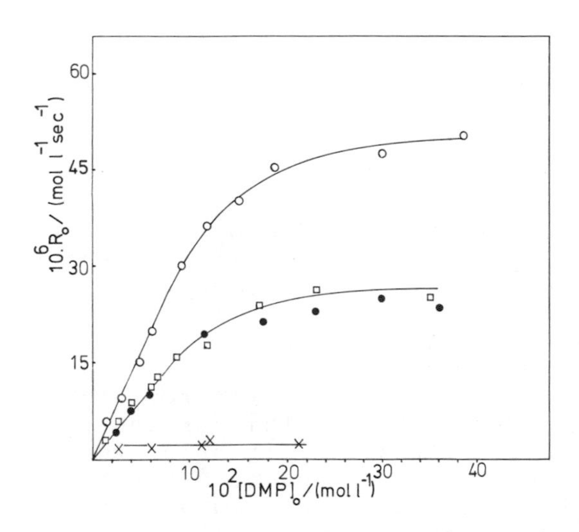

Die Makromolekulare Chemie

Figure 5. Initial rate of oxygen consumption R_o *vs. initial DMP concentration for various ligands.* (×) *DMBA;* (□) *polymer ligand (I) with* $\alpha = 0.10$ *and* $[\eta] = 1.00$; (●) *polymer ligand (I) with* $\alpha = 0.10$ *and* $[\eta] = 0.17$; (○) *polymer ligand (I) with* $\alpha = 0.18$ *and* $[\eta] = 0.17$. $[CuCl_2]_o = 3.33$mM; *N/Cu = 1; temp 20°C; solvent: 1,2-dichlorobenzene/methanol (13:2, v/v)* (17).

An estimation of the local ligand concentration, $[N]_{coil}$, could be achieved by assuming free movement of the ligands in the interior of a sphere with radius $<s^2>^{\frac{1}{2}}$, the root mean square radius of gyration of the polymer chain:

$$[N]_{coil'} = \frac{1000P\alpha}{N_A} \; / \; \frac{4}{3}\pi <s^2>^{3/2} \qquad (1)$$

Here P denotes the degree of polymerization and α the degree of substitution.
This expression is related to eq. (2) used by Morawetz (1) for the effective local concentration of one chain end in the neighbourhood of the other, which is relevant for ring closure kinetics:

$$c_{eff.} = \frac{1000}{N_A} \; / \; (\frac{2}{3}\pi <h^2>)^{3/2} \qquad (2)$$

The mean square end-to-end distance $<h^2>$ of a freely jointed chain without excluded volume is known to be equal to $6<s^2>$. The radius of gyration can be derived from light scattering or from the intrinsic viscosity (19):

$$[\eta] = \Phi' \frac{\overline{<s^2>}^{3/2}}{\overline{M}_n} \qquad (3)$$

Φ' is a universal constant and $\overline{M}_n = m.\overline{P}$. Under the conditions applied in Figure 5 with a constant overall ligand concentration of 3.3 mM, we found, indeed:

$$[N]_{coil} \gg [DMBA] = 3.3 \text{ mM}$$

Substitution of eq.(3) in eq. (1) reveals that $[N]_{coil}$ should be proportional to $\alpha/[\eta]$ for the polymeric catalysts. However, the activities of two polymeric catalysts with the same value of α but a nearly 6-fold difference in [η], i.e. a 10-fold difference in $\overline{M}_v$, were practically equal to each other in Figure 5. This means that the local concentration concept does no longer sufficiently apply to comparison of activities of different polymeric catalysts (see last section).

In Fig.5 we also saw that the initial rates, R_0, of oxidative coupling showed a limiting value with increasing substrate concentration, which resembles the

the saturation effect occurring in enzyme kinetics. This prompted us to describe our kinetics for medium substrate concentrations also in terms of the Michaelis-Menten scheme as Tsuchida et al. (20) did before for oxidative coupling with the electron transfer in the polymeric Cu(II)-substrate complex as rate-determining step (see also scheme 1 and Fig. 4):

$$\text{Cu(II)} + \text{DMP} \underset{k_{-1}}{\overset{k_1}{\rightleftharpoons}} [\text{Cu(II)-DMP}] \xrightarrow{k_2} \text{Cu(I)} + \text{DPQ/PPO}$$

$$\frac{1}{R_o} = \frac{1}{V_s} + \frac{K_s}{V_s[\text{DMP}]_o} \quad (4)$$

$$V_s = k_2[\text{Cu(II)complex}]_o \quad (5)$$

$$K_s = (k_{-1} + k_2)/k_1 \quad (6)$$

wherein R_o = initial reaction rate, V_s = limiting rate for $[\text{DMP}]_o = \infty$ and K_s = Michaelis constant. Eq. (4) denotes the so-called Lineweaver-Burk plot of reciprocal rate vs. reciprocal substrate concentration. This kind of analysis was successfully applied to both 2,6-dimethylphenol (DMP) and 2,6-diphenylphenol (DPP) and to the polymeric ligands (I), (II) and (III) listed in scheme 2 (15,17,18). Good Lineweaver-Burk plots derived from Figure 5 are shown in Figure 6.

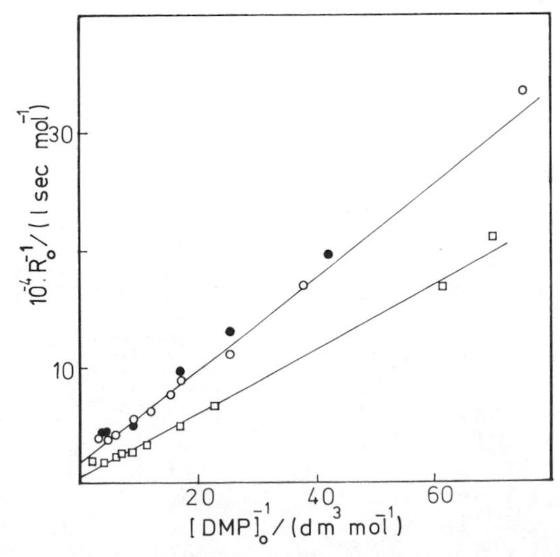

Die Makromolekulare Chemie

Figure 6. Lineweaver–Burk plots derived from Figure 5 for polymer ligands (I). (□) $\alpha = 0.18$, $[\eta] = 0.17$; (●) $\alpha = 0.10$, $[\eta] = 0.17$; (○) $\alpha = 0.10$, $[\eta] = 1.00$ (17).

From the intercepts and slopes V_s, k_2 and K_s could be calculated. For the analog DMBA the values of V_s and k_2 were directly taken from the constant maximum rate as shown in Figure 5. Most of our results are gathered in Table I (15,21,22), which demonstrates that the observed increase in rate with α is governed by an increase of the electron transfer rate constant, k_2, whereas the Michaelis constant, K_s, changed in the wrong way considering eq. (4).

Table I: Kinetic results on oxidative coupling of 2,6-disubstituted phenols at 25°C in the solvent mixture 1,2-dichlorobenzene/methanol (13:2 v/v).

ligand type	$10^2\alpha$	10^3k_2 s^{-1}	K_s^{-1} dm^3mol^{-1}	$\Delta H_2^{\ddagger}$ $kJmol^{-1}$	$\Delta S_2^{\ddagger}$ $JK^{-1}mol^{-1}$
substrate 2,6-dimethylphenol (DMP)					
DMBA	-	2.5	32.3	20	- 228
pol. I	6.5	3.1	11.4	11	- 259
pol. I	10.3	12.6	4.5	26	- 197
pol. I	18	39.5	1.7	36	- 151
pol. I	39	88.3	1.1	52	- 96
substrate 2,6-diphenylphenol (DPP)					
pol. I	5	2	4.2	27	- 209
pol. I	12	7	3.1	44	- 138
pol. I	18	19	1.5	73	- 33
pol. I	39	32	0.8	87	+ 21
substrate 2,6-dimethylphenol (DMP)					
pyridine	-	3.9	10.2	13	- 267
pol. II	3.6	6.0	8.4	29	- 198
pol. II	7.2	11.5	7.0	49	- 120
pol. II	12.7	15.4	5.6	111	+ 126

Effect of the Degree of Substitution on Catalytic Activity

The increase in rate of oxidative coupling when applying polymeric ligands with higher α is once more presented in Figure 7 for different substrates and polymeric ligands (15,21,22). So, while keeping all overall concentrations and conditions unaltered, the rate can be enhanced simply by concentrating the catalytic sites in a smaller number of polymeric microphases. Since this enhancement did not arise when the number of microphases was lowered by increasing

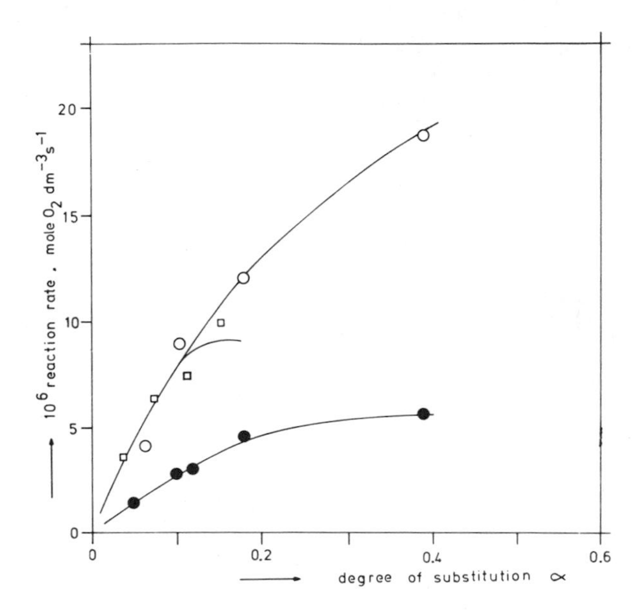

Figure 7. Initial rate of oxygen consumption R_o *vs. degree of substitution* α. *Reaction conditions: temp 25°C;* $[CuCl_2]_o = 3.3mM$; $N/Cu = 1$; *[substrate]*$_o$ = *0.06M.* (●) *substrate DPP, polymer ligand (I);* (□) *substrate DMP, polymer ligand (II);* (○) *substrate DMP, polymer ligand (I).*

the molecular weight of the polymer ligand with constant α, it must be concluded that a decreasing intermediate chain length between neighbouring ligand groups exerts an extra positive effect on catalytic activity.

In order to analyze this kinetic effect of α, we determined the activation parameters of k_2 from the temperature dependencies of V_{max} as derived from Lineweaver-Burk plots at different temperatures. In Figure 8 this procedure is shown for DPP as substrate and polymeric ligand (I) with α = 0.39. The finally resulting values of the activation enthalpy $\Delta H_2^\ddagger$ and activation entropy $\Delta S_2^\ddagger$ were already presented in Table I. It is peculiar to note that both $\Delta H_2^\ddagger$ and $\Delta S_2^\ddagger$ increase with α, i.e. shorter intermediate chain length between neighbouring ligand groups. This means that the increase of k_2 is caused by the relatively stronger increase of $\Delta S_2^\ddagger$ which compensates the retarding effect of increasing $\Delta H_2^\ddagger$.

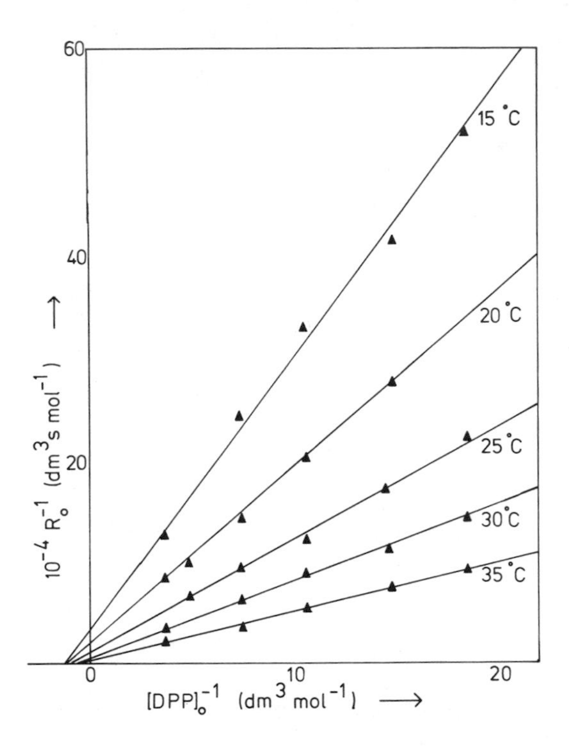

Figure 8. Lineweaver–Burk plots for oxidative coupling of DPP catalyzed by copper complexes of polymer ligand (I) with α =0.39 at 5 different temperatures. $[CuCl_2]_o$ = 3.3mM; N/Cu = 1; solvent: 1,2-dichlorobenzene/methanol (13:2, v/v).

Figure 9 demonstrates this compensation effect by the linear relationship between $\Delta S_2^{\ddagger}$ and $\Delta H_2^{\ddagger}$. This indicates that both activation parameters depend equally on α and that the isokinetic temperature, i.e. the slope of the line, amounts to 256°K. Thus, at -17°C the rate would become independent of α, whereas it increases with α at higher temperatures.

For a possible quantitative description of typical polymer effects we made the assumption that the values of $\Delta H_2^{\ddagger}$ and $\Delta S_2^{\ddagger}$ found for the low molecular weight catalysts stand for the activation process of the naked catalyst-substrate complex and are independent of α. So, after subtracting these values the separate polymer effects are found. Then we have to explain why more entropy is gained and more enthalpy is needed for adaptation of the intermediate chains to

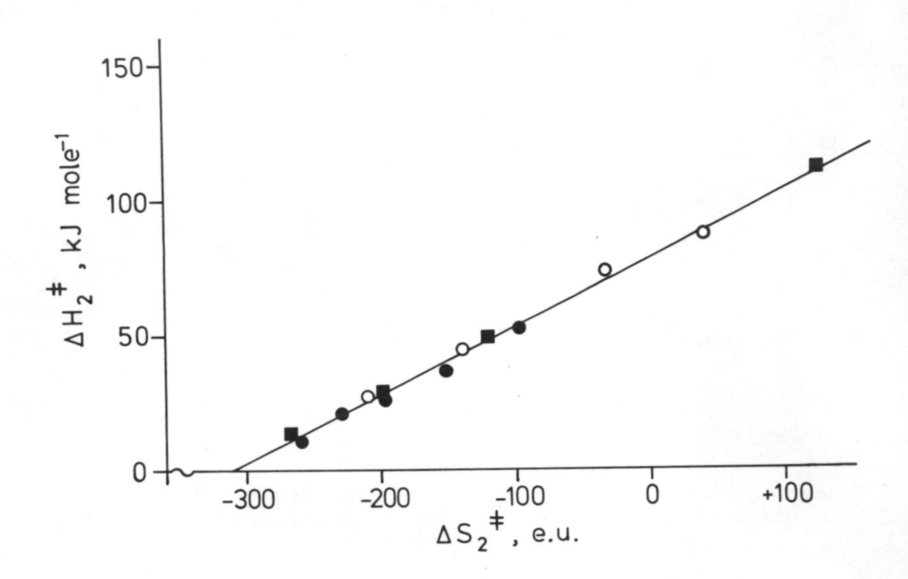

Figure 9. Compensation plot of activation parameters for the electron transfer rate constant k_2 *taken from Table I*

the transition state, when those chains become shorter. In principle, such trends were found in the same way by Sisido et al. for intramolecular hydrolysis between two chain ends of polysarcosine (23).

Interpretations of electron transfer reactions within normal transition metal complexes are based on the Franck-Condon principle, thus indicating that the metal-substrate complex has to be deformed before electron transfer takes place (24). This means that in our case the whole polymeric catalyst-substrate complex is deformed into a transition state which resembles more or less the final configuration of a tetrahedral Cu(I) complex. Building molecular models of octahedral and trigonal bipyramidal copper complexes we noticed that the tertiary NMe_2 groups of polymer ligand (I) are almost fixed in the former case, predominantly due to steric interaction of the Me groups. In a trigonal bipyramid, however, these amine groups can rotate almost freely (Figure 10). The consequences for the chains between adjacent aminated styrene units are drastic. It follows that prior to activation the end-to-end distances fall within a very small range, ∂, whereas in the activated state they

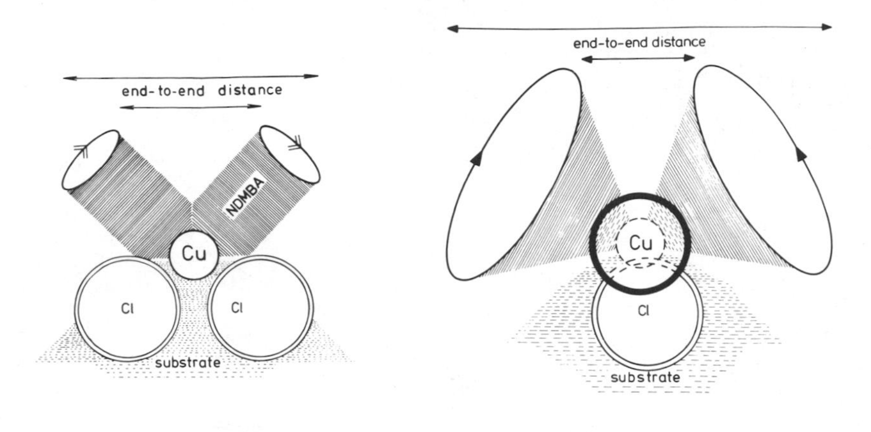

Figure 10. Schematic of the proposed catalyst–substrate complex, before and in the activated state. The increase of the chain end-to-end distance possibilities is represented by the bases of the cones.

can take any value in a much larger interval, Δ. Hence the number of conformations increases with a factor λ given by:

$$\lambda = \int_{\Delta} W(h)dh \; / \int_{\partial} W(h)dh \qquad (7)$$

where W(h) denotes the end-to-end distribution. This procedure is illustrated in Figure 11.

Clearly, λ is an increasing function of α, for not too small values of α. This is in conformity with the increase of the activation entropy $\Delta S_2^{\ddagger}$ with α. Generally, the conformational energy increases with decreasing end-to-end distance (25). As most of the additional conformations of the activated state have shorter end-to-end distance (see Figure 11), the contribution to the activation enthalpy $\Delta H_2^{\ddagger}$ is also positive. Moreover, it appears that this contribution increases with α. The stronger increase of $\Delta S_2^{\ddagger}$ and $\Delta H_2^{\ddagger}$ with α for DMP located at the bridge position of the catalytic complex of polymer ligand (II) (see Table I and Figure 4), is in line with the above views, since one should expect additional steric interaction in that case.

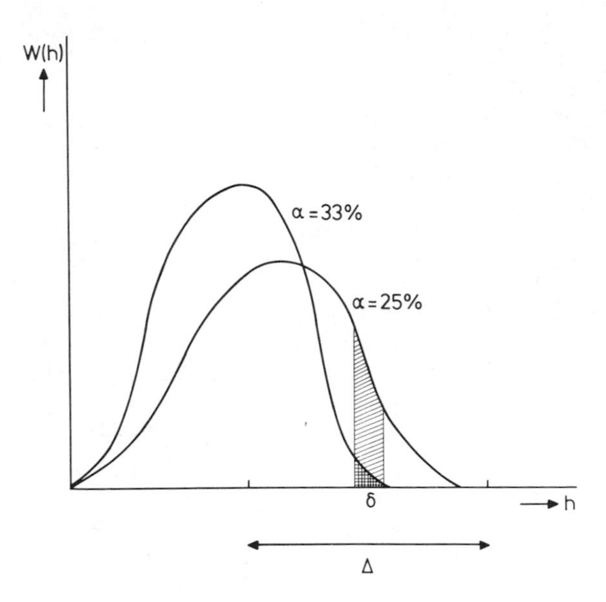

Figure 11. Illustration of Equation 7 for the calculation of the increase in number of intermediate chain conformations accompanying deformation and activation of the polymeric catalyst–substrate complexes

Anyhow, our study has demonstrated the benefit of "strained" polymeric catalyst-substrate complexes, a phenomenon well-known in enzymology (26) and once indicated by the term "entatic state" (16).

Literature Cited

1. Morawetz, H. "Macromolecules in Solution", John Wiley, New York, 1965, 1975, ch. IX.
2. Overberger, C.G. J. Polym. Sci., Polym. Symp. Ed., 1975, 50, 1.
3. Kunitake, T.; Okahata, Y. Adv. Polym. Sci., 1976, 20, 159.
4. Ise, N. in "Reactions on Polymers", Moore, J.A., Ed., Reidel, Dordrecht-Holland, 1973, p. 27.
5. Tsuchida, E.;Nishide, H. Adv.Polym.Sci., 1977, 24, 1.
6. Kabanov, V.A., Intern. Symp. Macromolecules, Dublin, 1977.
7. Hay, A.S. Polym. Eng. Sci., 1976, 16, 1.
8. Malkin, R.; Malmström, B.G. in "Advances in Enzymology", Nord, F.F., Ed., Interscience, New York, 1970, p. 177.

9. Brown, B.R. in "Oxidative Coupling of Phenols", Taylor, W.I.; Battersby, A.R., Eds., Marcel Dekker, New York, 1967, p. 167.
10. Ochiai, E.I. "Bioinorganic Chemistry", Allyn & Bacon, Boston, 1977, ch. 9.
11. Galeazzi, L., Ger. Pat. 2,455,946, June 1975.
12. Kevelam, H.J.; de Jong, K.P.; Meinders, H.C.; Challa, G. Makromol. Chem., 1975, 176, 1369.
13. Meinders, H.C.; van Bolhuis, F.; Challa, G. J. Mol. Catal., 1979, 5, 225.
14. Schouten, A.J.; Wiedijk, D.; Borkent, J.; Challa, G. Makromol. Chem., 1977, 178, 1341.
15. Meinders, H.C.; Challa, G. J. Mol. Catal., submitted.
16. Williams, R.J.P. Pure and Appl. Chem., 1974, 38, 249.
17. Schouten, A.J.; Prak, N.; Challa, G. Makromol. Chem., 1977, 178, 401.
18. Meinders, H.C., thesis, Groningen, 1979.
19. Flory, P.J. "Principles of Polymer Chemistry", Cornell University Press., Ithaca, 1953, p. 661, 616.
20. Tsuchida, E.; Kaneko, M; Nishide, H. Makromol. Chem., 1972, 151, 221.
21. Schouten, A.J.; Noordegraaf, D.; Jekel, A.P.; Challa, G. J. Mol. Catal., 1979, 5, 331.
22. Breemhaar, W.; Meinders, H.C.; Challa, G., to be published.
23. Sisido, M.; Mitamura, T; Imanishi, Y.; Higashimura, T. Macromolecules 1976, 9, 316.
24. Tsuchida, E.; Nishide, H.; Nishiyama, T. Makromol. Chem., 1974, 175, 3047.
25. Primilat, S.; Hermans, J. J. Chem. Phys. 1973, 59, 2602.
26. Jenks, W.P. "Catalysis in Chemistry and Enzymology", McGraw-Hill, New York, 1969, ch. 5.

RECEIVED July 12, 1979.

3

Kinetics of Intramolecular Cross-Linking and Conformational Properties of Cross-Linked Chains

N. A. PLATÉ, O. V. NOAH, I. I. ROMANTZOVA, and YU. A. TARAN
Moscow State University, Moscow, USSR

The reactions of intramolecular cross-linking is a rather poorly investigated area in the field of macromolecular reactions. However, the problems of regularities of such processes are related to such important problems of polymer chemistry as chemical modification of polymers, networks formation, sorption of low molecular reagents by polymers, intramolecular catalysis, conformational transitions and so on. In spite of the great importance of the study of regularities of cross-linking reactions, the experimental and theoretical analysis of such processes is complicated by many difficulties.

Complexities of the experimental study are due to the difficulty of the isolation of the intramolecular reaction properly said, as even for the reaction in very dilute solution the probability of cross-linkage formation between different molecules is not negligible. Another difficulty: the measuring of the characteristic parameters of the reaction /kinetics, hydrodynamic properties/.

The first problem of the theory of the intramolecular reactions is a calculation of dimensions of the intramolecularly cross-linked coils as a function of the degree of cross-linking. For the analytical calculation of such dependence one needs to know all possible topological structures for any number of cross-linkages and to have the calculation algorithm for each of

0-8412-0540-X/80/47-121-025$05.00/0

them. However the number of structures is rapidly increased with an increase of cross-linkages number /there are three possible structures for the chain with two cross-linkages; for three cross-linkages there are eight structures, and so on. Fig. 1/. The complexity of the algorithm of the calculation for each of them is increased even more rapidly.

The second important problem is a kinetical description of intramolecular cross-linking. Kinetic characteristics can vary in wide range depending on the nature of cross-linking agents, properties of the polymeric chain and experimental conditions.

The determination of the kinetic regularities for different systems is important in the first turn for understanding of the process of the networks formation and for the study of sol and gel properties. On the other hand, the solution of the kinetic problem is of the great importance from the viewpoint of the further development of the general theory of macromolecular reactions.

The kinetic problem for the intramolecular cross-linking reactions in general form was not yet solved. Only some particular cases, i.e. the cyclization of macromolecules, the intramolecular catalysis and diffusion-controlled collision of two reactive groups were studied theoretically by Morawetz, Sisido and Fixman [1-4].

Here we'll consider a more general case assuming the possibility of the cross-link formation between any two sites of the molecule rapproaching one to another to some critical distance /we'll call such pairs "contacts"/ and assuming that the rate constant of the elementary act does not depend on the chain conformation as a whole and the nearest environment. Besides we'll assume that the reaction is a kinetically-controlled one, i.e. the system reaches the state of the conformational equilibrium between two consequent cross-links formations but the elementary act is irreversible and so fast that the chain conformation remains constant during it [5-6].

Such model corresponds to many real chemical reactions of cross-linking with and without cross-linking agent.

In this case the r ate of cross-linkages formation must be proportional to the number of reactive contacts in each particular chain, z_j /where j is a number of cross-links/. Assuming the independence of the average contacts number $\bar{z}_j$ on cross-linkages configuration one can describe the reaction by the following system of kinetic equations:

$$dc_j/dt = k_o(\bar{z}_{j-1}c_{j-1} - \bar{z}_j c_j) \qquad (1)$$

where c_j is a number of chains with j cross-linkages.

Then the calculation of equilibrium values $\bar{z}_j$ is the only problem of kinetic description.

The average number of cross-linkages in a chain

$$\bar{n}(t) = 1/c \sum_{j=1}^{M} j\, c_j \qquad (2)$$

/where $c = \sum c_j$ is a total number of chains/ is determined by the solution of following equation

$$d\bar{n}/dt = k_o/c \sum_{j=0}^{M-1} \bar{z}_j c_j \qquad (3)$$

The exact analytical approach to the estimation of $\bar{z}_j$ now is practically impossible because of the reasons mentioned above. Therefore to solve the problem we used the method of mathematical experiment, Monte Carlo method. Our aim was:

1. to calculate the equilibrium values $\bar{z}_j$;
2. to calculate the cross-linking kinetics;
3. to calculate the cross-links number distribution;
4. to calculate dimensions of partially cross-linked macromolecular coils;
5. to consider the influence of the reactive groups distribution and MMD.

Besides we have shown the possibility to apply the results of model calculations to some experimental data and consider a simple approximate approach to the calculation of dimensions of partially cross-linked coils and cross-linking kinetics. The accuracy of this approximation is evaluated by comparison with Monte Carlo results.

The model and computation procedure

The linear macromolecule was simulated by the chain of N sites on the volume-centered lattice allowing the self-intersection with minimum loop of 4 chain units.

The procedure of the simulation included the following steps: the random conformation was built in the computer, the number of reactive contacts /i.e. non-cross-linked self-intersections/ was calculated and then one of the contacts was cross-linked with a probability

$$w_j = \beta z_{j-1} \qquad (4)$$

/where β is a normalization coefficient being equal to 1/N/.

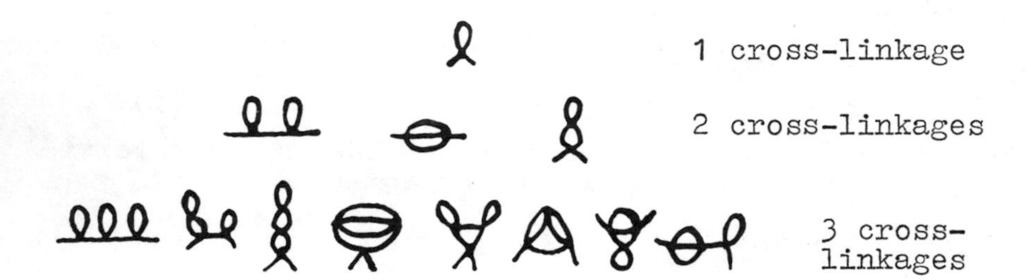

Figure 1. Topological structures of cross-linked chains

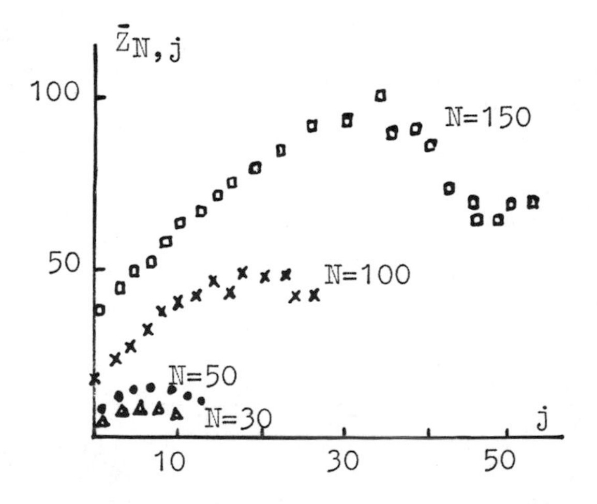

Figure 2. Average number of reactive contacts vs. the number of cross-linkages (□) N = 150; (×) N = 100; (●) N = 50; (△) N = 30

If the chain was not cross-linked the new conformation was built with the same number of cross-linkages [j-1] and so on up to the cross-link formation. The time between formation of two consequent cross-linkages was determined as

$$t = \beta / k_o(m + \xi) \qquad (5)$$

where m is a number of conformations built between /j-1/-th and jth cross-link formation, ξ is a random number equally distributed betwee 0 and 1 [7].

The procedure was repeated up to the formation of given number of cross-linkages or up to the given time [t_{max}].

The reaction was simulated for chains of 30, 50, 100 and 150 units with reactive groups being in each site of the chain and for the chains of 100 units with the degree of occupation of reactive groups [ω] being equal to 0.5 and 0.25. For the last case two types of reactive groups distribution were considered: the Bernoullian and regular distributions. Besides that the reaction for the polymolecular sample with Flory's MMD was simulated [$\bar{N}$ = 100, ω = 0.5].

Now we shall discuss some results obtained.

1. Statistics of intrachain contacts

The change of the contacts number with increase of the cross-links number is shown in the Figure 2. The contacts number is increased due to the decrease of the effective volume of macromolecular coil during cross-linking. The decrease of the contacts number at the high degrees of cross-linking can be explained by exhaustion of free reactive groups. It is not surprising that the dependence of $\bar{z}$ upon j is changed with the change of the model chain length. The longer the chains, the greater the effect of coils compression and more rapidly the average contacts number is growing.

It should be pointed out, that the initial part of the curve $\bar{z}_j(j)$ can be represented in the linear form. Below we discuss the possibility of the usage of this linear dependence.

2. Kinetics of intramolecular cross-linking

One can expect that the kinetic curves reach the saturation at high degrees of cross-linking due to the exhaustion of reactive groups. However in computer experiment one can obtain only the initial part of kinetic curves. The kinetic results of computer experiment are shown in the Figure 3. It can be seen that the reaction of intramolecular cross-linking is an autoaccelerated one, and the initial rate and the degree of autoacceleration is increased with an increase in chain length.

It was mentioned above that the kinetics of intramolecular cross-linking is determined by the solution of the kinetic equations system supposing the independence of the average number of contacts on the cross-linkages configuration. The validity of this assumption can be checked by comparison of results of numerical solution of this system [with $\bar{z}_j$ obtained from the computer experiment] with kinetic results obtained directly by simulation. It can be seen from the Figure 3, that two approaches to the calculation give the same results. Thus, the kinetics of cross-linking can be completely described by equilibrium values of the number of reactive contacts in a macromolecule. This seems to be the important result of the computer calculation.

The linear dependence of $\bar{z}_j$ on j in the initial stage of the reaction mentioned above

$$\bar{z}_j = Aj + B \qquad (6)$$

permits to write the kinetic equation in the simple form

$$d\bar{n}/dt = k_o(A\bar{n} + B) \qquad (7)$$

The results of the solution of this equation

$$\bar{n}(t) = B/A(e^{k_oAt} - 1) \qquad (8)$$

are shown in the same Figure 3 too. It can be seen that the kinetics of intramolecular cross-linking at the initial stage can be described by linear approximation indeed.

3. Cross-linkages number distribution

The exact cross-linkages number distribution of the chains at any moment in time is determined by the solution of the system (1).

If the average number of contacts in the chain were constant during the reaction, the process would be a random one with the Poisson cross-links number distribution. The dispersion of such process is

$$DP = \bar{z}_o k_o t \qquad (9)$$

The calculation of the dispersion in the linear approximation, which is valid at the initial stage of the reaction, gives the equation

$$D_L = e^{k_o A t} - \bar{n}(t) \qquad (10)$$

In the Figure 4 the values of the dispersion obtained from a computer experiment are compared with Poisson and linear dispersions. It can be seen from that figure that the true distribution is much wider than the Poisson distribution and that the width is increased with an increase in chain length. In the initial stage, the dispersion follows the linear approximation. For short chains at high degrees of cross-linking the distribution becomes narrower due to the accumulation of chains with many cross-linkages [close to the maximum value] and the dispersion tends to that of a Poisson.

4. Dimensions of partially cross-linked coils and their shape

The dimensions of polymer coil are usually characterized by the value of the mean square radius of gyration $\overline{R^2}$ or mean square and-to-end distance $\overline{h^2}$.

The dependence of relative values $\overline{R^2}/\overline{R_o^2}$ on the degree of cross-linking is presented in the Figure 5. It can be seen that chain dimensions are essentially decreased with cross-linking, and that this effect becomes greater with an increase of the chain length.

The same is with the ratio $\overline{h^2}/\overline{h_o^2}$.

Now consider the change of the shape of the polymer coils during cross-linking. Usually the shape of the macromolecule is represented by the rotation ellipsoid. For the Gaussian coil the ratio of the axes of this ellipsoid [p] is equal to 3 [for the sphere p= 1]. The dependence of the anisotropy factor p on the degree of cross-linking for different chain lengths is shown on the Figure 6. In despite of the considerable scattering of the points it can be seen that the conformation of partially cross-linked coils is changed tending to spherical form.

It is well known that for Gaussian coils the mean square dimensions $\overline{R^2}$ and $\overline{h^2}$ are related by the ratio $\overline{h^2}/\overline{R^2}= 6$. It was of interest to check the validity of this ratio for cross-linked coils. It can be seen from table for the long chains [N =100, 150] the ratio $\overline{h^2}/\overline{R^2}= 6$ is practically constant at different cross-linkages number, while for the short chains [N= 30, 50] this ratio is decreased with the degree of cross-linking.

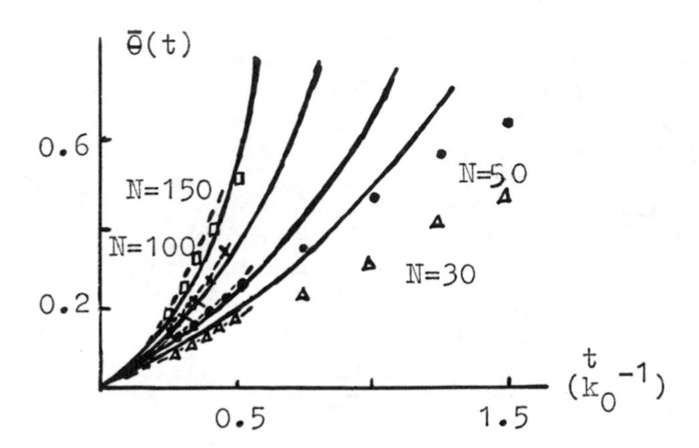

Figure 3. Average degree of cross-linking vs. time; (– – –) the solution of Equation 1; (——) linear approximation; points, Monte Carlo calculation (□) N = 150; (×) N = 100; (●) N = 50; (△) N = 30

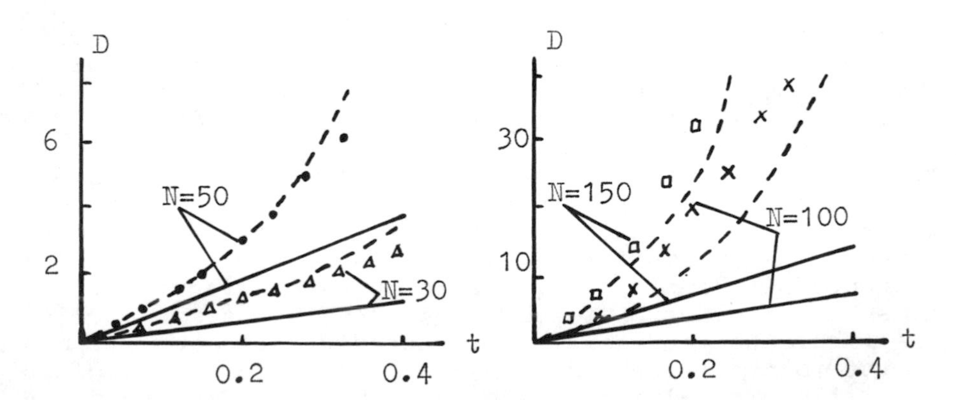

Figure 4. Dispersion of cross-linkage number distribution vs. time; (——) Poisson distribution; (– – –) linear approximation; points, Monte Carlo calculation (□) N = 150; (×) N = 100; (●) N = 50; (△) N = 30

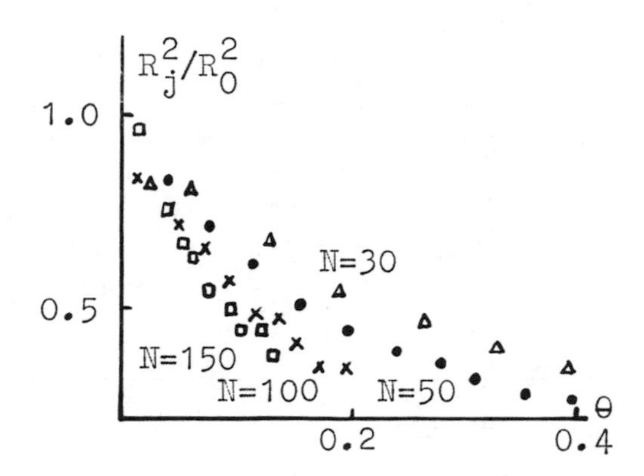

Figure 5. Relative coil dimension vs. the degree of cross-linking (□) N = 150; (×) N = 100; (●) N = 50; (△) N = 30

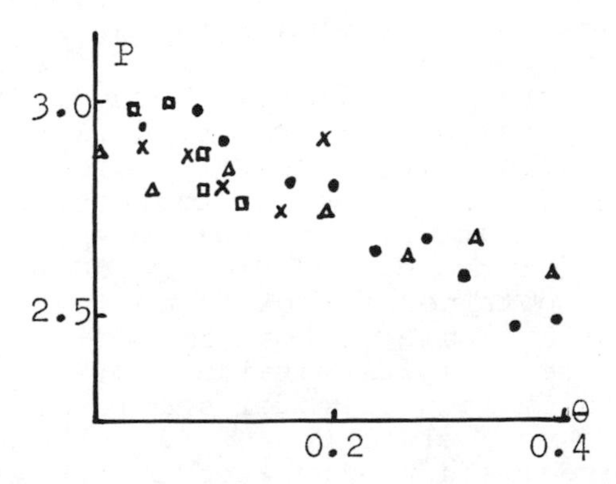

Figure 6. Form anisotropy factor vs. the degree of cross-linking

Table

$\overline{h^2}/\overline{R^2}$ for cross-linked coils

N	30	50	100	150
0	6.0	6.0	6.0	6.0
1	5.3	5.7	6.0	5.9
2	5.1	5.8	6.2	5.9
3	5.1	5.5	6.4	5.4
4	4.9	5.7	5.8	5.8
5	5.0	5.5	5.6	6.2
6	4.6	5.3	5.6	6.2
7		5.5	5.8	5.8
8		5.2	5.5	5.9
9		5.1	5.8	6.2
10		5.4	5.2	5.6

5. Influence of reactive groups distrbution and MMD

In the Figure 7 the kinetic curves for the model chains with some distribution of reactive groups along the chain are shown. The reaction was simulated for the degree of occupation of reactive groups $\omega = 0.5$ and 0.25 and for the Bernoullian and regular distributions [8]. The results of such procedure are compared with results obtained for molecules with the same number of reactive groups situated on each site of the chain /$\omega=1$/ and the chain is evidently shorter. It is rather evident that the rate of the reaction is decreased with a decrease in the number of reactive groups at the constant length of the chain. At the constant number of reactive groups the rate for the case $\omega = 1$ /$N = 50, \omega = 1$/ is higher than for the case $\omega < 1$ /$N = 100, \omega = 0.5$/. However the results obtained for different types of reactive groups distribution are close to one another. The difference between results obtained for unimolecular /$N = 100$/ and polymolecular model samples /$\bar{N} = 100$/ with Flory's MMD are very close too.

Thus this model study permitted to elucidate the main regularities of the process of intramolecular cross-linking. The most interesting of them are, firstly, the existence of a uniform relationship between the kinetics of the reaction and the equilibrium properties of partially cross-linked chains and secondly, the independence of the kinetics of cross-linking on the character of reactive groups distribution on the initial stage of the reaction.

6. Applying of results obtained to experimental data

The direct comparison of results obtained with experimental data is now unfortunately impossible, because of the practical absence of such experiments. However the suggested model and calculation method can be used for interpretation of some data obtained during the study of the sol-fraction properties in the process of network formation. Irzhak, Enikolopyan et al. [9-10] found the sharp decrease of intrinsic viscosity of the sol-fraction with polymer concentration during cross-linking of polyvinylbutyral containing some unreacted hydroxyl groups with diisocyanates [at constant $\bar{M}$ and the average cross-linkages number].

There are two different explanations of this fact. One of them assumes the possibility of a compression of polymer coils at average concentrations down to the dimensions less than in the θ-solvent. The alternative is based on the existence of the wide distribution of macromolecule dimensions in any time. It is rather natural to assume an increase of the probability of intramolecular reaction with an increase of the dimensions of the macromolecule. .e. more extended conformations go to the gel-fraction and more coiled remain in the sol. With the increase of solution concentration the distances between coils are diminished and the critical dimensions, necessary for a transition into the gel are decreased too. This process will be accompanied by a decrease of the average dimensions of molecules in sol.

The hypothesis about the fractionation during cross-linking is consistent with experimental viscosity data.

The hypothesis can be checked by the mathematical experiment, if one takes into consideration the probability of the intermolecular cross-linking. Let's assume that sol fraction contains only intramolecularly cross-linked chains while the formation of even one intermolecular cross-linkage leads to the sol-gel transition. Because only the properties of sol fraction are of our interest we don't need to follow the intermolecularly cross-linked chains. It is rather natural to assume that the probability of the transition into the gel is proportional to the dimensions of the macromolecular coil.

$$w = c(R^2)^{\gamma}$$

The coefficient "c" has an analogy with a concentration, as more is "c" more coiled conformation can be intermolecularly cross-linked.

In this case the procedure of the cross-linking simulation has to include checking of intermolecular cross-

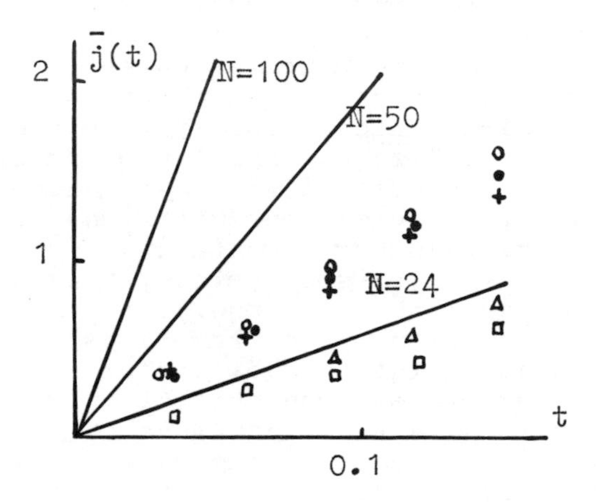

Figure 7. Kinetics of cross-linking for regular (△, ●) and Bernullian (□, ○) distributions of reactive groups, and for polymolecular sample with Flory's MMD (+)

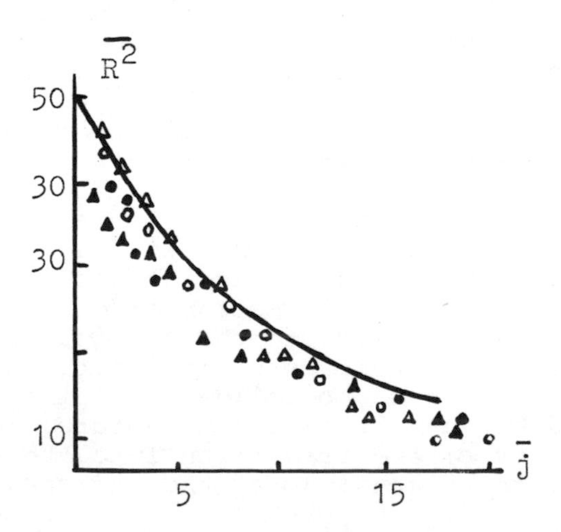

Figure 8. Monte Carlo calculation of the $R^2(j)$ *function with* C = *0.5 (△, ○); C = 1 (▲, ●) and γ = 1 (△, ▲); γ = 1.5 (○, ●) (solid curve,* C = *0)*

linking by the calculation of the correspondent probability. If the intermolecular cross-linkage is formed this chain mathematically is thrown away and the new conformation is being built. For checking of fractionation hypothesis dimensions of remained molecules having the same cross-linkages number and different c are compared.

We have accomplished such calculation for the chains of 50 and 100 units with $\gamma = 1$; 3/2 and 3 and $c = 0.5$ and 1. The dependences of $\bar{R}^2$ on average cross-links number are presented in the Figure 8. The results obtained at different c and γ show that there is no essential decrease of dimensions of sol molecules. The change of γ led only to the change of the yield of the sol-fraction and does not influence its properties.

I.e. the results of Monte Carlo experiment show that the effect of the fractionation is too small to explain the experimental fact of significant decrease of sol viscosity with an increase of the polymer concentration.

Thus, in spite of the model character of the approach described it can be applied to the estimation of the experimental data on the particular chemical cross-linking reactions.

7, Approximate approach

The method of mathematical simulation has many advantages, and is very close to the physical experiment. However the further development of this approach [a consideration of volume effects, reversible reactions and so on] can be rather difficult because it will require too much computer time. Therefore it is expedient to search some simple analytical or semianalytical approximate approaches to the calculation of cross-linking kinetics and conformational properties of cross-linked macromolecules. The results obtained by the Monte Carlo calculation can serve as criteria of the accuracy of such approximation.

We suggest here an approximation based on one of the results obtained in computer experiment. It was found that the mean square of the radius of gyration $\bar{R}^2$ and the total number of self-intersections of the partially cross-linked chain $\bar{\sigma}$ [it consists of the reactive contacts and "dead" contacts, cross-links] are related by the following relationship:

$$\bar{\sigma}_m (\bar{R}_m^2)^{3/2} = \text{const} \qquad (11)$$

which is valid for different chain lengths and different cross-links number.

As it was shown above the kinetics of intramolecular cross-linking is completely determined by the number of reactive contacts, which is equal to the number of self-intersections minus the number of cross-linkages [we do not consider the multiple self-intersections]. Then this relationship turns the kinetic problem to the average dimensions calculation.

The problem of the analytical representation of the dependence of the average dimensions of cross-linked coil on the number of cross-linkages was considered by some authors.

So Edwards et al. [11] suggested the following approximate relationship based on the thermodynamic consideration:

$$\bar{R}_m^2 = \frac{\bar{R}_0^2}{m + 1} \qquad (12)$$

This relation is too inaccurate and gives results very far from the Monte Carlo ones.

Gordon et al. [12] have modified this relationship:

$$\bar{R}_m^2 = \frac{R_o^2}{(m + 1)^q} \qquad q \sim 0,2 \qquad (13)$$

This one is much better, but the discrepancies are rather essential.

Here we propose for the $\bar{R}^2$ calculation the procedure including the random choice of the pairs of cross-linking units and the calculation after every cross-linkage formation the number of units in the elements of three types which compose the topological structure of every partially cross-linked coil. These elements are "the tails" [always two] arches between two cross-links and circles [loops].

$\bar{R}^2$ of the chain with "m" cross-linkages and definite number of elements of each type [the total number of elements is 2m+1] and definite number of units in each element is calculated as the sum of $\bar{R}^2$ of the linear part [the tails] and R^2 of the cross-linked part [the arches and circles] [13]:

$$\bar{R}_m^2 = \frac{\bar{n}_t l^2}{6} + \frac{(\bar{n}_c + \bar{n}_a) l^2}{12} \qquad (14)$$

[n - the number of units in each element].

On the $\bar{R}^2$ one can calculate $\bar{q}_m$ and $\bar{z}_m$

$$q_m = \text{const} / (\bar{R}_m^2)^{3/2} \qquad (15)$$

[const is calculated from the initial conditions]

$$\bar{z}_m = \bar{q}_m - m \qquad (16)$$

The solution of the system of the kinetic equations with coefficients $\bar{z}_m$ gives the dependence of the average number of cross-linkages on the time.

In the Figures 9 and 10 the results of the approximate calculation of $\overline{R^2}/\overline{R_0^2}$ and $\bar{z}$ are compared with results of mathematical experiment. It can be seen, that the results are rather close.

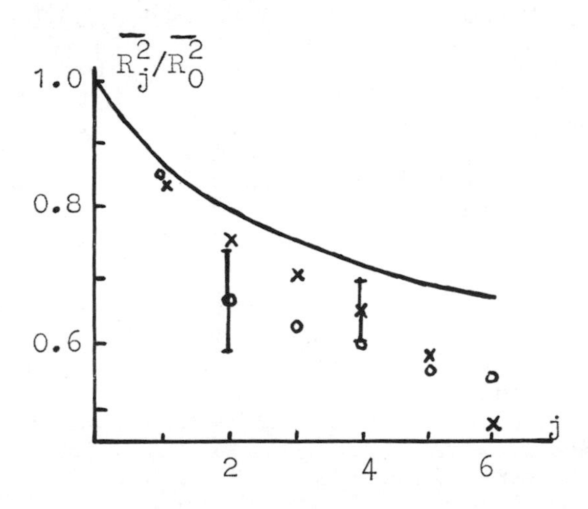

Figure 9. Relative dimensions vs. the number of cross-linkages obtained by the Monte Carlo simulation (×), approximate calculation (○), and by the solution of Equation 13 (N = 100)

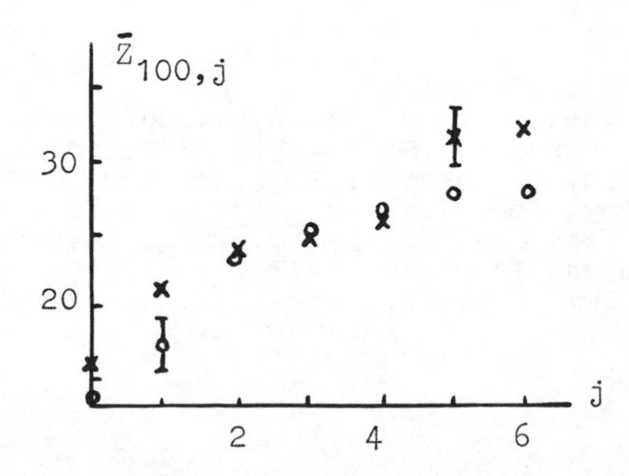

Figure 10. Average number of reactive contacts calculated by the Monte Carlo simulation (×) and by the approximate approach (○) (N = 100)

Thus the approach described can be applied to the description of the kinetics of intramolecular cross-linking and the conformational properties of cross-linked products.

It should be pointed out in conclusion that the results obtained can be used for the further development of the theory of reactions of functional groups of macromolecules in different processes including both intra- and intermolecular interactions of polymer chains.

On the other hand this is the necessary step to go from the theoretical description of the macromolecular reactions in isolated macromolecules which is the ideal rather than more practical case with concentrated solutions when one can't neglect the intermolecular interaction and its competition with the intramolecular one.

The comparison of predicted calculated results with the experimental ones in kinetics of the cross-linking [when the latter will be available] will allow also to choose one or another pathway of cross-linking processes during chemical modification of polymers.

REFERENCES

1. H. Morawetz, Pure Appl.Chem., 38, 267, 1974.
2. M. Sisido, Macromolecules, 4, 737, 1971.
3. M. Sisido, T. Mitamura, Y. Imanishi, T. Higashimura, Macromolecules, 10, 125, 1977.
4. G. Welmski, M. Fixman, J.Chem.Phys., 58, 4009, 1973.
5. I.I. Romantzova, Yu.A. Taran, O.V. Noah, N.A. Platé, Dokl.AN SSSR, 234, 109, 1977.
6. I.I. Romantzova, O.V. Noah, Yu.A. Taran, A.M. Elyashevich, Yu.Ya. Gotlib, N.A.Platé, Vysokomol.Soed. A19, 2800, 1977.
7. A.M. Elyashevich, Vysokomol.Soed., A20, 951, 1978.
8. I.I. Romantzova, Yu.A. Taran, O.V. Noah, N.A.Platé, Vysokomol.Soed., A21, 1176, 1979.
9. V.I. Irzhak, L.I. Kuzub, N.S. Enikolopyan, Dokl.AN SSSR, 214, 1340, 1974,
10. L.I. Kuzub, V.I. Irzhak, L.M. Bogdanova, N.S. Enikolopyan. Vysokomol.Soed., B16, 431, 1974.
11. G. Allen, J. Burgess, S.F. Edwards, D.Y. Walsh, Proc.Roy. Soc. London, A334, 453, 465, 477, 1973.
12. M. Gordon, J.A. Torkington, S.B. Ross-Murphy, Macromolecules, 10, 1090, 1977.
13. B.H. Zimm, W.H. Stockmayer, J.Chem.Phys., 17, 1301, 1949.

RECEIVED October 31, 1979.

4

Chemical Modification of Polyvinyl Chloride and Related Polymers

M. OKAWARA
Research Laboratory of Resources Utilization, Tokyo Institute of Technology, Nagatsuta, Midori-ku, Yokohama 227, Japan

Y. OCHIAI
Oji Paper Co., Ltd., Shinonome, Koto-ku, Tokyo 135, Japan

Widespread chlorine-containing polymers would include, 1) stable molding material for practical use such as polyvinyl chloride (PVC), polyvinylidene chloride and poly(epichlorohydrin)(PECH) and, 2) reactive polymers capable to introduce additional functional groups via their active chlorines such as chloromethyl polystyrene, poly (β-chloroethyl vinylether) and poly (vinyl chloroacetate). While the latter, especially the chloromethyl polystyrene, has been widely used recently for the synthesis of variety of functional polymers, we should like to talk in this article about the chemical modification of the former, mainly of PVC and PECH, which was developed in our laboratory.

Retardation of Discoloration of PVC and Decolorization of Discolored PVC

PVC has long been utilizing as a representative engineering plastics with low cost and stable properties, while the toxicity of the monomer and plasticizer included has given rise to public discussion recently. Nevertheless, the improvement of the thermal stability is one of the most important points in practical modification of PVC. The deterioration of PVC is known to proceed through a rapid and sequential elimination of hydrogen chloride along a length of polymer chain giving a chromophoric and easily oxidizable polyene structure. The ease of formation of the conjugated structure (2), once a double bond has

0-8412-0540-X/80/47-121-041$05.00/0

formed at a weak point of PVC (1), is readily understood in terms of allylic effect of an olefinic group upon an adjoining chlorine-bearing carbon atom. Thus, the polyene structure might be developed by this zipperlike elimination (PVC → 1 → 2).

$$\text{-CH(Cl)-CH}_2\text{-CH(Cl)-CH}_2\text{-} \quad \text{(PVC)}$$

$$\downarrow \text{-HCl}$$

$$\underset{2}{\text{-CH=CH-CH=CH-}} \xleftarrow{\text{-HCl}} \underset{1}{\text{-CH=CH-CH(Cl)-CH}_2\text{-}} \xrightarrow{+X_2} \underset{3}{\text{-CH(X)-CH(X)-CH(Cl)-CH}_2\text{-}}$$

While the numerous kinds of stabilizers were empirically developed and effectively used in practice, we proposed a device for stabilization logically derived from a viewpoint of reaction scheme (1). Namely, if the double bond was saturated (3) simultaneously when it formed, the retardation of the development of conjugated polymer structure might be expected. We chose the reduction with diimide (NH=NH) for the saturation of double bond. Thus, p-toluenesulfonyl hydrazide (PSH) was used which acts as a diimide source under the condition (130-150°C) of PVC-processing. Under the optimum condition (130°C in dimethylformamide (DMF)), the dehydrochlorination was distinctly suppressed as shown in equation 1 and Figure 1.

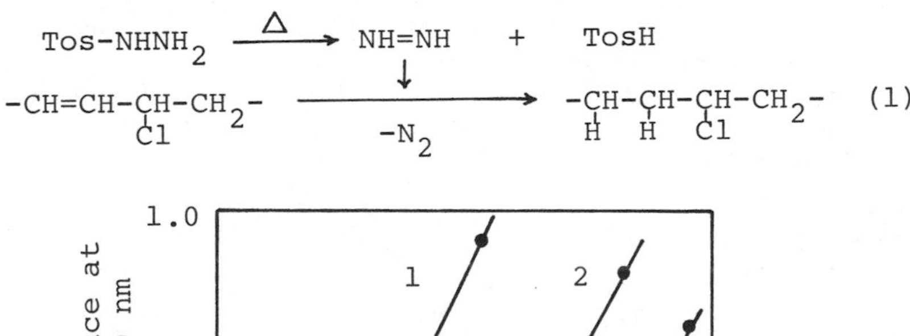

$$\text{Tos-NHNH}_2 \xrightarrow{\Delta} \text{NH=NH} + \text{TosH}$$

$$\text{-CH=CH-CH(Cl)-CH}_2\text{-} \xrightarrow[\text{-N}_2]{\text{NH=NH}} \text{-CH(H)-CH(H)-CH(Cl)-CH}_2\text{-} \qquad (1)$$

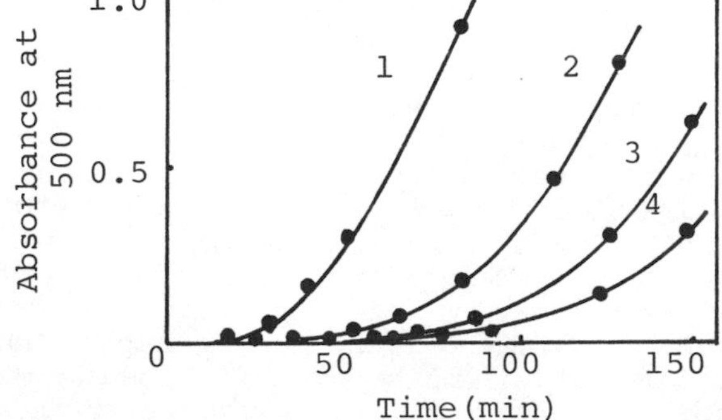

Journal of Polymer Science

Figure 1. Effect of PSH on the rate of discoloration of PVC—1.0 g of PVC, 5 mL of DMF at 130°C: (1) without PSH; (2) 4.5 g of PSH; (3) 0.16 g of PSH; (4) 0.75 g of PSH (1)

PSH also could reduce (decolorize) the deteriorized (discolored) PVC as well as retard the discoloration of PVC. Decolorization rate and the degree of decolorization in the final stage according to the per cent transmission of various discolored polymers are found to depend on the histories of the discolored PVC (heating in o-dichlorobenzene (ODB) at 180°C), the effect of temperature on the rate of decolorization was examined to result in Figure 2. Alternately, the decolorized PVC thus obtained was thermally stable compared with that obtained by oxidative method. That is, the thermal stability of two types of PVC, decolorized by the diimide reduction at 100°C for 4 h (A) and decolorized by oxygen bubbling in dioxane at 100°C for 4 h (B) were compared with that of the original PVC at 130°C in DMF. As shown in Figure 3, the induction period of the discoloration for (A) was longer than that for the original PVC. In the PVC decolorized by the diimide reduction, the C=C double bonds formed by dehydrochlorination at the most labile chlorine atoms of PVC (tertiary or allylic) were saturated with hydrogen and so were the most stabilized.

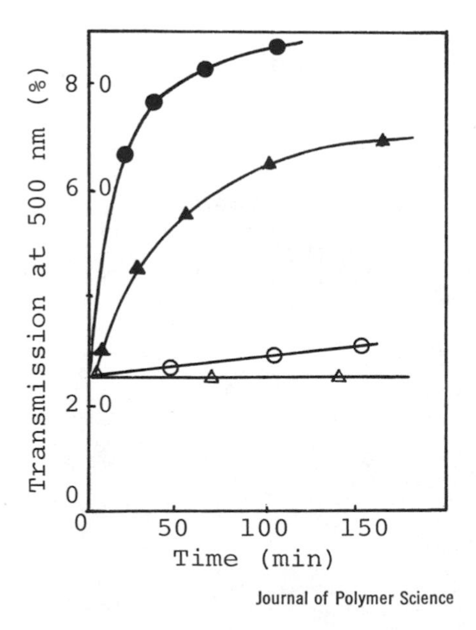

Figure 2. Decolorization in various temperatures with PSH—0.3 g of PVC, 20 mL of ODB: (△) without PSH at 100°C; (▲) 0.6 g of PSH at 100°C; (○) without PSH at 130°C; (●) 0.6 g of PSH at 130°C (1)

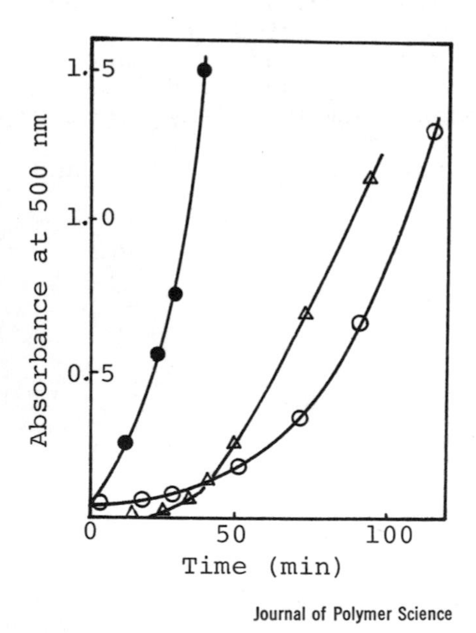

Figure 3. Rediscoloration of decolorized PVC in DMF at 130°C—0.30 g of PVC, 15 mL of DMF: (○) decolorized PVC with diimide; (●) decolorized PVC with oxygen; (△) original PVC (1)

This PSH technique was later utilized as an effective method of reductive modification for the unsaturated polymers such as polydiene rubbers and polypentenamer by Mango (2) and Samui et al (3).

Nucleophilic Substitution of PVC with Dithiocarbamate

It has been long believed that the nucleophilic substitution (a) of chlorine in PVC is extremely difficult in contrast to the ease of elimination (b) (equation 2).

$$-CH{=}CH- \xleftarrow[-HCl]{(b)} -CH(H\cdots Nu)-CH(Cl)(\leftarrow Nu)- \xrightarrow[-Cl^-]{(a)} -CH_2-CH(Nu)- \quad (2)$$

We found that the nucleophilic substitution of PVC with N,N-dialkyldithiocarbamate ($R_2N-CS-S^-$, DTC) was feasibly achieved at low temperature as 50-60°C in DMF without accompanied dehydrochlorination (equation 3)(4).

$$-CH(Cl)-CH_2- \xrightarrow[DMF,\ 50-60°C]{R_2N-CS-SNa} -CH(S-CS-NR_2)-CH_2- \quad (3)$$

These results were assumed to be ascribed to the neighboring group participation via the intermediary cyclic carbocation (4) as shown in equation 4, as well as the high nucleophilicity of DTC and acceleration by use of the dipolar aprotic (DA) solvent. The formation and reactivity (stability) of such carbocation were examined at length with the corresponding model compounds of low molecular weight (5).

(4) [scheme: Cl-substituted chain attacked by DTC → S-C(NR₂)=S substituted chain, Cl leaving → cyclic carbocation **4** (S +S ring with C–NR₂, Cl⁻) attacked by DTC → product with $R_2N-C(=S)-S-$ and cyclic S–C(NR₂)=S group, Cl]

Dithiocarbamoylated PVC (PVC-DTC) thus obtained was easily crosslinked by irradiation, and could photograft the monomer coexisted (equation 5) (6). The mechanism of these photolytic cleavage was deduced from the detailed researches (product examination and kinetic study) on the photolysis of model compounds such as dithiocarbamate, $RS-CSNR_2$ (7) and xanthates, RS-CSOR (8). These reactions are utilized to design photosensitive polymers and recently in Japan the technique, photografting with water-soluble monomer, was applied to prepare the materials for artificial organs.

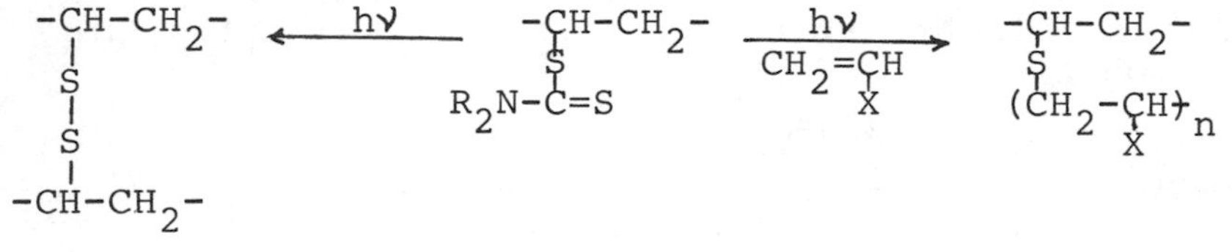

$$-CH-CH_2-S-S-CH-CH_2- \xleftarrow{h\nu} -CH(S-C(=S)-NR_2)-CH_2- \xrightarrow[CH_2=CHX]{h\nu} -CH(S-(CH_2-CHX)_n)-CH_2- \quad (5)$$

Further, N-methyl dithiocarbamate (MDTC) (9) and N-metylglycine (sarcosine) (10) were similarly incorporated into PVC matrices resulting in the derivatives usable to chelate forming and introduction of thiol-function as shown in following scheme. Of the two main purpose of modification of commercial polymers; 1) improvement of original property of each polymer and 2) incorporation of new function into the polymeric materials, our studies would be served from the view-point of the latter as the fundamental design for PVC and PECH with specific functions.

In addition, Nakagawa et al. (11) have shown that dehydrochlorination of PVC was suppressed extremely by incorporation of DTC or MDTC group, while the mechanism was not fully obvious.

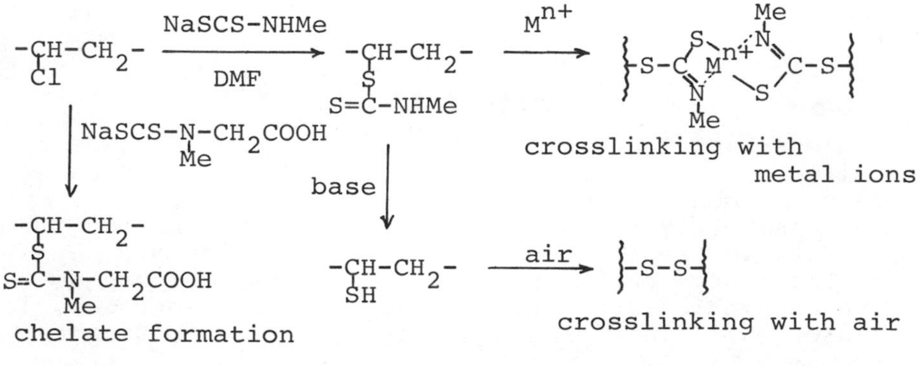

$$-CH(Cl)-CH_2- \xrightarrow[DMF]{NaSCS-NHMe} -CH(S-C(=S)-NHMe)-CH_2- \xrightarrow{M^{n+}} \text{crosslinking with metal ions}$$

$$-CH(S-C(=S)-NHMe)-CH_2- \xrightarrow{base} -CH(SH)-CH_2- \xrightarrow{air} -S-S- \quad \text{crosslinking with air}$$

$$-CH(Cl)-CH_2- \xrightarrow{NaSCS-N(Me)-CH_2COOH} -CH(S-C(=S)-N(Me)-CH_2COOH)-CH_2- \quad \text{chelate formation}$$

Nucleophilic Substitution of PVC with Thiolates

Thiolate is another class of effective nucleophile. Nakamura et al. (12) have proved that thiols reacted with PVC efficiently in the presence of ethylenediamine to give the thioetherificated PVC and PECH. Especially, PECH with triazine-3,5-dithiol substituent (5) is interesting in their characteristic behavior towards metal ions and metal surface.

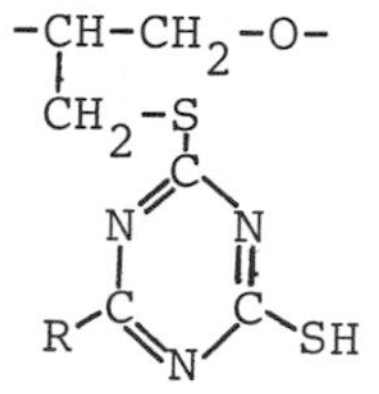

-CH-CH$_2$-O-
CH$_2$-S
N N
R C C SH
N

5

We also investigated the thioetherification of PVC with thiolate or analogous system (13):

1) The reaction by use of sodium thiolate proceeds easily in DMF at room temperature and the degree of substitution (DS) goes up to over 80% for sodium thiophenoxide (equivalent).
2) In the case of thiol and inorganic salts, the best result was obtained for thiophenol and sodium cyanide, and DS of 90% was attained (60°C) with minimum amounts of C=C formation even for p-thiocresol (equivalent) and potassium carbonate.
3) Even when allyl iso-thiuronium salt (and NaOH) is used tractably as a precursor of thiol (DMF, 40°C), the extent of thioetherification is nearly equal to that for the case 1).
4) Substitution up to 20% is possible at 80°C even in water, in which PVC is quite insoluble, by use of thiophenoxide together with a surfactant such as quarternary ammonium salt.

The reaction of the thiolated PVC with Chloramine T was carried out and the structure (6) of the resultant polymer was examined (14). In the case of phenylthioether, the sulfilimine structure (7) was mainly produced accompanied with, in part, the sulfenamide structure (8) binded to the main chain with N atom. In the case of allyl thioether derivative containing C=C moiety in the pendent group, the sulfilimine (9) formed rearranged exclusively in Claisen type to the sulfenamide structure (10) connected to the main chain with S atom.

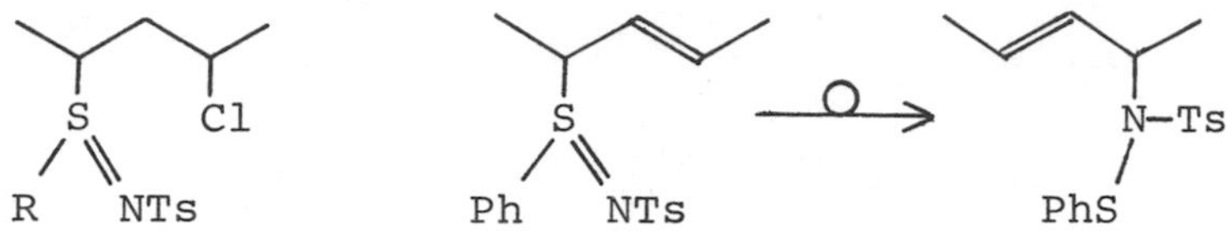

6 7 8

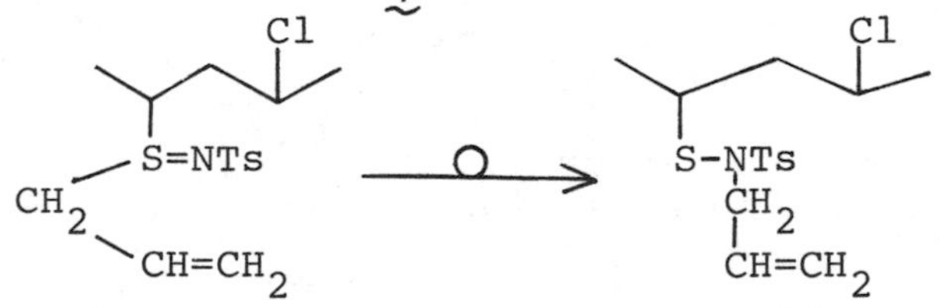

9 10

<u>Nucleophilic Substitution of PVC with Azide</u>

From the results mentioned above, it was ascertained that the chlorine in PVC is easily replaced by nucleophiles in contrast to the traditional notion. Thus, PVC was substituted with xanthate (EtO-CSS$^-$), dithiophosphate ($(EtO)_2P(=S)S^-$), thiourea, thiosulphate ($^-S-SO_3^-$), and even with carbanion such as dimethyl malonate/DBU to some extents. In the last case, PVC was reacted in tetrahydrofuran (THF) with a small excess of dimethylmalonate in the presence of equimolar amounts of 1,5-diazabicyclo[5,4,0]undec-5-ene (DBU). After 5 h at room temperature, 6 mol% of chlorine was replaced by the malonate accompanied with ca. 6% of C=C structure to result in a pale brown powdery product (15). Temperature of decomposition of the product, however, fell to ca. 200°C (DTA) compared with that of PVC (270°C).

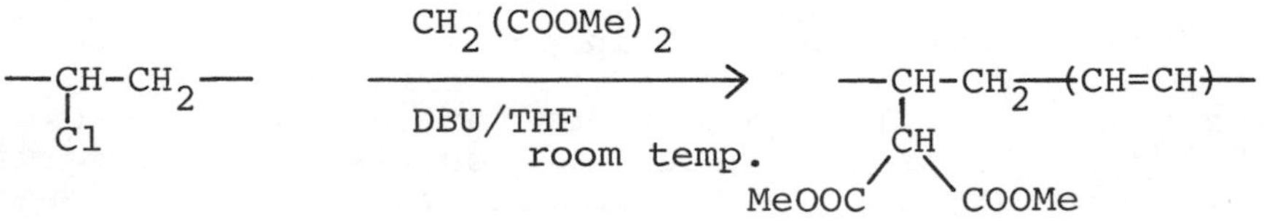

$$\text{—CH(Cl)—CH}_2\text{—} \xrightarrow[\text{DBU/THF, room temp.}]{CH_2(COOMe)_2} \text{—CH(CH(COOMe)}_2\text{)—CH}_2\text{—(CH=CH)—}$$

Further interestingly, azide anion (N_3^-), rather weak nucleophile, could easily react with PVC in DA-solvents to give the azide derivatives (equivalent NaN_3, 60°C, DS (degree of substitution): 64% (hexamethylphosphortriamide, HMPA), 33% (DMF) and 0% (THF, H_2O)) as shown in equation 6 (16).

$$\text{-}\underset{\underset{\text{Cl}}{|}}{\text{CH}}\text{-CH}_2\text{-} \xrightarrow{\text{NaN}_3} \text{-}\underset{\underset{\text{N}_3}{|}}{\text{CH}}\text{-CH}_2\text{-} \qquad (6)$$

Azide group in PVC was further transformed to various types of derivatives as shown below and might be used for the modification and functionalization of PVC (<u>17</u>).

$$\text{-}\underset{\underset{\text{N}_3}{|}}{\text{CH}}\text{-CH}_2\text{-}(\underset{\underset{\text{Cl}}{|}}{\text{CH}}\text{-CH}_2) \xrightarrow{\text{PhC≡CH}} \text{-CH-CH}_2\text{- (triazole, N–N=N, Ph)}$$

$$\xrightarrow{\text{LiAlH}_4} \text{-}\underset{\underset{\text{NH}_2}{|}}{\text{CH}}\text{-CH}_2\text{-} \qquad \xrightarrow{\text{PhMgBr}} \text{-}\underset{\underset{\text{N=N-NHPh}}{|}}{\text{CH}}\text{-CH}_2\text{-}$$

$$\xrightarrow{h\nu} \text{-CH-CH}_2\text{-NH-CH-CH}_2\text{-} \qquad \xrightarrow{\text{salicyl aldehyde}} \text{-}\underset{\underset{\text{N=CH-C}_6\text{H}_4\text{-HO}}{|}}{\text{CH}}\text{-CH}_2\text{-} \xrightarrow{\text{Cu}^{+2}} \text{(CH=N, O, Cu, N=CH complex)}$$

$$\xrightarrow[0\,°\text{C},\,\sim 100\%]{\text{PPh}_3} \text{-}\underset{\underset{\text{N=PPh}_3}{|}}{\text{CH}}\text{-CH}_2\text{-}$$

$$\text{-CH-CH}_2\text{-(N=PPh}_3) \xrightarrow{\text{CS}_2} \text{-}\underset{\underset{\text{N=C=S}}{|}}{\text{CH}}\text{-CH}_2\text{-} \qquad \xrightarrow{\text{CO}_2} \text{N=C=N} \qquad \xrightarrow{\text{DMSO}} \text{Cl}^- \, \text{N}^+\text{PPh}_3 \;\; \text{Cl}^- \, \text{N}^+\text{PPh}_3$$

All the above reactions of PVC were performed homogeneously in DA-solvents such as HMPA, DMF and dimethylsulfoxide (DMSO). For the practical modification of PVC, the reaction must be conducted under more commercial conditions as in slurry water. As mentioned before, azidation of PVC did not occur in water. However, the reaction proceeded feasibly in water by addition of some cationic surfactant to give, e.g. 8-20% (DS) of azidated PVC at 80°C by use of tetra-n-butyl ammonium chloride (<u>18</u>). The use of cationic surfactant was also effective in organic solvents and attracted increased attention as the conception of "phase transfer catalyst" in organic chemistry developed.

<u>Reaction of Poly(epichlorohydrin) with Nucleophiles</u>

Above experience in PVC modification was recently applied to PECH which seems to have more labile (reactive) primary chlorine atom. PECH would be useful in the preparation of poly(propylene oxide) substituted on the methyl groups with various kinds of functional groups. Thus, there has been a number of studies (19) on the nucleophilic substitution on PECH, but many of them are patents and ambiguous in details. We investigated the reaction of PECH with various nucleophiles (vide infra) to develope the useful methods for modification of PECH and compare the reactivity of PECH with PVC.

Thioetherification and Related Derivation.

Thioetherification of PECH is feasibly performed in DA-solvents as already described in the patent (20). For example, the highest substitution was obtained by the reaction of P(ECH-EO)(1:1 copolymer of epichlorohydrin and ethylene oxide) and equimolar thiophenoxide in HMPA at 100°C for 10 h as DS 83% for sodium and 93% for potassium salts. The DS in our nucleophilic substitution was estimated by the elemental analysis as well as the titration of liberated chloride ion with mercuric nitrate (21). In the latter method, reacted medium was pretreated with hydrogen peroxide when the reductive nucleophiles which can react with mercuric ion were used. As described before for PVC, thiolation was also achieved conveniently with iso-thiuronium salt followed by alkaline hydrolysis without the direct use of ill-smelling thiolate. The thiolated PECH obtained are rubbery solids, soluble in toluene, methylene chloride, ethyl methyl ketone and DMF and insoluble in water, acetone, dioxane and methanol.

11 → 12: H_2O_2/xylene, room temp., 2 days

11 → 13: Chloramine T, DMF-AcOH, 60°C, 6 h

13 → 12: 1) H_3O^+-dioxane 2) OH^-

13 ⇌ 14: $BF_3 \cdot MeOH/CH_2Cl_2$; Et_3N/CH_2Cl_2

11: CH_2-S-Ph 12: CH_2-S(→O)-Ph 13: CH_2-S(=NTos)-Ph 14: CH_2-S(=NTos)-Ph · BF_3

The thiophenylated PECH (11) thus obtained was converted to the sulfoxide structure (12, ν(SO), 1040 cm^{-1}) by treating with hydrogen peroxide (35%) in xylene at room temperature. When the polymer 11 was treated with Chloramine T at 60°C in DMF, sulfilimine type polymer (13, ν(S=N), 965 cm^{-1}) was obtained, which also affords the method to obtain a sulfoxide polymer without sulfone moiety on heating a dioxane solution containing small amounts of hydrogen chloride. A donor-type polymer 13 gave complex polymer (14) by the reaction with BF_3-methanol in methylene chloride, which reversibly returned to the sulfilimine 13 with triethylamine. The complex 14 exhibited the IR spectrum (1035-1080 cm^{-1}) of BF_3 but not ESR signal which was observed in the dark violet BF_3-complex of polystyrene-type sulfilimine, indicating the electron transfer did not occur. BF_3-included polymer 14 are interesting as a polymeric hardner for epoxy resin and a polymeric Lewis acid-catalyst.

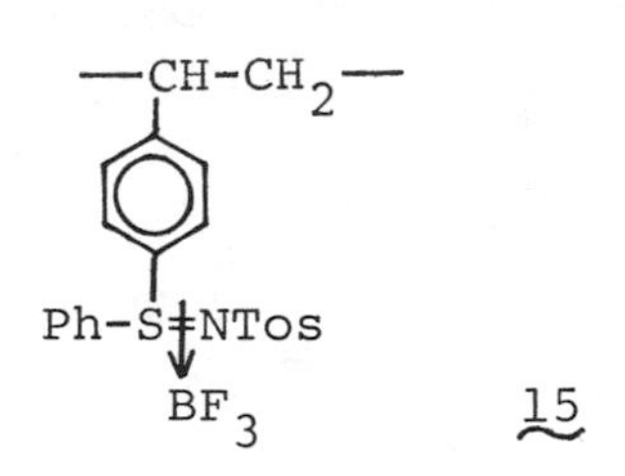

Reaction with Dithiocarbamate and Thiocyanate.

PECH reacted smoothly with sodium dithiocarbamate (NaDTC), as PVC, in DA-solvents affording a photosensitive polymer 16. The most interesting is the reaction of epichlorohydrin and NaDTC.

It has been reported (22) that the low polymerization of ethylene oxide proceeded through a step-by-step addition mechanism due to the enhanced nucleophilicity of the dithiocarbamate (DTC) in DA-solvents. For example, a DMF solution (35 ml) of sodium N,N-dimethyldithiocarbamate (0.08 mol) was added dropwise to a DMF solution (15 ml) of epichlorohydrin (0.05 mol) at room temperature; the mixture was poured into a large amounts of water to give a precipitate (softing point, 68-70°C) in good yield, which showed [η], 0.04 (30°C, DMF) after purification by reprecipitation (DMF-H_2O). The polymer (17) obtained exhibited characteristic IR bands in the range 1490-1375 cm^{-1} (-S-CS-N<) and ether linkage (C-O-C), essentially identical with that of the polymer obatined by the reaction of PECH with NaDTC in DMF-toluene, except in the range 1700~1650 cm^{-1}. Various lines of evidence indicated the mechanism involving initial polymerization of the epichlorohydrin catalyzed by NaDTC followed by the nucleophilic substitution on the chloromethyl carbon atom as shown in the following scheme. The IR absorption at 1700 cm^{-1} can not be assigned, but that of 1650 cm^{-1} would be ascribed to the thiolcarbamate structure (-S-CO-N<, 19), which can reasonably be explained by the mechanism, involving the 1,3-dithianylium ion (18) through anchimerism by the DTC function.

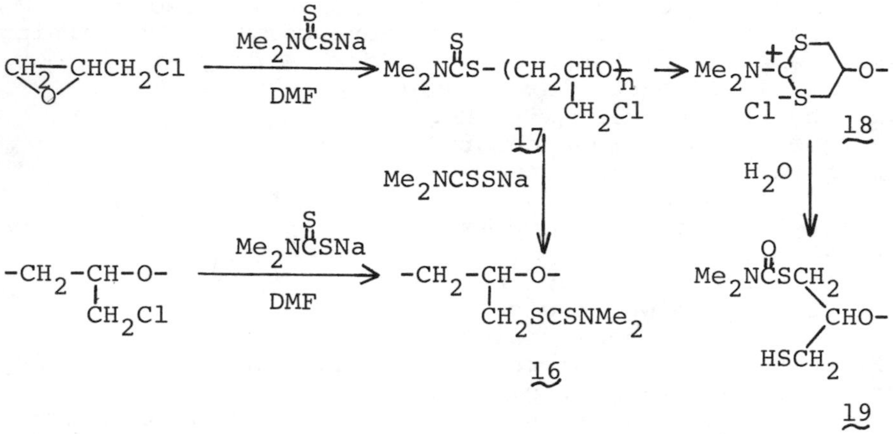

$$CH_2\text{-}CHCH_2Cl \ (\text{epoxide}) \xrightarrow[DMF]{Me_2NC(=S)SNa} Me_2NC(=S)S\text{-}(CH_2CH(CH_2Cl)O)_n \ (17) \rightarrow Me_2N\text{-}C^{+}(\text{-S-},\text{-S-})\text{-O-} \ Cl^{-} \ (18)$$

$$17 \xrightarrow{Me_2NCSSNa} 16; \quad 18 \xrightarrow{H_2O} 19$$

$$\text{-}CH_2\text{-}CH(CH_2Cl)\text{-O-} \xrightarrow[DMF]{Me_2NC(=S)SNa} \text{-}CH_2\text{-}CH(CH_2SCSNMe_2)\text{-O-} \ (16)$$

$$Me_2NC(=O)SCH_2\text{-}CHO\text{-}(CH_2SH) \ (19)$$

PECH did not react with potassium cyanate but reacted with equimolar potassium thiocyanate in DMF (90°C, 16 h) to give the thiocyanated polymer (20, IR, 2180 cm^{-1}) in 53% of DS. Comparing the IR spectrum with those of model compounds, Me_2CHCH_2SCN (2180 cm^{-1}) and Me_2CHCH_2NCS (2200, 2125 cm^{-1}), the isothiocyanate moieties are scarcely existed in the polymer 20. Since the -SCN is a protecting form of thiol likewise the -SCl, the polymer 20 are insolubilized with aqueous alkali presumably due to the S-S crosslinking (21). Further, absorption at 2180 cm^{-1} in 20 was completely disappeared treating it with two equivalents of triethyl phosphite at 90°C for 16 h in DMF probably due to the formation of phosphonate structure (22).

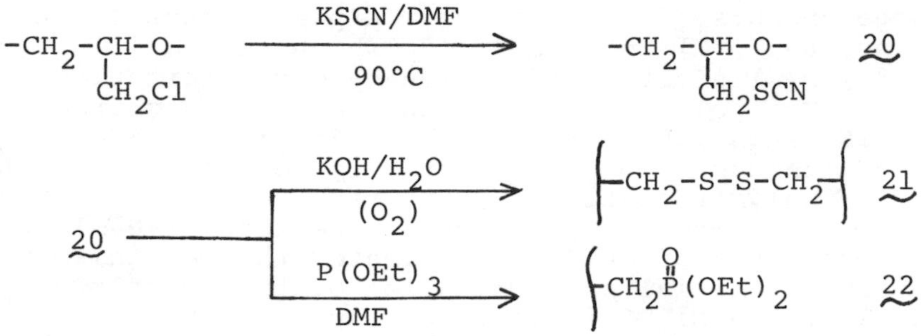

$$\text{-}CH_2\text{-}CH(CH_2Cl)\text{-O-} \xrightarrow[90°C]{KSCN/DMF} \text{-}CH_2\text{-}CH(CH_2SCN)\text{-O-} \quad 20$$

$$20 \xrightarrow[(O_2)]{KOH/H_2O} \text{-}CH_2\text{-S-S-}CH_2\text{-} \quad 21$$

$$20 \xrightarrow[DMF]{P(OEt)_3} \text{-}CH_2P(=O)(OEt)_2 \quad 22$$

Reaction with Thiosulfate-Formation of Bunte Salt.

The formation of S-alkyl thiosulfate (Bunte salt) by the reaction of alkyl halide and sodium thiosulfate has been well known. Whereas a patent claimed the formation of Bunte salt from PECH and sodium thiosulfate (23), the reaction hardly proceeded in DMF owing to low solubility of sodium salt. On the other hand, both ammonium thiosulfate and PECH were soluble in HMPA-H_2O (7:1 vol/vol) and the reaction proceeded homogeneously. Water soluble Bunte salt (23, ν(SO), 1200, 1020 cm^{-1}) was isolated by pouring the reaction mixture into water and salting out with ammonium chloride. The DS based on the mercuric nitrate titration was in nearly accord with that on elemental analysis. The DS values depended on the thiosulfate concentration were shown below.

$$-CH_2-\underset{\substack{|\\ CH_2Cl}}{CH}-O- \xrightarrow[HMPA/H_2O]{(NH_4)_2S_2O_3,\,90°C,\,16hr} -CH_2-\underset{\substack{|\\ CH_2S-SO_3^-NH_4^+}}{CH}-O- \qquad \mathbf{23}$$

$(NH_4)_2S_2O_3$/PECH	1.0	0.15	0.05	0.01
DS (%)	94	12	4.6	0.77

The Bunte salt 23 in aqueous dope decomposed gradually resulting an increase of acidity as; initial (pH, 5.47), after 7 days (4.81), 14 days (4.76), 28 days (4.64). An aqueous dope of 23 forms gel with protonic acid, alkali, oxidizing and reducing agents as formulated below. The crosslinking of S-S type would be suggested since the formation of -SH or $-S^-$ was confirmed for the model compound under the similar condition.

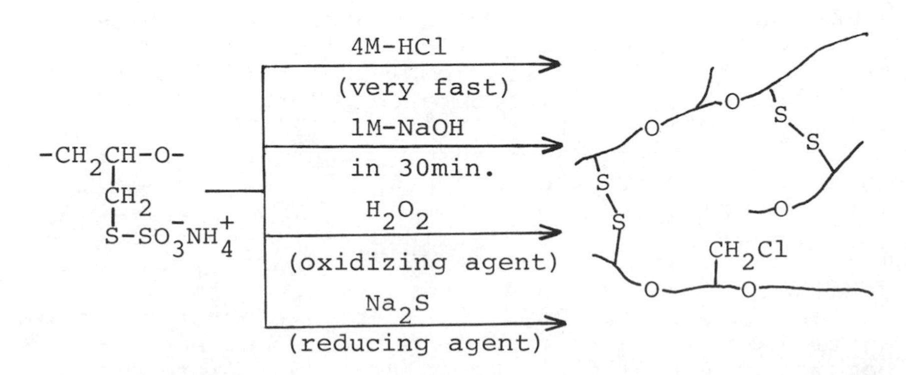

Further, a film was obtained from an aqueous or methanol solution of 23. The film containing 1 wt% of water-soluble bis-acrylate ($CH_2=CMeCOO(CH_2CH_2O)_{14}COCMe=CH_2$) insolubilized rapidly on UV-irradiation. The result would be ascribed to the photolytic cleavage ($P-SSO_3Na \rightarrow PS\cdot$) and successive radical coupling assisted by bis-acrylate as has been discussed by Tsunooka et al (24). Therefore, PECH-Bunte salt is promising as a water-soluble photoresist or photo-hardening paint. Polymeric Bunte salt, 23 also could bind heavy metal ions such as mercury and cadmium from their aqueous solution. In addition, when the aqueous dope of 23 is added to an aqueous solution of polymeric cation as 24 or 25, polymer complex was obtained which are insoluble in common solvents suggesting the formation of tight, saltlike structure between polyanion and polycation.

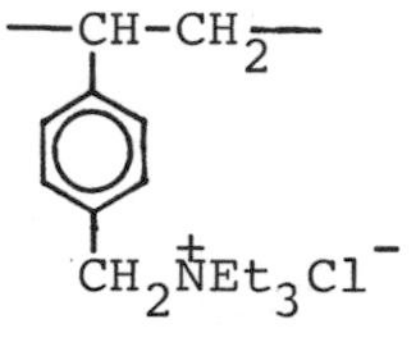

$$\text{—CH-CH}_2\text{—} \quad (\text{p-C}_6\text{H}_4\text{-}CH_2\overset{+}{N}Et_3Cl^-) \qquad \textbf{24}$$

$$\text{piperidinium ring }(\overset{+}{N}Me_2)\text{—}SO_2\text{—} \; Cl^- \qquad \textbf{25}$$

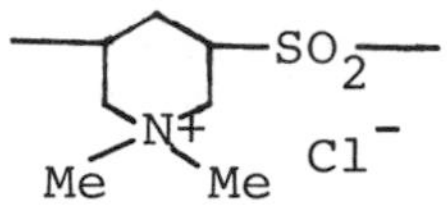

Reaction with Alcoholate.

Though the PECH decomposes to indefinite fragments with n-butyl lithium or sodium hydride in THF at room temperature, it reacts with sodium methoxide with liberation of Cl^- in which the β-elimination of hydrogen chloride predominates instead of nucleophilic substitution. For instance, PECH in DMSO was reacted with double the molar quantity of sodium methoxide at room temperature for 24 h to give the unsaturated polyether 26 (DS 92.3%, ν(C=C) 1630, $\delta(=CH_2)$ 795 cm^{-1}) after purification by dissolution (DMF)-precipitation (H_2O) technique. A similar unsaturated polyther was obtained by the pyrolysis of the sulfilimine 13 (110-130°C) but not of sulfoxide 12 (100-150°C). When the polymer 26 was heated to 90°C, the absorption of C=C and $=CH_2$ decreased and a new absorption at 1720 cm^{-1} appeared and increased. This is explained as a result of [3.3] sigmatropic rearrangement of 26 to afford 27 including $C=CH_2$ and C=O structure as shown in equation 7. The formation of carbonyl structure was observed in the reaction of PECH at higher temperature but not for the reaction of P(ECH-EO). The polymer 26 is interesting because it crosslinked rapidly by the Lewis acid without any hetero atoms, N or S and additional cross-linking agents.

$$-CH_2-\underset{CH_2Cl}{\underset{|}{CH}}-O- \xrightarrow[24\ h,92\%]{\substack{MeONa\\DMSO,25°C}} -CH_2C(=CH_2)\ldots O\ldots CH_2 \ldots C(-O-)=CH_2\ \ (\underset{\sim}{26}) \xrightarrow{90°C} -CH_2-C(=O)-CH_2-CH_2-C(=CH_2)-O-\ \ (\underset{\sim}{27})$$

Survey of Nucleophilic Substitution on PECH.

To compare the reactivities of various nucleophiles, the reactions of PECH with equimolar amounts of nucleophiles were carried out at 90°C for 16 h in DMF and the conversion was estimated by titration of chloride ion liberated. The results were summarized in Table 1. The reactivity of S-nucleophile is high as in general. The xanthate obtained was soluble in DMF, but insolubilized gradually on drying. Photosensitive PECH-N_3 is obtained in good yield notwithstanding the low solubility of sodium azide in DMF.

The products with ammonium diethyl dithiophosphate and thiourea were soluble in water, but the structure of the expected product (PECH-S-P(=S)$(OEt)_2$) for the former was not ascertained at present. The formation of acetate with AcOH/KF expecting hydrogen bonding assist was failed. No definite product was obtained in the case of potassium cyanide and sodium nitrate Contrary to the reports using tertiary amine (25) and triphenyl phosphine (26), there was no definitive clue for such reactions in our case. Comparison of the substitution reaction of PECH and PVC against the same nucleophiles shows the former is more reactive as shown in the first two cases of Table 2. On the other hand, whereas sodium cyanate does not react with PECH (in HMPA), it induces gelation of PVC liberating chloride ion. That is, when the basicity of nucleophiles predominates over nucleophilicity, β-elimination of hydrogen chloride takes place in preference, and then, PVC would show high reactivity in which the conjugation of C=C is effected more easily.

Thus, it was concluded that, combined the medium reactivity of chloromethyl group and flexible polyether-backbone, PECH is one of the most promising halogenated polymers for the developement of various new functional materials.

Table 1. Reactivities of PECH with Nucleophiles in DMF

Nucleophiles	Conv.(%)	Notes
PhSNa	100	formation of P-SPH
MeONa	100	formation of C=C (elimination)
$Me_2NCSSNa$	87	formation of $P\text{-}SCSNMe_2$
EtOCSSK	86	P-SCSOEt, drying ⟶ insolub.
NaN_3	77	heterogeneous, ⟶ $P\text{-}N_3$
$(EtO)_2P(=S)SNH_4$	64	folubilized in H_2O
KSCN	52	formation of P-SCN
BzONa	30	formation of P-OBz
$(NH_4)_2S_2O_3$	23	heterogeneous, solub. in H_2O
$(NH_2)_2C=S$	16	$P\text{-}SC(=NH)NH_3^+Cl^-$, solub. in H_2O
AcONa	16	formation of P-OAc
AcOH/KF	-	decomp. of polymer
KCN	-	reddish, decomp. of polymer
$NaNO_2$	-	yellow precipitate
KF, NaOCN NEt_3, PPh_3	0	no reaction

Table 2. Comparison of Reactivity of PECH and PVC[a)]

Nucleophiles	PECH	PVC
PhSK	100%	75.8%
$(EtO)_2P(=S)SNH_4$	64%	28%
NaOCN[b)]	no reaction	53% (gelation)

a) equimolar amounts of polymer and nucleophile, in DMF, 90°C, 16 h.
b) in HMPA/H_2O(7/2 vol/vol), 90°C, 18 h.

References

1) T.Nakagawa,M.Okawara,J.Polym.Sci.,A-1,6,1795(1968).
2) L.A.Mango,R.W.Lenz,Makromol.Chem.,163 1,13(1973).
3) K.Snui,W.J.Macknight,R.w.Lenz,Macromolecules,7,952 (1974).

4) M.Okawara,K.Morishita,E.Imoto,Kogyo Kagaku Zasshi, 69,761(1966).
5) cf. M.Okawara,K.Hiratani,Kagaku no Ryoiki,26,25 (1966)(review).
6) Ref.4) and,T.Nakai,M.Okawara,High Polymers,Jap., 18,2(1969)(review).
7) M.Okawara,T.Nakai,E.Imoto,Kogyo Kagaku Zasshi,69, 973(1966).
8) M.Okawara,T.Nakai,U.Otsuji,E.Imoto,J.Org.Chem.,30, 2025(1965).
9) T.Nakagawa,U.Taniguchi,M.Okawara,Kogyo Kagaku Zasshi,70,2382(1967).
10) T.Nakagawa,M.Okawara,ibid.,71,2076(1968).
11) T.Nakagawa,H.B.Hopfenberg,V.Stannett,J.Appl.Polym. Sci.,15,231,747(1971) and related papers.
12) For example,Y.Nakamura,K.Mori,M.Kaneda,Nippon Kagaku Kaishi,1976,1620.
13) T.Yamamoto,M.Imaura,Y.Naito,M.Okawara,Kobunshi Ronbunshu,31,164(1974).
14) T.Yamamoto,M.Imaura,M.Okawara,ibid.,31,171(1974).
15) T.Yamamoto,M.Okawara,unpublished data.
16) M.Takeishi,M.Okawara,J.Polym.Sci.,Polym.Lett.Ed., 7,201(1969).
17) M.Takeishi,M.Okawara,Ibid.,8,829(1970).
18) M.Takeishi,Y.Naito,M.Okawara,Angew.Makromol.Chem., 28,111(1973).
M.Takeishi,R.Kawashima,M.Okawara,Makromol.Chem.,167, 261(1973).
19) e.g.E.Schacht,D.Bailey.O.Vogl,J.Polym.Sci.Polym. Chem.Ed.,16,2343(1978) and references cited therein.
20) D.S.Breslow,U.S.Pat.,3417060(1968).
21) K.Hagino,Anal.Chem.Jap.,5 428(1956).
22) T.Nakai,M.Okawara,Bull.Chem.Soc.Jap.,41,707(1968).
23) E.J.Vandenberg,U.S.Pat.,3706706(1972).
24) M.Tsunooka,M.Fujii,N.Ando,M.Tanaka,N.Murata,Kogyo Kagaku Zasshi,73,805(1970).
25) T.Nishikubo,Y.Toyama,K.Maki,Y.Imamura,Japan Pat., 7303219(1973).
26) D.Redmore,U.S.Pat.,3664807(1972).

RECEIVED July 12, 1979.

5

Low-Temperature Modification of Polymers

CHARLES E. CARRAHER, JR.

Department of Chemistry, Wright State University, Dayton, OH 45435

We have been active in modifying commercially available vinyl polymers which contain polar groups utilizing techniques developed by us through condensation reactions of Lewis diacids with Lewis dibases. Most of these modifications can be depicted in general as follows

$$\text{-(CH}_2\text{-CH(A-H))-} + R_3MX \longrightarrow \text{-(CH}_2\text{-CH(A-MR}_3\text{))-} \quad \text{I}$$

$$\text{-(CH}_2\text{-CH(A-H))-} \xrightarrow{R_2MX_2} \text{-(CH}_2\text{-CH(A-MR}_2\text{-A)-CH-CH}_2\text{)-} \quad \text{II}$$

where A-H can be $-CO_2H$, $-NH_2$, -OH, -N-OH

Reaction with monofunctional metal containing reactants leads to the formation of linear products (form I.) which are soluble in a number of polar solvents, whereas reaction with di-functional reactants leads to formation of cross-linked products (form II.) which are insoluble in all attempted solvents.

0-8412-0540-X/80/47-121-059$05.00/0

We have, and continue to be actively engaged in the modification of such industrially available polymers for a number of reasons. First, such modifications yield products which generally exhibit dramatically different physical properties many of the new or enhanced properties being potentially useful. For instance, the modification of poly(ethyleneimine) with tin-containing moieties yielding products III and IV gave products which a. exhibit generally greater thermal stabilities; b. are semiconductors exhibiting bulk resistivities in the general range of 10^5 to 10^{11} ohm-cm; c. are hydrophobic (useful for situations requiring hydrolytic stability and/or water repellency); and d. are active at the 40 ppb level against fungi which are responsible for much of the natural mildew and rot. Potential uses for such products range from medical bandages to water repellent rainwear (1-3).

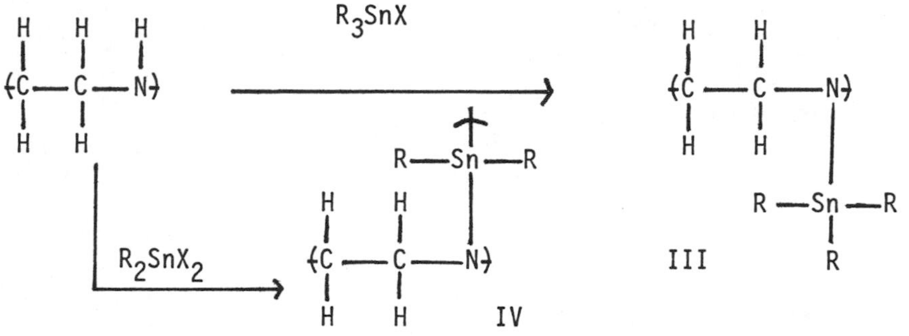

$$-(CH_2-CH_2-NH)- \xrightarrow{R_3SnX} -(CH_2-CH_2-N(SnR_3))- \quad \text{III}$$

$$-(CH_2-CH_2-NH)- \xrightarrow{R_2SnX_2} -(CH_2-CH_2-N(\text{-}SnR_2\text{-}))- \quad \text{IV}$$

Second, the modifications are intended to act as models to others that similar modifications can be effectively carried out utilizing similar reaction systems and reactants. It is significant that similar condensation between salts of poly(acrylic acid) with organic acid halides does not occur under typical low temperature condensation conditions presumably due to a combination of two interrelated factors - reaction with fully charged Lewis bases typically occur within the aqueous phase or very near the interface where rapid hydrolysis of the organic acid occurs. A number of already utilized organometallic halides and metal containing cations exhibit better water stability in comparison with organic acid halides or form water stable moieties (such as $Cp_2TiCl_2 + H_2O \rightarrow Cp_2Ti(H_2O)_x^{+2}$). Thus there exists a distinct advantage in utilizing such metal containing reactants in the modification of fully charged polymers.

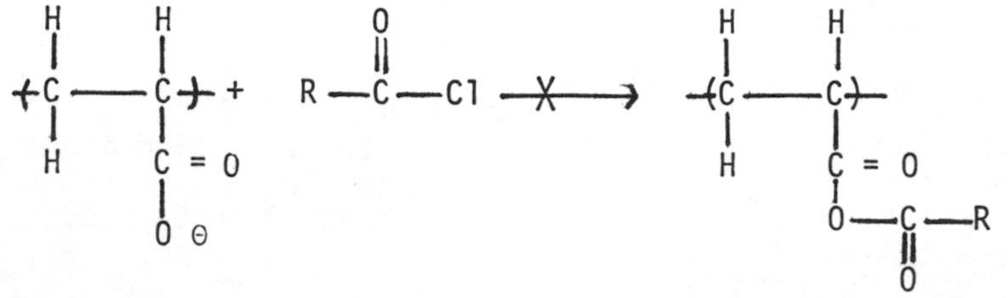

$$-(CH_2-CH(C(=O)O^{\ominus}))- + R-C(=O)-Cl \not\longrightarrow -(CH_2-CH(C(=O)-O-C(=O)-R))-$$

Third, the modifications are being utilized by use as models for the modification of naturally occurring materials including renewable resource materials such as celluloses, polyamides and natural oils.

Structure

As previously noted products from monofunctional reactants are linear whereas those which form difunctional reactants are crosslinked. Extent of inclusion appears to be generally more dependent on the nature of the condensing agent than the functionality (i.e. mono or difunctional reactants; for a given polymer and reactants containing a given metal, yield and inclusion appear to be more dependent on the actual chemical and physical nature of the condensing reactant than on the factor of functionality). Extent of inclusion varies from the low (20% and less) to high (80% to 100%, for instance Table 1). Fisher-Hirschfelder-Taylor space saving models of products approaching 50% and greater inclusion (assuming alternating or random inclusion) require elongation of the vinyl chain backbone resulting in an "extended" structure for the modified products. Further products from monofunctional reactants probably exhibit some short range helical segmental structural. It would be useful to investigate the actual stereo products for such properties as dynamic bulk properties (the suspected extended structures might offer some useful compression related properties) and optical properties. Optical activity could result from either the assymetric carbon present in the polymer chain now being randomized due to steric constraints or from an imposed helical structure of the products. Further crosslinking of the products under compressed conditions might produce products with potentially useful bulk properties.

For the products formed utilizing the interfacial technique in particular, the fact that high inclusion of the monomeric portion within polymer chains even for "low overall yield" systems may be due to the polymer chains being drawn, or remaining within the critical reaction zone until modification is essentially complete rather than to some neighboring group assistance - their position rather than increased chemical reactivity may be the essential aspect. The relative amount of modification of polymer chains which are in the critical reaction zone may be enhanced

Table 1. Sample Results for the Modification of Polymers Through Reaction with Acid Chlorides.

Acid Chloride	Product Yield (%)	Inclusion Yield (%)	Reaction Cond.	Ref.
Thionyl Chloride	17	56	a	4
Methanesulfonyl Chloride	15	79	a	4
p-Bromobenzenesulfonyl Chloride	10	69	a	4
Dibutyltin dichloride	76	98	b	5
Triphenyltin dichloride	48	66	b	5
Triphenylsilicon Chloride	42	40	b	5
Phenylphosphonic Dichloride	2	75	a	6
Phenylthioshosphonic Dichloride	6	70	a	6
Triphenyltin Chloride	26	24	c	7
Diphenyltin Dichloride	20	16	c	7
Tri-n-butyltin Chloride	21	14	c	7
Triphenyltin Chloride	97	83	d	8
Dibutyltin Diiodide	94	88	d	8
Diethyltin Dichloride	16	33	d	8
Diethyl Chlorothiophosphate	29	56	e	9
Phenyl Dichlorophosphate	72	100	e	9
Benzenesulfonyl Chloride	67	54	e	10
p-Bromobenzenesulfonyl Chloride	64	39	e	10
Diphenyltin Dichloride	97	98	e	11
Tri-n-propyltin Chloride	98	98	e	11
Cp_2TiCl_2	96	82	c	12
Cp_2HfCl_2	81	74	c	12
Cp_2TiCl_2	35	80	e	13
Cp_2HfCl_2	34	80	e	13
Cp_2TiCl_2	99	100	c	14

Reaction Conditions - General - 25-28°C, 17,500 to 20,500 rpm no load stirring rate.

a. PVA (23 mmoles) in 50 ml H_2O and acid chloride (23 mmoles) in 50 ml CCl_4 for 2 mins. stirring time.

b. Poly(sodium acrylate) (3 mmoles) in 50 ml H_2O and acid chloride (3 mmoles) in 50 ml $CHCl_3$ for 30 secs. stirring time.

c. PVA (3 mmole) in 50 ml H_2O and acid chloride (3 mmole) in 50 ml diethylether with an equivalence of NaOH added, 30 seconds stirring time.

d. Polyethyleneimine (1 mmole) in 50 ml H_2O and acid chloride (1 mmole) in 50 ml hexane with an equivalence of NaOH, 30 secs. stirring time.

e. Polyacrylamideoxime (2.5 mmole) in 50 ml H_2O, acid chloride (2.5 mmole) in 50 ml $CHCl_3$, equivalence of NaOH, 20 secs. stirring time.

through the initial inclusion of the somewhat polar comonomers since most interfacial condensation occurs near the interface which is a mixture of polar (aqueous) and essentially nonpolar (organic phase) with modified chains containing both the organic, nonpolar backbone and the polar "modified" attachment from the modifying organometallic comonomer. Thus some assistance may be due to the generation of modified and "modifying" polymer chains which are more compatible with the "active reaction zone" than the unmodified polymer chains.

In summary, little is actually known regarding sequence of inclusion and modified chain structure. There are enough potentially interesting points regarding both questions as to justify studies aimed at answering these questions. The presence in many cases of heavy metals may allow such studies to be carried out utilizing techniques not particularly useful with more classical polymers. Results from such study may assist the understanding of structures of other products derived from modifications where large, bulky moieties are included into the polymer.

Extent of Reaction

It appears that most of the Lewis acids, noted below, which have been successfully condensed to give polymeric or oligomeric chains with typical Lewis bases can be successfully included onto polymer chains containing similar nucleophilic reaction conditions similar to those employed for the analogous difunctional reactions (for instance Table 1).

```
 O   O        O        R
 ||  ||       ||       |
X-S-R-S-X   X-P-X    X-M-X    where M=Si, Ge, Sn, Pb, Ti, Zr,
 ||  ||       |        |            Hf, Pt, Mn
 O   O        R        R

  R R
  \/
X-M-X    where M=As, Sb, Bi,   UO2+2
  |
  R
```

There are some differences though. For instance the failure of poly(sodium acrylate) to condense with a number of Group IV A reactants (5,15-19) stands against the more general synthesis of of Group IV A polyesters. This is probably due to the large size of the polymer (greater steric requirements) and not to a difference in the intrinsic reactivity of the particular functional group.

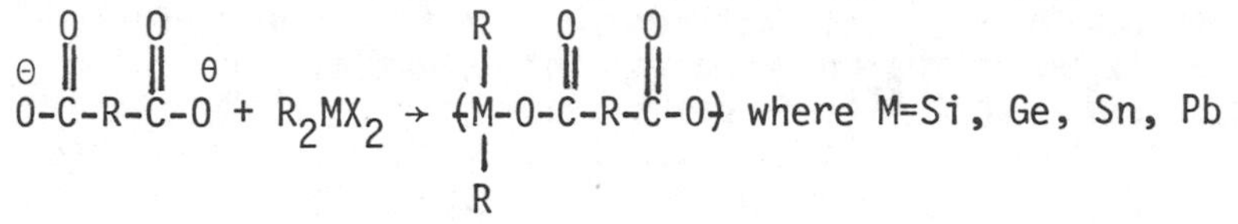

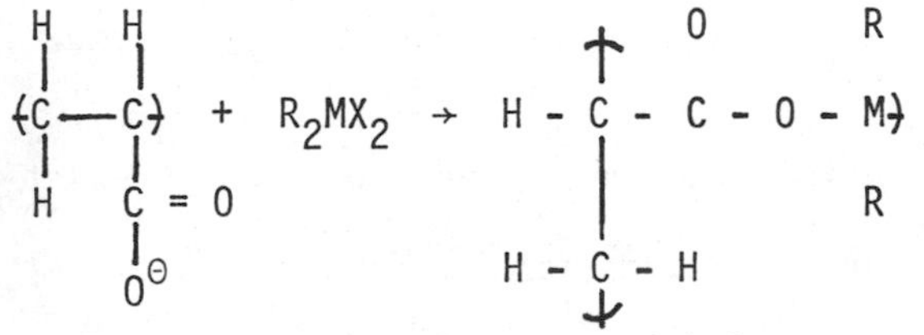

$$^{\ominus}O-\overset{\overset{O}{\|}}{C}-R-\overset{\overset{O}{\|}}{C}-O^{\ominus} + R_2MX_2 \rightarrow -\!\!\left(\underset{\underset{R}{|}}{\overset{\overset{R}{|}}{M}}-O-\overset{\overset{O}{\|}}{C}-R-\overset{\overset{O}{\|}}{C}-O\right)\!\!- \text{ where M=Si, Ge, Sn, Pb}$$

$$-\!\!\left(\underset{\underset{H}{|}}{\overset{\overset{H}{|}}{C}}-\underset{\underset{\underset{O^{\ominus}}{|}}{\underset{C\,=\,O}{|}}}{\overset{\overset{H}{|}}{C}}\right)\!\!- + R_2MX_2 \rightarrow \underset{\underset{\underset{\downarrow}{H\,-\,C\,-\,H}}{|}}{\overset{\uparrow}{H-C}}-\overset{\overset{O}{\|}}{C}-O-\underset{\underset{R}{|}}{\overset{\overset{R}{|}}{M}}\!\!\left.\right)\!\!-$$

In theory any polymer containing a condensable Lewis base (or acid) site is modifiable through condensation with a Lewis acid (or base). In actuality many possibilities have failed for both inclusion into polymer chains and the bifunctional reactions leading to linear product formation. We have not accomplished the inclusion of phosphonate and phosphate moieties into poly-(sodium acrylate) or with diacid salts such as disodium adipate. Hydrolysis of the phosphorus acid chloride appears to be critical since only hydrolyzed phosphorus containing moieties and unreacted poly(sodium acrylate) have been recovered from their attempted condensation. The analogous polythiophosphate esters, polythiophosphonate esters, polyphosphate esters and polyphosphonate esters have been synthesized utilizing analogous routes (for instance 20-22). It is possible that reaction with diols and thios occurs near the interface or within the organic phase whereas because of the greater ionic and polar nature of the organic salt reaction with the salt occurs clearly within the aqueous phase as previously noted. This is consistent with other findings (for instance 23-26).

Physical Appearance and Properties

The products generally exhibit poor to moderate low temperature stabilities (i.e. inception of degradation) and moderate to good high temperature stabilities (Table 2). For instance PVA itself exhibits less than 20% residue in air and nitrogen at 500°C whereas the tributyltin (14% substitution) retains 80% of its weight to 800°C (7). Degradation of the metal containing products occurs through a series of oxidations in air. Representative TGA thermograms appear in Figures 1 and 2.

Table 2.

Thermogravimetric Data For Polymers Modified Through Condensation With Group IV A and B Reactants

Polymer Modified[a.]	Modifying Agent (% Inclusion)	Initial Loss Temp. (°C)[b.]	10% Loss Temp. (°C)	20% Loss Temp. (°C)
PEI	$(ØCH_2)_2SnCl_2$(72)	225	225	250
PEI	$ØSnCl_2$ (61)	250	325	375
PEI	$Ø_3SnCl_2$ (62)	250	250	250
PEI	$(C_4H_9)_3SnCl$	175	225	250
PEI	$(C_2H_5)_2SnCl_2$ (33)	250	300	300
PEI	$(C_3H_9)_2SnCl_2$	275	275	275
PEI	$(C_8H_{17})_2SnCl_2$ (32)	250	250	250
PEI	$(C_4H_9)_2SnCl_2$ (31)	250	250	250
PVA	$Ø_2SnCl_2$ (16)	75	250	300
PVA	$(C_4H_9)_2SnCl_2$ (30)	150	275	300
PVA	$Ø_3SnCl$ (24)	175	250	275
PVA	$(C_4H_9)_3SnCl$ (14)	250	300	550
PVA	Cp_2TiCl_2 (82)	100	275	325
PVA	Cp_2ZrCl_2 (74)	100	250	400
PVA	Cp_2HfCl_2 (74)	75	275	450
PAA	$Ø_2SiCl_2$ (98)	200	300	325
PAA	$Ø_3SiCl$ (40)	175	200	225
PAA	$Ø_3SnCl$ (66)	175	350	375
PAA	$Ø_2SnCl_2$ (62)	150	200	225
PAA	$(C_4H_9)_2SnCl_2$ (98)	75	300	375
PAO	Cp_2TiCl_2 (95)	50	300	500
PAO	Cp_2ZrCl_2 (95)	50	250	375
PAO	Cp_2HfCl_2 (95)	50	300	375
PAA	----	200	275	300
PVA	----	225	275	300
PAA	Cp_2TiCl_2 (100)	250	275	300
PAA	Cp_2ZrCl_2 (100)	350	400	450
PAA	Cp_2HfCl_2 (100)	250	375	425

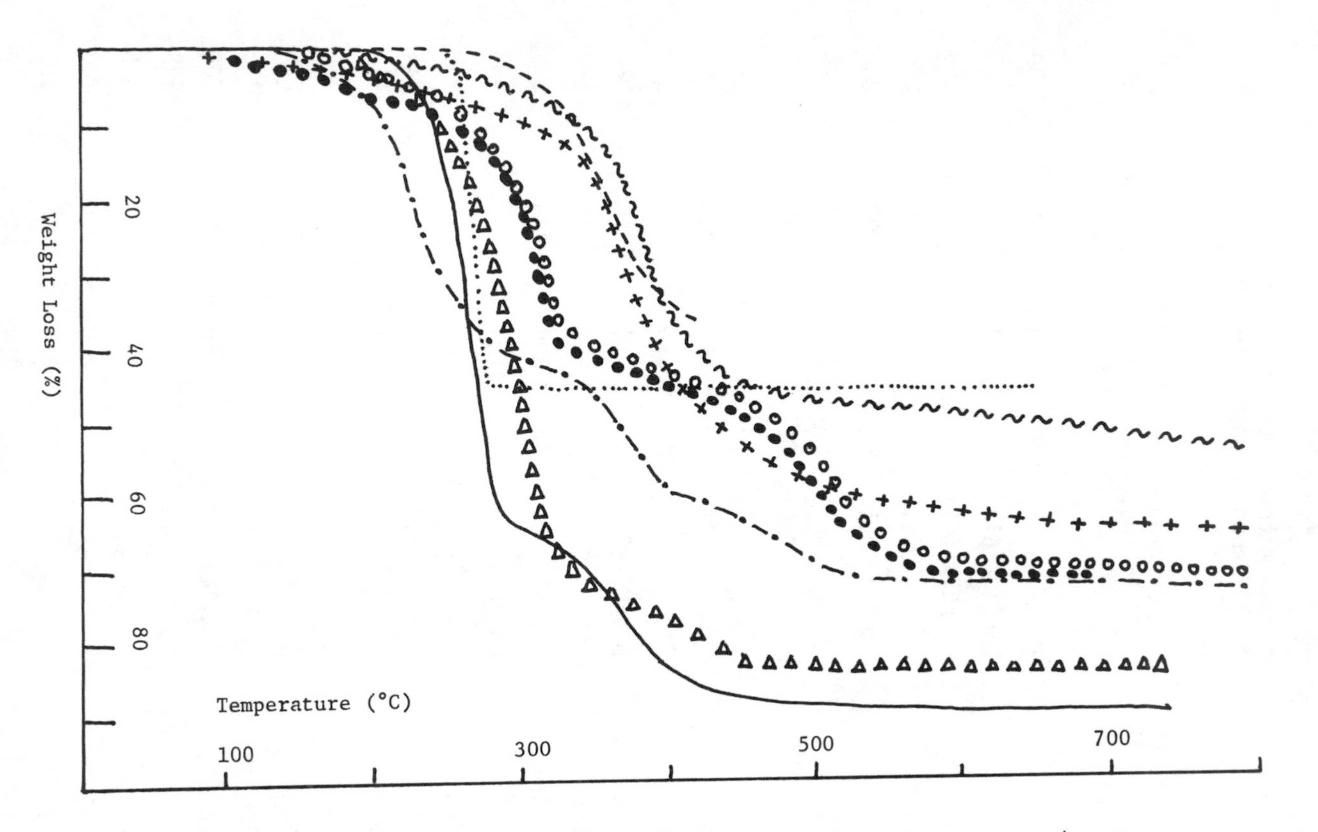

Figure 1. TGA thermograms of condensation products of PEI with Bu_2SnCl_2 (···); $Ø_2SnCl_2$ (– – –); $Ø_3SnCl$ (——); PVA with $(Bu)_2$ $SnCl_2$ (○); $Ø_2SnCl_2$ (●); $Ø_3SnCl$ (△); PAA with Bu_2SnCl_2 (+); $Ø_2SnCl_2$ (– · – ·); and $Ø_3SnCl$ (∿∿) in air at a heating rate of 30°C/min

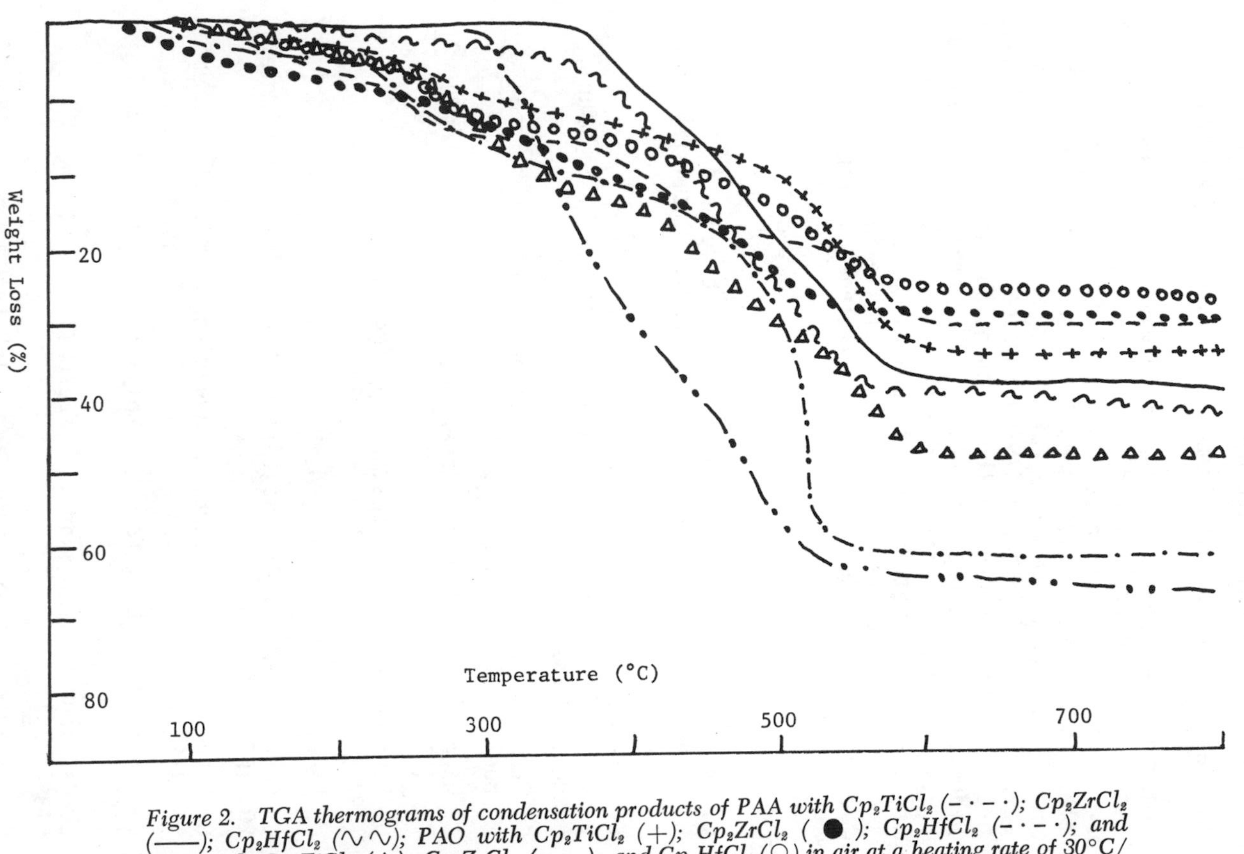

Figure 2. TGA thermograms of condensation products of PAA with Cp_2TiCl_2 (— · — ·); Cp_2ZrCl_2 (——); Cp_2HfCl_2 (∿∿); PAO with Cp_2TiCl_2 (+); Cp_2ZrCl_2 (●); Cp_2HfCl_2 (— · — ·); and PVA with Cp_2TiCl_2 (△); Cp_2ZrCl_2 (— — —); and Cp_2HfCl_2 (○) in air at a heating rate of 30°C/min

a. PEI = polyethyleneimine, PVA = polyvinyl alcohol, PAA = polyacrylic acid, PAO = polyacylamideoximes.

b. All results are for air, temperatures are to the nearest 25C°.

Many of the linear products can be cast to give tough films. Both the linear and crosslinked products can be rubbery and flexible with the crosslinked products inclined towards being more brittle. As expected, brittleness tends to increase with substitution. The products are hydrolytically stable to boiling water. for several days. This stability is physical, not chemical, i.e. the products are hydrophobic, thus preventing intimate contact of the hydrolyzable portions with water. This stability is overcome by addition of dipolar aportic solvents such as acetone which appear to act as a "wetting" agent towards the polymer.

The linear products are generally soluble in such dipolar aprotic solvents as DMSO, DMF, HMPA, Sulfolane, acetone, and triethylphosphate.

References

1. C. Carraher, D. Giron, W. Woelk, J. Schroeder and M. Feddersen, Organic Coatings and Plastics Chemistry, 38, 122 (1978).

2. C. Carraher, D. Giron, J. Schroeder, M. Feddersen and W. Woelk, "Additives in Polymers" (Edited by R. Seymour), Academic Press, 1978.

3. C. Carraher, M. Feddersen, J. Schroeder, D. Giron, and W. Woelk, CHEMISTRY, 51(5), 36 (1978).

4. C. Carraher and L. Torre, Angew. Makromolekulare Chemie, 21, 207 (1972).

5. C. Carraher and J. Piersma, J. Applied Polymer Sci., 16, 1851 (1972).

6. C. Carraher and L. Torre, J. Polymer Sci., A-1, 9, 975 (1971).

7. C. Carraher and J. Piersma, Angew. Makromolekulare Chemie, 28, 153 (1973).

8. C. Carraher and M. Feddersen, Angew. Makromolekulare Chemie, 54, 119 (1976).

9. C. Carraher and L-S. Wang, J. Polymer Sci., A-1, 9, 2893 (1971).

10. C. Carraher and L-S. Wang, J. Macromol. Sci.-Chem., A7(2), 513 (1973).

11. C. Carraher and L-S. Wang, Makromolekulare Chemie, 152, 43 (1972).

12. C. Carraher and J. Piersma, J. Macromol. Sci.-Chem., A7(4), 913 (1973).

13. C. Carraher and L-S. Wang, Angew. Makromolekulare Chemie, 25, 121 (1972).

14. C. Carraher and J. Piersma, Makromolekulare Chemie, 152, 49 (1972).

15. C. Carraher, Inorganic Macromolecules Reviews, 1, 271 (1972).

16. C. Carraher and R. Dammeier, Makromolekulare Chemie, 135, 107 (1970); 141, 245 (1971) and 141, 251 (1971).

17. M. I. Skenderov, K. Plekhanova and N. Adigezalora, Uch. Zap. Azerb. Gas. Univ. Ser. Khim. Nauk., 4, 71 (1965).

18. F. Frankel, D. Gertner, D. Wagner and A. Zilkha, J. Applied Polymer Sci., 9, 3383 (1965).

19. C. Carraher and R. Dammeier, J. Polymer Sci., 8, 3367 (1970).

20. C. Carraher, "Interfacila Synthesis," Vol. II, Chpt. 19 (Edited by F. Millich and C. Carraher), Marcel Dekker, N.Y., 1977.

21. F. Millich and C. Carraher, Macromolecules, 3, 253 (1970).

22. F. Millich and C. Carraher, J. Polymer Sci., A-1, 9, 1715 (1971).

23. C. Carraher, "Interfacial Synthesis," Vol. 2, Chpt. 20 (Edited by F. Millich and C. Carraher), Marcel Dekker, N.Y., 1977.

24. C. Carraher and J. Lee, J. Macromol. Sci.-Chem., A9(2), 191, 1975.

25. C. Carraher and J. Lee, Coatings and Plastics, 34(2), 478 (1974).

26. C. Carraher and S. Bajah, Polymer, 15, 9 (1974).

RECEIVED July 12, 1979.

6

Incorporation of Metal Ions into Polyimides

L. T. TAYLOR, V. C. CARVER, and T. A. FURTSCH[1]
Department of Chemistry, Virginia Polytechnic Institute and State University, Blacksburg, VA 24061
A. K. ST. CLAIR
NASA Langley Research Center, Hampton, VA 23665

The effect of ionic groups on the properties of bulk polymers(1) has normally referred to studies on polyelectrolytes in which an ionic group is covalently attached to the polymer chain which is usually neutralized by a metallic counterion. Studies of systems consisting of neutral polymers with dissolved inorganic salts are only beginning to receive considerable attention.

The results of the incorporation of ions in polymers and their effect on the glass transition temperature (T_g) has been reviewed through 1969.(2) Therefore, only the more recent and pertinent reports will be mentioned here. Large increases have been produced in the T_g of poly(propylene oxide) by dissolving $LiClO_4$ in the polymer.(3) Crystallization adducts of poly(ethylene oxide) that have been treated with $HgCl_2$ or $CdCl_2$ have also been reported.(4) Solutions of $Ca(NCS)_2$ and "Phenoxy" polymer have significantly different physical properties compared to the pure polymer. Increased water absorption, T_g and electrical conductivity are results of salt incorporation.(5) Mechanical properties of these glassy polymers are also affected by the presence of dissolved salt. Investigations(6) regarding the interaction of inorganic nitrate salts with cellulose acetate, poly(vinyl acetate), poly(vinyl alcohol), poly(methyl methacrylate) and poly (methyl acrylate) have been cited. In addition to observing large effects in T_g, large shifts in the infrared spectrum of both nitrate and polymer carbonyl frequencies have been observed. These observations have been interpreted in terms of complex formation between polymer and salt in the solid state. The change in T_g was shown to be an unusual function of metal ion concentration (i.e. T_g increases with increasing metal concentration up to

[1] Current address: Department of Chemistry, Tennessee Technological University, Cookeville, TN. 38501

0-8412-0540-X/80/47-121-071$05.00/0

a maximum point then T_g begins to decline).

The effects of various metal salts on the T_g and crystallinity of some polyamides have been noted.[7] Mixtures of Nylon 6 and either LiCl, LiBr or KCl were prepared by melting _in vacuo_ at 260°C an intimate blend of the two components. The equilibrium melting temperature of pure Nylon 6 was continuously depressed by increasing salt content with KCl being the most effective. In these studies the salt could be extracted from the polymer with hot water with complete recovery of the fusion temperature which is characteristic of pure Nylon 6. The occurrence of a strong interaction possibly between the amide group of the amorphous polymer and salts such as LiCl and LiBr was predicted. In a later study[8] these workers investigated the effect of these salts on the glass transition, T_g. T_g is not affected by type and content of metal salt. Mechanical data for unoriented specimen of Nylon 6-salt mixtures reveal that below T_g the shear modulus is not affected by the salt. Whereas above T_g the modulus of Nylon 6 is increased by KCl and decreased by LiCl and LiBr. The Newtonian melt viscosity was consistently higher for the mixtures than for pure Nylon 6 in each case. It should be noted, however, that LiBr and KCl were more effective than LiCl in causing the viscosity increase.

The employment of transition metal salts with neutral polar polymers is noticeably lacking. Recently, the addition of $ZnCl_2$ and $CoCl_2$ to high and to low molecular weight poly(propylene oxide) has been reported.[9] $ZnCl_2$ was found to increase the T_g of both high and low molecular weight polymer but $CoCl_2$ only increased the T_g of the low molecular weight polymer. A single phase system in the $ZnCl_2$ case was indicated; while a two phase system with $CoCl_2$ acting as a filler was suggested. In the zinc case, the elevation of T_g is believed to result from the formation of five-membered chelate rings by coordination of two adjacent oxygen atoms in the polymer chain with a $ZnCl_2$ molecule. In an analogous situation, $ZnCl_2$ was added to poly(tetramethylene glycol) with similar results albeit the T_g was raised less for a given amount of metal chloride. Intermolecular coordination with ether oxygen atoms from two neighboring chains was postulated since intramolecular bonding to zinc(II) would involve the formation of a less stable seven-membered chelate ring.

Angelo[10] has briefly reported in a patent the addition of metal ions to several types of polyimides. The object of the invention was a process for forming particle-containing (<1μ) transparent polyimide shaped structures. Unlike the work discussed previously, all of the metals were added in the form of coordination complexes rather than as simple anhydrous or hydrated salts. The properties of only one film (e.g. cast from a N,N-dimethylformamide(DMF) solution of 4,4'-diaminodiphenyl methane, pyromellitic dianhydride and bis(acetylacetonato)-

copper(II)) were given. These properties included: %Cu = 3.0%, dielectric constant = 3.6 and volume resistivity = 8 x 10^{12} ohms-cm. No further patents or published work in this area are apparently available(11). Since aromatic polyimides have proven to be excellent high temperature adhesives and limited data suggest that addition of metal or metalloid material as a filler may enhance(12) adhesive properties, we wish to report our results relating to the incorporation of a series of metal ions into various polyimides.

Results and Discussion

Polyimides derived from 3,3',4,4'-benzophenone tetracarboxylic acid dianhydride (BTDA), I, and 3,3'-diaminobenzophenone (m,m'-DABP), IIA, 4,4'-diaminobenzophenone (p,p'-DABP), IIB, or 4,4'-oxydianiline, IIC, (and to which have been added numerous metal compounds) have been prepared. The synthetic procedure em-

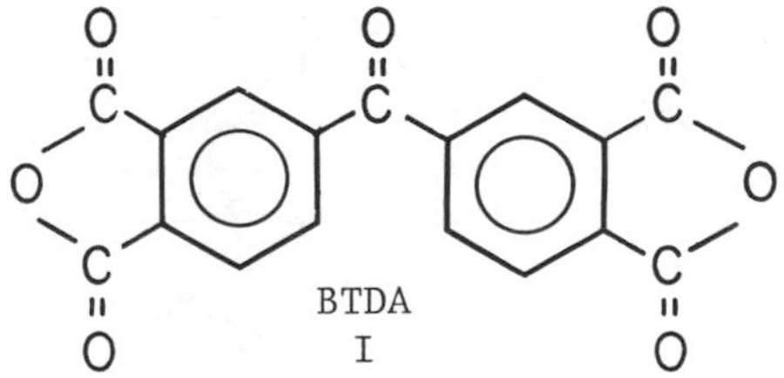

BTDA
I

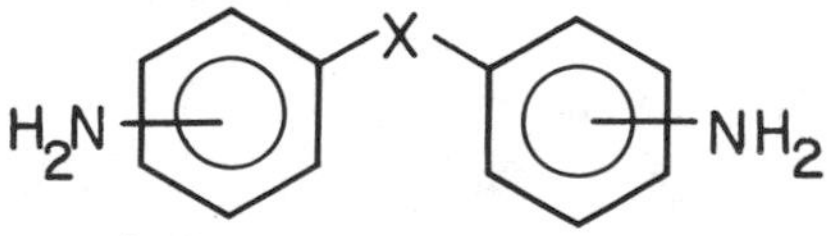

H_2N — X — NH_2

m,m'-DABP, X = C=O, IIA
p,p'-DABP, X = C=O, IIB
p,p'-ODA, X = O, IIC

ployed involved (1) formation of the polyamic acid (20% solids) in either DMF, N,N-dimethylacetamide(DMAC) or diethylene glycol dimethyl ether (Diglyme) (2) addition of the metal complex to the polyamic acid, III, in a 1:4 ratio (3) fabrication of a film of the polyamic acid-metal compound mixture and (4) thermal conversion (300°C) to the metal containing polyimide. Approximately twenty metals in a variety of forms were added to the polyamic acid solutions. Several experimental problems were encountered in following this procedure

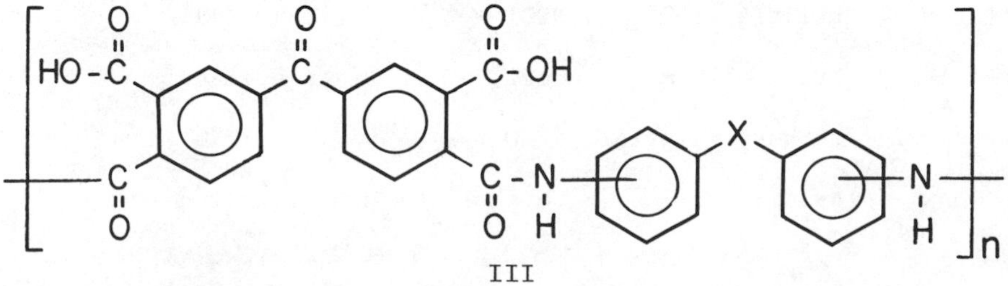

III

which limited the number of good quality films obtained. These problems include: (1) metal complex not dissolving in appropriate solvent, (2) gel formation or cross-linking of the polymer occurring upon interaction with the metal, (3) polyamic acid precipitating when the metal complex is added and (4) metal promoting thermal oxidative degradation of the polymer film upon curing.

In general polyimides derived from BTDA + p,p'-DABP yielded rather poor quality, very brittle films which were due in part to the low viscosity of the resulting polyamic acid solution. This fact, coupled with the observation that the adhesive properties of the p,p'-DABP isomer are poorer than the adhesive properties of the m,m'-DABP isomer (13) dictated that any extensive physical measurements should be carried out on the latter.

Table 1 lists some of the metal compounds employed and the results obtained when attempts were made to cast films of the resulting metal ion filled polyimide derived from BTDA + m,m'-DABP. Brittle films were produced in most cases regardless of whether the added metal ion was hydrated or anhydrous. The relatively low viscosities of the resulting polyamic acid-metal ion solutions no doubt accounted for this. Addition of $AlCl_3 \cdot 6H_2O$ or any simple aluminium salt to the polyamic acid produced immediately a rubbery material that could not be cast into a film. A similar result was obtained with $Ti(OEt)_4$ and $Ni(acac)_2$.

Table 1
Films Cast
BTDA+m,m'-DABP Polymer and Metal Ions*

Metal Compound	Results	Metal Compound	Results
$Al(acac)_3$	flexible film	$Mn(acac)_3$	very brittle film
$AlCl_3 \cdot 6H_2O$	rubbery material	$Ti(OEt)_4$	rubbery material
LiCl	surface DMAC	$SnCl_2 \cdot 2H_2O$	brittle film
$CrCl_3 \cdot 6H_2O$	brittle film	$MgCl_2 \cdot 6H_2O$	brittle film
$Fe(acac)_3$	brittle film	$NiCl_2 \cdot 6H_2O$	flexible film
$Cr(acac)_3$	brittle film	$CuCl_2 \cdot 4H_2O$	brittle film
$Ni(acac)_2$	rubbery material	$AgNO_3$	very brittle film
$CaCl_2$	brittle film	$Co(acac)_3$	brittle film

*Solvent = DMAC

Lithium containing films were unusual in that the film after curing was damp on the surface with what appeared to be the solvent, DMAC. No other films exhibited this property. The $AgNO_3$ containing film had the appearance of a silver mirror but the film was exceedingly brittle and "flaky-like". Only two truly flexible films were produced from BTDA + m,m'-DABP. These contained $Al(acac)_3$ and $NiCl_2 \cdot 6H_2O$ respectively.

A representative sample of many of the polyimide films that were produced were subjected to thermo-mechanical analysis (TMA), torsional braid analysis (TBA), thermal gravimetric analysis (TGA), infrared spectral analysis and weight loss on prolonged heating (e.g. isothermal studies), TABLE II. The softening temperature as measured by TMA and TBA are in general

TABLE II
TMA, TBA and TGA of Metal Ion Filled Polymers[a] of BTDA + m,m'-DABP

METAL ION	TMA(°C)	TBA(°C)	TGA(°C)
$Fe(acac)_3$	292	293	412
$Cr(acac)_3$	287	279	
$NiCl_2 \cdot 6H_2O$	279		495
$MnCl_2 \cdot 4H_2O$	279		495
$Al(acac)_3$	271	270	555
$CaCl_2$	264	266	518
$Co(acac)_3$	268	296	480
$CrCl_3 \cdot 6H_2O$	260		460
LiCl	252	264	480
$MgCl_2$	252	267	520
No Metal	251	252	570
$SnCl_2 \cdot 2H_2O$	237		550

[a]Solvent = DMAC; 0.1 g of metal complex (salt) per 4 g of polymer (20% solids)

[b]Polymer Decomposition Temperature

increased when metal ions are added with the exception of $SnCl_2 \cdot 2H_2O$. On the other hand, some thermal stability has been sacrificed as evidenced by the TGA data. Softening temperatures were more dramatically increased with p,p'-DABP polyimide than with the m,m'-DABP but at the expense of considerable thermal stability loss. No trend is apparent in the changes brought about by each metal ion. In fact, each metal is almost a case unto itself. This observation is further dramatized by some rather limited isothermal measurements on selected films (TABLE III). This data is typical of the metal ion filled BTDA + p,p'-DABP polyimides which we have examined. No changes in chemical functionality in the polyimide-metal film were apparent as judged by infrared spectral comparisons of polyimide alone and polyimide plus metal regardless of the metal employed.

Table III
Isothermal Studies[a]

MATERIAL	%Weight Loss
m,m'-DABP[b]	3
m,m'-DABP[c]	4
Polymer + $Al(acac)_3$[b,d]	5
Polymer + $CaCl_2$[b,d]	7
Polymer + LiCl[c,d]	13
Polymer + $Cr(acac)_3$[c,d]	52
Polymer + $Co(acac)_3$[c,d]	13

[a] 65 hours @ 316°C

[b] Solvent = DMAC

[c] Solvent = Diglyme

[d] 0.1 g of metal complex (salt) per 4 g of polymer (20% solids)

The best system studied in regard to enhancement of polymer properties while maintaining excellent film quality involves tri(acetylacetonato)aluminum(III) addition to the m,m'-DABP polyimide. An inspection of Tables I-III reveals that with $Al(acac)_3$ the softening temperature is increased without the loss of any polymer thermal stability, Figure 1. The remaining flexible film, $NiCl_2 \cdot 6H_2O$, has a slightly lower decomposition temperature than the "polymer-alone" film although the softening point has again been increased.

Since the m,m'-DABP polyimide is known to be an outstanding adhesive, lap shear strength tests employing titanium-titanium adherends and metal ion filled polyimides were conducted. Tests were performed at room temperature, 250°C and 275°C employing either DMAC or DMAC/Diglyme as the solvent. At room temperature regardless of the metal ion employed adhesive strength is de-

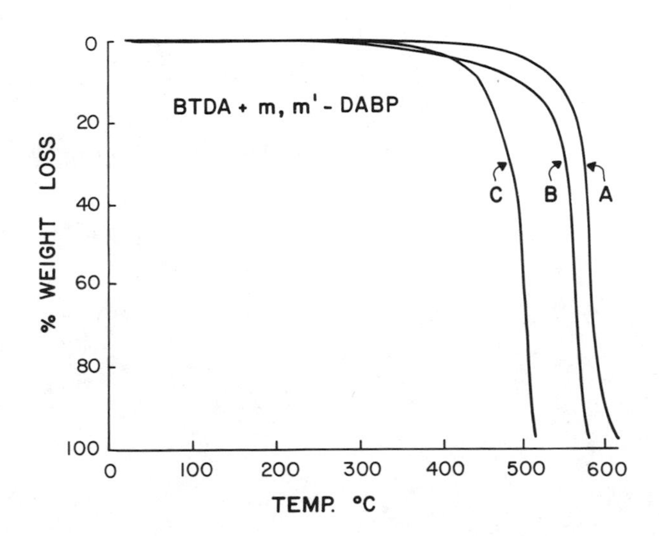

Figure 1. Thermogram (TGA) of (A) polyimide, (B) polyimide plus Al(acac)$_3$, and (C) polyimide plus $NiCl_2 \cdot 6\,H_2O$

creased relative to the "polymer-alone" case. Again the choice of metal ion is critical. The two best cases, i.e., $Al(acac)_3$ and $NiCl_2 \cdot 6H_2O$ were subjected to adhesive testing at elevated temperatures. Under these conditions the metal ion filled polyimides were superior. The $Al(acac)_3$ case proved to be exceptional in that it exhibited approximately four times the lap shear strength of "polymer-alone" at 275°C. We feel that this enhanced adhesiveness is due in part to the increased softening temperature of the $Al(acac)_3$ filled polyimide. These results are somewhat analogous to data collected by St. Clair and Progar earlier (14) regarding the use of aluminum metal as a filler with various polyimides. Lap shear strength was found to double at 250°C with 79% Al filled polyimide versus the unfilled polyimide.

Table IV
Lap Shear Tests[a]
Titanium-Titanium Adherend
BTDA-m,m'-DABP

Metal Ion Added	Lap Shear Strength[b] (psi)		
	25°C	250°C	275°C
No Metal	2966 (3138)	1573	438 (496)
$Al(acac)_3$[c]	2378 (2400)	1891	1641 (1348)
$NiCl_2 \cdot 6H_2O$[c]	1800	1332	608
LiCl[c]	781	--	--
$Cr(acac)_3$[c]	830	--	--

[a]Numbers in parenthesis correspond to DMAC/Diglyme solvent mixture.

[b]Average of four tests.

[c]0.1 g of metal complex (salt) per 4g of polymer (20% solids)

Surface and volume resistivity measurements have been performed on films of polymer alone and polymer with $Al(acac)_3$ added. Special care was taken to insure that only films of uniformly high quality were measured. Regardless of the film pretreatment, volume resistivities on two independently cast films of polymer alone fall in the 10^{16} ohm-cm range. No previously published resistivity is available on this particular polymer although upon surveying several ether polyimides from independent sources we measured similar volume resistivities.

Incorporation of $Al(acac)_3$ into the polyimide disappointingly shows no significant reduction in volume resistivity relative to the polymer alone. Replicate measurements (1.59×10^{15} and 1.12×10^{16} ohm-cm) on two independently cast films support this conclusion. Reorientation of the same film in the electrode assembly yielded identical results suggesting uniform behavior throughout the film containing $Al(acac)_3$. Similar results were obtained on $NiCl_2 \cdot 6H_2O$ filled polyimides. (Table V)

Table V

Resistivity Data on BTDA-m,m'-DABP Polyimide Films[a]

Film	Volume Resistivity (ohms-cm)	Metal Content
Polymer	1.36×10^{16}	-
	2.32×10^{16}	-
Polymer + $Al(acac)_3$	1.59×10^{15}	1.8% Al
	1.92×10^{16}	2.9% Al
Polymer + $NiCl_2 \cdot 6H_2O$	1.66×10^{16}	0.9% NI
	1.27×10^{16}	-

[a]Electrification Period = 1 minute @ 500 volts

[b]Solvent = DMAC

Surface resistivity measurements were carried out, but there is considerable scatter in the measurements. Other workers have also noted greater uncertainties associated with surface resistivity measurements relative to volume resistivy measurements Again, the $Al(acac)_3$ and $NiCl_2 \cdot 6H_2O$ containing films exhibit a surface resistivity similar to the average of the three data points obtained for "polymer alone" films. Numerous efforts to prepare high quality films incorporating other metal ions into BTDA + m,m'-DABP or p,p'-DABP were not satisfactory because metal ion addition resulted in a decrease in solution viscosity leading to unsatisfactory films insofar as resistivity measurements are concerned.

These results suggest that during the film curing process, the non-conducting $Al(acac)_3$ and $NiCl_2$ maintain their integrity rather than being converted to the more conductive aluminum or nickel metal as originally envisioned. X-ray photoelectron spectroscopic (XPS) data on the $Al(acac)_3$ containing polyimide employing a magnesium anode suggests one type of aluminum species, Figure 2. A binding energy of 118.4 eV for Al $2s_{1/2}$ is determined based on an internal carbon calibration C $1s_{1/2}$ assumed to be 284.0 eV. This value of 118.4 eV coupled with the facts that (1) no change in binding energy is found for

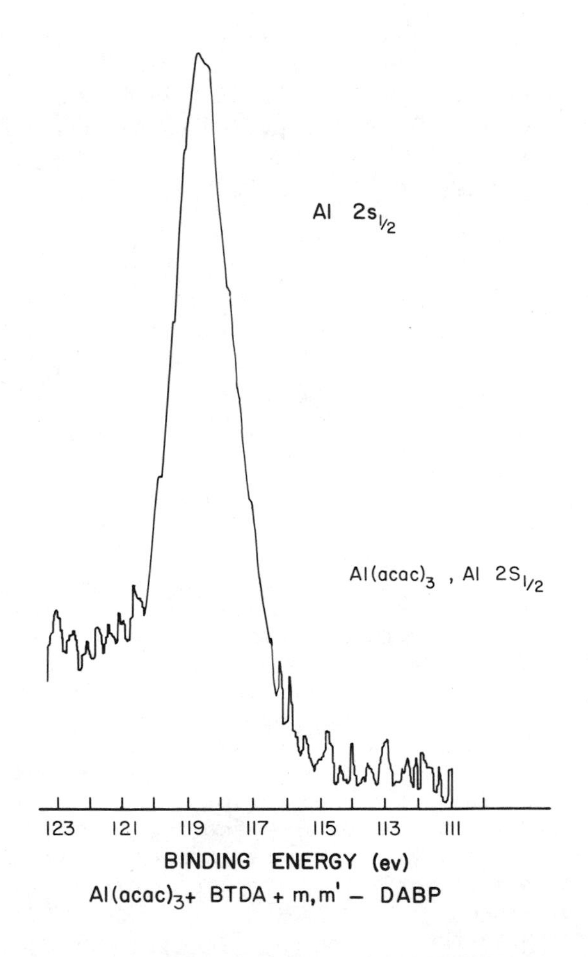

Figure 2. X-ray photoelectron spectrum (Al $2s_{1/2}$) of BTDA + m,m'-DABP + Al(acac)$_3$ polymer film (B.E. = 118.4 eV)

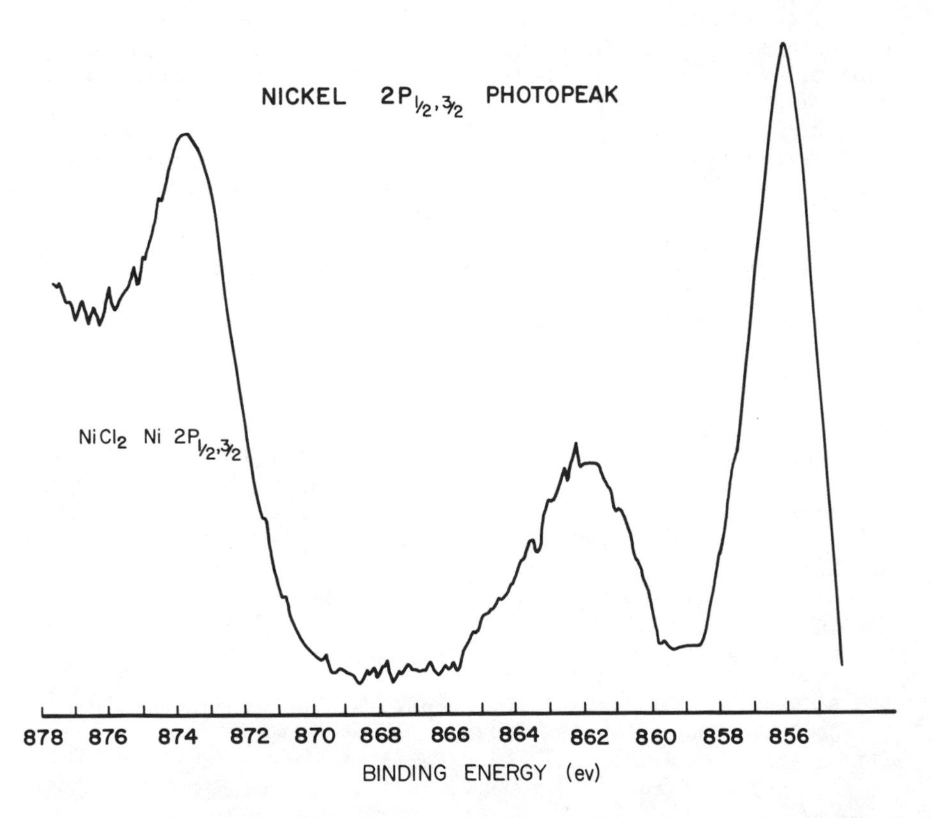

Figure 3. X-ray photoelectron spectrum (Ni $2p_{1/2,3/2}$) of BTDA + m,m'-DABP + $NiCl_2 \cdot 6\,H_2O$ polymer film (B.E. = 855.7 eV, 873.3 eV)

oxygen or nitrogen in the polyimide and (2) the binding energy of Al $2s_{1/2}$ in $Al(acac)_3$ is 117.9 eV indicates that aluminum continues to be bound to acetylacetone in the polyimide. An examination of XPS data obtained on the $NiCl_2$ containing polyimide, Figure 3, leads to a similar conclusion (Ni $2p_{3/2}$ = 854.2 eV in the polyimide, Ni $2p_{3/2}$ = 855.0 eV in $NiCl_2$. The smaller intensity peak centered around 862 eV is the $2p_{3/2}$ satellite photopeak which again is indicative of nickel(II).

Further work in this area is underway employing polyamic acid systems which are known to produce higher viscosity solutions (e.g. polyimides derived from 4,4'-oxydianiline and either BTDA or pyromellitic dianhydride). This is being carried out in the belief that higher viscosity solutions will give rise to higher quality, less brittle films and will, thereby, enable a broader spectrum of metal systems to be studied regarding the adhesive and electrical conductance properties of metal ion filled polyimides.

Acknowledgment - Informative discussions regarding this work with T. L. St. Clair are gratefully appreciated. We also wish to thank Robert Ely for technical assistance.

References

1. E. P. Otocha, J. Macromol. Sci., Sci., C5, 275 (1971).
2. A. Eisenbert, Macromolecules, 4 125 (1971).
3. J. Moacanin and E. F. Cuddihy, J. Polym. Sci. C, 14, 313 (1966).
4. M. Yokoyama, H. Ishihara, R. Iwamoto and H. Tadokoro, Macromolecules, 2, 184 (1969).
5. M. J. Hannon and K. F. Wissbrun, J. Polym. Sci. Polym. Phys. Ed., 13, 113 (1975).
6. M. J. Hannon and K. F. Wissburn, J. Polym. Sci. Polym., Phsy. Ed., 13, 223 (1975).
7. B. Valenti, E. Bianchi, G. Greppi, A. Tealdi and A. Ciferri, J. Phys. Chem. 77, 389 (1973).
8. D. Acierno, E. Bianchi, A. Ciferri, B. DeCindio, C. Migliaresi and L. Nicolais, J. Polym. Sci., Symposium No. 54, 259 (1976).
9. R. E. Wetton, D. B. James and W. Whiting, Polym. Letters Ed., 14, 577 (1976).
10. R. J. Angelo and E. I. duPont deNemours & Co., U.S. Patent, No. 3, 073, 785 (1959).
11. Private Communication, R. J. Angelo.
12. H. A. Burgman, J. H. Freeman, L. W. Frost. G. M. Bower, E. J. Traynor and C. R. Ruffing, J. Appl. Polym. Sci., 12, 805 (1968).
13. T. L. St. Clair and D. J. Progar, Polymer Preprints, 16(1) 538 (1975).
14. D. J. Progar and T. L. St. Clair, 7th National SAMPE Technical Conference, Albuquerque, NM Oct., 1975.

RECEIVED July 12, 1979.

7

Biologically Active Modification of Polyvinyl Alcohol: The Reaction of Phenyl Isocyanate with Polyvinyl Alcohol

CHARLES G. GEBELEIN and KEITH E. BURNFIELD

Department of Chemistry, Youngstown State University, Youngstown, OH 44555

Numerous papers have appeared in the literature describing the reactions of various compounds with polymers in order to change the properties of the polymers (1-5). Often these reactions have resulted in significant changes in such properties as flammability, solubility, thermal degradation, photodegradation, strength and biological activity. In this present paper, we will briefly review some of the reactions that have been run on poly(vinyl alcohol), especially in regards to their potential biomedical activity, and then we will describe the reaction of phenyl isocyanate with poly(vinyl alcohol) and consider some of the potential utility of this modified polymer.

PVA Formation Reaction. Poly(vinyl alcohol) is itself a modified polymer being made by the alcoholysis of poly(vinyl acetate) under acid or base catalysis as shown in Equation 1 (6,7). This polymer cannot be made by a direct polymerization because the vinyl alcohol monomer only exists in the tautomeric form of acetaldehyde. This saponification reaction can also be run on vinyl acetate copolymers and this affords a means of making vinyl alcohol copolymers. The homopolymer is water soluble and softens with decomposition at about 200°C while the properties of the copolymers would vary widely. Poly(vinyl alcohol) has been widely utilized in polymer modification because: (1) it is readily available, (2) it is inexpensive and (3) it has the readily reacted hydroxyl group present.

$$-\!\!\left(CH_2\underset{\underset{\displaystyle CH_3-C=O}{|}}{\underset{O}{\underset{|}{CH}}}\right)_{\!n}- \xrightarrow[H^+ \text{ or } OH^-]{CH_3OH} -\!\!\left(CH_2\underset{OH}{\underset{|}{CH}}\right)_{\!n}- + n\ CH_3COOCH_3$$

(Equation 1)

Esterification Reactions. Possibly the simplest reaction on poly(vinyl alcohol) would be the acetylation to regenerate the poly(vinyl acetate), Equation 2. While this reaction occurs readily, with acetic anhydride in pyridine solution, the result-

0-8412-0540-X/80/47-121-083$05.00/0

ing poly(vinyl acetate) often has a structure (and properties) different from the orginal vinyl acetate polymer. This change occurs because the original poly(vinyl acetate) often has chain branching occurring from the acetate group as well as the polymer backbone chain. On alcoholysis, these branches are cleaved and the polymer decreases both in molecular weight and the extent of branching (8-12). Other polymeric esters can be made from poly(vinyl alcohol) and this is illustrated in Equation 3 which shows the formation of poly(vinyl butyrate).

$$-\!\!(\!-CH_2\underset{\underset{OH}{|}}{CH}-\!)_n \xrightarrow[\text{pyridine}]{(CH_3CO)_2O} -\!\!(\!-CH_2\underset{\underset{\underset{O=C-CH_3}{|}}{O}}{\underset{|}{CH}}-\!)_n \qquad \text{(Equation 2)}$$

$$-\!\!(\!-CH_2\underset{\underset{OH}{|}}{CH}-\!)_n \xrightarrow[\text{pyridine}]{(CH_3CH_2CH_2CO)_2O} -\!\!(\!-CH_2\underset{\underset{\underset{O=CCH_2CH_2CH_3}{|}}{O}}{\underset{|}{CH}}-\!)_n \qquad \text{(Equation 3)}$$

Radiation Induced Reactions. Graft polymers have been prepared from poly(vinyl alcohol) by the irradiation of the polymer-monomer system and some other methods. The grafted side chains reported include: acrylamide, acrylic acid, acrylonitrile, ethyl acrylate, ethylene, ethyl methacrylate, methyl methacrylate, styrene, vinyl acetate, vinyl chloride, vinyl pyridine and vinyl pyrrolidone (13). Poly(vinyl alcohols) with grafted methyl methacrylate and sometimes methyl acrylate have been studied as membranes for hemodialysis (14). Graft polymers consisting of 50% poly(vinyl alcohol), 25% poly(vinyl acetate) and 25% grafted ethylene oxide units can be used to prepare capsule cases for drugs which do not require any additional plasticizers (15).

When poly(vinyl alcohol) is irradiated with an electron beam above its Tg, a crosslinked hydrogel forms (16). Poly(vinyl alcohol) hydrogels have been found to be effective eye lens substitutes in albino rabbits since their properties were very similar to the intact eye lens (17). Poly(vinyl alcohol) ointments containing pilocarpine have been shown to be effective sustained release agents which last 4-5 hours longer than the usual solutions of pilocarpine and are nearly free of side effects such as eye irritation (18). Another hydrogel application involves crosslinking poly(vinyl alcohol) and heparin together with glutaraldehyde and formaldehyde. These hydrogels are claimed to have low thrombogenicity and have been evaluated as a hemodialysis membrane. They show promise in this latter application since they are more permeable to molecules such as insulin than are the usual cellophane membranes. Radiation crosslinked poly(vinyl alcohol) gels have been proposed as synthetic cartilage in synovial joints (19).

Acetal Formation Reactions. Like other alcohols, poly(vinyl alcohol) undergoes an acid catalyzed reaction with various aldehydes to form acetals. Two of these products, the formal and the butyral, are of importance industrially and the reactions are shown in Equations 4 and 5, respectively. Poly(vinyl butyral) has been known since the 40's and is used as a plastic interlayer in safety glass. In this case, the reaction is stopped with about 25% of the OH groups remaining, to promote good adhesion to the glass, and the polymer is usually plasticized with about 30% dibutyl sebacate (20). This acetal reaction will not proceed beyond about 87.5% conversion of the OH groups since some of these groups become isolated and cannot undergo acetal formation.

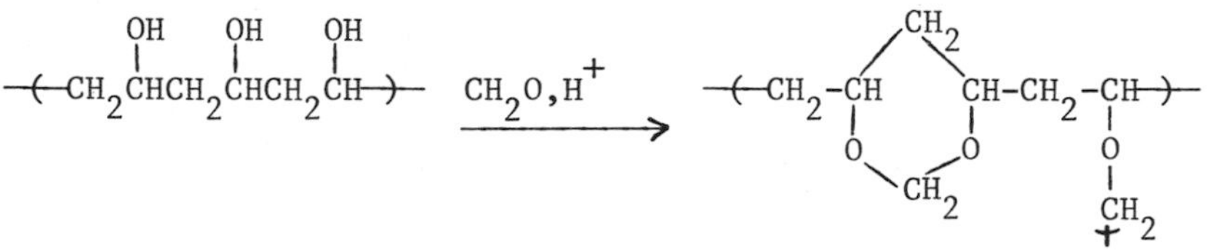

$$-(CH_2CH(OH)CH_2CH(OH)CH_2CH(OH))- \xrightarrow{CH_2O,\,H^+} -(CH_2-CH-O-CH_2-O-CH-CH_2-CH(-O-CH_2-))-$$

(crosslinked poly(vinyl formal)

(Equation 4)

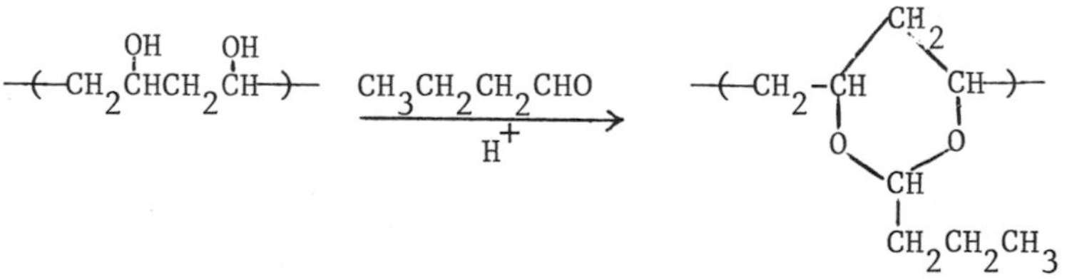

$$-(CH_2CH(OH)CH_2CH(OH))- \xrightarrow[H^+]{CH_3CH_2CH_2CHO} -(CH_2-CH-O-CH(CH_2CH_2CH_3)-O-CH)-$$

(Equation 5)

Poly(vinyl formal) (as in Equation 4 but not crosslinked) is a tough plastic material and is sometimes used for fuel tank coatings. The reaction of poly(vinyl alcohol) with formaldehyde quickly renders the product water insoluble and these materials can be used as "poly(vinyl alcohol)" fibers which have much higher water absorption than most other fibers. Normally these have about a third of the OH groups reacted (21). A crosslinked version of poly(vinyl alcohol) called Ivalon has been used in the form of a sponge in various soft tissue replacements, corneal implants, as a plasma extender and as a bone material in human and animal bodies. Unfortunately, this material tends to harden with time when implanted and has generally been found to be unsuitable. In addition, it also appears to retard bone healing (22-24). A recent acetal reaction, shown in Equation 6, has been used to introduce sulfonamide groups into this type of polymer. The acetal reaction is carried out to 33-67% substitution and the final product is water soluble (after base treatment) (25).

$$\text{-(CH}_2\text{CH(OH)CH}_2\text{CH(OH)CH}_2\text{CH(OH))-} + \text{OHC-C}_6\text{H}_4\text{-SO}_2\text{NHR} \xrightarrow{H^+} \text{-(CH}_2\text{CH(OH)CH}_2\text{-CH-CH}_2\text{-CH)-} \text{ (cyclic acetal: O-CH(C}_6\text{H}_4\text{SO}_2\text{NHR)-O)}$$

(insoluble)

$$\xrightarrow{NaOH} \text{-(CH}_2\text{CH(OH)CH}_2\text{-CH-CH}_2\text{-CH)-} \text{ (cyclic acetal: O-CH(C}_6\text{H}_4\text{SO}_2\text{-}\overset{-}{N}\text{-R } Na^+\text{)-O)}$$

(water soluble)

($R = H, C_6H_5, n\text{-}C_4H_9$)

(Equation 6)

Complexes with Iodine. One of the simplest "reactions" of poly(vinyl alcohol) is the formation of a blue complex with iodine. This complex formation, which requires the presence of KI, has been studied extensively by many workers (26-31). This complex also forms with partially hydrolyzed poly(vinyl acetates) (26) and is known to be affected by the 1,2-glycol content and the isotacticity of the polymer both of which tend to reduce complex formation (31). The complex also depends on the molecular weight of the poly(vinyl alcohol) and the iodine concentration. A helical structure has been proposed (28,29) for this complex but has not been firmly proven (30). One thing that is certain, however, is that these complexes exhibit a high degree of biological activity (3). The PVA-iodine complex showed no apparent toxic effects when injected into mice at the level of 1-5 mg/kg body weight. Higher dose levels did cause muscle contraction, inhibit locomotor activity and decrease appetite (32). When a PVA-I_2 complex was administered to guinea pigs which were infested with *Ascaris lumbricoids*, decreased larvae content resulted and the peroxide content of the brain increased (33). PVA-I_2 complexes were found to be effective in the treatment of gastrointestinal diseases (enteritis) in swine (dose level of 10-15 ml twice daily) and cattle (dose level of 25-30 ml twice daily) for a three to four day period, in the treatment of coccidiosis and streptococcosis in rabbits (2-3 ml, three times daily) and in the treatment of atrophic rhinitis in swine (34). Ram sperm has been disinfected by incubating 24 hours at 37°C with a 10% PVA-I_2 solution which was nontoxic to the sperm. Iodinated starch was toxic under these conditions (35). Finally, iodine complexes of poly(vinyl alcohol), poly(vinyl pyrrolidone) and amylose were found to possess bactericidal activity against eleven bacteria stains with the PVA complex being the most effective. The minimum inhibitory concentration range for the strains tested were 477, 683 and 874 g/ml for the iodine complexes of poly(vinyl alcohol), poly(vinyl pyrrolidone) and amylose, respectively (36).

Phosphorus Containing Modifications. Many workers have modified poly(vinyl alcohol) in order to include the phosphorus atom in the molecule and a full survey of these would be beyond the scope of this paper. Much of this work was done in the hope of imparting flame resistance to the polymers. The reaction of poly(vinyl alcohol) with P_2O_5 and H_3PO_4 resulted in crosslinked products with up to 20% phosphorus (37,38). Likewise the treatment of poly(vinyl alcohol) with $POCl_3$ resulted in a crosslinked polymer which contained primary, secondary and tertiary phosphate groups. This modified polymer could be used as an ion exchange resin and had a capacity of 9.7 meq/g resin (39). When poly(vinyl alcohol) was reacted with H_3PO_4 and urea, about 80% of the hydroxyl groups were converted to primary phosphate ester groups and the polymer was water soluble. This polymer also contained up to 21.5% nitrogen (as ammonium groups) (38).

Organophosphorus compounds have been reacted with poly(vinyl alcohol) by many workers. In one case, about 80% of the hydroxyl groups were converted to an acetal by the reaction with butyraldehyde and the remainder were reacted with $(C_6H_5)_2POCl$ to give a polymer with 5.2% phosphorus (40). The reaction of poly(vinyl alcohol) with $((CH_3)_2CHO)_2POCl$ followed by crosslinking with diisocyanates is reported to give a material useful for flameproofing textiles (41). The reaction of $RPOCl_2$ or $RPSCl_2$ (R = C_6H_5,$-CH_2Cl$ or $-C_2H_5$) with poly(vinyl alcohol) gives crosslinked products with some primary phosphate groups present (42).

Organometallic Modifications. Various organometallic groups have been incorporated into the poly(vinyl alcohol) polymer chain. For example, $(cyclopentadienyl)_2^-$ MCl_2 (M = Zr, Ti or Hf) reacts with PVA to give crosslinked polymers with 37-41% of the organometallic product present (43). Modified poly(vinyl alcohols) with tin groups were prepared by reacting the polymer with R_2SnCl_2 (R = C_6H_5 or n-C_4H_9), which gave crosslinked products, or R_3SnCl which gave linear polymers with 7-12% organometallic (44). Poly(vinyl alcohol) also reacts with RSO_2Cl compounds to give linear products (45). In general, where the organometallic, etc., contains only one reactable halogen, the resulting modified poly(vinyl alcohol) is linear rather than crosslinked.

Modifications with Thiol Groups. Mercapto or thio groups often have powerful physiological activity and have been claimed to be useful as radiation protective groups. These groups have been introduced into poly(vinyl alcohol) in a variety of ways. The hydroxyl group of PVA has been converted to the thiol group by the sequence of reactions shown in Equation 7 (46,47). A single thiol group was introduced at the bottom of an acetal ring by the reaction sequence shown in Equation 8 (48,49). Two thiol groups have been introduced into the poly(vinyl alcohol) base molecule by three different routes which are outlined in Equations 9, 10 and 11 (48,49).

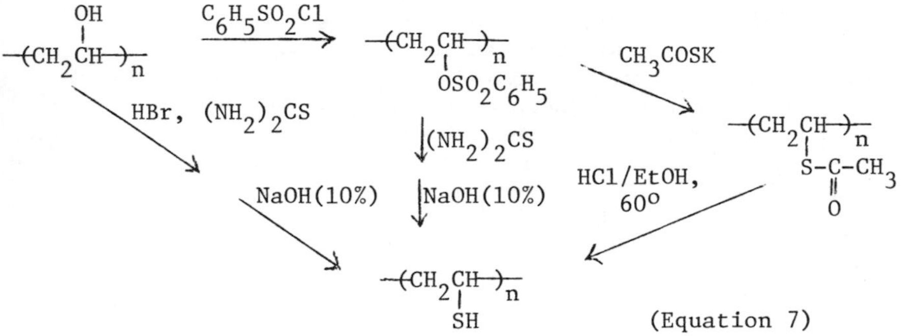

$$-(CH_2CH(OH))_n- \xrightarrow{C_6H_5SO_2Cl} -(CH_2CH(OSO_2C_6H_5))_n- \xrightarrow{CH_3COSK} -(CH_2CH(S-C(=O)-CH_3))_n- \xrightarrow{HCl/EtOH,\ 60^o} -(CH_2CH(SH))_n-$$

$$-(CH_2CH(OSO_2C_6H_5))_n- \xrightarrow{(NH_2)_2CS} \xrightarrow{NaOH(10\%)} -(CH_2CH(SH))_n-$$

$$-(CH_2CH(OH))_n- \xrightarrow{HBr,\ (NH_2)_2CS} \xrightarrow{NaOH(10\%)} -(CH_2CH(SH))_n-$$

(Equation 7)

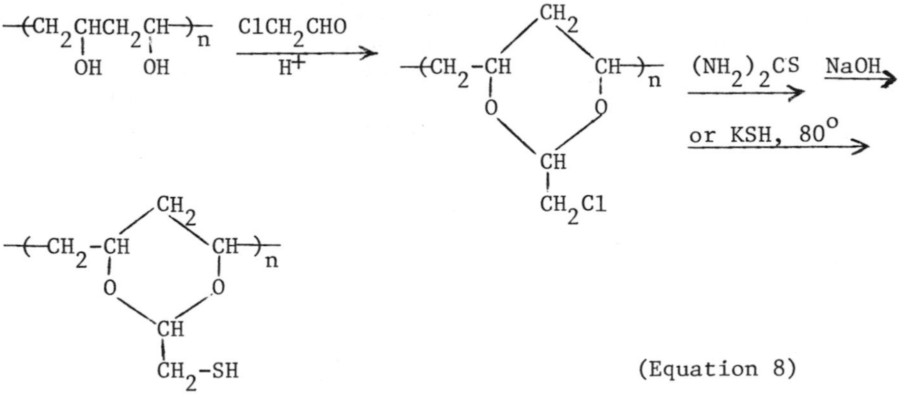

$$-(CH_2CH(OH)CH_2CH(OH))_n- \xrightarrow[H^+]{ClCH_2CHO} -(CH_2-CH-CH_2-CH)_n- \text{ (acetal ring: } O-CH(CH_2Cl)-O\text{)} \xrightarrow{(NH_2)_2CS} \xrightarrow{NaOH} \text{ or } \xrightarrow{KSH,\ 80^o}$$

$$-(CH_2-CH-CH_2-CH)_n- \text{ (acetal ring: } O-CH(CH_2-SH)-O\text{)}$$

(Equation 8)

$$-(CH_2CH(OH))_n- \xrightarrow{Na} \xrightarrow{CH_2=CHCH_2Br} \xrightarrow{Br_2,CCl_4}$$

$$\xrightarrow[(NH_2)_2CS]{CH_3COSK\ or\ KSH\ or} -(CH_2CH(O-CH_2-CH(SH)-CH_2(SH)))_n-$$

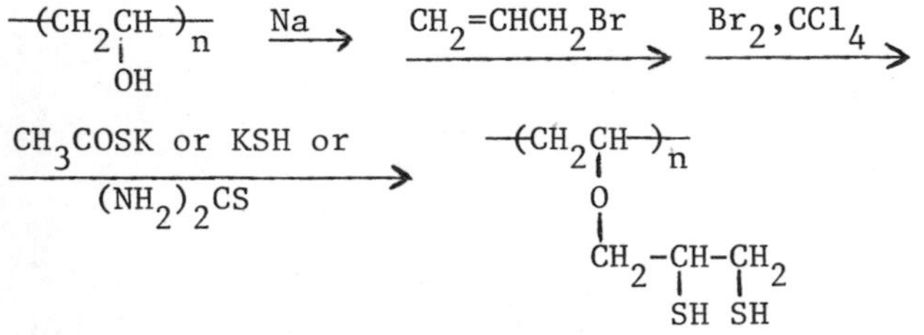

(Equation 9)

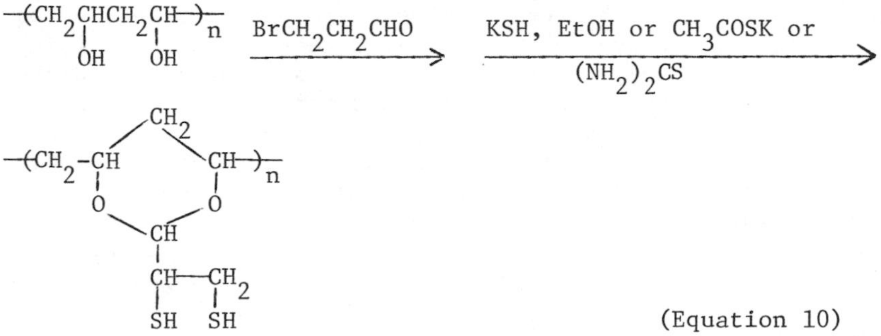

$$-\!\!(CH_2\underset{OH}{CH}CH_2\underset{OH}{CH})_n- \xrightarrow{BrCH_2CH_2CHO} \xrightarrow[(NH_2)_2CS]{KSH,\ EtOH\ or\ CH_3COSK\ or}$$

$$-\!\!(CH_2-\overset{}{CH}\ \ CH_2\ \ CH)_n- \quad (\text{1,3-dioxane ring: O–CH–O, with } CH-CH(SH)-CH_2(SH))$$

(Equation 10)

$$-\!\!(CH_2\underset{OH}{CH}CH_2\underset{OH}{CH})_n- \xrightarrow[H^+]{BrCH_2CHO} \xrightarrow{CH_2=CHCH_2ONa} \xrightarrow{Br_2,CCl_4}$$

$$\xrightarrow[(NH_2)_2CS]{KSH\ or\ CH_3COSK\ or} -\!\!(CH_2-CH\ \ CH_2\ \ CH)_n- \quad (\text{1,3-dioxane ring: O–CH–O, with } CH-CH_2-O-CH_2CH(SH)-CH_2(SH))$$

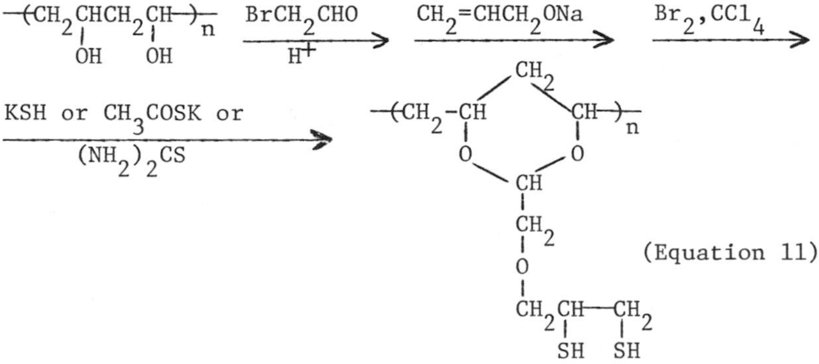

(Equation 11)

Modifications Containing Biologically Active Groups. Many examples of modified poly(vinyl alcohols) containing biologically active groups have been made in recent years. Nucleic acid models were prepared by reacting 2-pyridone-5-carboxylic acid with N,N'-carbonyldiimidazole in dimethylformamide and then reacting with poly(vinyl alcohol) (50). Aspirin, and some other salicylic acid derivatives, have been added to poly(vinyl alcohol) and cellulose acetate by melt esterification as illustrated in Equation 12. The PVA-bound aspirin showed a longer duration of the antiinflammatory and analgesic activities than does free aspirin (51).

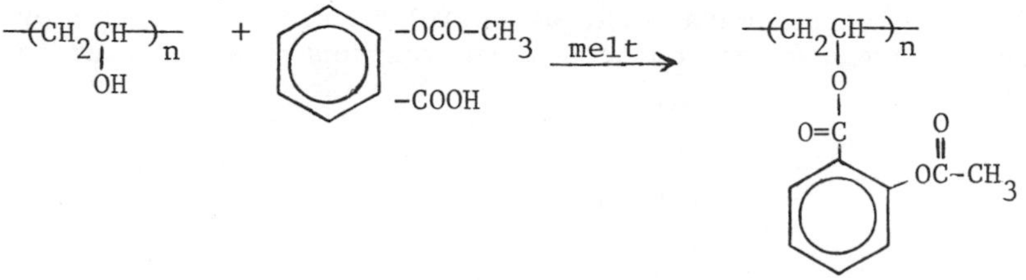

$$-\!\!(CH_2\underset{OH}{CH})_n- \ + \ C_6H_4(-OCO-CH_3)(-COOH) \xrightarrow{melt} -\!\!(CH_2\underset{O-C(=O)-C_6H_4-OC(=O)-CH_3}{CH})_n-$$

(Equation 12)

Several examples of the binding of enzymes to poly(vinyl alcohol) are in the literature. These could possibly be used to treat enzyme deficiency diseases. In a recent example, trypsin was immobilized on poly(vinyl alcohol) fibers using maleic dialdehyde or bromal. While the reaction was more complete with bromal, the reaction with maleic dialdehyde gave a better support which showed decreasing activity with increasing enzyme content. The activity of the bromal activated system was independant of the enzyme content (52). Trypsin and papain were attached to poly(vinyl alcohol) by the reaction sequence shown in Equation 13. In this case, the crosslinked poly(vinyl alcohol) is treated by the 1,3-dioxalone derivative and then converted to either the isothiocyanate or the diazonium salt for coupling with the enzyme. The bound enzymes showed significant, altho reduced, activity in each case (53).

$-(CH_2CH)-$ (OH) + OHC–C$_6$H$_4$–CHO → [1,3-dioxolane-substituted aniline, NH_2] $\xrightarrow{HONO \text{ or } CSCl_2}$ $\xrightarrow{(R=N_2^+ Cl^- \text{ or } -NCS)}$ [poly(vinyl alcohol) acetal with CH–C$_6$H$_4$–R and crosslinking CH acetal units, OH]

(Equation 13)

Imidazole units have been substituted on poly(vinyl alcohol) chains to the extent of 19-51% as shown in Equation 14 (54). This reaction was run 1-5 days as a 1% solution in DMF at 85°C. Using the sequence of reactions outlined in Equation 15, 6-methylthiopurine units have been placed on the poly(vinyl alcohol) backbone (55). These polymers might be useful in treating cancer or leukemia since the parent compound, 6-mercaptopurine, is used in this way.

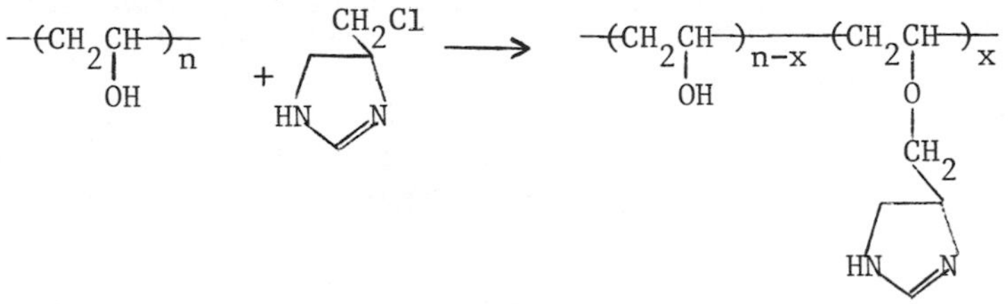

$$-\!(CH_2CH(OH))_n- \; + \; \text{4-(chloromethyl)imidazole} \longrightarrow -\!(CH_2CH(OH))_{n-x}\!-\!(CH_2CH(O\!-\!CH_2\!-\!\text{imidazolyl}))_x-$$

(Equation 14)

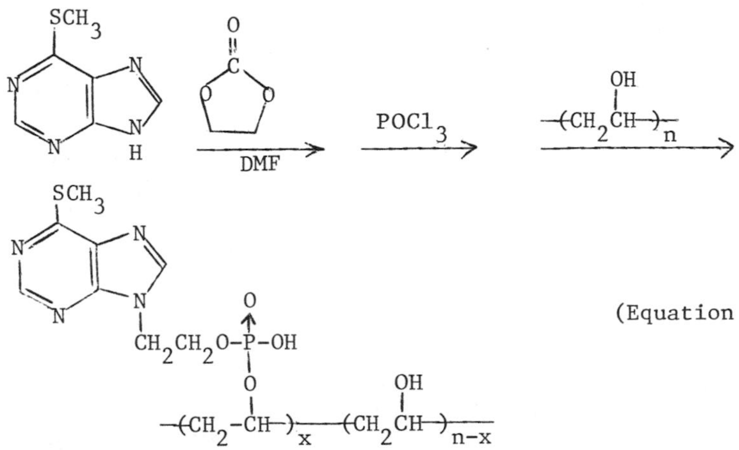

$$\text{6-(methylthio)purine} \; + \; \text{ethylene carbonate} \xrightarrow{DMF} \xrightarrow{POCl_3} \xrightarrow{-(CH_2CH(OH))_n-} -\!(CH_2\!-\!CH(O\!-\!P(\rightarrow O)(OH)\!-\!OCH_2CH_2\!-\!\text{N9-(6-methylthiopurinyl)}))_x\!-\!(CH_2CH(OH))_{n-x}-$$

(Equation 15)

A potential polymeric herbicide and/or pesticide has been made from poly(vinyl alcohol) and 2,6-dichlorobenzaldehyde, which is a known pesticide with strong herbacidal and moderate fungicidal activity, as outlined in Equation 16. This polymer shows negligible hydrolysis at room temperature and only 2% hydrolysis after three days at 60°C. The extent of substitution ranged from 18-68% (56).

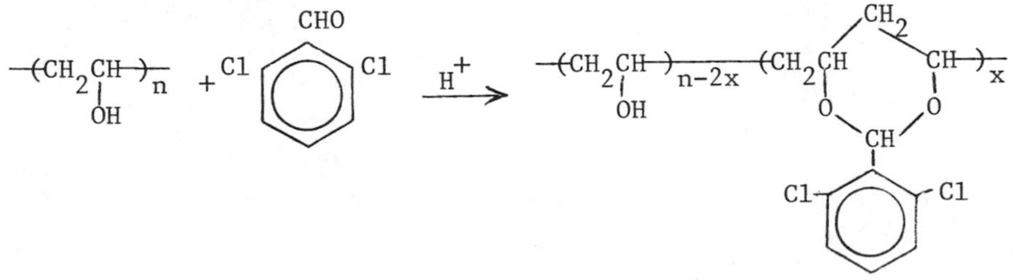

$$-\!(CH_2CH(OH))_n- \; + \; \text{2,6-}Cl_2C_6H_3CHO \xrightarrow{H^+} -\!(CH_2CH(OH))_{n-2x}\!-\!(CH_2CH\!-\!CH_2\!-\!CH)_x- \; [\text{1,3-dioxane ring: } O\!-\!CH(C_6H_3\text{-2,6-}Cl_2)\!-\!O]$$

(Equation 16)

Reactions with Isocyanates. The reaction of alcohols with isocyanates to form carbamates is well known and similar reactions with poly(vinyl alcohol) would be expected. Until recently, the only available reaction conditions were to use a heterogeneous reaction mixture or to run the reaction in a poor solvent for poly(vinyl alcohol). The best poly(vinyl alcohol) solvents, water and formaide derivatives, react rapidly with isocyanates. Nevertheless, several such reactions have been run in the past and we will cite only a few of them. A potentially photosensitive polymer was made by the reaction of allyl isocyanate with poly(vinyl alcohol) (57) and several workers have crosslinked poly(vinyl alcohol) with hexamethylene diisocyanate (58,59). These latter systems were examined as membranes (58) and water resistant wood adhesives (59).

Recently, poly(vinyl alcohol)-isocyanate reactions have been run successfully in dimethylsulfoxide solutions by several laboratories, including ours. While there are reports of solvent derived products from the reactions of isocyanates in DMSO (60), these reactions do not seem to pose a problem in the reaction with poly(vinyl alcohol). This solvent has been used to react PVA with methoxymethyl isocyanate (61,62) and a series of isocyanates including methyl, ethyl, isopropyl, phenyl and 1-naphthyl (63). This reaction is illustrated in Equation 17 where R is CH_3^-, $C_2H_5^-$, $(CH_3)_2CH^-$, $CH_3OCH_2^-$, $C_6H_5^-$ or $C_{10}H_7$. Degrees of substitution of 47-96% have been reported and the polymers are claimed to have increased viscosity and an increased weight loss in thermogravimetric analysis. The polymer elongation increases slightly with the increasing degree of substitution but the tensile strength appears to reach a maximum at about 50% substitution with the phenyl isocyanate modified poly(vinyl alcohol). With this polymer, the softening point increases from about 50°C at 15.3% to about 130°C at 78.1 mole% substitution (63).

$$-\!\!(CH_2\underset{\displaystyle OH}{CH})_n- \xrightarrow[DMSO]{R-NCO} -\!\!(CH_2\underset{\displaystyle OH}{CH})_{n-x}\!-\!(CH_2\underset{\displaystyle \underset{\displaystyle O=\underset{\displaystyle NHR}{C}}{O}}{CH})_x- \qquad \text{(Equation 17)}$$

A controlled release herbicide based on Metribuzin attached to poly(vinyl alcohol) has been reported via an isocyanate reaction as shown in Equation 18. Data on the release rate has been published (54).

$$(CH_3)_3C\text{-triazinone-}N\text{-}NH_2\ (SCH_3) \xrightarrow{OCN\text{-}R\text{-}NCO} (CH_3)_3C\text{-triazinone-}N\text{-}NH\text{-}\overset{O}{\overset{\|}{C}}\text{-}NH\text{-}R\text{-}NCO\ (SCH_3)$$

$$\xrightarrow{\text{-}(CH_2\underset{OH}{CH})_n\text{-}} \text{-}(CH_2\underset{O\text{-}C(=O)\text{-}NH\text{-}R\text{-}NH\text{-}C(=O)\text{-}NH\text{-}N(\text{triazinone})}{CH})_x\text{-}(CH_2\underset{OH}{CH})_{n-x}\text{-}$$

(Equation 18)

General Comment. The foregoing examples clearly show that poly(vinyl alcohol) can be modified readily and that some of these derivatives have potential (and/or actual) biological activity. This survey is definitely not encyclopedic in scope but rather illustrative. No doubt many more examples of potentially biologically active poly(vinyl alcohol) derivatives will be developed in the future and it is entirely possible that some of these may become of value in chemotherapy and other areas where biologically active polymers are now being studied.

Experimental Section

Materials Used. The poly(vinyl alcohol) used in this study was a commercial (Borden chemical) grade of fully hydrolyzed material which had an aqueous intrinsic viscosity of 0.762 which corresponds to a molecular weight of about 59,900. This material was dried in a vacuum oven for several days at about 100°C and 10 torr before it was used in the modification experiments. Dry, analytical grade dimethyl sulfoxide (DMSO) was used as supplied.

Modified Polymer Preparation. The general procedure was the same for all cases. The dried poly(vinyl alcohol) was dissolved in DMSO at a concentration of about 7.5 g/dl and a small, catalytic quantity of triethylamine (2-3 drops) was added to each solution to increase the reaction rate. (Altho higher concentrations could be used, the solutions were too viscous to permit easy handling and stirring.) Next, the amount of phenyl isocyanate calculated to give the desired degree of substitution was added, while stirring rapidly. These homogeneous solutions were stirred magnetically for one week at room temperature and a small quantity of methanol was added to destroy any residual trace of phenyl isocyanate. (In actual fact, there was no evidence of residual isocyanate except in those cases where a large excess of reagent was used.) The polymer was then recovered by pouring into water, separating and drying. Each polymer was further purified by redissolving in a solvent and precipitating it by pouring into an excess of a non-solvent.

In a typical example, 5.2 g poly(vinyl alcohol) was dissolved in 70 ml DMSO, two drops triethylamine was added and the solution was placed on a magnetic stirrer in a stoppered flask. Next, 11.1 ml phenyl isocyanate was added slowly while stirring to produce a solution which should give a theoretical 86.5% substitution of the hydroxyl groups. After stirring for a week at room temperature, 25 ml methanol was added and the entire sample was slowly poured into 1500 ml water. The precipitated polymer was collected by filtration and vacuum dried. This polymer was then dissolved in acetone and reprecipitated by pouring into water. The sample was collected by filtration and, after drying for several days under vacuum at 110°C, gave 14.4 g modified polymer (88% of theory). Elemental analysis showed this sample contained 7.67% N (theory is 7.42% N, assuming 86.5% substitution).

Polymer Physical Properties

Infrared Spectra. Films of the polymer samples were prepared by casting from a DMSO or a dimethylacetamide solution unto a salt plate. These were then examined for their infrared spectral properties in a Beckman AccuLab 4 spectrophotometer. Basically, as the degree of substitution increased, the OH peak (3300-3400 cm^{-1}) diminshed in intensity and the C=O peak (1710 cm^{-1}) increased in intensity. Peaks due to the NH (3120-3140 cm^{-1}) and phenyl groups (3030-3050 cm^{-1}) also developed as the substitution increased and the relative intensity of the aliphatic CH (backbone chain) (2910-2940 cm^{-1}) decreased at the same time. Some of these results are summarized in Table I where the ratios of the C = O/OH and phenyl/aliphatic CH peaks are reported for various degrees of substitution.

Nuclear Magnetic Resonance Spectra. NMR spectra were run on the original poly(vinyl alcohol) and some modified samples as solutions in deuterated DMSO using a Varian EM 360 spectrometer. The details of the NMR spectroscopy are complicated by tacticity considerations as noted in the literature (3). In the present case, this situation is complicated further by the presence of two different types of repeat units in the polymer chain. This results in variable amounts of a minimum of at least eleven different types of protons (ignoring any tacticity or nonequivalent magnetic environment problems) which will interact with each other along the polymer chain. In this study, we are not attempting to resolve these details but are merely reporting what new peaks arise on substitution. A group of peaks due to the phenyl group arise at about 7.3δ and increase with the degree of substitution. At the same time, the peak due to the hydroxyl group decreased in intensity. Numerous other small peaks at 5.0, 8.6 and 9.3δ arise as the degree of substitution increases but peak assignments have not been made at the present time.

Density. The density of the polymer samples was determined by the solvent floatation method described in the literature (65). All these samples ranged from 1.24 to 1.27 and these results are reported in Table I.

Polymer Solubility. The modified polymers were soluble in DMSO, dimethylacetamide, dimethylformamide and formic acid. They were insoluble in water, methanol and xylene. Above about 57% degree of substitution, the polymers were also soluble in butyrolactone and acetic acid. Solubility parameters were determined for each polymer by the titration procedure as described in the literature (65). The polymer was dissolved in DMSO and titrated with xylene for the low end of the solubility parameter and a second DMSO solution was titrated with water for the high end of the solubility parameter range. These solubility parameters and some other solubility data are summarized in Table II.

Solution Viscosity Studies. Ths polymer solution viscosity was run on two modified polymers and the original poly(vinyl alcohol) at 30°C in DMSO solutions using a series 100 Cannon-Fenske viscometer. The observed specific viscosities and the intrinsic viscosity for each of these samples are summarized in Table III.

The original poly(vinyl alcohol) was studied in both aqueous and DMSO solutions. Viscosity-molecular weight relationships have been reported for each of these solutions at 30°C as shown in Equations 19 and 20 (3).

$$[\eta]_{water} = 6.67 \times 10^{-4} M^{0.64} \quad \text{(Equation 19)}$$

$$[\eta]_{DMSO} = 1.58 \times 10^{-4} M^{0.84} \quad \text{(Equation 20)}$$

For our sample, these equations give a molecular weight of 59,900 and 69,740 for the water and DMSO solutions, respectively, with an average value of 64,850. The molecular weight of the fully substituted polymer would be 240,240.

Results and Discussion

The reaction between an isocyanate and an alcohol to form a carbamate or urethane has been known for many years but has been applied to the poly(vinyl alcohol) system only recently. This is due largely to the fact that the heterogeneous reaction between PVA and an isocyanate is difficult to control reproducibly.

TABLE I. THE EFFECT OF THE DEGREE OF SUBSTITUTION ON THE INFRARED SPECTRA AND DENSITY OF THE POLYMERS.

% SUBSTITUTION	INFRARED SPECTRA C=O/OH (a)	INFRARED SPECTRA AROMATIC/ALIPHATIC (b)	DENSITY
0.0	0.00	0.00	-
14.8	0.34	-	1.27
28.8	0.98	0.60	1.24
43.7	1.16	0.71	1.25
57.7	1.26	0.81	-
72.5	1.63	1.00	1.26
86.5	2.60	1.00	1.25
100.0 (c)	2.60	1.07	1.24

(a) Ratio of the C=O and OH absorption peaks.
(b) Ratio of phenyl and aliphatic CH absorption peaks.
(c) 116% of theory of phenyl isocyanate used.

TABLE II. THE EFFECT OF THE DEGREE OF SUBSTITUTION ON THE POLYMER SOLUBILITY.

% SUBSTITUTION	SOLUBILITY PARAMETER (a) MEAN	LOW	HIGH	SOLUBILITY (b) IN BUTYRO-LACTONE	ACETIC ACID
0.0	17.2	11.1	23.4	-	-
14.8	13.8	11.0	16.6	I	I
28.8	13.4	10.7	16.0	I	P
43.7	12.9	10.7	15.1	I	P
57.7	12.7	10.3	15.1	S	S
72.5	12.4	10.2	14.5	S	S
86.5	12.7	10.2	15.3	S	S
100.0	12.8	10.2	15.4	S	S

(a) Determined by titration procedure.
(b) S soluble; P partially soluble; I insoluble

TABLE III. A STUDY OF THE SPECIFIC VISCOSITY OF SOME MODIFIED POLY(VINYL ALCOHOL) POLYMERS IN DMSO SOLUTION AT 30°C

% SUBSTITUTION	CONCENTRATION g/dl 0.10	0.20	0.33	0.50	0.67	0.80	INTRINSIC VISCOSITY
0 (a)	0.08	0.17	0.29	0.47	0.68	0.86	0.762
0	0.20	0.40	0.58	1.14	1.59	2.05	1.86
43.7	0.11	0.22	0.34	0.64	0.93	1.24	0.96
86.5	0.07	0.14	0.17b	0.29	0.43	0.51	0.70

(a) Aqueous solution viscosity.
(b) Concentration was 0.27 g/dl.

Recently it has been observed in this laboratory and others (60-63) that this reaction can be run under homogeneous conditions in dimethyl sulfoxide solution to obtain the desired modified polymers. These modified polymers would have several potential uses and exhibit some unusual properties.

Polymer Structure. The reaction studied here is summarized in Equation 21. As shown in the experimental section, it is possible to prepare these polymers at various degrees of substitution. As the degree of substitution increases, the ratios of the infrared C=O/OH absorption peaks and the phenyl/aliphatic C-H absorption peaks increase in a linear manner (Table I). (It would be possible to determine the degree of substitution from such calibrated curves.) At the same time, the intensity of the OH band in the NMR spectra diminishes while a strong set of peaks due to the phenyl group forms. Elemental nitrogen analysis values for the modified polymers agree closely with the calculated values. In addition, the infrared spectra show the necessary carbamate N-H bands. These factors enable us to have confidence that the polymer structure is as shown in Equation 21.

$$-(CH_2\underset{OH}{CH})_n- \xrightarrow[DMSO,\ Et_3N]{C_6H_5-NCO} -(CH_2\underset{OH}{CH})_{n-x}-(CH_2\underset{O-C(=O)-NH-C_6H_5}{CH})_x-$$

(Equation 21)

There are, however, things about the polymer structure which are not known for certain. We assume that the reaction occurs in a random manner along the polymer backbone but there is little evidence at all concerning this problem and a detailed analysis must await future research. In addition, we know very little about the effects of polymer tacticity on the reaction shown in Equation 21. This also remains to be studied. On the other hand, we are confident that this reaction does not lead to a novel crosslinking reaction sequence since these polymers are soluble in a number of different solvents (Table II).

Polymer Properties. The modified polymers do exhibit some interesting properties. Water solubility is lost at 15% or less degree of substitution altho the polymers are at least partially soluble in acetic acid above about 25% degree of substitution. Solubility in butyrolactone also occurs above about 50% degree of substitution but the polymers are not soluble in xylene or methanol regardless of the degree of substitution. All polymers studied were soluble in DMSO, dimethylacetamide, dimethylformamide and formic acid. The experimentally determined solubility parameters for all these modified samples show a lower limit between 10.2 and 11.0 with the upper limit ranging from 14.5 to 16.6 (Table II). The actual value of the solubility parameter decreases as the degree of substitution increases. This is due largely to the more hydrophobic nature of the N-phenylcarbamoyl group being added to the polymer backbone. Most of these modified polymers are soluble in solvents having solubility parameters ranging from 11 to 13 (66).

The effect of the degree of substitution on the polymer viscosity is striking. Poly(vinyl alcohol) is more viscous in DMSO than in aqueous solutions (Table III). As the free OH groups are replaced by N-phenylcarbamoyl groups, the specific viscosity decreases sharply. DMSO is considered as a better solvent for poly(vinyl alcohol) than is water (3). We therefore envision these DMSO solutions as consisting of fairly extended chains of PVA which interact with each other by entanglement and by hydrogen bonding. As the OH groups are replaced by bulky N-phenylcarbamoyl units, the hydrogen bonding interactions would drop sharply because the NH would interact less than the OH and the bulky N-phenylcarbamoyl group would hinder this interaction sterically. The increased presence of the bulky N-phenylcarbamoyl units would also tend to increase chain stiffness which would also tend to reduce interchain interactions due to entanglements. Thus, the presence of the N-phenylcarbamoyl groups would reduce both chain entanglement and hydrogen bonding interactions. This, in turn, would result in the observed reduction of viscosity as the degree of substitution increases.

It is worth noting that this viscosity reduction occurs even though the polymer molecular weight increases. Assuming the initial poly(vinyl alcohol) molecular weight to be 64,850, the molecular weight of the 43.7% and the 86.5% substituted polymers would be 141,500 and 216,600, respectively. In the 86.5% degree of substitution case, the molecular weight is about 3.3 times as great while the intrinsic viscosity is only 0.38 that of the unmodified polymer.

Potential Uses of These Polymers. We have studied the phenyl isocyanate modification of poly(vinyl alcohol) as a model system. Many uses exist for carbamates as medicines, pesticides and herbicides (67,68). For example, ethyl carbamate has been used to treat leukemia and multiple myeloma. Ethyl carbamate has also been used as an antidote for central nervous system poisoning by strychnine. The tranquilizer Meprobamate is a carbamate derivitive. Numerous pesticides and herbicides, such as Sevin and Propham, are also carbamate derivatives. Propham is isopropyl N-phenylcarbamate which bears a strong resemblence to the polymers of Equation 21, and this compound is used as a preemergence herbicide. Numerous other close analogs could be cited also. We might note also that the N-phenyl carbamoyl unit bears a close resemblence to acetanilide which has been used in headache remedies.

In recent years there has been a growing interest in the use of polymeric herbicides, pesticides and drugs. Several reviews have appeared on this general area (69-71) and we earlier noted several examples of such potential behavior with poly(vinyl alcohol) modifications. These included modifications containing 6-methylthiopurine (an antileukemia drug) (55), 2,6-dichlorobenzaldehyde (a herbicide) (56), various enzymes (52,53), aspirin (analgesic)(51) and mercapto groups (46-49).

Our study has clearly shown that carbamate groups can be attached to the poly(vinyl alcohol) backbone. Further studies are in progress to ascertain whether the N-phenylcarbamoyl modified poly(vinyl alcohol) samples will show any utility as a herbicide or a drug. These modified polymers would, of course, have potential use as a new plastics or fibers but we would not expect great thermal stability from this system (63).

Summary

Polymers containing pendant carbamate functional groups can be prepared by the reaction of phenyl isocyanate with poly(vinyl alcohol) in homogeneous dimethylsulfoxide solutions using a triethylamine catalyst. These modified polymers are soluble in dimethyl sulfoxide, dimethylacetamide, dimethylformamide and formic acid but are insoluble in water, methanol and xylene. Above about 50% degree of substitution, the polymers are also soluble in acetic acid and butyrolactone. The modified polymers contain aromatic, C=O, NH and CN bands in the infrared and show a diminished OH absorption. Similar results were noted in the NMR spectroscopy. These modified polymers show a lower specific and intrinsic viscosity in DMSO solutions than does the unmodified poly(vinyl alcohol) and this viscosity decreases as the degree of substitution increases.

References

1. Fettes, E.M., ed., Chemical Reactions of Polymers, Wiley-Interscience, New York, 1964
2. Moore, J.A., ed., Reactions on Polymers, D. Reidel Publ. Co., Dordrecht, Holland, 1973
3. Pritchard, J.G., Poly(vinyl alcohol): Basic Properties and Uses, Gordon & Breach, New York, 1970
4. Gebelein, C.G., J. Macromol. Sci.-Chem., A5, 433 (1971)
5. Gebelein, C.G., & Baytos, A. in reference 2, p 116
6. Herrmann, W.O. & Haehnel, W., Ber., 60, 1658 (1927)
7. Minsk, L.M., Priest, W.J., & Kenyon, W.O., J. Am. Chem. Soc., 63, 2715 (1941)
8. Bevington, J.C., Guzman, G.M. & Melville, H.W., Proc. Roy. Soc. (London), A221, 437 (1954)

9. Melville, H.W. & Sewell, P.R., Makromol. Chem. 32, 139 (1959)
10. Clark, J.T., Howard, R.O. & Stockmayer, W.H., Makromol. Chem., 44/46, 427 (1961)
11. Wheeler, O.L., Ernst, S.L. & Crozier, R.N., J. Polymer Sci., 8, 409 (1952)
12. Wheeler, O.L., Lavin E. & Crozier, R.N., J. Polymer Sci., 9, 157 (1952)
13. Danno, A., Atomic Energy Review, 9, 399 (1971)
14. Yamashita, S., Takakura, K. Imai, Y. & Masuhara, E, Kobunshi Ronbunshu, 35, 283 (1978); Chem. Abstr. 89, 111592b (1978)
15. Hoechst, A.-G., Neth. Appl. 74, 16,362 (1975); Chem Abstr. 84, 79,741h (1976)
16. Peppas, N.A. & Merrill, E.M., J. Polymer Sci.-Chem., 14, 441 (1976)
17. Hara, Y., Nishioka, K., Kamiya, S., Yamauchi, A. & Matsuzawa, Y., Nippon Ganka Kiyo, 28, 1522 (1977); Chem. Abstr., 88, 141,661v (1978)
18. Kamiya, S., Hara, Y., Matsushima, S., Nishioka, K., Matsuzawa, Y., & Yamauchi, A., Nippon Ganka Kiyo, 29, 420 (1978); Chem. Abstr., 89, 94,957e (1978)
19. Ratner, B., & Hoffman, A.S., in Hydrogels for Medical and Related Applications, Andrade, J.D., ed., Am. Chem. Soc., Washington, DC, 1976, pp 1-36
20. Stamatoff, G.S., U.S. Patent 2,400,957 (1946)
21. Kranzlein, G. & Reis, H., German Patent 765,265 (1954)
22. Block, B., & Hastings, G.W., Plastic Materials in Surgery, Thomas, C.C., Springfield, IL, 1972, p 102
23. Kronenthal, R.L., in Polymers in Medicine and Surgery, Kronenthal, R.L., Oser, Z. & Martin, E., ed., Plenum Press, New York, 1975, p 119
24. Hulbert, S.F., & Bowman, L.S., in Polymers in Medicine and Surgery, Kronenthal, R.L., Oser, Z. & Martin, E., ed., Plenum Press, New York, 1975, p 161
25. Taylor, L.D., Fitzgerald, M., MacLaughlin, P. & Plohar, M., J. Polymer Sci., Pt. B, 5, 73 (1967)
26. Miller, S.A. & Bracken, A., J. Chem. Soc., 1951, 1933
27. Imai, K. & Matsumoto, M., J. Polymer Sci., 55, 335 (1961)
28. Zwick, M.M., J. Appl. Polymer Sci., 9, 2393 (1965)
29. Zwick, M.M., J. Polymer Sci., Pt. A1, 4, 1642 (1966)
30. Hayashi, S., Nakabayashi, T., & Yoshida, K., Bull. Chem. Soc. (Japan), 43, 3292 (1970)
31. Kikukawa, K., Nozakura, S. & Murahashi, S., Polymer J., 2, 212 (1971)
32. Gizatullin, N.I., Uch, Zap. Kazan Vet. Inst., 111, 177 (1971); Chem. Abstr., 81, 72,571g (1974)
33. Gevonydyan, S.A., Gevondyan, V.S., & Babayan, B.L., Tr. Staurop. S-Kh. Inst., 36, 206 (1973); Chem. Abstr. 82, 11,251b (1975)

34. Sytdykov, A.K., Burlutskii, I.D., & Turakulov, B., Tr. Uzb. Nauchno-Issled. Inst. Vet., 23, 207 (1973); Chem. Abstr., 82, 188,424g (1975)
35. Aliev, Ya.N., Alferov, V.V. & Pulatov, T., Tr. Uzb., Nauchno-Issled. Inst. Vet., 21, 238 (1973); Chem. Abstr., 82, 52,180w (1975)
36. Wojciak, L., Uchman, G. & Kucharski, P., Med. Dosw. Mikrobiol., 27, 19 (1975); Chem. Abstr., 83, 38,189v (1975)
37. Ferrel, R.E., Olcott, H.S. & Fraenkel, N., J. Am. Chem. Soc., 70, 2101 (1948)
38. Daul, G.C., Reid, J.D. & Reinhardt, R.M., Ind. Eng. Chem., 46, 1042 (1954)
39. Ashida, K., Chem. High Polymers (Japan), 10, 17 (1953); Chem. Abstr. 48, 1402 (1954)
40. Kosolapoff, G.M., U.S. Patent 2,495,108 (1950); Chem. Abstr., 44, 7091 (1950)
41. Koalnes, D.E. & Brace, N.O., U.S. Patent 2,691,567 (1954); Chem. Abstr., 49, 2090 (1955)
42. Carraher, C.E. & Torre, L., J. Polymer Sci., Pt. A1, 9, 975 (1971)
43. Carraher, C.E. & Piersma, J.D., J. Macromol. Sci.-Chem., A7, 913 (1973)
44. Carraher, C.E. & Piersma, J.D., Angew. Makromol. Chem., 28, 153 (1973)
45. Carraher, C.E., in reference 2, p 126
46. Nakamura, Y., Kogyo Kagaku Zasshi, 58, 269 (1955); Chem. Abstr., 49, 14,376h (1955)
47. Cerny, J. & Wichterle, O., J. Polymer Sci., 30, 501 (1958)
48. Okawara, M., Nakagawa, T. & Imoto, E., Kogyo Kagaku Zasshi, 60, 73 (1957); Chem. Abstr., 53, 5730d (1959)
49. Okawara, M. & Sumitomo, Y., Kogyo Kagaku Zasshi, 61, 1508 (1958); Chem. Abstr., 56, 1330i (1962)
50. Hoffmann, S., Witkowski, W. & Schubert, H., Z. Chem., 14, 154 (1974); Chem. Abstr., 81, 136,565t (1974)
51. Weiner, B.Z., Havron, A. & Zilkha, A., Isr. J. Chem., 12, 863 (1974)
52. Khorunzhina, S.I., Khokhlova, V.A., Shamolina, I.I. & Vol'f, L.A., Zh. Prikl. Khim. (Leningrad), 51, 651 (1978); Chem. Abstr., 88, 148, 195e (1978)
53. Manecke, G. & Schlünsen, J., in Polymeric Drugs, Donaruma, L.G., & Vogl, O., ed., Academic Press, New York, 1978, p 39
54. Kida, M. & Nakano, H. Polymer J., 10, 117 (1978)
55. Seita, T., Kinoshita, M. & Imoto, M., J. Macromol. Sci.-Chem., 7, 1297 (1973)
56. Schacht, E.H., Desmarets, G.E., Goethals, E.J. & St. Pierre, T. in Polymeric Drugs, Donaruma, L.G. & Vogl, O. ed., Academic Press, New York, 1978, p 331
57. Pande, K.C. & Kallenbach, S.E., U.S. Patent 3,776,889 (1973); Chem. Abstr., 80, 83,922c (1974)

58. Caro, S.V., Jr., Paik Song, C.S., & Merrill, E.W., J. Appl. Polymer Sci., 20, 3241 (1976)
59. Yamakawa, Y., Tashiro, T., Miyazaki, Y. & Sakurada, S., Japan Kokai 78 37,739 (1978); Chem. Abstr., 89, 76,074k (1978)
60. Carleton, P.S. & Farrissey, W.J., Jr., Tetrahedron Letters, 40, 3485 (1969)
61. Sikorski, R.T., Hadrowicz, B., Kokocinski, J. & Kowalczyk, M., Pr. Nauk. Inst. Technol. Org. Twerzyro Sztweznych Polstech. Wroclaw, 16, 101 (1976); Chem. Abstr., 83, 44,761s (1975)
62. Sikorski, R.T., Hadrowicz, B., Kokocinski, J. & Kowalczyk, J., Pol. 86,452 (1976); Chem. Abstr., 86, 191,429e (1977)
63. Sastre, R., Garcia Perez, M. & Acosta, J.L., Rev. Plast. Mod., 34, 76 (1977); Chem. Abstr., 87, 136,586z (1977)
64. McCormick, C.L., & Fooladi, M. in Controlled Release Pesticides, Scher, H.B., ed., ACS Symposium Series No. 53, Am. Chem. Soc., Washington, DC, 1977, p 112
65. McCaffery, E.M., Laboratory Preparation for Macromolecular Chemistry, McGraw-Hill, New York, 1970
66. Barton, A.F.M., Chem. Revs., 75, 731 (1975)
67. Adams, P. & Baron, F.A., Chem. Revs., 65, 567 (1965)
68. Neumeyer, J., Gibbons, D. & Trask, H., Chem. Week, April 12, 1969, pp 38-68 & April 26, 1969, pp 38-68
69. Gebelein, C.G., Polymer News, 4, 163 (1978)
70. Paul, D.R. & Harris, F.W., ed., Controlled Release Polymeric Formulations, Am. Chem. Soc., Washington, DC, 1976
71. Colbert, J.C., Controlled Action Drug Forms, Noyes Data Corp., Park Ridge, NJ, 1974

RECEIVED July 12, 1979.

Vinyl Polymerization (383): Radical Polymerization of Vinyl Monomer with an Aqueous Solution of Polystyrenesulfonate or Polyvinylphosphonate

M. IMOTO, T. OUCHI, M. SAKAE, E. MORITA, and T. YAMADA

Department of Applied Chemistry, Faculty of Engineering, Kansai University, Suita, Osaka 564, Japan

In 1962, Kimura, Takitani and Imoto (1) found that an aqueous solution of starch could easily polymerize methyl methacrylate (MMA) and about a half of polymerized MMA grafted on starch. This novel polymerization was called as "uncatalyzed polymerization". Since then, a lot of macromolecule was applied, instead of starch, and many of them were effective to initiate the radical polymerization of MMA.

Effective macromolecules were found to be divided into two groups. The macromolecules which belong to Group I are effective only in the presence of some metal ion, particularly Cu(II) ion. The macromolecules of Group II require no metal ion. They are listed in Tables 1 and 2.

As can be seen, the effective macromolecules are water-soluble or at least somewhat hydrophilic. Strongly hydrophobic macromolecules and low molecular compounds were always ineffective. The ineffective substances which were tested are listed in Table 3.

Table 1. Effective Macromolecular Substances: Group I (Effective in the presence of Cu(II) ion)

Starch (1), Cellulose (2), Cellulose Methyl Ether (3), Oxycellulose (4), PVA (5), Partially Hydrolyzed PVAc (6), Silk (2), Wool (7), Hide-Powder (8), Natural Rubber Latex (9), Synthesized Poly-(α-Amino Acids) (10), Nylon-6 (11), Nylon-3 (12), α-Amylase (13), Lysozyme (14), RNA (15), Polyacrylonitrile (16), Polyvinylsulfonate (17).

Table 2. Effective Macromolecular Substances: Group II (Effective in the absence of Cu(II) ion)

Polymethallylsulfonate (18), Polyallylsulfonate (19), Polystyrenesulfonate (20), Crosslinked Polystyrenesulfonate (Ion Exchange Resin) (21), Chondroitin Sulfate (22), Polyvinylphosphonate (23).

0-8412-0540-X/80/47-121-103$05.00/0

Table 3. Ineffective Substances (24)

Macromolecules: Polyvinylchloride, Polyethylene, Polypropylene, Styrene-Butadiene Rubber.

Low molecular compounds: Glucose, Sucrose, ATP, $CH_3CH_2SO_3Na$, α-Amino Acid, Polyphosphonic Acid, etc.

The present paper deals with the uncatalyzed radical polymerization initiated with the water-soluble macromolecule in the absence of Cu(II) ion. Using polystyrenesulfonate (PSS-Na) and polyvinylphosphonate (PVPA) as the macromolecules, a study on the process of polymerization was made. And a new concept on the "hard and soft hydrophobic areas and monomers" was proposed.

Experimental

Materials: PSS-Na (20) was prepared by the radical polymerization of p-styrenesulfonate. PVPA was obtained by the hydrolysis of poly-bis-(β-chloroethyl) vinylphosphonate and was concluded to have a following formula (23):

$$Bu-\left[CH_2-\underset{\underset{ONa}{|}}{\overset{}{CH}}\!\!\left(O{=}P{-}ONa\right)\right]_{16.8}\left[CH_2-CH\left(O{=}P(-ONa)-O-CH_2CH_2OH\right)\right]_{7.0}\left[CH_2-CH\left(O{=}P(-ONa)-O-CH_2CH_2Cl\right)\right]_{4.2}-H$$

Procedures: Vinyl monomer and an aqueous solution of the macromolecule were placed in a tube and sealed under vacuum after thawing with nitrogen. The tube was shaken or allowed to stand at 85°C. In the case of shaking, the contents were poured into methanol to precipitate the polymer. In the case of standing, the upper MMA phase and the lower water phase were pipetted out separately and poured into methanol.

It was confirmed by IR and elemental analysis that poly-MMA contained neither sulfonate group nor phosphonate group. Therefore, any grafted copolymerization of MMA onto macromolecule was not observed.

Results and Discussion

1. Polymerizations by PSS-Na and PVPA.

We have repeatedly reported that a coexistense of water is indispensable for the uncatalyzed polymerization. Also in the present cases of PVPA and PSS-Na, the polymerization of MMA proceeded only in the presence of water, as shown in Fig. 1.

Figure 2 showed the effect of the dissolved mass of PSS-Na or PVPA on the rate of polymerization of MMA.

When the mass of PSS-Na or PVPA was less than a certain limit, the rate of polymerization of MMA increased with the mass of PSS-Na or PVPA. However, passing a certain mass, the conversion became to decrease or to be independent of the mass of feeded

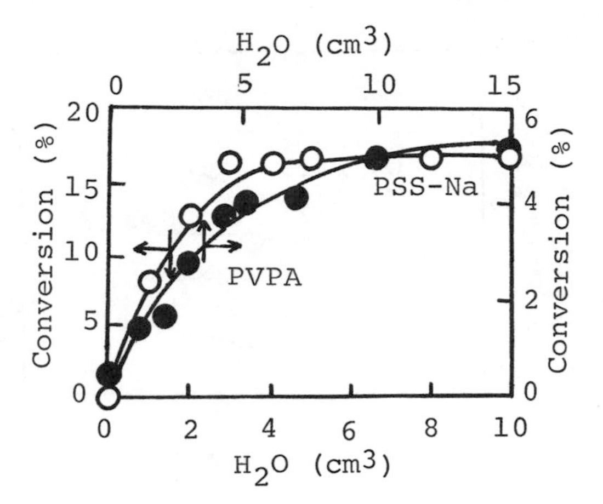

Figure 1. Conversion of MMA vs. mass of H_2O (PSS-Na(P_n 450) or PVPA 0.1 g, MMA 3 cm^3; 85°C, 3 hr, under shaking)

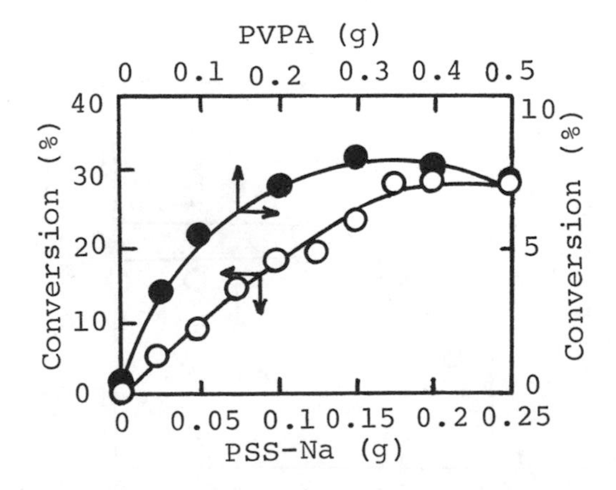

Figure 2. Conversion of MMA vs. mass of PSS-Na(P_n 450) or PVPA (MMA 3 cm^3, H_2O 5 cm^3 and 10 cm^3; 85°C, 3 hr, under shaking)

macromolecule. These results suggested that when the concentration of dissolved macromolecule was high, the macromolecules entangled with each other and became difficult to form the adequate hydrophobic areas, into which the monomer was incorporated. Accordingly, the conversion of MMA decreased, when the concentration of macromolecule was too high.

The effects of the mass of styrene (St) and MMA on polymer yields can be seen in Figs. 3 and 4. The concentration of PVPA or PSS-Na was kept constant and added mass of St or MMA was varied.

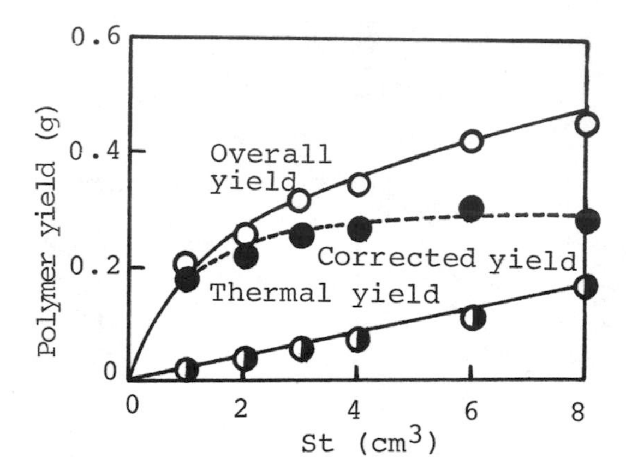

Figure 3. Conversion of styrene vs. mass of styrene (PSS-Na(P_n 450) 0.1 g, H_2O 5 cm^3; 85°C, 3 hr. under shaking)

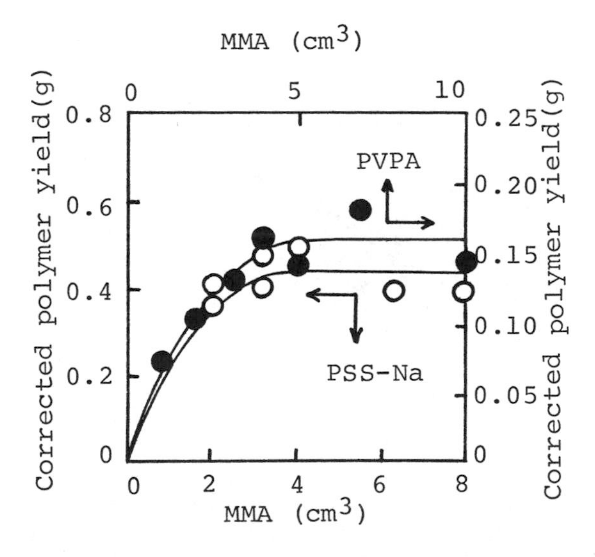

Figure 4. Conversion of MMA vs. mass of MMA (PSS-Na(P_n 450) 0.1 g, H_2O 5 cm^3; PVPA 0.1 g, H_2O 10 cm^3; 85°C, 3 hr, under shaking)

By substracting the thermal yield from the overall yield, the corrected yield was calculated. Beyond a certain mass of MMA or St, the yields became to be independent of the mass of the monomer. This is explained by the following consideration: the first step of the polymerization is the incorporation of monomer into the hydrophobic areas. When a sufficient mass of the monomer is added, the areas may be saturated with the monomer. Thus, the excess of the monomer becomes useless.

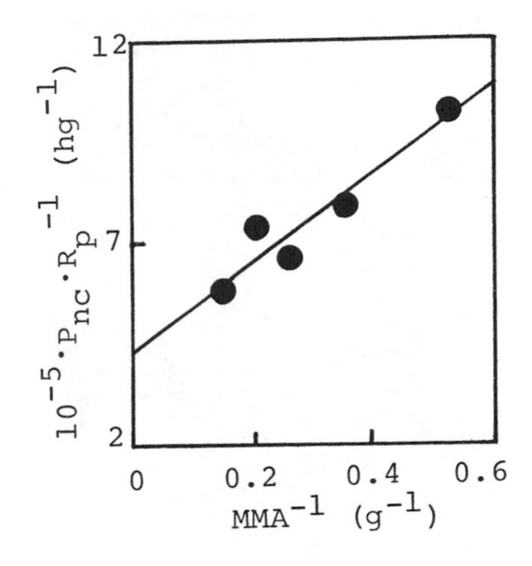

Figure 5. Application of Michaelis–Menten–Lineweaver–Burk's equation (PVPA 0.1 g, H_2O 10 cm^3; 85°C, 3 hr, P_{nc} indicates the degree of polymerization of poly-MMA)

Such a relationship between the polymer yield and the mass of feeded MMA is similar to that in the enzymatic reaction. Therefore, the result was applied to Michaelis-Menten equation and in the case of PVPA, the result shown in Fig. 5 was obtained. Such a good agreement with the Michaelis-Menten-Lineweaver-Burk's equation was always observed in the uncatalyzed polymerization.

2. Confirmation of the Formation of Hydrophobic Areas.

The direct evidence of the formation of hydrophobic areas (HA) was obtained by scanning electron microscopy; cf. Figs. 6 and 7.

As mentioned above, the presence of water is indispensable. Now the reason is clear. Water is necessary for the formation of HA in which the polymerization starts. Figure 8 verified this conclusion. When DMSO was mixed, PVPA became difficult to form HA.

Thus, the process of polymerization could be concluded to be as follows: (i) PVPA or PSS-Na forms HA in the aqueous phase, (ii) Vinyl monomer is incorporated into the HA, (iii) In the HA, the polymerization starts.

3. Effect of the Degree of Polymerization of PSS-Na on the Vinyl Polymerization.

As Table 4 showed, the conversions of MMA and St decreased with the increase of the degree of polymerization (P_n) of PSS-Na. This was due to the difficulty of the formation of HA when the large PSS-Na was dissolved in such a quantity of water. In other words, when PSS-Na with a large P_n were dissolved in water, the molecules of PSS-Na entangled with each other and became difficult to form the adequate HA. This assumption was verified by the

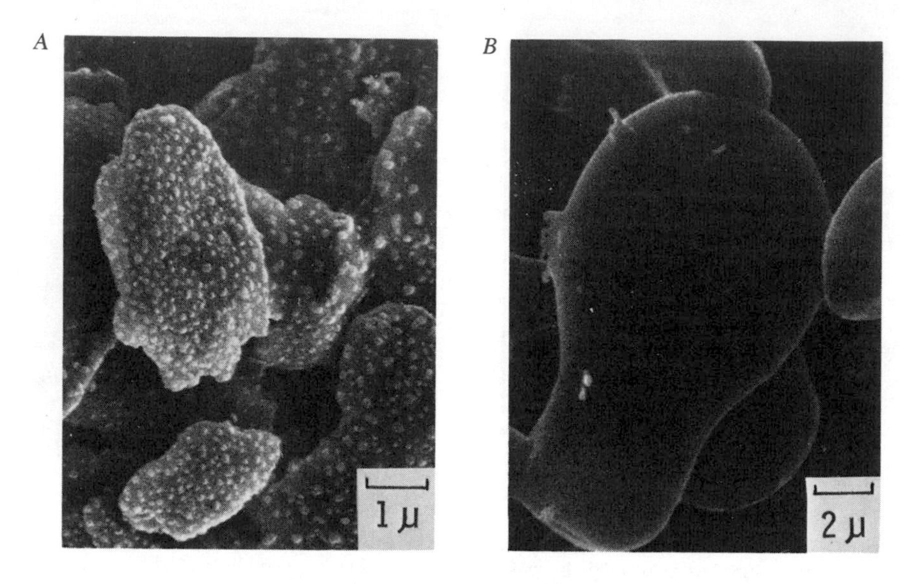

Figure 6. Surface views of PVPA: (A) 0.1 g of PVPA was dissolved in 1 dm^3 of H_2O; (B) PVPA 0.01 g, H_2O 1 cm^3, MMA 0.3 cm^3; 85°C, 3 hr. After the polymerization the system was diluted with 100 cm^3 of H_2O

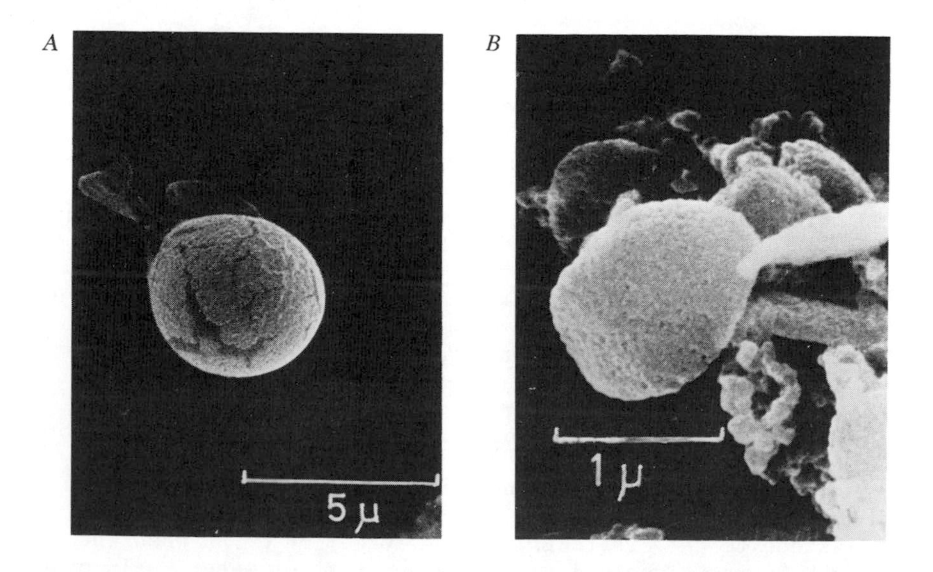

Figure 7. Surface views of PSS-Na (PSS-Na(P_n 450) 0.1 g, MMA 3 cm^3, H_2O 5 cm^3, diluted to 0.5 dm^3) (A) before polymerization; (B) after polymerization (85°C, 3 hr)

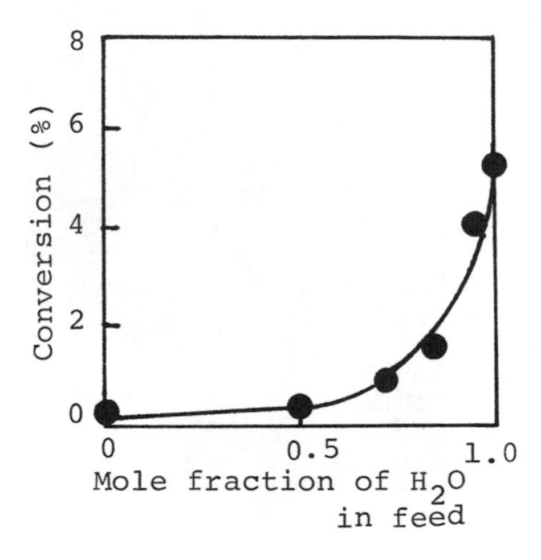

Figure 8. Conversion of MMA vs. fraction of H_2O in the mixed solvent of H_2O and DMSO (MMA 3 cm^3, PVPA 0.1 g, (H_2O + DMSO) 10 cm^3; 85°C, 3 hr)

Table 4
Effect of P_n of PSS-Na on the Vinyl Polymerization
(PSS-Na 0.1 g, Monomer 3 cm^3, H_2O 5 cm^3; 85°C, 3 h)

PSS-Na		Conversion (%)		
P_n	$[\eta]^{**}$	MMA	AN	St
1*		0.0	0	0.0
85		41.0	0	20.0
107	0.062	54.0	0	3.8
450	0.260	16.7	0	6.4
870	0.500	4.0	0	3.1
3090	1.780	1.7	0	1.8

* Sodium ethylbenzene sulfonate.
** Measured in 0.5 N-NaCl aqueous solution at 30°C.

scanning electron microscopic method, using PSS-Na having P_n of 3090.

As shown in Fig. 9 (A), when PSS-Na having a P_n of 3090 was dissolved in 500 cm^3 of water, the figures were alike to assembled fibers. It is clear that PSS-Na did not form HA to incorporate the monomer. On the contrary, when the same PSS-Na solution was diluted to 5000 cm^3 with water, the commencement of formation of HA was observed, as shown in Fig. 9 (B). Accordingly, when very diluted solution is applied, the polymerization should be taken place, even if P_n is very high as 3090.

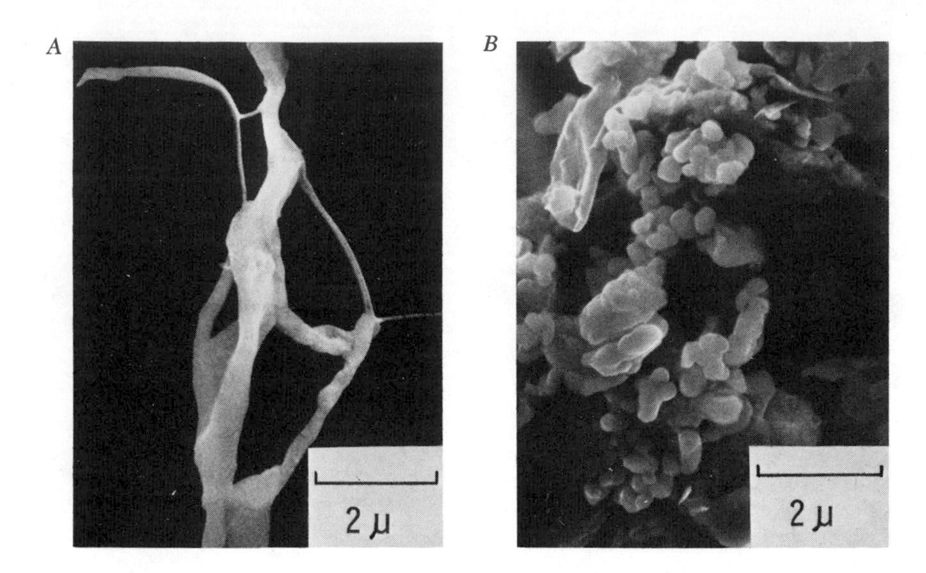

Figure 9. Surface views of PSS-Na(P_n 3090): (A) 0.1 g of PSS-Na(P_n 3090) dissolved in 5 cm^3 of H_2O, heated at 85°C for 3 hr, and diluted with 500 cm^3 of H_2O; (B) same sample as (A) diluted with 2500 cm^3 of H_2O

We carried out the polymerization of MMA. The results, which agreed well with the expectation, were obtained, as shown in Table 5.

Table 5
Effect of Dilution of the Aqueous PSS-Na Solution on the Polymerization of MMA
(MMA 3 cm^3; 85°C, 3 h, under shaking)

PSS-Na(P_n 3090) g	H_2O cm^3	Conversion* %
0.1	5	1.7
0.01	5	1.7
0.01	10	1.8
0.01	15	2.8
0.01	20	6.3

* Including the thermal conversion of 0.8 ± 0.2 %.

4. Hard and Soft Hydrophobic Areas.

Following to the concept of hard and soft acids and bases, we would like to propose a concept of hard and soft HA (micelles) and hard and soft monomers.

The micelles formed by dodecylbenzenesulfonate (ABS) in water are called to be "hard", because the interior is alike to an assemble of hydrocarbon molecules and strongly hydrophobic. However, HA formed by the water-soluble macromolecules are not hydrophobic in a strict meaning. It is rather alike an agglomerate of the macromolecules. Thus, the interiors of HA formed by PSS-Na or PVPA are not so hydrophobic, but rather somewhat hydrophilic and called to be "soft". The order of hardness of HA may be as follows:

$SO_3^-Na^+$ / $C_{12}H_{25}$ (Hard) $\gg$ Starch, PVPA $\sim$ $-(CH_2-CH)-$ / $SO_3^-Na^+$ (Soft (PSS-Na))

Similarly, vinyl monomers can be put in order from hard to soft monomer, according to their hydrophobicities. As a scale of hydrophobicity of vinyl monomer, the solubility in water may be adopted. Figure 10 showed some examples. Styrene is the most hard monomer and AN is the most soft monomer. Butyl acrylate and butyl methacrylate are more hard by one order than methyl or ethyl acrylate and methacrylate.

And the concept is realized as follows: A vinyl monomer having a certain hardness or softness for its hydrophobicity can be incorporated the most easily into the HA having a corresponding hardness or softness.

The validity of this concept could be observed in the following experimental results.

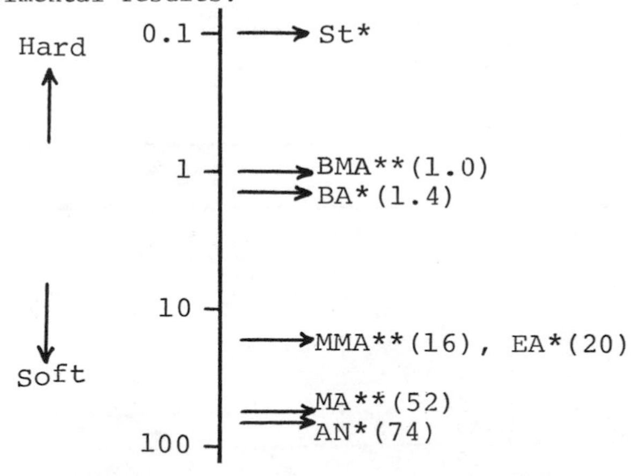

Figure 10. Solubilities of vinyl monomers at 20°C. Numbers indicate the solubilities of the monomers in water (g dm^{-3}) (BA) butyl acrylate; (BMA) butyl methacrylate; (EA) ethyl acrylate; (MA) methyl acrylate. (, 27); (**, 28)*

5. Verification of the Concept of Hard and Soft HA and Monomers.

(1) Selectivity of Vinyl Monomer for the Uncatalyzed Polymerization.

Among the methacrylates, methyl and ethyl ester can be the most easily polymerized by the uncatalyzed polymerization. This specificity was a conclusion (25) which was obtained in the uncatalyzed polymerization initiated with silk or cellulose. Also in the cases of PVPA (23) and starch (26) the same specificity was observed, as shown in Table 6. n-Butyl ester was always hardly polymerized.

These specific polymerizabilities of MMA and ethyl methacrylate are not due to their reactivities. According to literature (29), the propagating and terminating reaction constants, k_p and k_t, are shown in Table 7. Butylester is usually more reactive than methyl or ethyl ester.

Table 6
Polymerizations of Vinyl Monomers with Watersoluble Macromolecules
(PVPA, Starch, PSS-Na(P_n 1000), 0.1 g, H_2O 10 cm^3 (PVPA, Starch) 5 cm^3 (PSS-Na), Monomer 3 cm^3, $CuCl_2 \cdot 2H_2O$ 0 g (PVPA, PSS-Na), 0.5 mg (Starch); 85°C, 3 h, under shaking)

Monomer	Conversion (%)		PSS-Na(P_n 1000)
	PVPA(23)	Starch	
$CH_2=C(CH_3)-COOR$			
R=CH_3	4.1	6.0	4.9
C_2H_5	5.5	5.3	—
i-C_3H_7	—	2.2	1.6
n-C_4H_9	0	1.2	0.8
$CH_2=CH-COOR$			
R=CH_3	13.6	55	—
C_2H_5	11.8	33	—
n-C_4H_9	0		
AN	0	0	0
St	0	1.3	1.2

Table 7
k_p and k_t of Methacrylates at 30°C (29)

Ester Group	k_p dm^3 mol^{-1} s^{-1}	k_t dm^3 mol^{-1} s^{-1}
CH_3	143	12.2×10^6
C_2H_5	126	7.35×10^6
n-C_4H_9	369	10.2×10^6

Here, the concept of the hard and soft HA and monomers may be reasonably applied. As Fig. 10 shows, n-butyl esters are much harder than methyl or ethyl ester. Accordingly, n-butyl methacrylate and acrylate were too hard to be incorporated into the soft HA formed by PVPA or starch.

St is too hard to be easily incorporated into the soft HA formed by PVPA or starch. Therefore, the conversion of St was very low. However, Asahara et al. (32) polymerized St easily with the initiating system of ABS and water. The micelles or HA formed by ABS is very hard. Therefore, St could be easily incorporated in the micelles and easily polymerized.

PSS-Na could polymerize St, as shown in Table 4. This exceptional results may be explained as follows: PSS-Na contains the positively charged phenyl group which can adsorb the negatively charged phenyl group of St. Accordingly, regardless of the softness of the HA, St can be incorporated into the HA formed by PSS-Na, thereby takes place polymerization.

AN is too soft to be incorporated into the HA formed by starch, PVPA or PSS-Na.

(2) Copolymerization of MMA with St in HA having Various Hardnesses.

The composition curve of the copolymers of MMA with St by the copolymerization initiated with PSS-Na was shown in Fig. 11. The copolymer was isolated from the water phase. The contents of MMA in the copolymer are larger than those of St. This was due to the easier incorporation of soft MMA into the soft HA formed by PSS-Na.

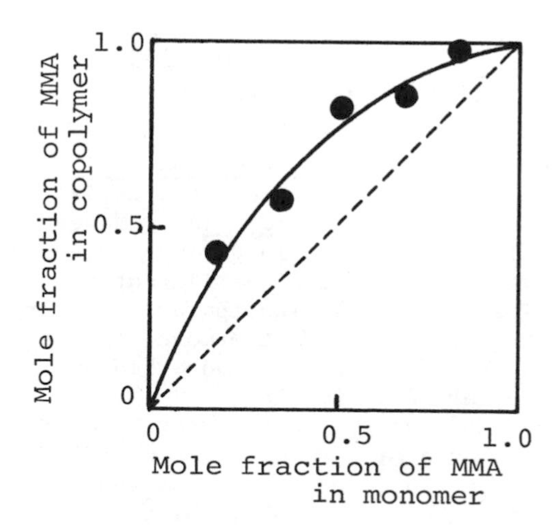

Figure 11. Copolymer composition curves of MMA and styrene (PSS-Na(P_n 85) 0.1 g, H_2O 5 cm^3, (MMA + styrene) 3 cm^3; 85°C, 4 hr on standing)

Second example was obtained from the copolymerization initiated with starch. The results were shown in Fig. 12. The copolymer isolated from the monomer phase was produced by the thermal polymerization and the composition curve was completely similar to the ordinary curve of the radical copolymerization product. The copolymer isolated from the water phase differed from the usual copolymer. The upper curve indicated that the HA formed by starch were soft, and soft MMA was much more easily incorporated than hard St.

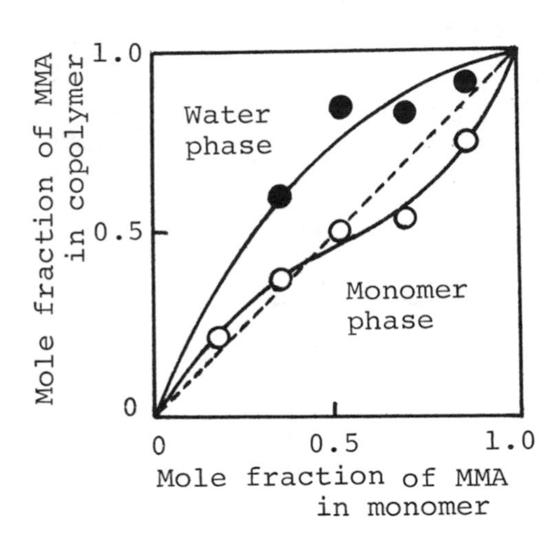

Figure 12. Composition curves of the copolymers of MMA and styrene by the copolymerization initiated with starch on standing (starch 0.1 g, $CuCl_2 \cdot 2H_2O$ 0.5 mg, H_2O 10 cm^3, (MMA + styrene) 3 cm^3; 85°C, 3 hr)

(3) The Rates of Polymerization of MMA and St.

As known well, the rate of radical polymerization (R_p) of MMA is laways larger than that of St (30, 31). However, according to Asahara et al. (32), R_p of St was larger by 3 times or more than MMA, when the polymerizations were carried out in the micelles formed by ABS. The hard St was incorporated easier in the hard HA by ABS, than the soft MMA. Furthermore, the negatively charged phenyl group of St could be easily adsorbed on the positively charged phenyl group of ABS molecule.

Inversely, PSS-Na gave the much larger R_p of MMA than that of St. For example, when a mixture of 0.1 g of PSS-Na(P_n 85) dissolved in 5 cm^3 of H_2O and 3 cm^3 of monomer was shaken at 85°C for 3 h, the conversion of MMA was 41.0 %, while that of St was only 20.2 %.

(4) Inhibition of the Polymerization with Pottasium Fluoride.

As can be seen in Fig. 13, pottasium fluoride could inhibit the polymerization of MMA initiated with PSS-Na. This is due to the adsorption of fluoride anion on the positively charged part of MMA. The loose complex of MMA with F^- was too soft to be incorporated in the HA formed by PSS-Na. Furthermore, the complex could not be adsorbed on the sulfonate group, even if the incorporation into the HA was possible in a small extent. Accordingly, the initiation reaction did not take place.

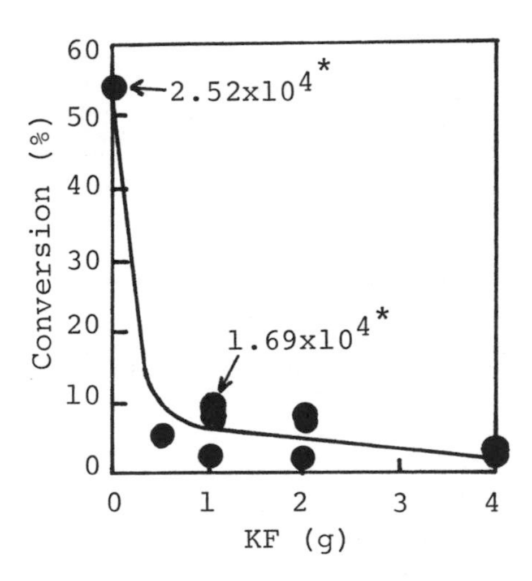

Figure 13. Conversion of MMA vs. amount of KF (PSS-Na(P_n 107) 0.1 g, MMA 3 cm^3, H_2O 5 cm^3; 85°C, 3 hr) (*) numbers indicate P_n of poly-MMA isolated

6. A Proposed Mechanism of Initiation.

The initiation mechanism in the system of PSS-Na and MMA was considered to proceed as shown in Scheme 1 (20).

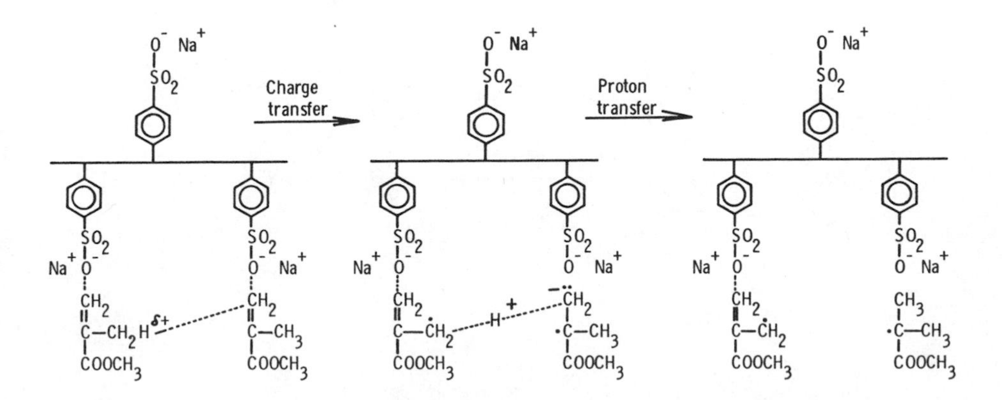

Scheme 1. Proposed mechanism of the initiation reaction of the polymerization of MMA with PSS-Na

PSS-Na could initiate the polymerization of St. The mechanism is assumed as Scheme 2 (20).

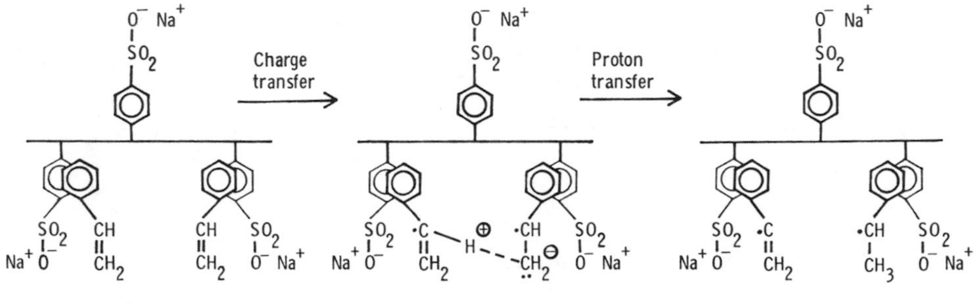

Scheme 2. Proposed mechanism for the initiation of the polymerization of styrene by PSS-Na

The initiation mechanism with PVPA was proposed as follows (23):

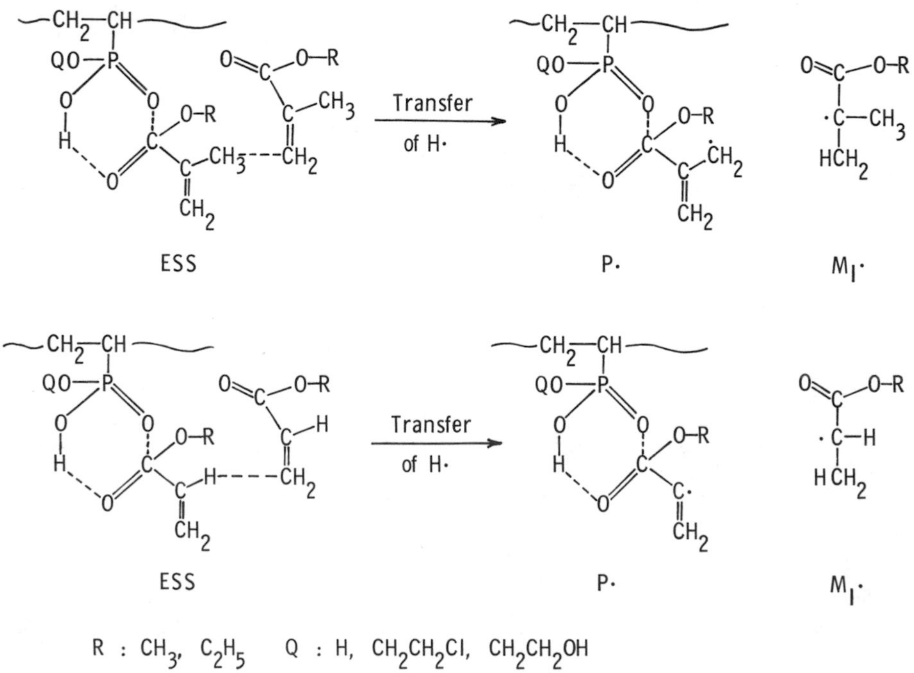

R : CH_3, C_2H_5 Q : H, CH_2CH_2Cl, CH_2CH_2OH

Scheme 3. Proposed mechanism for the initiation with PVPA

Conclusion

A study on the process of the uncatalyzed polymerization was made. The conclusion was as follows; (1) The hydrophilic macromolecules form hydrophobic areas (HA) in the water phase; (2) Into the HA, vinyl monomers are incorporated; (3) Then, the radical polymerization starts in the HA. A new concept was proposed; HA are put in order from hard to soft HA, according to their hydrophobicities, and hard or soft vinyl monomer can be the most easily incorporated into the HA having corresponding hardness or softness. Experimental verifications for this concept were obtained.

Literature Cited

1. Kimura, S., Takitani, T. and Imoto, M., Bull. Chem. Soc. Jpn. (1962), 35, 2012
2. Imoto, M., Kondo, M. and Takemoto, K., Makromol Chem. (1965), 89, 165
3. Simionescu, C. I., Feldman, D. and Ciubotariu, M., J. Polym. Sci. Part C (1972), 37, 173
4. Imoto, M., Kushibe, S. and Ouchi, T., J. Macromol. Sci.-Chem. (1977), A11, 321
5. Imoto, M., Takemoto, K. and Otsuki, T., Makromol. Chem. (1967), 104, 244
6. Imoto, M., Takemoto, K., Okuro, A. and Izubayashi, M., Makromol. Chem. (1968), 113, 111
7. Tanaka, Z., Kogyo Kagaku Zasshi (1971), 74, 1683
8. Okamoto, K., Yamamoto, T., Kogyo Kagaku Zasshi (1971), 74, 527
9. Kondo, M., Yamada, K., Takemoto, K. and Imoto, M., Bull. Chem. Soc. Jpn. (1966), 39, 536
10. Fujie, A. and Kawai, T., Makromol. Chem. (1975), 176, 629
11. Hayashi, S. and Imoto, M., Angew. Makromol. Chem. (1969), 6, 46
12. Ouchi, T., Nishimura, T. and Imoto, M., Kobunshi Ronbunshu (1975), 32, 196
13. Imoto, M., Nishimura, T., Sakade, N. and Ouchi, T., Chem. Lett. (1975), 1119
14. Ouchi, T., Yoshikawa, T. and Imoto, M., J. Macromol. Sci.-Chem. (1978), A12, 1523
15. Sugiyama, K. and Lee, S. W., J. Polym. Sci. Polym. Lett. Ed. (1977), 15, 17: Imoto, M., Nishida, Y., Yoshikawa, T., Ouchi, T., Bull. Chem. Soc. Jpn. (1978), 51, 1456
16. Tang, H.-S., Kinoshita, M. and Imoto, M., J. Macromol. Sci.-Chem., (1973), A7, 831
17. Imoto, M., Suzuki, H. and Ouchi, T., J. Macromol. Sci.-Chem. (1976), A10, 1585
18. moto, M., Ouchi, T., Nakamura, Y. and Ogushi, H., J. Polym. Sci., Polym. Lett. Ed. (1975), 13, 131

19. Imoto, M., Yamada, T., Tatsumi, E. and Ouchi, T., Nippon Kasaku Kaishi (1977), 1883
20. Nakamura, Y., Ouchi, T. and Imoto, M., Kobunshi Ronbunshu (1976), 33, 36: Ouchi, T., Suzuki, H., Yamada, T. and Imoto, M., J. Macromol. Sci.-Chem. (1978), A12, 1561
21. Ouchi, T., Tatsumi, A. and Imoto, M., J. Polym. Sci., Polym. Chem. Ed., (1978), 16, 707
22. Ouchi, T., Yamada, T. and Imoto, M., Chem. Lett. (1977), 1371
23. Imoto, M., Sakae, M. and Ouchi, T., Makromol. Chem., in press
24. Imoto, M., et al., unpublished
25. Imoto, M., Takemoto, K., Azuma, A., Kita, N. and Kondo, M. Makromol. Chem. (1967), 107, 188: Imoto, M., Kondo, M., Takemoto, K., ibid. (1965), 89, 165
26. Imoto, M., Morita, E. and Ouchi, T., J. Polym. Sci., Symposia, in press
27. Windholz, M., Ed., "Merck Index, 9th Ed." Merck and Co. New Jercy 1976
28. Chem. Soc. Jpn. Ed., "Kagaku Benran, Oyo-Hen", Maruzen, Tokyo 1973
29. Bradrup, J., Immergut, E. H., Ed. "Polymer Handbook, 2nd Ed." John Wiley and Sons, New York 1975, p. II48
30. Yokota, K., Kani, M. and Ishii, Y., J. Polym. Sci. (1968), A1, 6, 1325
31. Fагдасарян, "Теория радикальной Ролимеризации" НАУК, МОСКВА, 1966
32. Asahara, T., Seno, M., Shiraishi, S. and Arita, Y., Bull. Chem. Soc. Jpn. (1970) 43, 3895: Arita, Y., Shiraishi, S., Seno, M. and Asahara, T., Bull. Chem. Soc. Jpn. (1973), 46, 249, 2599

RECEIVED July 12, 1979.

9

Chemical Modifications of Polymers—Mechanistic Aspects and Specific Properties of the Derived Copolymers

J. C. GALIN

Centre de Recherches sur les Macromolécules (CNRS) 6, rue Boussingault, 67083 Strasbourg-Cedex, France

1. Introduction

Chemical modification of polymers (1) still remains a field of continuously increasing importance in macromolecular chemistry. In spite of its high diversification, it may be divided into 2 distinct but complementary main research lines : a) the fundamental study of the chemical reactivity of macromolecular chains ; b) the synthesis of new homopolymers and copolymers, and the functionalization of linear or crosslinked polymers. Some of these facets have been reviewed in the last years (2-6), and the purpose of this presentation is to illustrate a number of characteristic topics both from fundamental and applied points of view, through some literature data and through our own studies on nucleophilic substitution of polymethylmethacrylate (PMMA).

2. Theoretical framework of the analysis of the chemical reactivity of macromolecular chains

Quite recently, Plate and Noah[7] have critically discussed all the theoretical aspects of macromolecular chain reactivity, and we summarize below the main conclusions related to the processes of interest in our studies : kinetically controlled irreversible reaction of a low molecular weight compound R on an homopolymer A_n, involving a single monomeric unit A.
In most cases the chemical transformation of a polymer $A_n+R \rightarrow A_{n-x}B_x$ in homogeneous solution cannot be identified with the same reaction on a monomeric model compound $A+R \rightarrow B$, and the main differences may be classified into 3 categories : a) Neighbouring group effects implying various types of interactions - often depending on tacticity (8) - between the reaction site A and its 2 vicinal units ; they lead to different kinetic constants : $k_0(A\overset{*}{A}A) \neq k_1(A\overset{*}{A}B) \neq k_2(B\overset{*}{A}B)$. A limiting case is intramolecular cyclization in AB or BB dyads leading to new reaction products(8,9a). b) Solvation effects: the polarity of the microenvironment of the reaction site A(10) is a complex function of the nature of its vicinal units, of the concentration and of the polymer-solvent interactions.

0-8412-0540-X/80/47-121-119$05.00/0

c) Conformational effects: as the reaction progresses, the intrinsic properties of the chain (steric hindrance, rigidity, expansion in the solvent...) are continuously modified, leading to possible kinetic perturbations, as a result of conformational transition for instance.

Since the pionneering work of Keller (11) in 1962, theoretical studies of the reactivity of macromolecules have been steadily developed. They take into account solely neighbouring group effects, in most cases restricted to the vicinal units (8b, 12-15), but recently broadened to cooperative effects at longer distance (15,16). In spite of this oversimplification, it can be assumed that a good theoretical framework emphasizing the major role of the ratios of the 3 kinetic constants $k_0:k_1:k_2$, has now been elaborated ; it allows a quantitative description of reactions on polymers from 3 related points of view, a) kinetic analysis and limiting yields (9b) ; b) compositional heterogeneity (13,17) ; c) unit distribution (13,18,19) of the modified polymers, as illustrated in table 1.

Table 1 - Overview of cooperative effects on macromolecular chain reactivity

Neighbouring group effects	one single rate constant $k_0=k_1=k_2$	three different rate constants $k_0<k_1<k_2$	$k_0>k_1>k_2$
Kinetics	pure random processes	autoacceleration	autoretardation
Conversion	may be quantitative	may be quantitative	may be limited $k_2=0 \rightarrow \overline{DS}_m=0.666$ $k_1=k_2=0 \rightarrow \overline{DS}_m=0.432$
Compositional heterogeneity	moderate $\overline{\sigma}^2=\overline{DP}_n^{-1}(\overline{DS}_m-\overline{DS}_m^2)$	may be quite high	may be very low
Distribution of A and B units	Bernouilli statistics	Markov statistics - formation of B_n blocks	Markov statistics isolated B units between A_n blocks

. $\overline{DS}_m$ = substitution degree in mole. or molar fraction of B units in the copolymer
. $\overline{\sigma}^2$ = mean square standard deviation to the average composition.

Exhaustive studies on well-defined systems are rather scarce (4) ; nevertheless 3 systems thoroughly analyzed by independant research groups are of outstanding interest: a) the quaternization of polyvinylpyridines by alkyl halogenides (20-25) ; b) the chlorination of polyethylene (13,26-28) ; c) the basic or acid hydrolysis of PMMA (29-31). On the other hand, neighbouring groups effects have been quantitatively taken into account for the kinetic analysis of periodate oxidation of amylose (32,33).

3. Nucleophilic substitution of primary organolithium reagents on PMMA (34-36)

SN_2 reactions of primary organolithium compounds on PMMA in dilute homogeneous solution may be considered as a model system where all the important reaction parameters may be controlled ; they allow both a quantitative analysis of PMMA chain reactivity and the synthesis of well defined ketonic copolymers within a wide range of possible structural variations. The two homologous series of organolithium compounds and the corresponding reaction conditions we selected are given below :

Sulfur stabilized carbanions

I $C_6H_5SCH_2Li$ a) 25°C
II CH_3SOCH_2Na b) 25,60°C
III $CH_3SO_2CH_2Li$ b) 25,60°C
IV $(CH_3)_2NSO_2CH_2Li$ c) 25,60°C
V $CH_3OSO_2CH_2Li$ a) -78°C

a) THF ; b) DMSO ; c) THF + HMPA

Heterocycle stabilized carbanions

VI 2-pyridyl-CH_2Li c) 25°C
VII 2-quinolyl-CH_2Li c) 25°C
VIII 4,4-dimethyl-2-oxazolinyl-CH_2Li c) -78 to -15°C
IX 2-thiazolinyl-CH_2Li c) -78 to -15°C

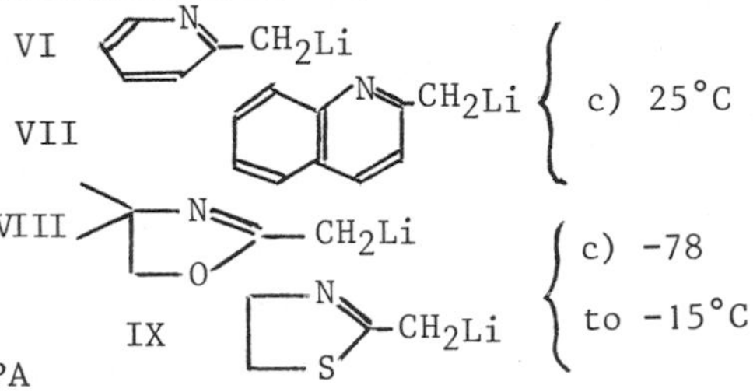

The reactions were carried out in dilute homogeneous solution in dipolar aprotic solvents ($[ester]_0$=0.2-0.4 mole.1^{-1}) using stereoregular (pure I or S) or predominantly syndiotactic radical (R) PMMA, polymethylacrylate (PMA) and radical azeotropic styrene-MMA copolymer (PSMMA, MMA mole.fraction = 0.47) as well as model monomeric (methylpivalate) and dimeric (dimethylglutarate) compounds. The overall reaction is outlined in the simplified scheme :

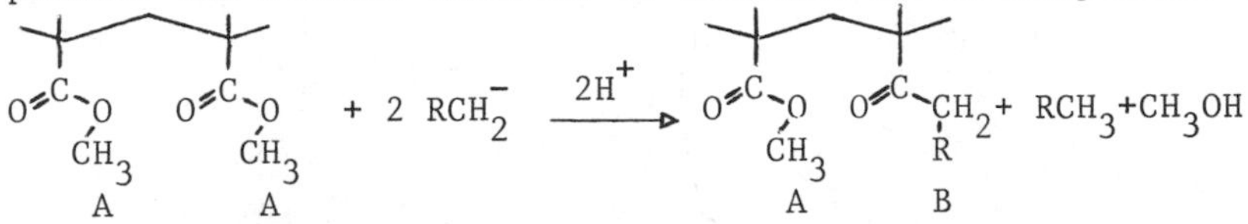

$$\text{[A: } C(=O)OCH_3\text{] [A: } C(=O)OCH_3\text{]} + 2\,RCH_2^- \xrightarrow{2H^+} \text{[A: } C(=O)OCH_3\text{] [B: } C(=O)CH_2R\text{]} + RCH_3 + CH_3OH$$

3.1. General survey of the reaction process.

The main conclusions of a previous study (34) may be summarized as follows :

- In no case does the process stop at the keto stage B as a result of the initial SN-2 step (reaction 1), but it proceeds further to keto-enolate B^- formation through proton abstraction by the more basic RCH_2Li species (reaction 2) ; carbonyl addition (reaction 3) was never observed : see reaction scheme. This behaviour was not expected for the reaction $C_6H_5SCH_2Li$/PMMA taking into account that of the model system $C_6H_5SCH_2Li/(CH_3)_3C\ CO_2CH_3$; it has to be considered as a specific feature of the PMMA chain. Moreover, keto-enolate formation gives to the copolymer an increasing anionic character as the subsitution progresses and restricts its solubility to highly dipolar aprotic solvents(DMSO, HMPA).

$$\text{PMMA (two } C(=O)OCH_3 \text{ units)} \xrightarrow[(1)]{RCH_2^-} CH_3O^- + \text{polymer with } C(=O)OCH_3 \text{ and } C(=O)CH_2R \text{ units} \xrightarrow{RCH_2^-} \begin{cases} (2)\ C(=O)OCH_3 \text{ and } C(O^-)=CHR \text{ units} + RCH_3 \\ (3)\ C(=O)OCH_3 \text{ and } C(O^-)(CH_2R)CH_2R \text{ units} \end{cases}$$

- In smooth reaction conditions ($t°C < 25$, $[RCH_2Li]_0/[ester]_0 < 1.5$) the process is remarkably free from side reactions, the substitution occurs selectively on the ester function without any racemization, degradation or crosslinking, leading after acidification to keto-β-functionalized $COCH_2R$ units.
- In more drastic reaction conditions ($t°C > 25$, $[RCH_2Li]_0/[ester]_0 > 1.5$), intramolecular cyclization on B^-B^- diads (see reaction (4) in 3.2.3) does appear to a very low extent (0.02 molar fraction), in sharp contrast to nearly quantitative cyclization on dimethylglutarate. Carboxylate formation (OCH_3 scission) may occur to an extent as high as 0.15 molar fraction for syndiotactic copolymers at 60°C, but the $-CO_2H$ units may be selectively and quantitatively methylated by CH_2N_2 into the original ester function.
- The limiting molar substitution degree ($\overline{DS}_m$ obtained for $[RCH_2Li]_0/[ester]_0 = 3$ and reaction time = 20 hours) depends on the RCH_2Li / polymer or model system as shown in table 2.

Table 2 - Limiting molar substitution degree for various RCH_2Li / PMMA or model systems

$R\text{-}CH_2Li$	t°C	model a)	PMMA b) I	PMMA b) R	PMMA b) S	PMA	PS+MMA
II, III, IV, VI and VII	25	1	0.85	0.59-0.66		0.98	1
	60		0.95	0.80	0.73		
V	-78 c)	1		0			
VIII	-15 c)	0.8		0.45			

a) $(CH_3)_3CCO_2CH_3$, $[RCH_2Li]_0/[ester]_0 = 1.1$

b) PMMA-I and S : >97 % I and S triads respectively. PMMA-R ($\overline{M}_n = 6.7x10^4$) : I = 0.05, H = 0.37 and S = 0.58.

c) The RCH_2Li species are not stable at t°C higher than -78 and -15 for V and VIII respectively.

SN_2 reactions of primary stabilized organolithium compounds on polyalkylmethacrylates affords a very versatile synthetic route to model ketonic copolymers : substitution is selective and quantitative up to $\overline{DS}_m$ of 0.60, and it is easily monitored by the initial

$[RCH_2Li]_0/[ester]_0$ ratio. A series of structurally related copolymers of the same $\overline{DP}_n$ and same tacticity as the polymeric precursor may be obtained. Variations on R allow clean introduction of complex $COCH_2R$ groups of specific properties (see 4) in a polymethacrylic backbone. Generalization of this process to the polyalkylacrylate series is not possible without simultaneous chain degradation (34) :

The use of sterically hindered secondary organolithium reagents, such as $(C_6H_5)_2CH\,Li$, drastically favors the O-alkyl scission of the ester group (37) ; finally, organolithium compounds leading to non enolizable keto functions in the first step of substitution, such as C_6H_5Li, promote competitive and consecutive reactions resulting in complex copolymers of poorly defined structure (38,39) :

$$r = k_1/k_2$$

$(CH_3)_3CCO\text{-}OCH_3$ $\quad r = 0.12\times10^{-3}\ e^{2.85/RT}$ $\quad\left\{\begin{array}{l} r = 0.016 \text{ at } 20°C \\ r = 0.14 \text{ at } -78°C \end{array}\right.$

S.M*MA.S $\quad r = 0.35$ at 20°C

MMA.M*MA.MMA $\quad r < 1$ at 20°C

On the other hand, the nitrile function may be a good precursor for the keto-β- functionalized group $CO\ CH_2R$; nevertheless this process variation is restricted to copolymers bearing isolated methacrylonitrile units to avoid degradation (no tertiary enolizable H atom) and cyclization of the CN bonds :

3.2. Reaction kinetics and structural characteristics of the modified PMMA.

3.2.1 - Reaction kinetics. Experimental measurements carried out at 25°C, as detailed elsewhere (35), lead to the values of the kinetic constants k_0 (AÅA), k_1(AÅB⁻) and k_2(B⁻ÅB⁻) given in table 3.

Table 3 - Reaction rates of SN_2 substitution for various RCH_2Li/PMMA systems

Reagent	Solvent	PMMA tacticity	k_0 x10^2	k_1 l.mole^{-1}	k_2 mn^{-1}	$k_0 : k_1 : k_2$
$CH_2SOCH_2^-Na^+$	DMSO	R	41	2.8	0.5	1:0.068:0.012
$CH_3SO_2CH_2^-Li^+$	DMSO	I	56	56	0.6	1:1:0.011
		S	10	0.6	≃ 0	1:0.06:0
$(CH_3)_2NSO_2CH_2^-Li^+$	THF + HMPA	R	1.55	0.32	≃ 0	1:0.21:0

$[Ester]_0$ = 0.3 mole.l^{-1}, $[RCH_2^-]_0/[Ester]_0$ = 2

In all cases the substitution process is characterized by autoretarded kinetics, and the tacticity of the PMMA precursor is the main factor determining the importance of these neighbouring group effects. For PMMA-S and R, $k_2 \sim 0$ implies a limiting $\overline{DS}_m$ value of 0.66 (9b) in fairly good agreement with the experimental data. The limiting $\overline{DS}_m$ of 0.45 obtained for the reaction of VIII on PMMA-R at t°C <-15 suggests that $k_1=k_2=0$ (calc. limiting $\overline{DS}_m$ = 0.43, 9b).

All these kinetic features may be readily taken into account within the 3 following assumptions : a) because of lower steric hindrance, isotactic triads exhibit a better accessibility than the syndiotactic ones : $k_0(A_m\overset{*}{A}_mA) > k_0(A_r\overset{*}{A}_rA)$; b) autoretarded kinetics arises from increasing steric hindrance around the reaction sites as the substitution proceeds further, and from electrostatic repulsion between the anionic reagent (RCH_2^-) and the modified negatively charged chain ($A\overset{*}{A}B^-$ and $AAB^-\overset{*}{A}B^-$) ; c) this electrostatic effect is partly canceled in isotactic triads by anchimeric assistance of the substituted B^- unit to the SN_2 step : $k_0(A_m\overset{*}{A}_mA)=k_1(A_m\overset{*}{A}_mB^-)$.

3.2.2 - Compositional heterogeneity of the substituted PMMA : For all systems investigated (36), the substituted PMMA are characterized by a fairly high chemical homogeneity over the whole range of $\overline{DS}_m$($\overline{DS}_m$<0.76), quite comparable to that of a radical azeotropic S-MMA copolymer (40) ($\overline{\sigma^2}$ = 1.6x10^{-4}). The mean square standard deviation $\overline{\sigma^2}$ related to two copolymers of $\overline{DP}_n$ = 700 derived from PMMA-R have been estimated by "cross fractionation" (36) : see table 4 and figure 1.

Table 4 - Compositional heterogeneity of substituted PMMA

B units	$\overline{DS}_m$	$\overline{\sigma^2}$ x 10^4 exp	calc. for $k_0 : k_1 : k_2$ 1:1:1	calc. for $k_0 : k_1 : k_2$ 1:0.2:0
$COCH_2SO_2N(CH_3)_2$	0.366	2.2	3.1	1.0
$COCH_2$-(oxazoline)	0.300	5.6	3.1	1.0

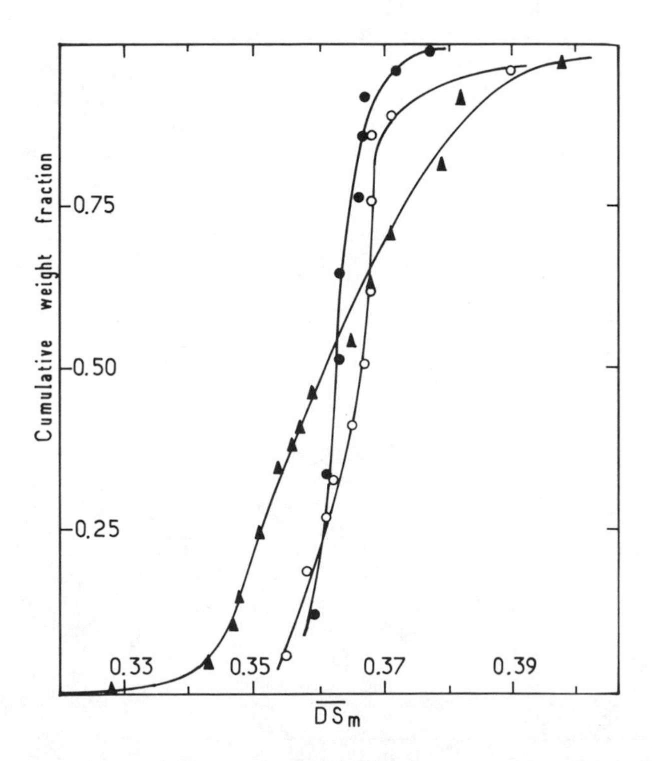

Figure 1. Compositional distribution of a predominantly syndiotactic copolymer bearing $COCH_2SO_2N(CH_3)_2$ units ($\overline{DP}_n = 650$, $DS_m = 0\ 366$) from precipitation fractionation data. System I (●): $CHCl_3$—Et_2O; system II (○): DMF—H_2O + O, 5% NH_4Cl; cross fractionation (▲): intermediate fractions obtained from system II are further fractionated according to System I.

A better agreement between experimental and calculated $\overline{\sigma}^2$ values cannot be reasonably expected because of the well known difficulties of accurate determination of low compositional heterogeneity on polydisperse copolymers (41). Both the high molecular weight of the PMMA precursor and the autoretarded kinetics contribute to the narrowing of the compositional distribution, but their relative influence cannot be estimated separately.

3.2.3 - Unit distribution in the substituted PMMA (35) was investigated by two independant methods : a) Direct analysis of copolymer microstructure by ^{1}H-NMR at 250 MHz ; the NMR spectrum (pyridine solution at 80°C) are sufficiently well resolved to allow a quantitative analysis of unit distribution, in terms of A centered triads and isolated B units in ABA triads. b) UV studies of the ionization and of the intramolecular cyclization of the B^-B and B^-B^- dyads in protic basic media ($NaOH$-H_2O 0.1N, NaOMe-MeOH 0.1N) : in such a medium the partially ionized copolymer chains are the site of a complex series of consecutive intramolecular reactions we have completely elucidated (35). The first step is of interest with respect to B unit distribution :

O=C–CH⁻(R) ... C(=O)–CH⁻(R) $\xrightarrow[(4)]{k_c}$ OH⁻ + O=C–C=C–CH⁻ (R, R) with R=SO_2CH_3, $SO_2N(CH_3)_2$

B^-, $\lambda_{max} \sim 250$ nm $\xrightarrow[\text{at 270 nm}]{\text{isosbestic point}}$ C^-, $\lambda_{max} \simeq 320$ nm

The cyclization rate k_c is higher for B^-B than for B^-B^- dyads and a drastic influence of tacticity is observed, the meso dyads being by far more reactive than the racemic ones, as expected (8). Some characteristic K_c values measured at 20°C in NaOH 0.1 N for different systems are collected in table 5.

Table 5 - Intramolecular cyclization rates of BB dyads in modified PMMA

B units	Tacticity	$\overline{DS}_m$	$k_c(B^-B)$, h^{-1}	$k_c(B^-B^-)\times10^2$, h^{-1}
$(CH_2)_3(COCH_2SO_2N(CH_3)_2)_2$				42.8
$COCH_2SO_2N(CH_3)_2$	I	0.856	2.35	8
$COCH_2SO_2CH_3$	I	0.688	4.1	38
	R	0.517	4.1	0.72
	S	0.398	0.24	0.65

The amount of uncyclized B units may be correlated with the B distribution, illustrated by the fraction F(ABA) of isolated B units for instance, allowing a direct comparison with ^{1}H-NMR data.

The experimental results of the two methods a) and b) applied to copolymers bearing $COCH_2SO_2CH_3$ and $COCH_2SO_2N(CH_3)_2$ groups are in fairly good agreement , as shown in Figure 2;for isotactic copolymers the distribution of B units is nearly bernouillian, as expected from the kinetics $k_0=k_1$; for syndiotactic copolymers B units tend to be isolated between A_n blocks, and their distribution is quite compatible with that calculated taken into account the corresponding autoretarded kinetics ($k_0>k_1>>k_2 \simeq 0$).

To conclude, kinetic measurements and structural analysis of the copolymers have allowed a quantitative and self-consistent description of the SN_2 reaction of RCH_2Li species on PMMA taking into account PMMA chain reactivity through the simplified model of the nearest neighbouring group effects. Two main features are particularly relevant : the definite influence of tacticity, and the independance of the reaction process on the total charge of the copolymer. In this sense, the $R\text{-}CH_2Li$/PMMA systems are closed to the PMMA basic hydrolysis in presence of excess base. (29,31).

4. Tautomerism of keto-β-heterocycles on macromolecular chains

Copolymers bearing keto-β-heterocyclic units $COCH_2Het$ are good systems for the comparative study of the specific tautomeric equilibrium of the B units (42) on model compounds $(CH_3)_3CCOCH_2Het$. and on well defined macromolecular chains :

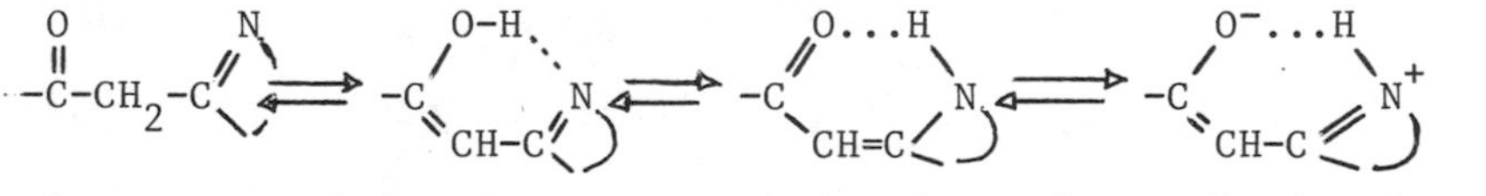

ketone ⇄ chelated enol ⇄ chelated enamine ⇄ dipolar form

In general, only one conjugated tautomer is in equilibrium with the keto form.

Tautomerism on polymer should be quite sensitive to neighbouring group effects (composition and unit distribution, steric hindrance and tacticity) and to the microenvironment polarity in solution (copolymer-solvent interactions, critical concentration c^* of coil interpenetration). The determination of the tautomerism constant K_T=(total conjugated forms)/(keto form) in dilute ($c<c^*$) and semi-dilute ($c>c^*$) solution from ^1H-NMR at 250 MHz and from UV spectroscopy has been reported elsewhere (39,43). The following spectrometric data related to keto-2-picolyl and keto-quinaldyl structures are quite illustrative :

a) keto-2-picolyl group (39) : ketone ⇄ chelated enol

b) keto-quinaldyl group (43) : ketone ⇄ chelated enamine

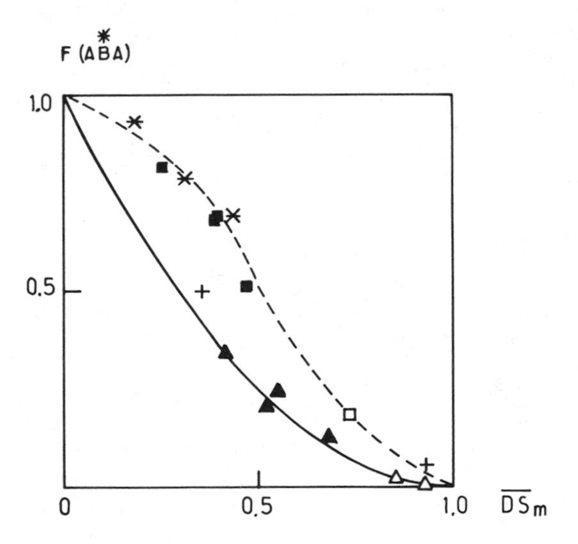

Figure 2. Distribution of isolated B units for various modified PMMA. Calculated for bernouillan ($k_0 = k_1 = k_2$) and for markovian ($k_0 : k_1 : k_2 = 1 : 0{,}2 : 0{,}01$) copolymers. B unit = $COCH_2SO_2CH_3$: I-UV (▲); S-¹H-NMR, (*); and S-UV (■). B unit = $COCH_2SO_2N(CH_3)_2$: I-¹H-NMR (+); I-UV (△); and S-UV (□)

a)

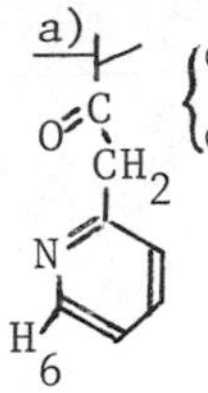

$\delta(CH_2CO)$=4.0-4.1 ppm
$\delta(H_6)$=8.2-8.3 ppm

In CF_3CO_2H sol. λ_{max}=263 nm, ε=7800 l.mole^{-1}.cm^{-1}

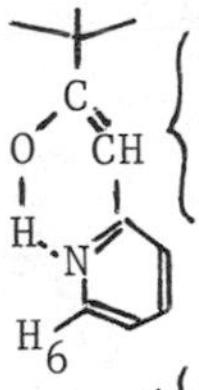

δ(CH=)=5.4-5.5 ppm
$\delta(H_6)$ =8.4-8.5 ppm
δ(OH) = 15.1 ppm

$\pi \rightarrow \pi^*$ λ_{max} = 320 nm, ε depends on copolymer structure

b)

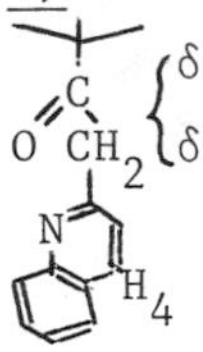

$\delta(CH_2CO)$=4.1-4.2 ppm
$\delta(H_4)$=8.1-8.2 ppm

In CF_3CO_2H sol. λ_{max}=319 nm, ε=10100 l.mole^{-1} cm^{-1}

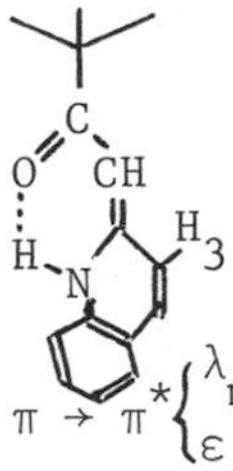

δ(CH=)=5.5-5.6 ppm
$\delta(H_3)$ =6.9-7.0 ppm
δ(NH) = 15.0 ppm

$\pi \rightarrow \pi^*$ λ_{max} = 420 nm, ε depends on copolymer structure

4.1 - Keto-β-oxazolines and keto-β-thiazolines (43)

At 25°C, the keto-β-oxazoline (from VIII) and keto-β-thiazoline (from IX) are nearly exclusively (>90 %) in the chelated enamine form both for model compounds and copolymers, either in bulk or in most solvents over a wide range of polarity (from $CHCl_3$ to DMSO). The keto form does appear only in very strong hydrogen bonding donor solvents like trifluoroethanol (TFE).

4.2. - Keto-2-picolines (P) (39) and a keto-quinaldines (Q) (43)

These keto-β-heterocyclic structures are well suited for tautomerism study , and the main conclusions of our studies are summarized below :

- For the model compounds studied at 30°C at a constant concentration of 0.3 mole.l^{-1} in 18 different solvents covering a very broad range of polarity from hexane to trifluoroethanol, K_T is systematically higher (about one order of magnitude) for the quinaldyl than for the 2-picolylketone, and L_nK_T is a linearly decreasing function of the solvent polarity parameter E_T defined by Dimroth (44) : fig. 3. The keto form is favoured in highly polar solvents and is the exclusive form in solvents able to complex the heterocycle nitrogen either by strong hydrogen bonding, like TFE, or by acid-base reaction, like HCO_2H or CF_3CO_2H.

t.butyl-2-picolyl ketone : $L_nK_T = -0.123\ E_T + 4.03$ (R_{18} = 0.991)

t.butyl-quinaldylketone : $L_nK_T = -0.0864\ E_T + 4.85$ (R_{13} = 0.972)

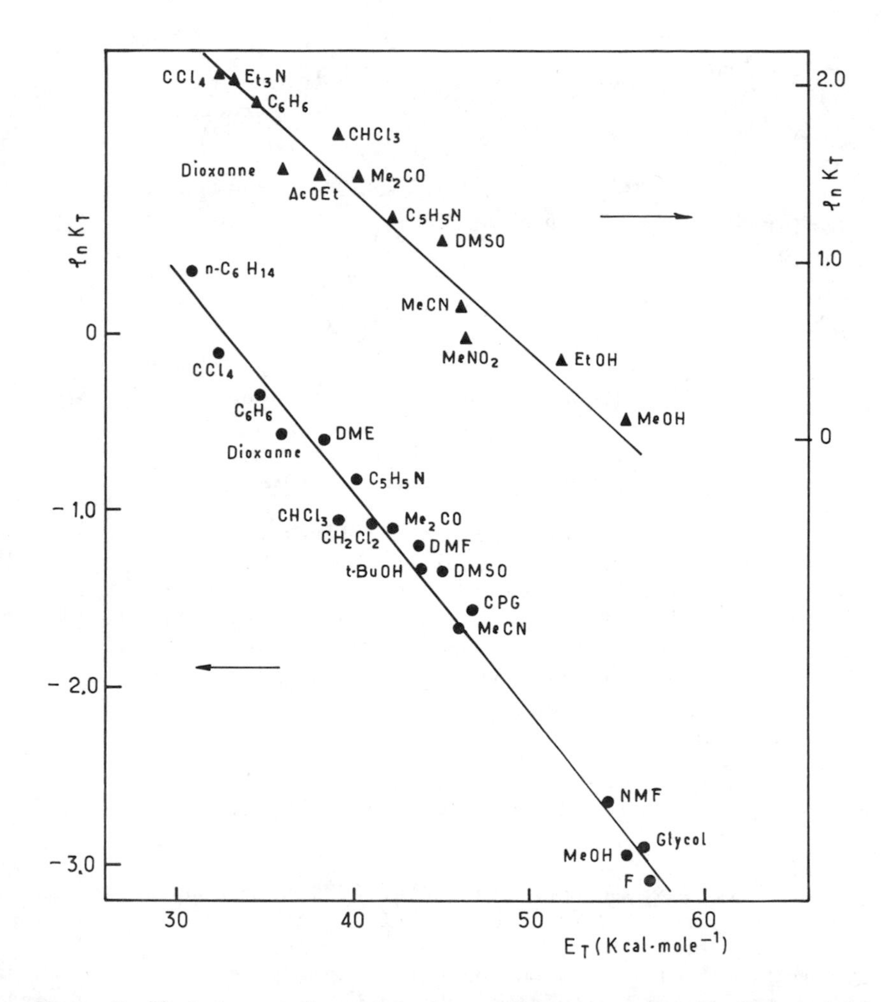

Figure 3. Variations of the tautomerism equilibrium constant K_T *([B] = 0.3 mole · l^{-1}, 30°C) of t-butyl-2-picolyl (●) and quinaldyl (▲) ketones vs. solvent polarity. (DME) dimethoxyethane; (CPG) propyleneglycol carbonate; (NMF) N-methylformamide; (F) formamide.*

- In semi dilute ($c > c^*$, [B] = 0.3-0.5 $mole.l^{-1}$) or dilute ($c < c^*$, [B] = 10^{-2} $mole.l^{-1}$) solution, K_T is significantly greater for copolymers than for the model compounds whatever the solvent is. For semi-dilute solution in a given solvent, the complex influences of composition, unit distribution and tacticity do not result in definite trends on K_T values, as illustrated in table 6 by some representative K_T data related to keto-2-picolyl structures at 25°C.

Table 6 - Tautomerism of keto-2-picolyl functions on PMMA-copolymers

$\overline{DS}_m$	Pyridine (E_T=40.2)		DMSO(E_T = 45)		
	% ketone	K_T	% ketone	K_T	ΔH $kcal.mole^{-1}$ a)
Model	69.4	0.44	79.3	0.26	- 1.10
R-0.129	50.0	1.0	48.6	1.06	
R-0.313	49.0	1.04	51.1	0.96	
R-0.538	51.0	0.96	51.2	0.99	
R-0.615	46.4	1.15	49.1	1.03	
S-0.298	50.0	1.0	46.4	1.15	- 3.68
I-0.311	45.0	1.22	34.8	1.87	- 0.82

a) ΔH enthalpy of enolization measured in dilute solution ([B] = 10^{-2} $mole.l^{-1}$) between 25 and 145°C.

In dilute solution, the dependance of L_nK_T on solvent polarity for copolymers is definitely measurable, but it is significantly reduced with respect to that of the model compounds, by a factor of about 3 for a predominantly syndiotactic chain bearing keto-2-picolyl functions in the form of isolated units ($\overline{DS}_m$ = 0.129, F(ABA) = 0.95 , $L_nK_T = -3.79 \times 10^{-2} E_T + 1.86$ (R_8 = 0.971)) : see figure 4. Finally, even in highly polar solvents like TFE where the model exists exclusively in the keto-form, the same copolymer has an enol content of about 16 % and, moreover, it shows an important specific absorption band at λ = 380 nm which may be tentatively attributed to new dipolar structures such as :

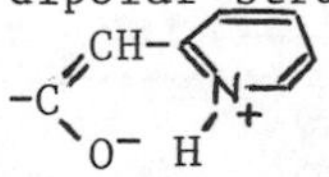

$$-C(-O^-)=CH-C_5H_4N^+-H$$

All these features are in contrast with the invariance of the extent of enolization of polyvinylacetoacetate ($K_T \simeq 0.07$) in solvents such as CH_3CO_2Et (E_T = 38.1), $CHCl_3$ (E_T = 39.1), CH_3COCH_3 (E_T = 42.2) and CH_3CO_2H (E_T = 51.2) (45).

In all cases, an increase in temperature shifts the equilibrium towards the keto form by disrupting the internal H-bond which is the main stabilization factor of the enol or enamine. The tautomerism is more temperature dependant for keto-quinaldines than for

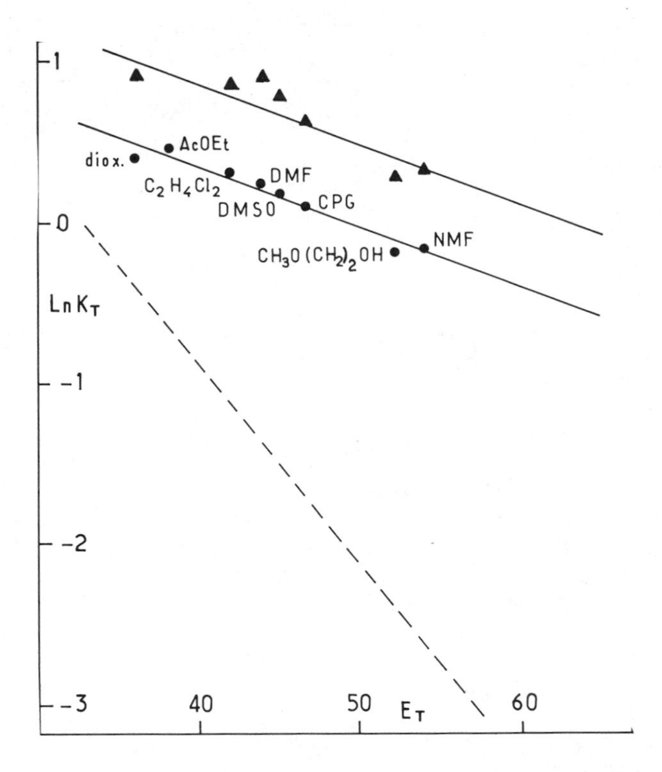

Figure 4. Variations of the tautomerism equilibrium constant K_T *([B] = 10^{-2} mole · l^{-1}, 20°C) of* t-butyl-2-picolyl *ketone (– – –) and of the corresponding PMMA-copolymers vs. solvent polarity. Predominantly syndiotactic (S = 0,58) copolymer of* DS_m *= 0,123 (●); syndiotactic copolymer of* $\overline{DS_m}$ *= 298 (▲)*

keto-2-picolines, and for syndiotactic chains than for isotactic ones.

Toconclude, the significant differences we have pointed out between the tautomerism of keto-2-picolyl and keto-quinaldyl structures on model compounds and on well defined MMA-copolymers strongly suggest that tautomerism of well choosen keto-β-heterocycles may be a valuable probe for the quantitative study of the specific characteristics of the microenvironment of a given site of a macromolecular chain. The increased enolization and its weaker dependance on solvent polarity we have noticed for copolymers may be reasonably attributed to neighbouring group and chain effects which promote higher steric hindrance and rigidity and tend to level off solvation and polarity variations around the enolizable probe.

5. Recent advances in chemical modification of polymers

5.1 - Mechanistic and theoretical aspects

A few years ago, Harwood[2] has experimentally studied the reactivity of copolymers in terms of their structure, focusing on the dependance of the reactivity of a given unit on the nature of its neighbouring stubstituents. More recently, quantitative analysis of cooperative effects during reactions on linear binary copolymers has been performed by González and Hemmer[46,47]. The copolymer sequential structure is obviously expected to influence the kinetics of cooperative reactions, and theoretical models have been developed both for periodic (alternating copolymers) and aperiodic (Bernouillian, first and second-order Markov chains) copolymers. These calculations allowed a quantitative interpretation of the kinetics of oxidation and repeated oxidation of various polysaccharides, leading to the sequential analysis of two important galactomannanns, guaran and locust bean gum[48].

5.2 - Chemical modification of polymers as a synthetic route to new polymers and copolymers or function alized materials

Reactions on macromolecular precursors are most often the key step in the synthesis of sophisticated polymers in various well documented fields of steadily increasing importance such as : a) linear or crosslinked polymeric reagents and catalysts (2,5,6, 49) ; b) polymers showing esterolytic enzyme-like properties (2, 49-52) ; c) polymeric drugs (53,54) and so on... Three more specific but still highly significant studies are outlined below.

5.2.1 - Polyorganophosphazenes - All the investigations carried out before 1965 to develop technological applications of polydichlorophosphazene (a) failed because of its unstability towards hydrolysis. Nevertheless complete halogen replacement by

SN_2 substitution in solution with a variety of nucleophiles (NuH) may be easily performed : this leads to a very broad range of new polyorganophosphazenes (b) of outstanding specific properties and high versatility[55]

$$\left[N = P(Cl)_2 \right]_n + 2n\ Nu^- \longrightarrow \left[N = P(Nu)_2 \right]_n + 2n\ Cl^-$$

(a) (b)

Nu H = ROH, ArOH, R_2NH

Some copolymers (Nu_1 and Nu_2 substituants) are already manufactured on an industrial scale as high performance elastomers

5.2.2. Chemical modification of polydienes - The studies of CAMERON et al.[56-59] may considered quite representative of recent trends in this very rich field[60]. Addition in solution of aryl or alkylsulfenyl chlorides accross the double bond is selective and may be quantitative ; furthermore it is regioselective and yields block copolymers for partial modification (R=H or CH_3) :

$$\left[CH_2-C(R)=CH-CH_2 \right]_n + m\ ArSCl \longrightarrow \left[CH_2-C(R)(Cl)-CH(SAr)-CH_2 \right]_m \left[CH_2-C(R)=CH-CH_2 \right]_{n-m}$$

5.2.3. Linear poly(ethylenimine) and poly(N-alkylethylenimines) - Linear poly(N-formyl or N-acylethylenimines) obtained by cationic polymerization of the corresponding oxazolines may be quantitatively hydrolyzed[61] or reduced[62] without degradation or crosslinking : the linear poly(ethylenimine) (PEI) and poly(N-alkylethylenimines) thus obtained cannot be prepared directly by the cationic polymerization of the corresponding aziridines which leads to highly branched structures and oligomer formation[63]

$$-\!\left(N(C(=O)R)-CH_2-CH_2 \right)_n \xrightarrow{\text{basic hydrolysis}} \left(NH-CH_2-CH_2 \right)_n$$

$$-\!\left(N(C(=O)R)-CH_2-CH_2 \right)_n \xrightarrow{\text{hydride reduction}} \left(N(CH_2R)-CH_2-CH_2 \right)_n$$

R=H, CH_3, C_6H_5

The hydrolysis of polyacylethylenimines to PEI is the key step of the preparation of well defined block (b)[64] and graft (g)[65-66] polymers where PEI blocks show interesting chelating properties for heavy metal ions: poly(butadiene-b-EI), poly(butadiene-g-EI) and poly(styrene-g-EI).

6. Perspectives and conclusions

6.1. Mechanistic and theoretical aspects of the reactivity of macromolecular chains in solution

Further quantitative studies related to both reaction kinetics and copolymer structure, and performed on well defined polymeric precursors, are still necessary in order to give, at the experimental level, a broader and more rigorous base to the theoretical calculations relying on cooperative effects restricted to the nearest neighbours. Moreover analysis of complex systems have to be considered to point out the necessary limitations of such a simplified theory, and some particular points may be of special interest, such as a) cooperative effects taking into account both configurational and compositional effects, which may lead to a maximum of 10 kinetic constants for a given function on a polymeric chain[7,67] ; b) cooperative effects at longer distance, for which calculations have been already carried out[15,16] ; c) specific solvation effects, like selective absorption of the reagent R on an previously reacted block B_n of the polymeric precursor[38,68] ; d) conformational transitions occuring as the reaction progresses.

On the other hand, two important factors of macromolecular chain reactivity in solution deserve more attention,namely the nature of the reaction medium and the polymer concentration.

6.1.1. Quantitative analysis of the influence of the reaction medium - It may be reasonably expected that steric and polar effects due to the chain backbone and to the neighbouring units contribute to an appreciable extent to the polarity of the microenvironment of a given reaction site of a polymeric chain. Thus the influence of the solvent on reaction process may be significantly weaker for polymers than for low molecular weight model compounds, but it may be still measurable. The concept of "microenvironment polarity" has promoted experimental approaches mainly through spectrometric methods such as fluorescence [69,70], photochromism[71], solvatochromism[10]. Chemical equilibrium such as tautomerism, as we have previously shown (see 4.), may be used as a sensitive probe in this field, and the quantitative analysis of the chemical reactivity of polymeric systems versus solvent polarity may also be of high value as pointed out by Morawetz et al.[72].

6.1.2. Polymer concentration effects on its chemical reactivity - At concentration higher than the critical concentration c^* of coil interpenetration (c^* is a function of both the polymer molecular weight and the polymer-solvent interactions) intermolecular interactions may become of increased importance, and their possible influence on chain reactivity remains an open field for future studies.

6.2. Chemical reactions on polymers as a synthetic route to new polymers and functionalized polymeric materials

Reactions on polymers may be now easily monitored and performed in a very selective way for an increasing number of cases. They may be thus considered as an efficient alternative to copolymerization, and they are indeed for some systems the best method for the synthesis of structurally well defined polymers and copolymers ($\overline{DP}_n$ and MWD, branching, tacticity, homogeneity...) On the other hand, the scope of functionalization of polymeric materials is practically infinite, even for very sophisticated structural units. In this field, the macromolecular chemist has to remain well aware of recent progress of organic chemistry which is continuously developing synthetic reagents and methods of outstanding efficiency and specificity.

Acknowledgments : The author is greatly indebted to Drs. J.J. Bourguignon, M. Oteyza de Guerrero, R. Roussel and P. Spegt for their decisive contribution to this work, and he gratefully acknowledges Dr. P. Rempp for his continuous interest and his critical discussions.

Literature cited

1. Chemical Reactions of Polymers, Ed. by FETTES, E.M., Interscience-Publishers, New-York, 1964.
2. Reactions on Polymers, Ed. by MOORE, J.A., D. Reidel Publishing Company, Dordrecht, Holland, 1973.
3. MORAWETZ, H., Macromolecules in Solution, 2nd.Ed., Interscience Publishers, New York, 1975, P. 439.
4. PLATE, N.A., Pure Appl.Chem. (1976), 46, 49.
5. MATHUR, N.K. and WILLIAMS, R.E., J.Macromol.Sci.-Rev.Macromol. Chem. (1976), C-15, 117.
6. HEITZ, W., Adv.Polym.Sci., (1977), 23, 1.
7. PLATE, N.A. and NOAH, O.V., Adv.Polym.Sci., (1979), 31, 133.
8. VAN BEYLEN, M.M., in the Stereochemistry of Macromolecules, Ed.by M. DEKKER, New-York, 1968, vol.3, p. 333.
9. BOUCHER, E.A., J.Chem.Soc., Faraday Trans. I a) (1972), 68, 2281 ; b) (1972), 68, 2295.
10. STROP, P., MIKES, F. and KALAL, J., J.Phys.Chem., (1976), 80, 694.
11. KELLER, J.B., J.Chem.Phys., (1962), 37, 2584 ; (1963), 38,325.
12. Mc QUARRIE, D.A., Mc TAGUE I.P. and REISS, H., Biopolymers, (1965), 3, 657.
13. FRENSDORFF, H.K. and EKINER, O., J.Polym.Sci., (1967), A-2, 5, 1157.
14. KLESPER, E., JOHNSEN A. and GRONSKI, W., Makromol.Chem., (1972), 160, 167.
15. GONZALEZ, J.J., HEMMER, P.C. and HØYE, J.S., Chem.Phys.,(1974) 3, 228.

16. KRISHNASWAMI, P. and VADAV, D.P., J.Appl.Polym.Sci., (1976), 20, 1175.
17. NOAH, O.V., TOOM A.L., VASIL'EV, N.B., LITMANOVITCH, A.D. and PLATE, N.A., J.Polym.Sci.- Polym.Phys.Ed., (1974) 12
18. Same authors as in (17), J.Polym.Sci.- Polym.Chem.Ed.,(1974) 12
19. GONZÁLEZ, J.J. and KEHR, K.W., Macromolecules, (1978), 11, 996.
20. FUOSS, R.M., WATANABE, M. and COLEMAN, B.D., J.Polym.Sci., (1960), 48, 5.
21. TSUCHIDA, E. and IRIE, S., J.Polym.Sci. - Polym.Chem.Ed., (1973), 11, 789.
22. NOA, O.V., TORCHILIN, V.P., LITMANOVITCH, A.D. and PLATE, N.A., Polym.Sci. USSR, (1974), A-16, 775.
23. MORCELLET-SAUVAGE, J. and LOUCHEUX, C., Makromol.Chem.,(1975) 176, 315.
24. BOUCHER, E.A., GROVES, J.A., MOLLETT, C.C. and FLETCHER, P.W. Trans.Farad.Soc., (1977) 73, 1629.
25. BOUCHER, E.A. and MOLLETT, C.C., J.Polym.Sci. -Polym.Chem.Ed. (1977), 15, 283.
26. SAITO, T., MATSUMARA, Y. and HAYASHI, S., Polym.J., (1970), 1, 639 ; (1973), 4, 124.
27. KRENTSEL, L.B., LITMANOVITCH, A.D., PASTUKHOVA, I.V. and AGASANDYAN, V.A., Polym.Sci. USSR, (1961), 13, 11.
28. LITMANOVITCH, A.D., PLATE, N.A., SERGEEV, N.M., SUBOTTIN, O. A. and USMANOV, T.I., Dokl.Acad.Nauk. USSR, (1973), 210,114.
29. LITMANOVITCH, A.D., PLATE, N.A., AGASANDYAN, V.A., NOA, O.V., YUN, E., KRYSHTO , V.I., LUK'YANOVA, N.A., LELYUSHENKO, N.V., KRESHETOV, V.V., Polym.Sci. USSR, (1975), 17, 1276.
30. JOHNSON, A., KLESPER, E., WIRTHLIN, T., Makromol.Chem., (1976), 177, 2397.
31. BARTH, V. and KLESPER, E.,, Polymer, (1976), 17, 777, 787 and 893.
32. GONZALEZ, J.J., Biophys.Chem., (1974), 2, 23.
33. HEMMER, P.C. and GONZALEZ, J.J., J.Polym.Sci.- Polym.Phys.Ed. (1977), 15, 321.
34. BOURGUIGNON, J.J. and GALIN, J.C., Macromolecules, (1977), 10, 804.
35. BOURGUIGNON, J.J., Thèse, Strasbourg, 1978.
36. BOURGUIGNON, J.J., BELISSENT, H. and GALIN, J.C., Polymer, (1977), 18, 937.
37. ROUSSEL, R. and GALIN, J.C., Unpublished results.
38. REMPP, P., Pure Appl.Chem., (1976), 46, 9.
39. ROUSSEL, R., Thesis, Strasbourg, 1979.
40. TERAMACHI, S. and KATO, Y., Macromolecules, (1971),4, 54.
41. INAGAKI, H., Adv.Polym.Sci., (1977), 24, 189.
42. ELGUERO, J., MARZIN, C., KATRITZKY, A. and LINDA, P., The Tautomerism of Heterocycles, Academic Press, New York, 1976.

43. OTEYZA DE GUERRERO, M., Thesis, Strasbourg, 1977.
44. REICHARDT, C. and DIMROTH, K., Fortschr.Chem.Forsch., (1968) 11, 1.
45. DAVYDOVA, S.L. and PLATE, N.A., Coordination Chem.Revs., (1975), 16, 195.
46. GONZALEZ, J.J. and HEMMER, P.C., J.Polym.Sci.- Polym.Lett.Ed. (1976), 14, 645.
47. GONZALEZ, J.J. and HEMMER, P.C., J.Chem.Phys., (1977), 67, 2496 and (1977), 67, 2509.
48. GONZALEZ, J.J., Macromolecules, (1978), 11, 1074.
49. MANECKE, G., STORCK, W., Angew.Chem.Int. Ed., (1978), 17, 657.
50. KUNITAKE, T. and OKANATA, Y., Adv.Polym.Sci., (1976), 20,159.
51. OKUBO, T. and ISE, N., Adv.Polym.Sci., (1977), 25, 135
52. SHIMIDZU, T., Adv.Polym.Sci., (1977), 23, 56.
53. DONARUMA, J.G., Progr.Polym.Sci., (1974), 4, 18.
54. BATZ, H.G., Adv.Polym.Sci., (1977), 23, 25
55. ALLCOCK, H.R., Angew.Chem.Int.Ed., (1977), 16, 147.
56. CAMERON, G.G., and MUIR, R.B., J.Polym.Sci.- Polym.Lett.Ed., (1976), 14, 661.
57. BRYDON, A., CAMERON, G.G. and MUIR, R.B., Makromol.Chem., (1977), 178, 1739.
58. BULLOCK, A.T., CAMERON, G.G. and MUIR, R.B., Europ.Polym.J. (1977), 13, 505.
59. BUCHAN, G.M. and CAMERON, G.C., J.Chem.Soc.- Perkin.Trans., (1978), 1, 783.
60. BRYDON, A. and CAMERON, G.G., Progr.Polym.Sci., (1974), 4, 209.
61. SAEGUSA, T., IKEDA, H. and FUJII, H., Macromolecules, (1972), 5, 108.
62. SAEGUSA, T., YAMADA, A., TAODA, H. and KOBAYASHI, S., Macromolecules, (1978), 11, 435.
63. GOETHALS, E.J., Adv.Polym.Sci., (1977), 23, 103.
64. SAEGUSA, T., and IKEDA, H., Macromolecules, (1973), 6,805.
65. SAEGUSA, T., YAMADA, A. and KOBAYASHI, S., Polymer,J., (1978) 11, 53.
66. SAEGUSA, T., KOVAYASHI, S. and YAMADA, A., Macromolecules, (1975), 8, 390.
67. PIS'MEN, L.P., Polym.Sci. USSR, (1972), 14, 2084.
68. LE MOIGNE, J., Thèse, Strasbourg, 1974.
69. KANAOKA, Y., Angew.Chem.Int.Ed., (1977), 16, 137.
70. TAZUKE, S., SATO, K. and BANBA, F., Macromolecules, (1977),
71. 10, 1224.
71. VANDEWIJER, P.H. and SMETS, G., J.Polym.Sci., (1968) C-22, 251.
72. CHIN-PAO SUN and MORAWETZ, H., J.Polym.Chem.- Polym.Chem.Ed. (1977), 15, 185 and (1978), 16, 1059.

RECEIVED July 12, 1979.

10

Conversion of Graft Polyacrylamide to Amines via the Hofmann and Mannich Reactions

R. J. ELDRIDGE

CSIRO, Division of Chemical Technology, South Melbourne, Victoria, Australia

Magnetic ion exchange resins have considerable advantages over conventional resins in their handling properties (1). They settle rapidly, permit high flow rates in fixed bed plant, and can be pumped continuously without damage in moving bed plant. By forming a shell of ion exchanger about a fine magnetic core, resins with very high reaction rates can be produced. Graft polymerization of monomers such as acrylic acid on core particles consisting of magnetic iron oxide embedded in crosslinked poly (vinyl alcohol) (PVA) has been described previously (2).

Magnetic weak base resins are of interest for a variety of water treatment applications. However, the attempted graft polymerization on magnetic PVA beads of diallylamine, methyldiallylamine, dimethylaminoethyl methacrylate and dimethylaminopropylacrylamide (as hydrochlorides) was unsuccessful. Attempts to prepare magnetic weak base resins by modification of grafted acrylic acid also failed. However acrylamide grafts readily to crosslinked PVA (2) and the Hofmann degradation of such grafts offers a possible alternative route. The Mannich reaction can also be used to introduce amino groups into magnetic polyacrylamide (PAM) beads.

Hofmann Degradation of Polymeric Amides

The preparation of (soluble) polyvinylamine by reacting PAM with hypohalite was attempted as early as 1944, but high conversions have been reported only recently (3,4,5,6). Early work at high temperature (7) resulted in a decrease in the nitrogen content of the polymer from 19.7% to <15% instead of the expected increase to 32.5%. Polymethacrylamide (PMAM) yielded a polymer containing <5.5% w/w amine units (8). The product from PAM treated at 25-30° for one hour contained up to 25% N (9) and that from PMAM at 21° for 3h up to 20% w/w amine (10). However, the carboxylic acid content of these products often exceeded 20% and other unwanted functional groups were also obtained. Mullier and Smets (10) found the amine content to increase slightly when the amount

0-8412-0540-X/80/47-121-139$05.00/0

of NaOC used was less than stoichiometric. PNAM gave higher yields of amine than polyacrylamide (10), and although Arcus (8) found hypobromite to react very slowly with PNAM, both he and Schiller and Suen (9) got higher yields of amine with OBr^- than with OCl^-.

In a study published after the present work commenced, Tanaka and Senju (3,4,5,6) reported conversion of 85-97% of the amide groups of PAM to amino groups (as determined by colloid titration). They used a very slight excess of NaOCl in 3.6-5.3 M NaOH solution at or below 0^oC for up to 24 hours. Formation of carboxyl groups under these conditions did not exceed 5-6%. The order of adding the reagents is significant, adding the polymer to hypochlorite being preferable to the reverse order. Prior hydrolysis of some of the amide groups increased the conversion to amine of the remainder.

At low temperatures formation of the N-chloroamide intermediate is much faster than hydrolysis of the amide groups. Conversion to the isocyanate is slow, and at high concentrations of NaOH, isocyanate groups are hydrolyzed to amine as fast as they are formed, preventing side reactions between isocyanate and amide or N-chloroamide groups. Viscosity measurements indicated that chain scission is also greatly reduced. Tanaka and Senju found OBr^- to be inferior to OCl^-, and disputed the claim of Sugiura *et al.* (11) to have obtained high conversions with OBr^-. St. Pierre *et al.* (12,13) were also unable to reporduce these results.

If Tanaka and Senju's findings apply also for grafted polyacrylamide, their procedure could be used to produce weak base resins consisting of polyvinylamine chains grafted to magnetic PVA beads. The capacity of these resins would be reduced by any chain scission during the Hofmann degradation, but could reach useful values. However, the high local concentration and reduced mobility of the grafted chains could well alter the course of the reaction, resulting in the optimum conditions being different from those for soluble polymer. Optimum conditions for grafted PAM were sought by treating grafts with NaOCl and NaOBr at various temperatures, NaOH concentrations and hypohalite:amide ratios.

Hofmann Degradation of Grafted PAM

Experimental: The preparation of ∿ 100μm beads of crosslinked PVA containing Fe_3O_4 or γ-Fe_2O_3 and grafted with PAM has been described previously (2). Graft beads containing 2.0-6.6 mmol/g PAM were stirred in 2.5M aqueous NaOH solution and commercial NaOCl solution (1.82M by iodometry) or aqueous Br_2/NaOH solution (0.175 - 0.50M/2.5M) was added. When no temperature control was applied the system reached temperatures of up to 60°. At the end of each reaction the product was filtered off, washed copiously with water, air dried to a free flowing powder, and weighed. It was then used to pack an ion exchange column, converted to the hydrochloride form and eluted with 2M $NaNO_3$ solution. Amine capacities were determined by titration with 1M $AgNO_3$ solution. Total acid-base capacities were determined in a few cases by adding excess acid to the base form of the resin or vice versa and back-titrating. Elemental analyses were performed by the Australian Microanalytical Service. Experimental details and analytical results are given in Tables I (hypobromite) and II (hypochlorite).

Results and Discussion: It is clear from the yields of amine shown in the Tables that OCl^- is superior to OBr^-. However the best capacities obtained were only 1.5-1.7 meq/g (23-25% conversion). This cannot be due entirely to chain scission since the loss in mass of the grafted particles was usually small. (In some cases there was actually an increase). Total base determinations indicate that a significant percentage of amide groups were hydrolyzed to carboxylate, even at 0°C. Conversion of all $\sim CONH_2$ groups to $\sim COONa$ would give a mass increase of 15% for a graft particle initially containing 6.6 mmol/g amide. This suggests that not all the original amide groups were converted to amino or carboxylate groups; probably some scission occurred. Crosslinking the grafted polymer with methylenebis-acrylamide did not prevent scission (H12).

In contrast to the behaviour of linear PAM (10), optium conversion to amine occurred at a hypochlorite:amide ratio of 1.5:1 (H21). At room temperature the reaction was essentially complete in 20 minutes (compare H10, 11,14). At 0°C the reaction is much slower (H17), but reasonable conversions are reached after about 120 minutes (H19). However, the very high conversions of Tanaka and Senju were not attained. Another difference from the behaviour of the linear polymer is that prior hydrolysis is disadvantageous (H13, H18).

Stirring NaOCl/NaOH solutions in open beakers in the absence of magnetic beads did not result in any decrease in the hypochlorite concentration. When ungrafted magnetic beads were added there was still no significant decrease over several hours. Thus the 1.5:1 ratio is not an artefact due to loss of Cl_2 or reaction of hypochlorite with the PVA core.

Microanalysis of several products showed the presence of 2.2-5.3% w/w of non-amine nitrogen (Table II). This suggests the presence of unreacted amide groups in the Hofmann products even when relatively high yields of amine were obtained. However, assuming the 4.9% non-amine nitrogen in H18 to be in the form of 3.5 mmol/g of unreacted amide leaves 0.3 mmol/g of the original amide unaccounted for, and the decrease in carboxyl content confirms that there has been some chain scission. The large amount of unreacted amide is presumably due to the inaccessibility of some monomer units, since use of excess hypochlorite does not increase conversion to amine (H9, H22). After standing for 18 hours in

TABLE I. TREATMENT OF PAM GRAFTS WITH NaOBr IN 2.5M NaOH SOLUTION

NO	GRAFT	mmol AMIDE	OBr^-/AMIDE [1]	$[OBr^-]$	TEMP	TIME	YIELD	% AMINE[2]	% -COOH[2,3]
H1	2.62g	8.74	1.1	0.33M	RT[4]	40 min.	2.36g	11[5]	16
H2	2.31g	4.85	3.0	0.13M	RT	40 min.	2.17g	8.9	20
H3	2.47g	6.63	1.0	0.11M	RT	120 min.	2.18g	1.6	20
H4	2.54g	8.03	0.5	0.08M	RT	40 min.	2.35g	7.9	11
H5	product of H4		0.5[6]	0.08M	RT	40 min.	2.06g	total 20%	
H6	2.60g	8.45	1.0	0.12M	50°	40 min.	2.30g	total 30%	
H7	0.72g	3.68	1.0	0.08M	RT	20 min.	0.70g	2.7	-

1. mol ratio
2. mol per 100 mol amide in starting material
3. by subtraction from total acid-base capacity
4. initially at room temperature of (20 ± 5)°C
5. 0.42 meq/g amine by titration; 0.46 mmol/g N by microanalysis
6. ratio calculated on original amide

TABLE II. TREATMENT OF PAM GRAFTS WITH NaOCl

NO.	GRAFT	mmol AMIDE	OCl⁻/AMIDE [1]	[OCl⁻]	[OH⁻]	TEMP.	TIME	YIELD	% AMINE
H8	2.70g	9.1	1.0	0.33M	2.0M	RT[2]	20 min.	2.44g	20[3]
H9	2.79g	11.1	3.2	0.81M	1.4M	RT	20 min.	2.00g	7.4
H10	2.42g	12.4	1.0	0.22M	2.2M	RT	20 min.	2.20g	23
H11	1.66g	10.9	1,0	0.20M	2.2M	RT	60 min.	1.40g	19[4]
H12	10.0g[5]	89.3	1.0	0.22M	2.2M	RT	40 min.	8.27g	6.8
H13	7.5g[6]	33.4	1.0	0.20M	2.2M	RT	40 min.	6.85g	6.2[8]
H14	10.0g	65.9	1.0	0.76M	1.5M	RT	40 min.	10.42g	20
H15	10.0g	65.9	1.1	0.81M	1.4M	RT	40 min.	9.75g	16
H16	10.0g	65.9	1.2	0.86M	1.3M	RT	40 min.	10.15g	23
H17	10.0g	65.9	1.0	0.76M	1.5M	0°	40 min.	10.15g	2.5[9]
H18	10.0g[7]	40.4	1.4	0.70M	1.5M	RT	40 min.	9.50g	5.9[10]
H19	10.0g	65.9	1.2	0.86M	1.3M	0°	120 min.	10.37g	23
H20	10.0g	65.9	1.2	0.86M	1.3M	60°	40 min.	8.92g	8.8
H21	10.0g	65.9	1.5	0.95M	1.2M	RT	40 min.	9.40g	25[11]
H22	10.0g	65.9	2.0	1.07M	1.0M	RT	40 min.	6.00g	9.5
H23	5.85g	38.6	0.5	0.48M	1.8M	RT	40 min.	5.55g	5.8
H24	18.0g	73.8	1.6	0.72M	1.5M	RT	40 min.	15.81g	24
H25	5.0g	30.8	0.64	0.41M	5.0M	0°	180 min.	4.45g	7.8
H26	5.0g	22.8	0.76	0.38M	5.2M	0°	24 hrs.	3.95g	19[12]

1. mol ratio
2. initially at room temperature of (20 ± 5)°C
3. 2.2% w/w non-amine N
4. 5.3% w/w non-amine N
5. crosslinked with methylenebisacrylamide
6. part hydrolyzed: 4.45 mmol/g amide + 2.14 mmol/g carbyxyl
7. Part hydrolyzed: 4.04 mmol/g amide + 2.21 mmol/g carboxyl
8. 0.30 meq/g amine; 3.5 meq/g total base
9. 0.16 meq/g amine; 0.72 meq/g total base
10. 0.25 meq/g amine; 1.76 meq/g total base; 4.9% non-amine N
11. 1.73 meq/g amine; 2.00 meq/g total base
12. 1.09 meq/g amine; 2.21 meq/g total base

2M NaOH solution, H10 contained 0.58% non-amine nitrogen. The residual non-amine nitrogen after hydrolysis may indicate the presence of other functional groups such as lactam rings (10), but the alkaline hydrolysis of PAM is very difficult to drive to completion (14, 15), and this nitrogen could equally well be present in residual amide groups. Product H18 also contained 2.0% nonionic Cl, attributed to α-chlorocarboxylic acid groups, which were reported by Mullier and Smets (10).

The stability of the base resins produced by the Hofmann reaction was tested by shaking 2.0g lots of H16 in 50 ml of 2% NaOH, 10% NaCl solution. The amine capacity decreased with time as shown in Table III. Because of this instability and the relatively low amine capacities attained, the products were not suitable for the intended applications.

Weak Base Resins via the Mannich Reaction

Amino-substituted (Mannich base) polymers can be prepared by reacting amide-containing polymers with formaldehyde and a suitable amine. Sugiyama and Kamogawa (16) treated PAM in aqueous solution with excess paraformaldehyde ($50^{o}C$, 1h) followed by excess dimethylamine ($50^{o}C$, 1h). This procedure gave 68% conversion to amine. Schiller and Suen (9) used a similar procedure with monomeric formaldehyde and various amines, but with excess PAM. Muller *et al.* (17) prepared monomeric amines from methacrylamide by reaction with formaldehyde and several primary and secondary amines (30 min at $60\text{-}70^{o}C$). These compounds were unstable in acid solution. Bartoli *et al.* (18) prepared a series of monomers by reacting acrylamide with paraformaldehyde and secondary amines in CCl_4.

Magnetic amine resins were prepared by a procedure based on that of Sugiyama and Kamogawa. A slurry of 4.25g of PAM graft beads (24.6 mmol) in 30 ml of water was heated to $50^{o}C$ and 2 ml of 37% H_2CO solution (24.7 mmol) was added. After one hour 1.8g (24.6 mmol) of diethylamine (b.p. $55.5^{o}C$) was added and heating at $50^{o}C$ under reflux continued for a further one hour.

Variations on this procedure are set out in Table IV. It can be seen that adding excess reagents does not improve the capacity obtained, but that sequential addition is of some benefit (M3, M4). Strong base resins cannot be prepared by this route since tertiary amines do not react (M5), and a very weak secondary amine also gives very little reaction (M6).

Products were converted to the hydrochloride salt with IM HCl. The initial capacity determination on M1 ($AgNO_3$ titration) gave a value of 2.48 meq/g, calculated on free base. 100% conversion would give 3.88 meq/g. However, repeat determinations on the same sample gave values of 2.01 and 1.80 meq/g. Conversion of this sample to the hydrochloride with 0.1M HCl resulted in a further drop to 1.66, and even eluting the nitrate form with 2M NaCl solution produced a further drop, to 1.57 meq/g. Evidently

TABLE III DEGRADATION OF BASE RESIN IN NaOH/NaCl SOLUTION

TIME/d	MASS/g	% DECREASE	mmol AMINE	% DECREASE
0	2.00	-	3.04	-
1	1.71	14.5	2.62	14
3	1.63	18.5	2.04	33
7	1.51	24.5	1.24	59
10	-	-	1.42	53

TABLE IV. REACTION OF PAM GRAFT WITH FORMALDEHYDE AND AMINES

NO	H_2CO	AMINE	TIME	YIELD	CAPACITY
M1	24.7 mmol	24.6 mmol diethylamine[1]	2h	5.44g	2.48 meq/g
M2	24.7 mmol	36.9 mmol diethylamine[1]	2h	5.33g	2.16 meq/g
M3	37.0 mmol	43.9 mmol diethylamine[1]	2h		2.42 meq/g
M4	37.0 mmol	43.9 mmol diethylamine[2]	2h	5.51g	2.16 meq/g
M5	37.0 mmol	36.6 mmol triethylamine[1]	2h	4.67g	0.0 meq/g
M6	37.0 mmol	37.2 mmol diethanolamine[1]	2h	5.18g	0.06 meq/g

1. added after 1h
2. added with H_2CO

the aminomethylamide linkage is highly labile under acidic conditions. Stability measurements were also carried out under alkaline conditions, by shaking 1.9g lots of fresh M1 in 50 ml of 2% NaOH/10% NaCl solution. After 24 hours, the capacity had fallen to 0.44 meq/g (82% loss) and after twelve days no capacity remained. The conversion of PAM grafts to useful anion exchangers clearly requires a much more inert linkage than that introduced by the Mannich reaction.

Abstract

Graft polyvinylamine has been prepared by the Hofmann degradation of polyacrylamide grafted to crosslinked polyvinyl alcohol particles containing magnetic iron oxide. Conversion of amide to amine groups was limited to c. 25%, and was accompanied by hydrolysis and chain scission. Hypochlorite was more effective than hypobromite, and the maximum yield of amine occurred at a hypochlorite:amide ratio of 1.5:1. The major factor limiting the yield appeared to be the inaccessibility of some monomer units to hypochlorite. Graft weak base resins were also prepared by treating the magnetic polyacrylamide grafts with stoichiometric amounts of formaldehyde and diethylamine. Both Hofman and Mannich products were labile, and unsuitable for use as ion exchangers.

Literature Cited

1. Bolto B.A., Dixon D.R., Eldridge R.J., Kolarik L.O., Priestley A.J., Raper W.G.C., Rowney E.A., Swinton J.E. and Weiss D.E., "Theory and Practice of Ion Exchange", Society of Chemical Industry, London, 1976, p. 27.1

2. Bolto B.A., Dixon D.R. and Eldridge R.J., J. Appl. Polym. Sci., 1978, 22, 1977

3. Tanaka H., and Senju R, Kobunshi Ronbunshu 1976, 33, 309 (English Edition 1976 5, 429)

4. Tanaka H. Suzuki K. and Senju R., Japan Tappi 1976, 30, 392

5. Tanaka H., and Senju R., Bull. Chem. Scc. Japan 1976, 49, 2821

6. Tanaka H., J. Poly. Sci. Polymer Letters Edition, 1978, 16, 87

7. Jones G.D., Zomlefer J. and Hawkins K., J. Org. Chem., 1944 9, 500

8. Arcus C.L. J. Polym. Sci., 1952, 8, 365

9. Schiller A.M. and Suen T.J. Ind. Eng. Chem, 1956, 48, 2132

10. Mullier M. and Smets G., J. Polym. Sci., 1957, 23, 915

11. Sugiura M., Ochi M., Tani Y. and Nagai Y., Kogyo Kagaku Zasshi, 1969, 72, 1926

12. St. Pierre T., Vigee G and Hughes A.R., in 'Reactions on Polymers', ed. J.A. Moore, Reidel, Dordrecht, 1973 p.61

13. Hughes A.R. and St. Pierre T., Macromolecular Syntheses 1977, 6, 31

14. Went P.M., Evans R. and Napper D.H., J. Polym Sci. Part C 1975, No. 49, 159

15. Higuchi M. and Senju R., Poly. J., 1972, 3, 370

16. Sugiyama H. and Kamogawa H., J. Poly. Sci. Part A-1 1966, 4, 2281

17. Muller E., Dinges K. and Graulich W., Makromol. Chemie 1962, 57, 27

18. Bartoli M., Sebille B., Audebert R. and Quivoron C., Makromol. Chemie 1975, 176, 2579

RECEIVED September 20, 1979.

Polymer-Copper Catalysts for Oxidative Polymerization of Phenol Derivatives

E. TSUCHIDA and H. NISHIDE

Department of Polymer Chemistry, Waseda University, Shinjuku, Tokyo 160, Japan

A polymer-metal complex is composed of a synthetic polymer and metal ions. Its synthesis represents an attempt to give an organic polymer with inorganic functions (1). For example, we bound metal chelates such as Fe-porphyrins or Co-Schiff-base chelates to a polymer-chain through coordinate or covalent bonds (Scheme 1, 2). These polymer-metal complexes carry molecular oxygen reversibly, as hemoglobin and myoglobin do. When a polymer ligand (3) is mixed directly with a "naked" metal ion, which generally has four or six coordinate bonding hands, another group of polymer-metal complexes is formed. This may be of the intra-polymer chelate type (Scheme 2) or of the insoluble inter-polymer chelate type.

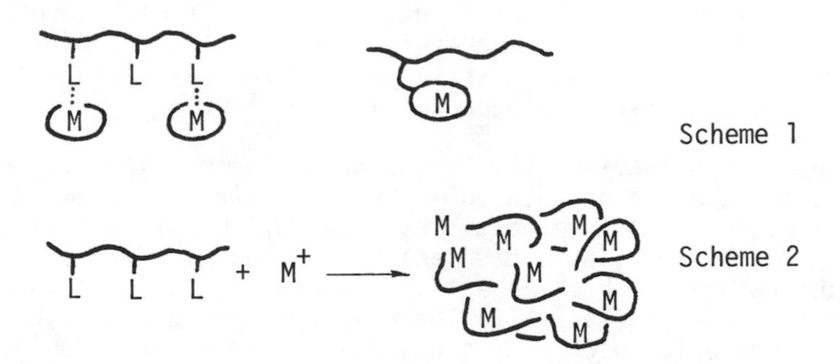

Scheme 1

Scheme 2

If one combines a catalytically active metal ion with a polymer via Scheme 2, a polymer to catalyze a reaction can be obtained. It is reasonable to assume that the metal catalyst bound to the polymer backbone will show a specific behavior compared with that of the corresponding monomeric complex, because the reactivities of the metal complex are sometimes strongly

0-8412-0540-X/80/47-121-147$05.00/0

affected by the polymer chain that exists outside the coordination sphere and surrounds the metal complex. Indeed some polymer-metal complexes have been found to exhibit high efficiency in catalysis (1).

The Cu-complex-catalyzed oxidative polymerization of phenol derivatives has been selected here as a model reaction in which a polymer-metal complex acts as a catalyst. The catalytic cycle is illustrated in Scheme 3, the example used being the oxidative

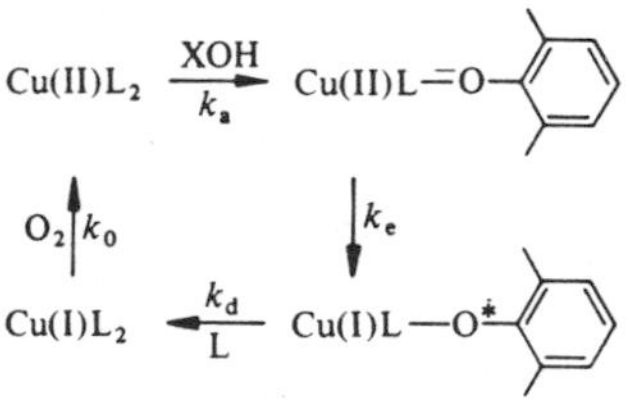

$$\begin{array}{ccc} Cu(II)L_2 & \xrightarrow[k_a]{XOH} & Cu(II)L{=}O{-}Ar \\ O_2 \uparrow k_0 & & \downarrow k_e \\ Cu(I)L_2 & \xleftarrow[L]{k_d} & Cu(I)L{-}O^{*}{-}Ar \end{array}$$

Scheme 3

dimerization of 2,6-dimethylphenol (XOH). In the first step, the substrate (phenol) coordinates to the Cu(II) complex and one electron transfers from the substrate to the Cu(II) ion. Then the activated substrate dissociates from the catalyst and the reduced catalyst is reoxidized to the original Cu(II) complex by oxygen. This oxidative reaction has some merits for the study of the catalytic effect of a polymer. (i) The complex homogeneously catalyzed the reaction. (ii) The reaction intermediate is the substrate-coordinated complex (substrate-metal ion -ligand mixed complex), so that the property of the ligand is clearly reflected in the catalysis. (iii) The reaction proceeds rapidly under mild conditions (at room temperature under an air atmosphere). (iv) The complex catalyst affects not only the reaction rate but also the characteristics of the resulting polymer.

Since the oxidative polymerization of phenols is the industrial process used to produce poly(phenyleneoxide)s (Scheme 4), the application of polymer catalysts may well be of interest. Furthermore, enzymic, oxidative polymerization of phenols is an important pathway in biosynthesis. For example, black pigment of animal kingdom "melanin" is the polymeric product of 2,6-dihydroxyindole 1, which is the oxidative product of tyrosine, catalyzed by copper enzyme "tyrosinase". In plants "lignin" is the natural polymer of phenols, such as coniferyl alcohol 2 and sinapyl alcohol 3. Tyrosinase contains four Cu ions in catalytically active site which are considered to act cooperatively. These Cu ions are presumed to be surrounded by the non-polar apoprotein, and their reactivities in substitution and redox reactions are controlled by the environmental protein.

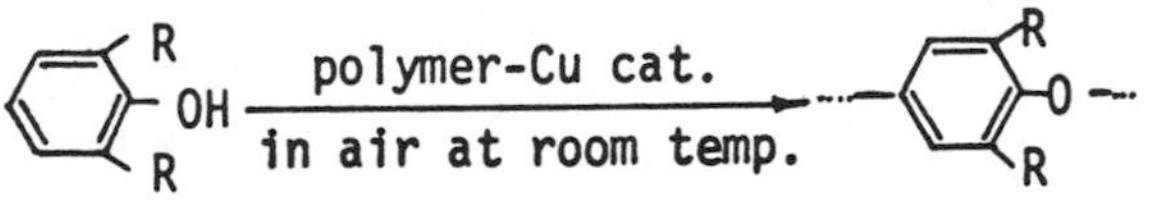

Scheme 4

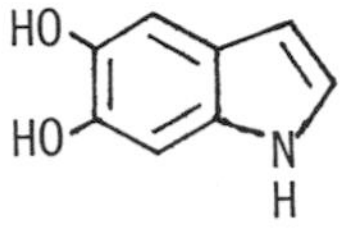

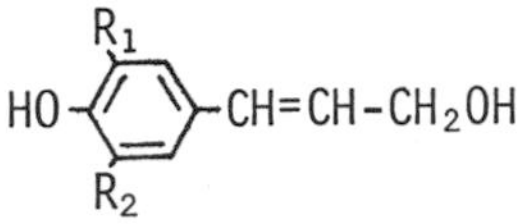

In the present paper we describe the catalytic mechanisms of synthetic polymer-Cu complexes: a catalytic interaction between the metal ions which attached to a polymer chain at high concentration and an environmental effect of polymer surrounding Cu ions. In the latter half, the catalytic behavior is compared with the specific one of tyrosinase enzyme in the melanin-formation reaction which is a multi-step reaction. To the following polymers Cu ions are combined.

PVP = poly(4-vinylpyridine)
QPVP = partially quaternized PVP with ethylbromide
DBQP = partially crosslinked PVP with dibromobutane
PSP = copolymer of styrene and 4-vinylpyridine
PVIm = poly(N-vinylimidazole)
PIPo = copolymer of N-vinylimidazole and N-vinylpyrrolidone
PSI = copolymer of styrene and N-vinylimidazole

Formation of Polymer-Cu Complexes

When a cupric salt is added to a polymer-ligand solution, a green~blue Cu complex is rapidly formed. Spectroscopic study indicated that complex formation between Cu and a polymer-ligand was not a step-by-step mechanism and that the composition of the complex, $Cu(ligand)_4$, in a polymer ligand remained constant throughout the course of the reaction (4).

In order to study the shape of a polymer-Cu complex, viscometric measurements of a homogeneous solution of QPVP were carried out (Fig. 1). At constant QPVP concentration, an increase in the added amount of Cu ions causes a decrease in viscosity, which reveals that the polymer-ligand chain is markedly contracted due to intra-polymer chelation. An intra-polymer chelate takes a very compact form and Cu ions are crowded within the contracted polymer chain (Scheme 2). The adsorption of Cu ions on the polymer ligand is sigmoidal, as can be seen in Fig. 1. At a low

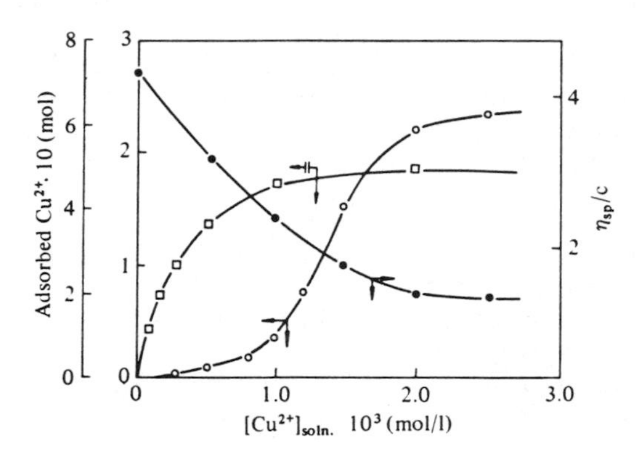

Figure 1. Adsorption of Cu ions on polymer ligand: adsorption of Cu(II) on (○) QPVP; (□) DBQP; (●) viscosity of QPVP solution in pH 5 CH_3COOH–CH_3-COONa buffer

Tab. I

Formation constant(K) of the PVP complexes

Complex	K
PVP-Cu	13,800
PVP-Mn	620
PVP-Cu,Mn	Cu 14,000 Mn 1,800
pyridine-Cu	320

in methanol, 30°

concentration of Cu ions, the conformation of the polymer is not sufficiently contracted to form stable Cu chelates, but at higher Cu concentrations, the tight packing of the polymer ligand, caused by intra-polymer chelation with a large number of Cu ions, progressively facilitates chelate formation. The conformational change in the polymer ligand from an extended shape to a contracted one, which occurs with the binding reaction, enhances the Cu ion-binding ability. This cooperative binding is illustrated by Scheme 5. On the other hand, the Cu complexation on the cross-linked polymer-ligand DBQP is of Langmuir's type, because its conformation is fixed throughout the binding reaction by pre-crosslinking.

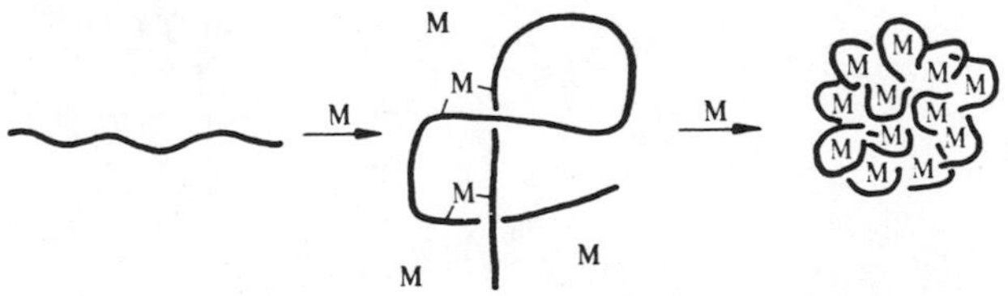

Scheme 5

The cooperative phenomenon of complexation was also observed for a heteronuclear polymer complex, e.g., the PVP-Cu,Mn mixed metal ion system (Tab. I). Although the complex forming ability of Mn ion itself is weak, the formation constant (K) for the Mn complex increases 3 fold in PVP-Cu,Mn. Cu ions assist Mn ions to coordinate by contracting the polymer chain.

The K values are greater in the PVP complexes than in the monomeric Cu complex system, as is also shown in Tab. I. This appeares to be general for polymer complex systems (5). This large stability of the polymer-Cu complex gives advantages as the catalyst. (i) The catalytically effective complex is formed, even if the polymer-ligand is not excess to Cu ion. Challa et al. reported that the maximum of the activity was observed at [ligand unit]/[Cu]$\simeq$1 for the aminated poly(styrene) ligand (6). (ii) The polymer complex is stable toward inactivation, e.g., alkaline hydrolysis during the polymerization.

Interaction between Metal Ions in the Catalysis

The polymer complex takes a very compact form and metal ions are crowded within the contracted polymer chain (Scheme 2), so that an interaction between the metal ions is expected in the catalysis of heteronuclear polymer complex.

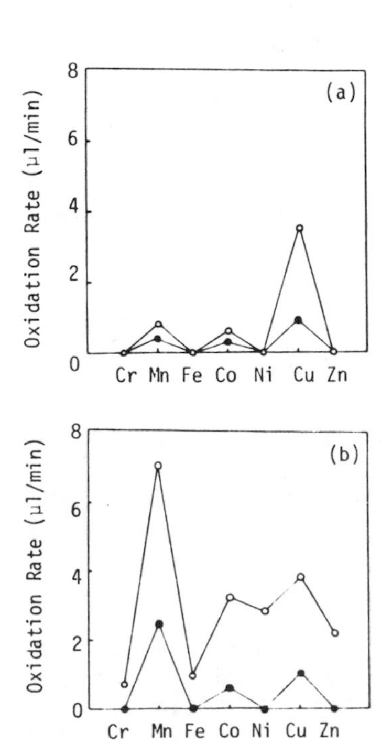

Figure 2. Catalytic activities of (a) the PVP complexes and (b) the PVP–Cu secondary metal mixed complexes (○) PVP catalyst; (●) pyridine catalyst, in methanol, 30°C

The oxidation rates of XOH were measured for the PVP complexes of the transition metal ions of the 4th series, i.e., Cr, Mn, Fe, Co, Ni, Cu and Zn ion. As can be seen in Fig. 2 (a), the Cu complexes exhibit the highest activity and the activity of the PVP-Cu catalyst is higher than that of the monomeric pyridine-Cu catalyst. To this Cu complex, equivalent amount of the second metal component was added i.e., the PVP-Cu, secondary metal ion mixed complexes were prepared. The activities of these mixed complexes are summarized in Fig. 2 (b). One notices that Mn ion increases the catalytic activity of the Cu ion although Cr and Fe ion inhibit the catalytic activity. Another important result in Fig. 2 (b) is that the effect of secondary metal ion is more clearly observed in the PVP system, comparing to the monomeric pyridine catalysts.

The PVP-Cu,Mn mixed complex gave not only the greatest oxidation rate but also the highest yield and molecular weight of the resulting XOH polymer. This higher efficiency of the Cu,Mn mixed complex was also recognized for other system, using other polymer-ligand and other solvent. Thus it is concluded that the combination of Cu ion and Mn ion on a polymer-ligand provides good catalysts for the oxidative polymerization of phenols.

The rate constants of elementary reactions (see Scheme 3) were estimated for the PVP-Cu,Mn catalyst. For example, the rate constant of electron-transfer (k_e) and of catalyst reoxidation (k_o) were determined by measuring the decrease and the increase in the d-d absorption of Cu(II). The k_e value for Cu(II) → Cu(I): 14 min^{-1} was much larger than that for Mn(III) → Mn(II). k_o were PVP-Mn (0.042 min^{-1}) ≃ PVP-Cu,Mn (0.040) > PVP-Cu(0.013), respectively. Furthermore the following rapid redox reaction was regocnized.

$$Cu(I) + Mn(III) \xrightarrow[\text{anaerobic conditions}]{PVP} Cu(II) + Mn(II) \qquad (4)$$

From these results, the schematic redox cycles are proposed for the PVP-Cu,Mn catalyst where two cycles geared each other (Scheme 6):

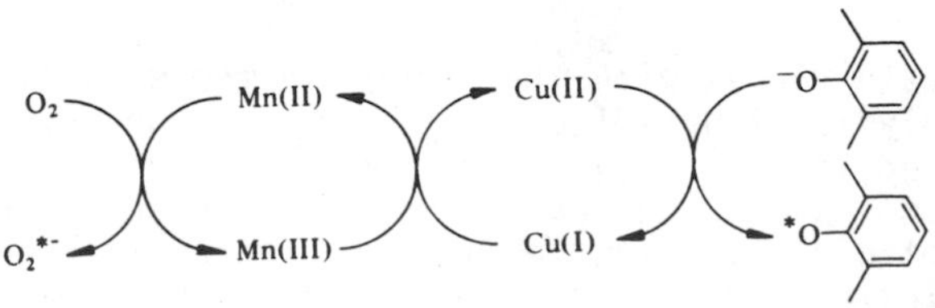

O_2 → $O_2^{\bullet -}$; Mn(II) ⇄ Mn(III); Cu(II) ⇄ Cu(I); $^{-}$O-Ar → $^{\bullet}$O-Ar

Scheme 6

i.e., the substrate XOH coordinates to the Cu(II) complex and is activated. The reduced Cu(I) catalyst is reoxidized to the original Cu(II) complex by rapid redox with Mn(III) ion. Thus formed Mn(II) complex smoothly reoxidized by oxygen. For the PVP complex the interaction between Cu and Mn ions occurs effectively because Cu and Mn ions coordinate on the polymer chain strongly and in high concentration.

This stoichiometric interaction between Cu and Mn on PVP is supported by the influences of the composition of the PVP-Cu,Mn catalyst. Fig. 3 (a) shows that the maximum of the catalytic activity appears at [Mn]/[Cu] = 1. This means that Cu ion and Mn ion interact each other equivalently. The ratio [pyridine unit of PVP]/[metal ion] is plotted as abscissa in Fig. 3 (b). The activity passes through a maximum at the moderately excess ratio. This result is explained as follows; the metal ions have to exist on the polymer chain at high concentration in order to interact each other and large excess of PVP decreases the ion concentration and retards the cooperative interaction.

Environmental Effect of Polymer-ligand

The catalytic activity of the Cu complex on the oxidative reaction in solution is much influenced also by the chemical environment around it. Tab. II shows the effect of reaction solvent. The highest rate is observed for the reaction in a benzene solvent.

Tab. II Polymerization rate of XOH catalyzed by pyridine-Cu complex in several solvents

Solvent	Benzene	Dichloro-benzene	Nitro-benzene	Methanol	DMSO
Rate$\cdot 10^3$ (mol/l min)	12	6.6	3.8	1.3	1.1

The reaction rate varies with the change in the solvent composition. The catalysis of pyridine-Cu in DMSO-benzene mixed solvent is summarized in Fig. 4 (a). The rate constant of the catalyst reoxidation (k_o) and the overall rate increase although the rate constant of electron-transfer (k_e) decreases with the benzene content. Instead of the benzene solvent, the copolymer of vinylpyridine with styrene (PSP) was used as a polymer ligand, as shown in Fig. 4 (b). The overall rate and k_o increase with the styrene content in the PSP ligand, just as the solvent effect of benzene. Only several times amount of styrene unit to Cu ion (as polymer concentration ca. 0.1 wt% of the solvent) affects

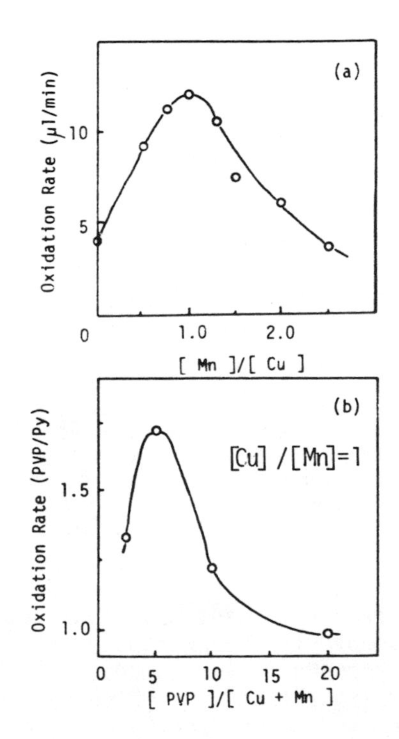

Figure 3. Influence of the composition of PVP–Cu,Mn mixed catalyst on activity

the reaction, that corresponds to the solvent effect caused by ca. 40 vol% of benzene. The local chemical environment of the polymer-Cu catalyst can be considered to be fairly different from that of the monomeric analogue due to the effect of the polymer chain and the neighboring group, even if part of the complex is the same in both.

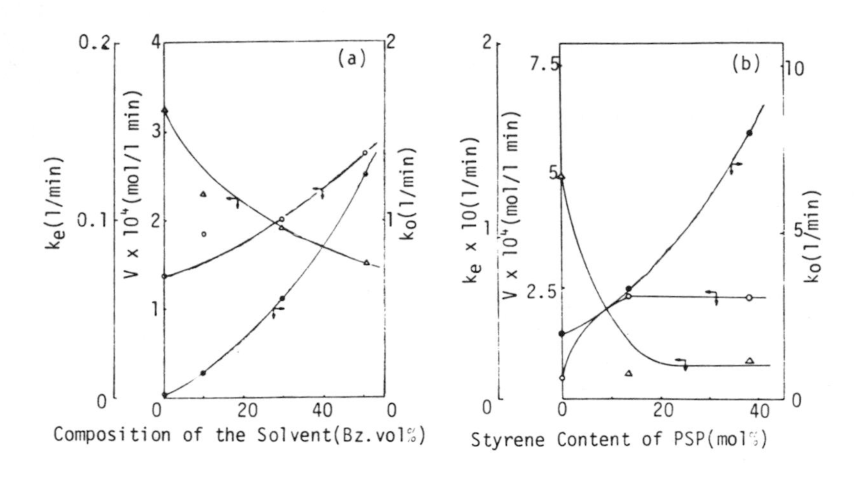

Figure 4. Catalytic activity of the pyridine–Cu catalyst in DMSO–benzene solvent (a) and activity of the PSP–Cu catalyst in DMSO (b): (○) oxidative polymerization rate of XOH; (△) rate constant of electron transfer step (k_e); (●) rate constant of catalyst reoxidation step

In aqueous solvent a hydrophobic environment was constructed by using a water-soluble and hydrophobic tri-block copolymer (Scheme 7). The central block is hydrophobic and composed of the copolymer of styrene and N-vinylimidazole (PSI), to which Cu ions can coordinate. This central block was synthesized by UV-irradiation polymerization by telechelic initiator of bis(4-carbomethoxyphenyl)-disulfide. The reaction of telechelic block with poly-(ethyleneoxide) gave the block copolymer PEO-PSI-PEO.

The block copolymer was complexed with Cu(II) ion in methanol, then the methanol was evaporated to dryness. The Cu complexed block copolymer was soluble in water. Its spectroscopic property showed that the structure and stability of the Cu complex were similar to those of the complex in a benzene solvent. But the reactivity of the Cu complex in the block copolymer could not be examined. The Cu complex was occuluded so tightly in the hydrophobic domain that it did not react with phenols and other reducing agents.

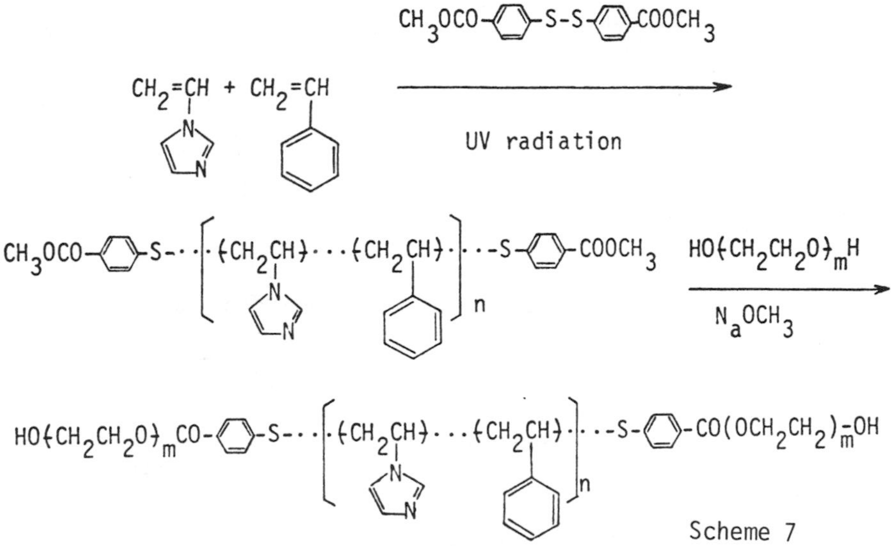

$$CH_2{=}CH(\text{N-imidazolyl}) + CH_2{=}CH(C_6H_5) \xrightarrow[\text{UV radiation}]{CH_3OCO\text{-}C_6H_4\text{-}S\text{-}S\text{-}C_6H_4\text{-}COOCH_3}$$

$$CH_3OCO\text{-}C_6H_4\text{-}S\cdots\left[\cdot(CH_2CH(\text{Im}))\cdots(CH_2CH(C_6H_5))\cdot\right]_n\cdots S\text{-}C_6H_4\text{-}COOCH_3 \xrightarrow[N_aOCH_3]{HO(CH_2CH_2O)_mH}$$

$$HO(CH_2CH_2O)_mCO\text{-}C_6H_4\text{-}S\cdots\left[\cdot(CH_2CH(\text{Im}))\cdots(CH_2CH(C_6H_5))\cdot\right]_n\cdots S\text{-}C_6H_4\text{-}CO(OCH_2CH_2)_{\overline{m}}OH$$

Scheme 7

Melanin Formation Catalyzed by Polymer-Cu Complexes

Oxidative polymerization of phenol derivatives is also important pathway in vivo, and one example is the formation of melanin from tyrosine catalyzed by the Cu enzyme, tyrosinase. The pathway from tyrosine to melanin is described by Raper (7) and Mason (8) as Scheme 8: the oxygenation of tyrosine to 4-(3,4-dihydroxyphenyl)-L-alanin (dopa), its subsequent oxidation to dopaquinone, its oxidative cyclization to dopachrome and succeeding decarboxylation to 5,6-dihydroxyindole, and the oxidative coupling of the products leads to the melanin polymer. The oxidation of dopa to melanin was attempted here by using Cu as the catalyst.

The Cu complexes with imidazole-containing ligands exhibit much higher catalytic activity among other complexes, i.e., Cu complexes of polyamine, polyaminocarboxylic acid or pyridine, and Mn, Fe, Co or Ni complexes of imidazole (Tab. III). The dopa oxidation with imidazoles-Cu complexes yielded a relatively stable intermediate with λ_{max} at 480 nm, which was assigned to dopachrome. Similar spectral change was observed for the tyrosinase-catalyzed oxidation. The formation rate of dopachrome (k_D) was determined from the increase in the absorbance at 480 nm and this rate corresponded to the oxidation of dopa to dopaqunone which was a rate-determining step of dopa-

Tab. III Melanin formation catalyzed by metal complexes

Catalyst	Melanin formation rate constant pH 7	pH 4 (l/mol sec)	Rate determining step
Cu(II)	0	0	Dopa→ (pH 4)
Im-Cu	23	0	
PVIm-Cu	23	0.17	
$PIPo_{20}$-Cu	40	0.16	Dopa→ , Dopachrome→ (pH 7)
$PIPo_{35}$-Cu	38	0.16	
PAA-Cu	0	0	
$NaIO_4$-PVIm-Cu		1.2	Dopachrome→
$NaIO_4$-PAA-Cu		0.4	
PVIm-Ni		0.04	
PVIm-Fe		0.04	Dopa→
Autoxid.		0	
Tyrosinase	900		Dopa→ ,Dopachrome→

pH 4=acetate buffer, pH 7=phosphate buffer, at room temperature under air

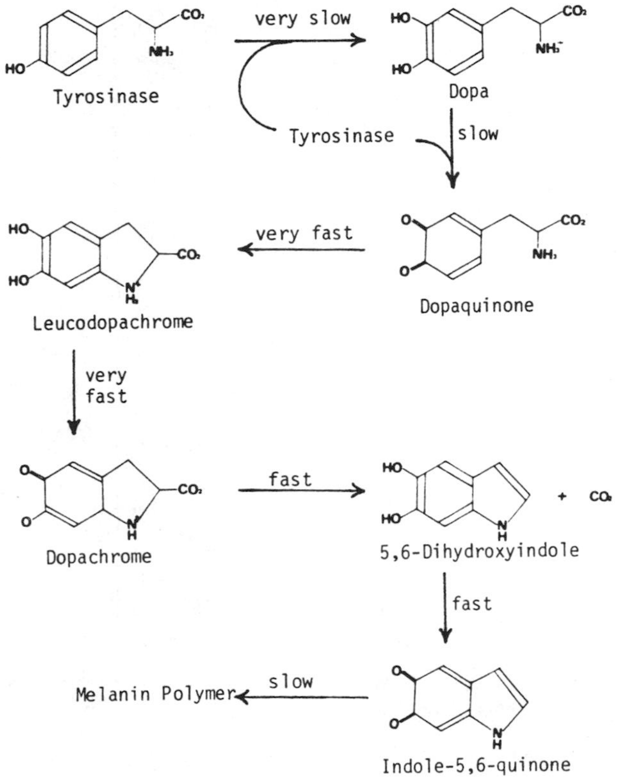

Tyrosinase → (very slow; Tyrosinase) → Dopa

Dopa → (slow) → Dopaquinone

Dopaquinone → (very fast) → Leucodopachrome

Leucodopachrome → (very fast) → Dopachrome

Dopachrome → (fast) → 5,6-Dihydroxyindole + CO_2

5,6-Dihydroxyindole → (fast) → Indole-5,6-quinone

Indole-5,6-quinone → (slow) → Melanin Polymer

Scheme 8

chrome formation. The succeeding increase in absorbance throughout visible region meant the melanin formation, and its formation rate (k_M) was estimated. When dopa was oxidized by the Cu complexes of polymeric imidazoles (PVIm and PIPo) or imidazole, both reactions of dopachrome formation and melanin formation proceeded at comparable rate.

The pH dependence of the rate constants were shown in Fig. 5. The k_D value increases and the ration k_M/k_D decreases with pH. The values with tyrosinase are also given in the rate per Cu equivalent, considering that the enzyme contains 4 Cu ions (Mol. wt. 12,800) (9). Much difference between the values with tyrosinase and those with other Cu complexes indicates that the rate of

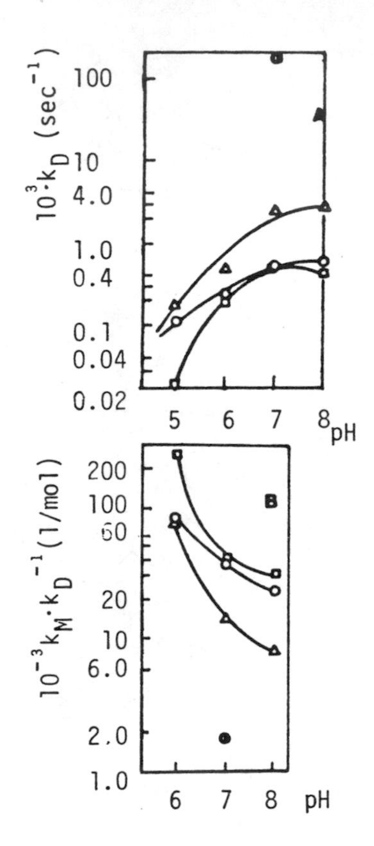

Figure 5. The pH dependence of rate constant of dopachrome and melanin formation: (□) imidazole–Cu; (○) PVIm–Cu; (△) PIPo–Cu; (◑) tyrosinase; 30°C, air, phosphate buffer

dopachrome formation is extremely accelerated with tyrosinase. In the synthetic catalysts, the behavior of the PIPo-Cu catalyst is most resemble to that of tyrosinase.

When dopa was oxidized using the PIPo-Cu catalyst, the distinguished acceleration was observed as compared with the PVIm-Cu or imidazole-Cu catalysts (Fig. 6). An increase in content of the N-vinylpyrrolidone residue in the PIPo copolymer caused higher activity of the Cu complex for the dopa oxidation. The similar acceleration was also produced when N-methyl-pyrrolidone was added to the system of PVIm-Cu. However, nearly 10^3 fold concentration of the pyrrolidone residue was necessary as compared with the PIPo copolymer. Addition of homopolymer of N-vinylpyrrolidone to PVIm-Cu caused no acceleration.

This acceleration by PIPo was restricted in the process of dopa oxidation to the quinone and was not observed in the melanin formation process. Thus the catalytic behavior of the PIPo-Cu

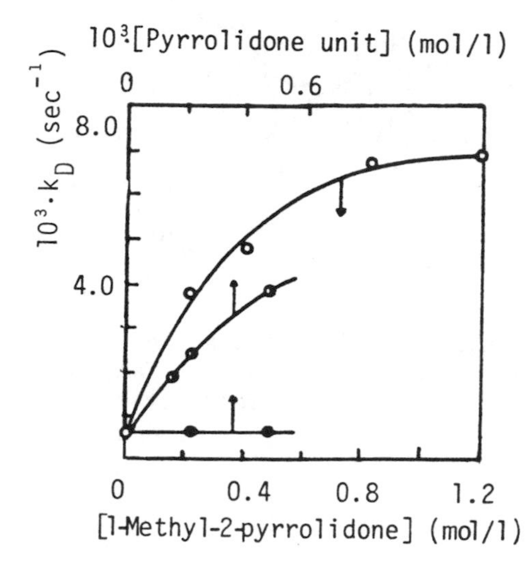

Figure 6. Effect of pyrrolidone residue: (○) PVIm–Cu + methylpyrrolidone; (◑) PIPo–Cu; (●) PVIm–Cu + poly-(vinylpyrrolidone)

complexes comes close to that of tyrosinase. From the following results, the effect of PIPo is concluded that the active species for dopa oxidation, i.e., molecular oxygen-bound Cu complex is formed with considerable ease in the PIPo-Cu catalytic system, which resulted in selective acceleration of dopa oxidation to dopaquinone in the course of melanin formation. (i) k_D increased till the catalyst composition [imidazole unit]/[Cu] = 2, which implied that the coordinated imidazole ligand necessary for the active species is less than two per Cu ion. (ii) Cu ion existed as Cu(II) in the steady state of the catalysis. (iii) Neither the oxidation of dopa nor the reduction of Cu(II) complex proceeded in the absence of oxygen. Molecular oxygen is necessary for the catalyst. (iv) Oxygen coordinated structure for the active intermediate Cu complex is approved from the result of acceleration induced by hydrogen peroxide. k_D was increased proportionally by adding H_2O_2, and the spectrum of the Cu complex with H_2O_2 resembled to that of oxy-tyrosinase. The active intermediate for the catalysis can therefore be presumed to be the binuclear μ-dioxo complex (Scheme 9). The dimeric structure is applied to facilitate the simultaneous two electron transfer. We speculate the active site of tyrosinase as Scheme 10.

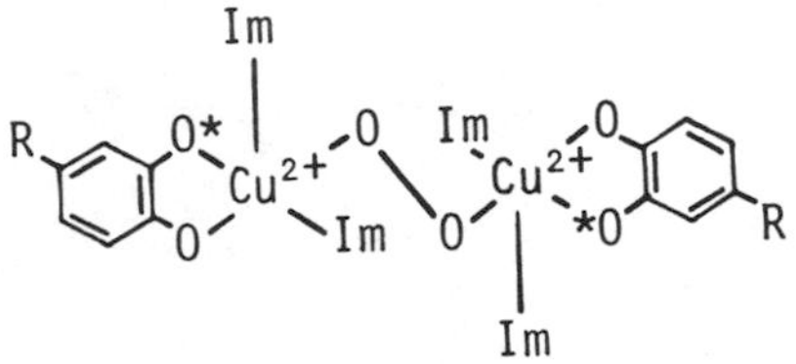

Scheme 9

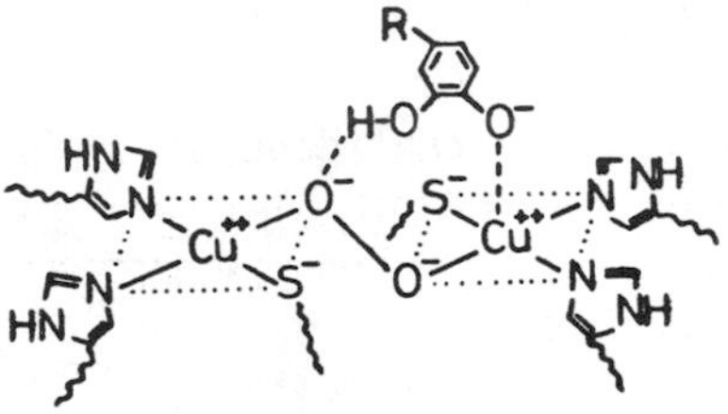

Scheme 10

The resulting, synthetic melanin were black powder and absorbed light over the ultraviolet and visible regions (Tab. IV). They contained very stable free radical, being of interest to physiology.

Tab. IV Properties of melanin polymers

Substrate	Catalyst	Yield(%)	Solubility	Color	ESR spectra
Catechol	PVIm-Cu	82	non	black	free radical
	imidazole-Cu	46	alkali aq.	"	"
Dopa	PVIm-Cu	86	non	deep balck	"
	imidazole-Cu	39	non	"	"
5,6-dihydroxy indole	PVIm-Cu	26	aq.	"	"

[Substrate]=10 mmol/l, [Cu]=2 mmol/l, 2 days in water, at room temperature under air

Conclusion

At present, it is not yet possible to predict whether a polymer influences a reaction as catalyst or an inhibitor, because the catalytic mechanism of one reaction differs from another and the rate-determining step varies with the reaction conditions. But at least in the catalysis of the oxidative polymerization of phenols, (i) ease in redox reaction of metal ion, and (ii) an interaction between metal ion and oxygen are important to design a catalyst which will exhibit high efficiency and selectively. Polymers are expected to give positive effects on this approach.

References

1) E.Tsuchida, H.Nishide, Adv. Polymer Sci., 24, 1 (1977)
2) E.Tsuchida, E.Hasegawa, T.Kanayama, Macromolecules, 11, 947 (1978)
3) A polymer ligand is a polymeric substance that contains co-ordinating groups or atoms (mainly N, O and S), obtained by the polymerization of monomers containing coordinating sites, or by the chemical reaction between a polymer and a low-molecular-weight compound having coordinating ability.
4) H.Nishide, J.Deguchi, E.Tsuchida, Bull. Chem. Soc., Japan, 49, 3498 (1976)
5) H.Nishikawa, E.Tsuchida, J. Phys. Chem., 79, 2072 (1975)
6) G.Challa, D.A.Noordegraaf, A.J.Schouten, Macrosymp. Dublin., 449, 455 (1977)
7) H.S.Raper, Physiol. Rev., 8, 245 (1928)
8) H.S.Mason, J. Biol. Chem., 172, 83 (1948)
9) S.Bouchilloux, P.McMahill, H.S.Mason, J. Biol. Chem., 238, 1699 (1963)

RECEIVED July 12, 1979.

Structure, Mechanism, and Reactivity of Organotin Carboxylate Polymers

K. N. SOMASEKHARAN and R. V. SUBRAMANIAN

Department of Materials Science and Engineering, Washington State University, Pullman, WA 99164

Tributyltin compounds are known to be toxic to marine organisms and have been incorporated as antifouling additives, e.g., tributyltin oxide (TBTO), fluoride (TBTF) or sulfide (TBTS), in marine biocidal paints (1). In recent years, efforts to develop longer-lasting and environmentally safer antifouling coatings have led to controlled release formulations in which the organotin group is chemically anchored to a polymer chain (2,3).

We have developed several such organotin polymer compositions that are suitable for antifouling applications (3). In these, the prepolymers were prepared by partial esterification by TBTO of linear polymers carrying carboxylic acid or anhydride groups. The free carboxylic acid or anhydride groups were then reacted with diepoxides to form thermoset organotin-epoxy polymers. Many variations of this scheme were investigated, including one which provided for simultaneous vinyl polymerization and carboxyl-epoxide reactions to form crosslinks.

Biotoxicity studies on some of these formulations have shown that the nature and degree of crosslinking have a profound influence on the size of the observed inhibition zones (4). More significantly, performance tests in marine environments reveal that polymer compositions that are composed of aromatic monomers resist fouling for longer periods (5). These observations suggest that the principles of antifouling performance are similar to those involved in plasticizer technology, environmental resistance of polymers, and related areas. Consideration of these principles, viz., transport mechanisms, boundary effects, etc., should be preceded by the identification of the chemical species involved in these processes.

Thus an essential step in the study of the mechanism of antifouling activity is the determination of the chemical nature of the tin compounds being released from the coatings. The identification of the chemical species will enable one to evaluate the change in bulk characteristics of the coatings as leaching proceeds. The knowledge of this chemical moiety is a prerequisite for a meaningful study of the rate of leaching and the mechanism of

0-8412-0540-X/80/47-121-165$05.00/0

release; and the factors controlling the rate and mechanism of leaching are the design parameters for newer and better coatings.

Since most of the above polymer compositions have tin anchored as tributyltin carboxylates, we have initiated a study of the structure and reactivity of tributyltin carboxylates.

Structure of Tributyltin Carboxylates

Tributyltin carboxylates act like weak acids, and can be titrated in hexane-ethanol (1:1) medium against alkali. Tributyltin chloride (TBTCl) is also a weak acid, and gives a sharp end point when similarly titrated against sodium hydroxide. TBTCl gives an instantaneous precipitate with silver nitrate in aqueous alcoholic medium.

These observations do not, however, mean that TBT carboxylates and TBTCl are ionic in nature. After detailed analysis of the physical evidence such as the low specific conductance and dipole moment of trialkyltin halides, Neumann has concluded that they have no "salt-like constitution" (6). Bonding in the trialkyltin carboxylates also is essentially similar to that in covalent alkyl esters, as evidenced by the low dipole moment of 2.2D for tributyltin acetate in benzene, as compared to 1.9D for alkyl acetates (7).

Neumann points out that the low dipole moment is probably the outcome of long bond lengths. Long bonds naturally are weak and reactive. The electron cloud being polarizable and the central atom less shielded, approaching nucleophiles should find the Sn atom an attractive target.

The electronic structure of tin is also conducive to such attacks. Tin atom has the electronic structure $1s^2 2s^2 2p^6 3s^2 3p^6 3d^{10} 4s^2 4p^6 4d^{10} 5s^2 5p^2$. As in the case of carbon, the valence electrons can undergo sp^3 hybridization, when tin is tetracovalent, and the geometry is tetrahedral. As the attacking nucleophile approaches, the hybridization transforms to sp^3d, with trigonal-bipyramid geometry. The picture is reminiscent of the transition state in S_N2 reactions. Not surprisingly, conventional techniques such as titration against alkali failed to yield information on the kinetics of the hydrolysis of TBT carboxylates.

NMR of Tributyltin Maleate. From the structure of TBT maleate, HO-OC-HC=CH-CO-O-$SnBu_3$, one would expect the vinyl protons to be different, giving rise to a doublet of doublets in the NMR spectrum. However, the spectrum of the compound, recorded on a JEOL MH-100 spectrometer, in carbon tetrachloride (0.25M) at room temperature, showed only a singlet absorption at 6.32 ppm downfield from TMS. This observation strongly suggests that the TBT group undergoes fast exchange in a 0.25M solution in carbon tetrachloride.

NMR of Tributyltin Acetate-Acetic Acid Mixture. Corroborating the above indication is our observation that a mixture of TBT acetate and acetic acid in carbon tetrachloride shows only one resonance for their acetyl protons. The position of this absorption is intermediate between the normal positions of the acetyl peaks in acetic acid-TBT acetate. Further, the position of this resonance depends on the molar ratios of acetic acid and TBT acetate in the mixture; in fact, there is a linear relationship between the chemical shifts of this new peak and the molar ratios of the two compounds (Figure 1).

The occurrence of an average CH_3-CO resonance line position from acetic acid and TBT acetate is a consequence of rapid chemical exchange of the tributyltin group. It represents the CH_3-CO protons in a time-averaged environment.

Reaction of Tributyltin Carboxylates with Sodium Chloride

Even the interfacial reaction between TBT carboxylates and chloride is very fast; the reaction is almost complete within one day. TBTCl is the product of the reaction (vide infra).

About 50 g of pure tributyltin acrylate (TBTA) was dissolved in 1 liter hexane. Evaporation of 20 ml of this solution yielded a *solid* residue, weighing 0.9796 g.

300 ml of the above solution of TBTA in hexane was placed on top of 3000 ml of 8% aqueous solution of soidum chloride taken in a bottle. The area of contact was about 180 cm^2. 20-ml samples were pipetted out from the hexane layer at the end of 24, 48, 72 and 96 hours. Evaporation of these aliquots yielded *liquid* residues weighing 0.8906 g (8-1), 0.8872 g (8-2), 0.8869 g (8-3) and 0.8866 g (8-4), respectively (Table I). The results of experiments repeated with 4%, 2% and 1% sodium chloride solutions, keeping all other conditions the same, are also given in Table I.

Each of the above liquid residues was tritrated against standard sodium hydroxide, using phenolphthalein as indicator. Identical titer values were obtained; the same titer value was also given by the original solid residue of unreacted TBTA (0-1). Such an observation of identical titer values should be expected if the conversion of TBTA, by reaction with sodium chloride, is solely to TBTCl. However, any side reaction leading to TBT hydroxide or TBTO will result in lower titer values since these tin compounds, unlike TBTA or TBTCl cannot be titrated like weak acids. Clearly, the side reactions are not noticeable in these experiments. Hydrolysis is not competitive under the conditions of this study, probably because chloride concentration never drops below 10^{-1} whereas hydroxide concentration is always below 10^{-5}. (It was noticed that the pH of the aqueous layer in each case had risen from 6.5 to 9.0.)

After complete conversion, 0.9796 g TBTA (0-1) is expected to yield 0.8830 g TBTCl (0-2). Based on this, the fraction of TBTA remaining unreacted in each case is calculated, and from that the equilibrium constant, K (Table I).

A similar set of experiments, where the volume of sodium chloride was varied but its initial concentration maintained the same in all cases, was also carried out. The results are summarized in Table II.

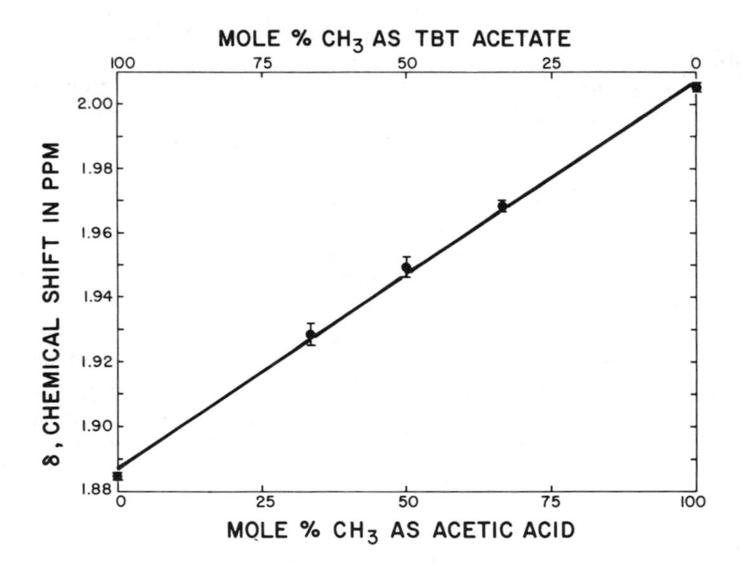

Figure 1. Variation of the position of acetyl proton resonance of TBT acetate–acetic acid mixtures with composition

TABLE I. Interfacial Reaction[1] Between TBTA and NaCl and Equilibrium Constant, K.

Sample	NaCl, wt%	Time, h	Residue[2], g	Unreacted % TBTA	K
8-1	8	24	0.8906	7.9	--
8-2	8	48	0.8872	4.3	--
8-3	8	72	0.8859	4.0	--
8-4	8	96	0.8866	3.7	0.249
4-1	4	24	0.8945	11.9	--
4-2	4	48	0.8934	10.8	--
4-3	4	72	0.8919	9.2	--
4-4	4	96	0.8900	7.2	0.240
2-1	2	24	0.8970	14.5	--
2-2	2	48	0.9016	19.3	--
2-3	2	72	0.9009	18.5	--
2-4	2	96	0.8978	15.3	0.192
1-1	1	24	0.9058	23.6	--
1-2	1	48	0.9029	20.6	--
1-3	1	72	0.9037	21.4	--
1-4	1	96	0.9032	20.9	0.253
0-1	-	0	0.9796	100.0	--
0-2	-	--	0.8830[3]	0.0	--

[1]The volume of NaCl solution (300 ml), volume of hexane solution (300 ml), concentration of TBTA (48.9800 g/liter), and area of contact (180 cm^2) were the same in all cases.
[2]On evaporation of 20 ml hexane solution.
[3]Calculated.

TABLE II. Interfacial Reaction Between TBTA and NaCl[1].

Sample	NaCl Soln., ml	Residue[2], g	Unreacted % TBTA[3]	K
1	1000	1.8908	15.4	0.200
2	2000	1.8826	11.4	0.146
3	3000	1.8753	7.8	0.153
4	4000	1.8697	5.0	0.187
0-1	--	2.0628	100.0	--
0.2	--	1.8594[3]	0.0	--

[1]Concentration of NaCl solution = 4%, Volume of hexane solution = 200 ml, Concentration of TBTA = 51.4700 g/liter, & area of contact = 180 cm^2 in all cases; all samples collected at the end of 7 days.
[2]On evaporation of 40 ml hexane solution.
[3]Calculated.

Identification of Reaction Products

IR Spectroscopy. Samples of the hexane layer (8-4, 4-4, 2-4, 1-4), after evaporation of the solvent, gave IR spectra of mixtures of TBTCl and TBTA. The amount of TBTA depended on the concentration of sodium chloride solution and its volume, as evidenced by the intensity of the broad peak at 1650 cm^{-1}.

Chloride Assay. 20-ml aliquots corresponding to sample 4-4 were evaporated. The residues were dissolved in aqueous alcohol (1:1), and silver nitrate in the same solvent was added to them. The precipitates were collected in sintered glass crucibles, washed with alcohol, and dried in vacuum. The weight of silver chloride was 0.3488 g. Authentic samples of TBTCl (M & T Chemicals), when gravimetrically analyzed under identical conditions, gave an assay of 95%. Thus the conversion from TBTA to TBTCl is 94% in sample 4-4.

Column Chromatography. The hexane layer corresponding to sample 4-4 was separated and the solvent evaporated. The residue was adsorbed on a silica gel column, and eluted with hexane. The IR spectrum of the first fraction was identical with that of an authentic sample of TBTCl. Frequencies in the finger-print region, in cm^{-1}: 1464 (s), 1416 (m), 1377 (s), 1359 (w), 1342 (m), 1293 (m), 1250 (m), 1180 (m), 1151 (m), 1074 (s), 1047 (w), 1021 (m), 1000 (w), 959 (m), 875 (s), 864 (m), 842 (w), 766 (w), 745 (w), 695 (s), 666 (s).

Gas Chromatography. TBTCl can be distilled at 145°C under 10 mm pressure, and it passes through nonpolar gas chromatographic columns. The flame ionization detector is not suitable, as tin oxide formed during combustion can deposit on the electrode causing excessive noise. However, electron capture detection is ideal; the sensitivity is very high. Analytical separation of the above fraction was attempted on a silicone oil column. Temperature of the column was varied only over a narrow range (130-160°C), but the flow rate over a wide range. There was only one peak under all conditions, the retention time of which corresponded to that of an authentic sample of TBTCl.

Mass Spectroscopy. Mass spectra were recorded on a Hitachi-Perkin-Elmer RMU-6G spectrometer.

The first fraction from column chromatography was analyzed. The sample has enough vapor pressure at room temperature to give a mass spectrum. At the voltages normally employed (70v), extensive fragmentation was observed. The most intense peak in the mass spectrum was at 57, corresponding to $C_4H_9^+$. Peaks corresponding to Cl^+ were also identified.

At low voltages (15v), organic compounds usually give only the M^+ peak, the energy being insufficient to rupture the molecule.

However, this compound undergoes fragmentation even at very low voltages, reflecting the low energy of some of its bonds.

The molecular ion of $C_{12}H_{27}SnCl$ is expected to produce a very complex pattern, because of the combination of the characteristic natural abundances of the isotopes of Sn, Cl and C. The theoretical calculation of the intensity pattern has taken into account all the ten isotopes of Sn, the two isotopes of Cl, three distinct contributions due to the 12 carbons (144, 145, 146), and the two significant contributions due to the 27 hydrogens (27 and 28). Of the 120 combinations, many overlap.

The calculated intensity pattern is compared with the observed spectrum in Table III. The matching is excellent; the root mean square deviation between the observed and calculated intensities is less than 2%.

The pattern computed for $C_{12}H_{27}Sn^+$ markedly differs from the observed intensity pattern. Only the presence of Sn and Cl together in the molecule can produce the complex pattern observed.

In fact, $C_{12}H_{27}Sn^+$ was identified as one of the fragmentation products; even at the low voltage (15v) used, the Sn-Cl bond undergoes scission. Other fragments are produced due to the successive loss of butyl groups.

TABLE III. Calculated and Observed Intensity Patterns in the Mass Spectrum

M/e	Observed[1]	Calc. for $C_{12}H_{27}SnCl^+$	Calc. for $C_{12}H_{27}Sn^+$
M-8	--	2.26	2.80
M-7	--	0.31	0.39
M-6	--	2.31	1.95
M-5	--	1.14	1.29
M-4	34.86	34.35	41.91
M-3	26.85	22.96	28.02
M-2	71.62	70.40	73.61
M-1	38.38	35.57	34.98
M	100.00	100.00[2]	100.00[3]
M+1	21.25	20.08	13.53
M+2	35.68	38.44	15.21
M+3	6.31	5.15	2.00
M+4	17.12	18.08	17.47
M+5	3.33	2.47	2.41
M+6	5.41	4.69	0.15
M+7	--	0.63	0.00

[1]For the reaction product of TBTA/hexane and $NaCl/H_2O$.
[2]M = 326
[3]M = 291

Models for Organotin Release

Antifouling performance of these organotin carboxylate polymers indicates that their mode of action corresponds to the bulk abiotic bond cleavage model proposed by Castelli and Yeager (8). The controlling factors to be considered here are:

(a) diffusion of water (and possibly chloride ions) into the polymer matrix from sea water;
(b) hydrolysis of tributyltin carboxylates to produce tributyltin oxide (or chloride);
(c) diffusion, from the matrix to the surface, of the mobile tin species produced;
(d) phase transfer of the organotin species;
(e) their migration across the boundary layer; and
(f) possible mechanical loss of the tributyltin species from the surface.

Conventional Systems. In the conventional antifouling compositions, the organotin compound (TBTO, TBTF, TBTCl, TBTOAc) is mechanically mixed into the paint vehicle. When the TBT species is completely soluble in the polymer matrix, factors (a) and (b) become unimportant in most cases. The mobile species is already present; its diffusion in the matrix, phase transfer and migration across the boundary layer into the ocean environment may be represented by Figure 2a. When the organotin compound forms a dispersed second phase, rate of its dissolution in the polymer matrix becomes another factor to consider.

Figure 2a represents the concentration profile of the tin species during the service life of the coating. The diffusion in the polymer matrix is represented by Fick's second law for non-steady state flow:

$$\partial c_1/\partial t = D_1 \; (\partial^2 c_1/\partial x^2) \qquad -\infty < x < 0$$

This refers to the depletion of the tin species at a given point in the matrix as a function of time. The diffusion across the boundary layer is given by Fick's first law for stationary state flow:

$$P = -D_2 \; (\partial c_2/\partial x) \qquad 0 < x < x_1$$

This refers to the permeation of the tin species through the boundary layer as a function of the concentration gradient. Steady state flow through normal unit area is in the opposite direction to the concentration gradient and is proportional to the absolute value of that gradient. (D_1 and D_2, the effective diffusivities, are expressed in cm^2/sec; c_1 and c_2, the concentrations, in g/cm^3; x, the distance, in cm; and t, the time, in s.) An equilibrium is usually assumed at the interface:

$$\text{at } x = 0, \qquad c_2 = m\, c_1 \qquad \text{for all } t > 0.$$

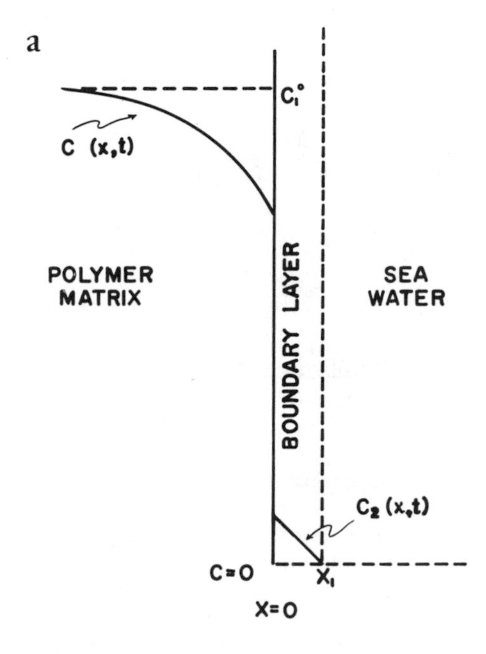

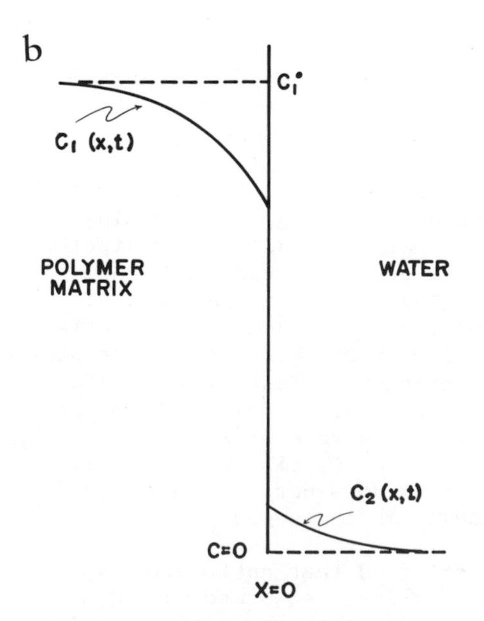

Figure 2. Release of organotin from polymer matrix, nonsteady state mass transport in (a) polymer matrix and stationary state mass transport in boundary layer and (b) both media

The boundary conditions may be set as follows:

$$\text{at } t = 0, \quad c_1 = c_1^{\circ} \quad -\infty < x < 0$$

$$\text{at } t = 0, \quad c_2 = 0 \quad +\infty > x > 0$$

$$\text{at } x = 0, \quad -D_1 \left(\frac{\partial c_1}{\partial x}\right) = -D_2 \left(\frac{\partial c_2}{\partial x}\right) \quad \text{for all } t > 0$$

$$\text{at } x = -\infty, \quad c_1 = c_1^{\circ} \quad \text{for all } t$$

$$\text{at } x > x_1, \quad c_2 = 0 \quad \text{for all } t.$$

Laplace transformation now yields the concentration profile. Substitution of the concentration gradient obtained for the boundary layer in Fick's first law will give the rate of release of tin into the environment.

Boundary Layer. There exists a quiescent boundary layer through which the organotin species must diffuse before being carried by the sea water flow past the surface of the coating. The boundary layer under laminar flow conditions and under turbulent flow conditions are quantitatively defined (9); the thickness of the layer (L) decreases as the fluid velocity increases.

The rate of mass transfer across the diffusion layer (in g/ cm^2/sec) may be given by:

$$\text{rate} = D_2 \ (C_2/L).$$

Here, C_2 represents the concentration of the organotin species in the diffusion layer at the surface of the coating (at $x = 0$); the concentration at the diffusion layer-sea water boundary (at $x = x_1$) approximates to zero.

Ketchum et al., who evaluated the leaching of copper from paint matrices, have reported substantial changes in leaching rate caused by agitation (10). However, Marson predicts less significant difference in the leaching rate of copper under laminar and turbulent flow conditions (9). (We have analyzed the data of Ketchum et al. in detail and found that the boundary layer effects are significant only at the beginning of leaching and become negligible as matrix effects become dominant.) Tests with antifouling rubber performed on an underwater rotating device showed no significant increase with velocity in the rate of loss of TBTO (11). We have observed that the rate of release of tin from controlled release epoxy coatings described later does not increase significantly when the laminar flow velocity is increased.

Phase Transfer. Cardarelli observed that antifouling vulcanized rubber in the partially toxicant-depleted state showed no measurable toxicant gradient (11), indicating that diffusion is very fast in the rubber matrix. Also, tests performed on an underwater rotating device showed no significant increase with velocity in the rate of loss of TBTO (11). Since the fast diffusion in the matrix is not a controlling factor, this observation suggests that the boundary layer effects are also not very significant. Thus controlling factor for rate of release of toxicant becomes the dissolution stage:

$$\text{rate} = -K\ C,$$

where the rate is expressed in $g/cm^2/sec$ and C, the concentration in the matrix, in g/cm^3.

As pointed out earlier, the conventional method of treating the problem is by assuming an interfacial equilibrium between C_2 & C_1. Based on the reported solubility, 50 ppm, of TBTCl in sea water (12), "m" may be assigned a value of 5×10^{-5}. However, an assumption is being made here that the equilibration is fast. Since Cardarelli has pointed out the possibility of a rate controlling interfacial transfer, we have decided to consider the phase transfer rate rather than interfacial equilibrium.

Diffusion in Matrix. The transport equation for a semi-infinite medium of uniform initial concentration of mobile species, with the surface concentration equal to zero for time greater than zero, is given by Crank (13). The rate of mass transfer at the surface for this model is:

$$\text{rate} = C_0\ (D_1/\pi)^{0.5}\ t^{-0.5}.$$

Since the matrix, interface and boundary layer are visualized as offering resistance in succession to mass transport (14), the resultant resistance is obtained by combining the individual resistance in series. On this basis, the expression for the overall rate thus becomes:

$$\text{Rate} = \frac{C_0 D_1^{0.5} K D_2}{K D_2 \pi^{0.5} t^{0.5} + D_1^{0.5} D_2 + D_1^{0.5} K L}$$

Our experimental data, when fitted to this equation, yield a value of 1.0×10^{-14} cm^2/sec for D_1. Least square fit of the same data to Crank's rate equation, which takes into account only the diffusion in the matrix, evaluates D_1 to be 2.2×10^{-14} cm^2/sec. The integral form of Crank's equation (13), giving the cumulative amount released per unit surface area,

$$Q = 2\ C_0\ (D_1/\pi)^{0.5}\ t^{0.5}$$

gives a value of 5.3×10^{-14} cm^2/sec for D_1. As can be seen from Figure 3, the experimental results agree with the model, though not very closely.

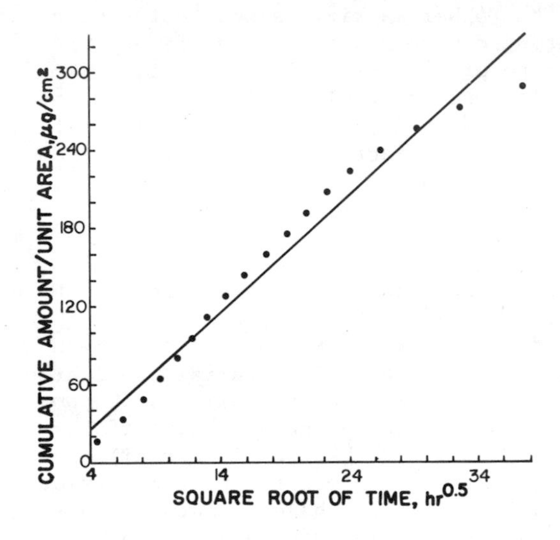

Figure 3. Cumulative amount of TBTCl released per unit area from the epoxy polymer: (●) experimental values; (——) best fit for the integral form of Crank's Equation

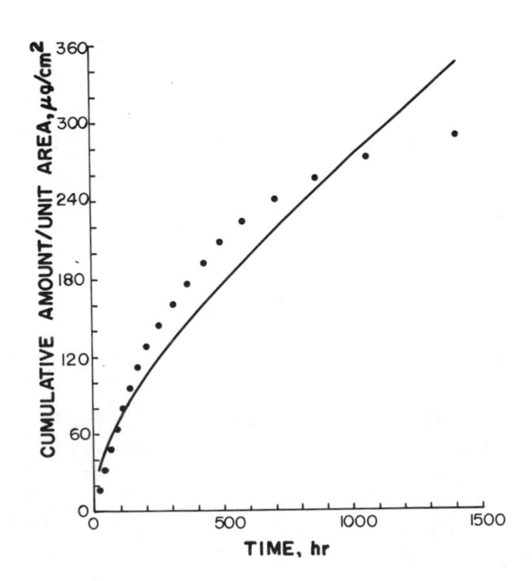

Figure 4. Cumulative amount of TBTCl released per unit area from the epoxy polymer: (●) experimental values; (——) curve calculated for the Godbee–Joy Equation with $D_1 = 3.38 \times 10^{-14}$ *cm²/sec and* $K_2 = 1.82 \times 10^{-7}$ *sec*$^{-1}$

Hydrolysis. NMR results show that TBT carboxylates undergo fast chemical exchange. Even the interfacial reaction between TBT carboxylates and chloride is shown to be extremely fast. The hydrolysis is thus not likely to be a rate determining step. Since the diffusivity of water in the matrix is expected to be much greater than that of TBTO, a hydrolytic equilibrium between the tributyltin carboxylate polymer and TBTO will always exist. As the mobile species produced diffuses out, the hydrolysis proceeds at a concentration-dependent rate. Godbee and Joy have developed a model to describe a similar situation in predicting the leachability of radionuclides from cementitious grouts (15). Based on their equation, the rate of release of tin from the surface is:

$$\text{rate} = CD_1^{0.5}K_2^{0.5}[\text{erf}(K_2^{0.5}t^{0.5}) + \frac{\exp(-K_2 t)}{(\pi K_2 t)^{0.5}}]$$

where K_2 is the concentration-dependent hydrolysis rate constant in sec^{-1}. Experimental data fitted to the integral form of the above equation,

$$Q = CD_1^{0.5}K_2^{0.5}[(t + \frac{1}{2K_2})\ \text{erf}(K_2^{0.5}t^{0.5}) + (\frac{t}{\pi K_2})^{0.5}\ \exp(-K_2 t)]$$

whith $D_1 = 3.38 \times 10^{-14}$ cm^2/sec and $K_2 = 1.82 \times 10^{-7}$ sec^{-1} is given in Figure 4. The agreement is not very good.

Laboratory Testing. The situation existing during service in marine environments is illustrated by Figure 2a. To simulate the conditions in the laboratory for the direct determination of the release rate, it is not just sufficient to maintain a flow; the concentration of TBTCl in the bulk will have to be kept near zero by constantly changing the sodium chloride solution or by constantly extracting TBTCl into hexane or carbon tetrachloride. In fact, any effort to maintain zero concentration in the bulk will ensure steady state mass transport in the boundary layer.

Static experiments approach the situation described in Figure 2b. The appropriate boundary conditions are set and Laplace transformation performed by Bird et al. (16). Differentiation of their equations, evaluation at $x = 0$, and substitution in Fick's first law will provide the mass transfer rate at the interface. Diffusivities in the matrix and in water can also be derived.

Loss of Tin. The preparation and characterization of organotin-epoxy polymers have been reported earlier (3). In an effort to determine the loss of tin from these controlled release formulations, 0.5-mm thick coatings were kept immersed in 4% sodium chloride solution under conditions approximating Figure 2a. The concentration of TBTCl in the aqueous phase was maintained low by continuously extracting it into hexane. Analysis of the coating at the end of 16 months revealed that not more than 2% tin was lost in any of the four cases studied.

This is in agreement with the results of Bennett and Zedler (12); they have pointed out that conventional urethane, vinyl and epoxy systems show widely divergent release rates. Whereas the vinyl system lost TBTO at a rate of 1-2 $\mu g/cm^2$/day and polyurethane at 25 $\mu g/cm^2$/day, the epoxy system virtually lost no tin.

Antifouling Performance. Compositions of several organotin-epoxy coatings, the method of preparing specimens for antifouling tests in marine environments and procedures for determining antifouling performance have been reported earlier (17). Fouling resistance up to 27 months have been observed. It is apparent from the laboratory tests that only a fraction of tin was released at the onset of fouling.

Miller has shown that TBTO will prevent fouling attachment at leaching rates as low as 1.25 $\mu g/cm^2$/day (18). It is thus reasonable to assume that fouling commences when the rate of release falls below 0.5 $\mu g\ Sn/cm^2$/day. Based on this, the effective diffusivities are calculated, using Crank's rate equation. The calculated effective diffusivities are then substituted in the integral form of Crank's equation to estimate the amount of Sn lost. The average effective diffusivity calculated for the 44 compositions is 1.7×10^{-13} cm^2/sec and the loss of tin 2.2%. These are in qualitative agreement with the results of laboratory testing. Cardarelli has also reported that antifouling rubber retains a considerable amount of organotin additives even after complete fouling (11).

Release Rate. Following the general procedure described earlier, a 60% TBT ester of 1:1 styrene-maleic anhydride copolymer (SMA-1000) was crosslinked by DGEBA (Ciba 6004) at an epoxy to anhydride ratio of 2:1, curing at 150°C for 36 hours. An epoxy coating of 406 cm^2 surface area and 3.17 mm thickness, containing 17.4% Sn, was immersed in 4% sodium chloride solution. Laminar flow conditions are maintained throughout, and pH between 6.9 and 7.1. TBTCl released was continuously extracted into carbon tetrachloride. The carbon tetrachloride layer was periodically analyzed for Sn by dithiol method.

The rate of release of TBTCl was 19.2 $\mu g/cm^2$/day (7.0 μg Sn/cm^2/day) on the first day of the experiment. It steadily dropped to 1.1 μg TBTCl/cm^2/day (0.41 $\mu g\ Sn/cm^2$/day) after 50 days. The total loss in 2 months was 0.16%. The experimental data are fitted to theoretical equations in Figure 3 and Figure 4.

Diffusion in Matrix. The rate of release drops faster than predicted by the models (Figure 3 and Figure 4). The epoxy compositions discussed here are highly crosslinked (3) and have glass transition temperatures around 140°C. Their free volumes and segmental mobilities are very low. It is known that these factors decrease the diffusivity in the matrix (19). Further, the magnitude of decrease is greater, the larger the diffusing molecule (19); TBTO is a relatively large molecule.

Conclusions

Tributyltin carboxylates undergo rapid chemical exchange, as evidenced by NMR. As a consequence, even the interfacial reaction between tributyltin carboxylate and chloride is fast. IR, mass spectra, gas chromatographic retention time and chloride assay show that the product of the reaction is tributyltin chloride.

Tributyltin carboxylates can undergo hydrolysis to form tributyltin hydroxide; the conversion is quantitative at high pH, as their titratability indicates. Tributyltin hydroxide readily loses water in organic media to form TBTO. However, in aqueous medium, the hydrated form prevails. Prince (20) has reported that trialkyltin chlorides establish an equilibrium with the corresponding hydroxides, in the presence of water. In marine environments, where the chloride concentration is over 0.5M and hydroxide concentration less than 10^{-5}M, any hydroxide formed would readily be converted to TBTCl, which, as shown by Aldridge et al., deranges mitochondrial function (21) and, therefore, can prove to be quite toxic to marine organisms.

Antifouling performance of organotin carboxylate polymers show that their mode of action corresponds to the "bulk abiotic bond cleavage" model. All the controlling factors are analyzed. It is found that the diffusion of the tributyltin species in the matrix is the rate limiting factor; the highly crosslinked epoxy matrix offers great resistance for the diffusion of the large molecule.

Abstract

Controlled release epoxy formulations in which tin is chemically anchored as tributyltin carboxylate to the polymer chain are discussed. NMR evidence is presented to establish that rapid exchange exists in tributyltin carboxylates. Consequently, even the interfacial reaction between tributyltin carboxylates and chloride is very fast; equilibrium constants are reported for the reaction between tributyltin acrylate in hexane and sodium chloride in water. IR spectra, gas chromatographic retention time, chloride assay, and the complex intensity pattern of the molecular ion peaks in the mass spectrum show that the product of the reaction is tributyltin chloride, suggesting that it is the chemical species responsible for antifouling activity in marine environment. The mode of action of the antifouling polymers thus conforms to the bulk abiotic bond cleavage model. All the controlling factors, viz., diffusion of water into the polymer matrix, hydrolysis of the tributyltin carboxylate, diffusion of tributyltin species from the matrix to the surface, phase transfer of the organotin species, and its migration across the boundary layer, are analyzed. It is found that the transport of the mobile tributyltin species in the matrix is the rate limiting factor.

Literature Cited

1. Phillip, A. T., Progr. Org. Coatings, 2, 159 (1973/74).

2. Montemarano, J. A. and Dyckman, E. J., J. Paint Technol., 47(600), 59 (1975).

3. Subramanian, R. V. and Anand, M., in "Chemistry and Properties of Crosslinked Polymers," S. S. Labana, Ed., Academic Press, N.Y., 1977, p. 1.

4. Subramanian, R. V., Garg, B. K., and Corredor, J., in "Organometallic Polymers," C. E. Carraher, Jr., J. E. Sheats, and C. U. Pittman, Jr., Eds., Academic Press, N.Y., 1978, p. 181.

5. Subramanian, R. V., Garg, B. K., and Somasekharan, K. N., Amer. Chem. Soc., Div. Org. Coat. Plast. Chem., Prepr., 39, 572 (1978).

6. Neumann, W. P., "The Organic Chemistry of Tin," Interscience Publishers, N.Y., 1970, Chapter 2.

7. Neumann, W. P., "The Organic Chemistry of Tin," Interscience Publishers, N.Y., 1970, Chapter 7.

8. Castelli, V. J. and Yeager, W. L., in "Controlled Release Polymeric Formulations," D. R. Paul and F. W. Harris, Eds., American Chemical Society, Washington, D.C., 1976, p. 239.

9. Marson, F., J. Appl. Chem., 19, 93 (1969).

10. Ketchum, B. H., Ferry, J. D., Redfield, A. C., and Burns, A. E., Jr., Ind. Eng. Chem., 37, 457 (1945).

11. Caradelli, N., "Controlled Release Pesticides Formulations," CRC Press, Cleveland, Ohio, 1976, pp. 35-36.

12. Bennett, R. F. and Zedler, R. J., J. Oil Colour Chem. Assoc., 49, 928 (1966).

13. Crank, J., "The Mathematics of Diffusion," 2nd ed., Clarendon Press, Oxford, 1975, p. 32.

14. Hershey, D., "Transport Analysis," Plenum Press, N.Y., Chapter 1.

15. Godbee, H. W. and Joy, D. S., "Assessment of the Loss of Radioactive Isotopes from Waste Solids to the Environment. Part I: Background and Theory," Oak Ridge National Laboratory, Oak Ridge, Tennessee, 1974.

16. Bird, R. B., Stewart, W. E., and Lightfoot, E. N., "Notes on Transport Phenomena, " John Wiley, N.Y., 1958, Chapter 19.

17. Subramanian, R. V. and Garg, B. K., in "Proceedings of the 1977 International Controlled Release Pesticide Symposium," R. L. Goulding, Ed., Oregon State University, Corvallis, Oregon, 1977, p. IV-154.

18. Miller, S. M., Ind. Eng. Chem., Prod. Res. Develop., 3(3), 226 (1964).

19. Stannett, V., in "Diffusion of Polymers," J. Crank and G. S. Park, Eds., Academic Press, N.Y., 1968, Chapter 2.

20. R. H. Prince, J. Chem. Soc., 1783 (1959).

21. Aldridge, W. N., in "Organotin Compounds: New Chemistry and Applications," J. J. Zuckerman, Ed., American Chemical Society, Washington, D.C., 1976, p. 186.

RECEIVED July 12, 1979.

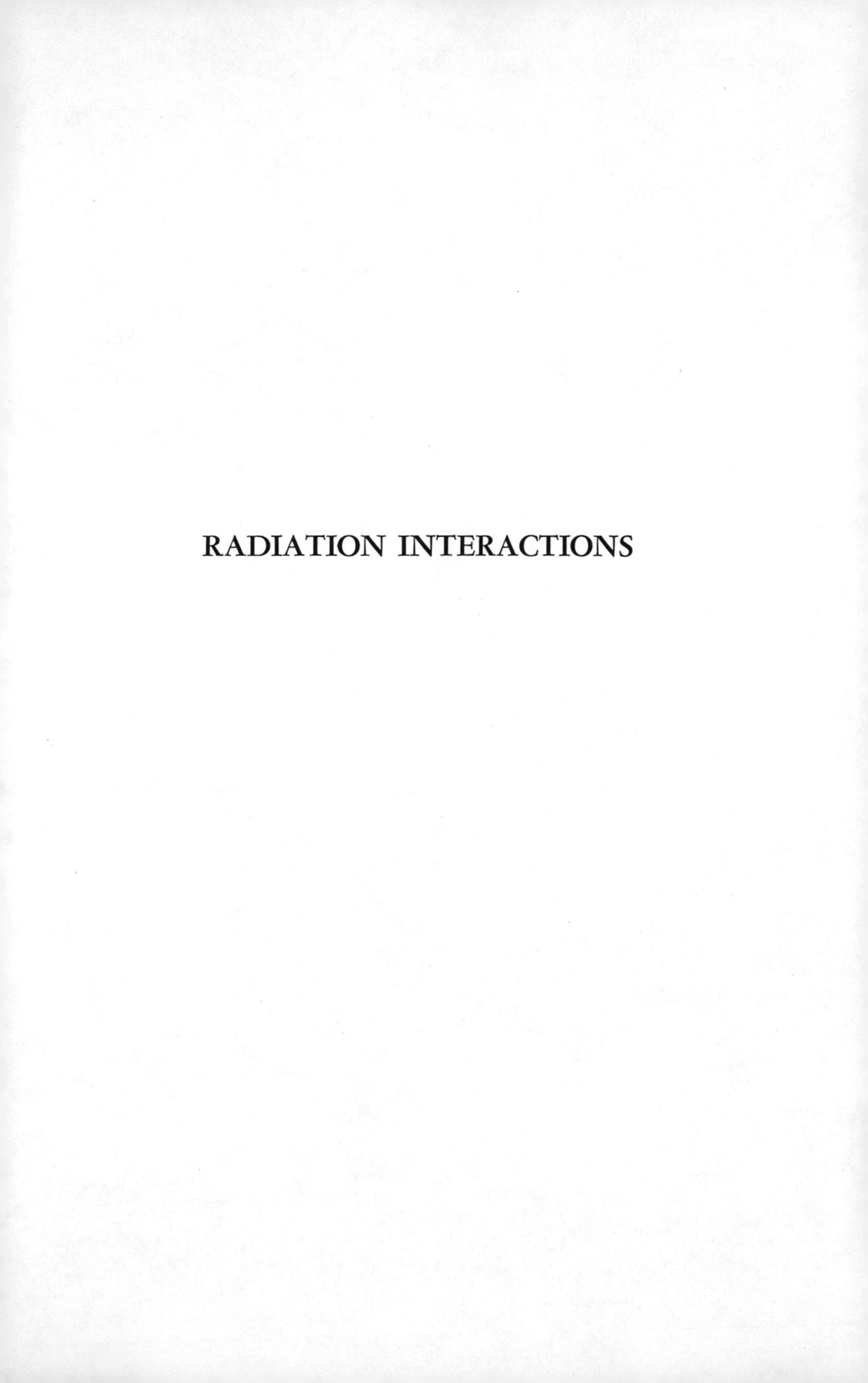

RADIATION INTERACTIONS

13

Photocross-Linking of 1,2-Polybutadiene by Aromatic Azide

TSUGUO YAMAOKA, TAKAHIRO TSUNODA, KEN-ICHI KOSEKI, and ISAO TABAYASHI

Chiba University, Faculty of Engineering, Chiba, Japan 260

Cyclized polyisoprene has been used as a photoresist by being sensitized with bisazides(1-3). Recently, H.Harada et al. have reported that a partially cyclized 1,2-polybutadiene showed good properties as a practical photoresist material in reproducing submicron patterns(4). S.Shimazu et al. have studied the photochemical cleavage of 2,6-di(4'-azidobenzal)cyclohexanone in a cyclized polyisoprene rubber matrix, and have reported that the principal photoreaction is the simultaneous cleavage of the both azido groups by absorption of a single photon with a 43% quantum yield(5). Their result does not support the biphotonic process in the photolysis of bisazide proposed by A.Reiser et al.(6).

Even if a same azide is used as the sensitizer, such properties of the photoresist as photosensitivity, photocurability and adhesion to base surfaces differ depending on the property of the base polymer. That is, degree of cyclization, content of the unsaturated groups and molecular weight of the polymer affect the photoresist properties mentioned above. H.L.Hunter et al. have discussed the dependence of the sensitivity of polybutadiene photoresist on the polymer structure, and have concluded that a higher sensitivity was obtained when 1,2- and 3,4-isomers were used(7).

It is known that aromatic azides are photodecomposed to give active nitrenes as the transient species, which react with the environmental binder polymers to crosslink them. However, the mechanism of these photocrosslinking polymers has not been studied in detail. L.S.Efros et al. have proposed that the rubber polymer is crosslinkes in such a way that the aromatic nitrene inserts into an unsaturated bond of the polymer to give an aziridine ring. The experimental evidence for this, however, has not been given (8).

In the present experiment, we have studied the mechanism of photocrosslinking of 1,2-polybutadiene by aromatic azide, based on the reaction of aromatic nitrene with unsaturated hydrocarbon monomeric compounds.

0-8412-0540-X/80/47-121-185$05.00/0

Experimental

Reagent

3-Methyl-1-butene The commercially available reagent was used after purifying by distillation.

Cyclohexene The commercially available reagent was washed with aqueous solution of ferrous sulfate, dried with anhydrous sodium sulfate and distilled(bp.83.0°C). The distilled cyclohexene was passed through an alumina column. This process of purification was repeated three times.

Phenylazide 5G of aniline was dissolved in 100ml of 10% hydrochloric acid and cooled to 5°C. To this solution, 1g of sodium nitrite in 10ml of water was added drop by drop with agitation under 5°C. After reacting for 30 minutes, 1g of sodium azide was added gradually with stirring. Oilly phenyl azide was extracted from this aqueous solution with diethylether. Phenyl azide was obtained by removing the diethylether with an evaporator. Phenyl azide thus obtained was purified by distillation at 2mmHg/50°C.

4,4'-Diazidodiphenyl 5G of 4,4'-diaminodiphenyl was dispersed in 200ml of 10% hydrochloric acid and 1.5g of sodium nitrite in 10ml of water was added drop by drop with stirring under 5°C. After reacting it for 30 minutes, the solution became transparent. 1.5G of sodium azide was added to this solution gradually. 4,4'-Diazidodiphenyl was precipitated. The precipitate was dried at room temperature after filteration, and purified by recrystallizing from methanol.

7-Phenyl-7-azabicyclo[4,1,0]heptane 20G of phenyl azide and 20g of cyclohexene in 50ml of tetrahydrofuran were refluxed for 8 hours. Triazoline was obtained by distilling off the solvent and the unreacted components. 3G of triazoline was dissolved in 100ml of benzene and irradiated with a 100 watts high pressure mercury lamp for 5 hours. After irradiation, benzene was distilled off. 7-Phenyl-7-azabicyclo[4,1,0]heptane was obtained as the residue. The residue was purified with column chlomatography using alumina (W-200, ICN Woelem Lab.Inc..) as the adsorbent and diethylether-n-hexane(1:1) as the developer.

N-(3-cyclohexenyl)aniline 2.5G of 3-chlorocyclohexene was dissolved in 20ml of diethylether. This solution was added to an excess amount of aniline drop by drop with stirring. A mixture of N-(3-cyclohexenyl)aniline, aniline hydrochloride and aniline was obtained. The mixture was washed with dilute hydroxyammonium and further washed with distilled water. N-(3-cyclohexenyl)aniline was obtained as the residue by distilling off aniline at 3mmHg/108°C.

3,3'-Bicyclohexenyl 0.5G of methylbromide was added slowly to 50ml of anhydrous diethylether containing 2.5g of magnesium powder. While the solution is bubbling, 11.6g of 3-cyclohexene in 100ml of anhydrous diethylether was added. This mixture was stirred for 10 hours at room temperature and washed with 1N hydrochloric acid. After the diethylether solution was dried with anhydrous sodium sulfate, diethylether was removed by distillation. 3,3'-bicyclohexenyl(bp.83°C at 2mmHg) was obtained by distillation.

Reaction of phenylnitrene with 3-methyl-1-butene

2G of phenylazide and 10g of 3-methyl-1-butene were dissolved in 300ml of benzene. This solution was irradiated with 100 watts high pressure mercury lamp in a Pyrex cell for 30 hours at 20°C. Benzene and unreacted 3-methyl-1-butene were removed by evaporation.

Measurement of gel fraction

A xyrene solution containing 10% of 1,2-polybutadiene(JSR RB-820, MW.150,000) and azide(the ratio of azide to vinyl group of 1,2-polybutadiene was adjusted to 0.05) was coated on a glass plate using a spinner and dried. The weight of the film on the plate(W_0) was measured. The film was irradiated with 500 watts high pressure mercury lamp from the back of the plate for one hour through a glass filter. The irradiated film was washed with xyrene, and then dried. The weight of the film remaining on the plate(W) was measured. W/W_0 was defined as the gel fraction.

Measurements

Gas chromatography was carried out utilizing a Shimazu GC-4C PTF(adsorbent:silicone SE-30, carrier:N_2 40ml/min., detector: FID).

Liquid chromatography was carried out utilizing a Hitachi model 634(adsorbent:Hitachi gel #3010, developing solvent:methanol, detecting wavelength:280nm).

Electronic, Infrared and Mass spectra were obtained using a Hitachi 200-20 spectrophotometer, a Hitachi EPI-S2 infrared spectrophotometer and a Hitachi RMU-6E mass spectrometer, respectively. NMR spectra were obtained using a Hitachi-Perkin-Elmer, model R-24 using carbontetrachloride as the solvent.

Results and discussion

Crosslinking of 1,2-polybutadiene by mono and dinitrene

Table 1 shows the gel fractions of 1,2-polybutadiene film containing mono or diazido compound which were irradiated by ultraviolet radiation. The results show that the gel fractions for diazides are 0.77 - 0.82, and 1,2-polybutadiene was crosslinked by dinitrene which was formed by the photodecomposition of diazide. The gel fractions for monoazides have lower values than those for diazides. This means that the crosslinking with monoazides is less effective than that with diazides.

Although mononitrenes do not act as crosslinking agents themselves, they can generate unpaired electrons in the polymer chains by the abstraction of hydrogen molecules from the polymer. As the result, the polymer chains are crosslinked by the recombinations of these unpaired electrons. Crosslinking by mononitrene in such a way, depends on the activity of the nitrene and may be not necessarily less efficient in some combination of monoazide with the polymer than that by dinitrene as is seen in the case for p-nitrophenylazide.

Table 1 Gel fractions of 1,2-polybutadiene containing azido compound after irradiated with the ultraviolet light

Azide	Gel fraction
p-Nitrophenylazide	0.38
p-(N,N-Dimethylamino)-phenylazide	0.06
1-Azidopyrene	0
4,4'-Diazidodiphenyl	0.77
4,4'-Diazidostilbene	0.80
2,6-Di(4'-azidobenzal)-cyclohexanone	0.82

Reaction of phenylnitrene with unsaturated olefines

With the purpose of understanding the crosslinking mechanism of 1,2-polybutadiene with aromatic nitrene, we studied the reaction of phenylnitrene with unsaturated olefine monomers such as 3-methyl-1-butene and cyclohexene. These monomers are structually similar to a unit segment of 1,2-polybutadiene.

Formation of triazoline by addition of phenylazide to 3-methyl-1-butene and its photodecomposition products

It has been reported by R.Scheiner that phenylazide forms triazoline compounds by 1,3-cyclic addition to unsaturated olefines such as n-butylethylene and norbornen(9). These triazolines are decomposed photochemically or thermally to give imine compounds and aziridine as is shown in scheme 1. These facts suggest that phenylazide may react with 3-methyl-1-butene to give triazoline in a similar reaction to that with norbornen.

Phenylazide and 3-methyl-1-butene were dissolved in n-hexane and stirred for 20 days in the dark. Then, unreacted phenylazide and 3-methyl-1-butene were removed by distillation. A liquid residue with a higher boiling point was obtained. The electronic spectrum of the residue differs from both components. It gives absorption peaks at 287nm and 303nm as is shown in Figure 1. These peak wavelengths are almost equal to those of the triazoline which was obtained by the addition of phenylazide to norbornen. IR spectrum of the residue shows the disappearance of absorption peaks at $2150cm^{-1}$ and $1650cm^{-1}$, $900cm^{-1}$ and $650cm^{-1}$ which are due to $\nu(N{=}N{=}N)$ of the azido group, $\nu(C{=}C)$ and $\delta(CH_2)$ of 3-methyl-1-butene, respectively. In the mass spectrum, the parent peak due

$$Ar\text{-}\overset{-}{N}\text{-}\overset{+}{N}{\equiv}N \; + \; >C{=}C< \;\longrightarrow\; \text{[triazoline transition state]} \;\longrightarrow\; \text{triazoline (N(Ar)-N=N ring)}$$

$$\xrightarrow[\text{or } \Delta]{h\nu} \; -\underset{\underset{Ar}{N}}{\overset{\|}{C}}-\overset{|}{C}- \;+\; -\overset{|}{C}\!-\!\overset{|}{C}-\ (\text{aziridine, N-Ar}) \;+\; N_2$$

to triazoline was not observed, but a peak was observed at m/e= 161. This number is equal to the molecular weight of the fragment where N_2 is removed from triazoline which is formed by the addition of phenylazide to 3-methyl-1-butene. These facts lead us to conclude that a triazoline having the structure [3] or [3'] was formed by the addition of phenylazide to 3-methyl-1-butene. It has not yet been determined which structure [3] or [3'] the triazoline has. The reaction mechanism of triazoline formation proposed by P.Scheiner, suggests that the structure may be 5-isopropyl-1-phenyl-Δ^2-1,2,3-triazoline(structure[3]). This conclusion was also supported by the fact that 3-methyl-2-butylideneaniline[5] was obtained by the decomposition of the triazoline (vide infra).

Refering to Figure 2, the triazoline showed a remarkable change in the IR spectrum after it was irradiated with ultraviolet radiation. The triazoline was dissolved in benzene and irradiated with ultraviolet radiation. After removing benzene by distillation, the photodecomposition products was isolated by distillation under reduced pressure. The NMR spectrum of the distillate gave peaks at τ=3.0 and τ=8.8 which are due to ring protons and methyl protons, respectively. IR spectrum of the distillate (Figure 3a) shows the absorption peak at 1210cm^{-1} which is attributed to symmetric stretching vibration of aziridine ring assigned by H.T. Hoffman and J.B.Patrick(10). The mass spectrum gave the parent peak at m/e=161 which is equal to the molecular weight of 1-phenyl-2-isopropylaziridine[4]. These results show that the aziridine compound is formed from the decomposition of the triazoline. The electronic spectrum of 1-phenyl-2-isopropylaziridine is close to that of aniline as is shown in Figure 4. The similarity of their spectra may be due to the common π-electronic structures of both compounds.

The residue which was obtained after the removal of the aziridine compound by distillation shows a strong IR absorption peak at 1670cm^{-1} which is assigned to ν(C=O). Since the aziridine could not be completely removed from the residue because of its close boiling point to the residue, the structure of the residue compound has not been determined in this step.

The residue is thermally unstable and a colorless transparent crystal was precipitated by standing at room temperature for a

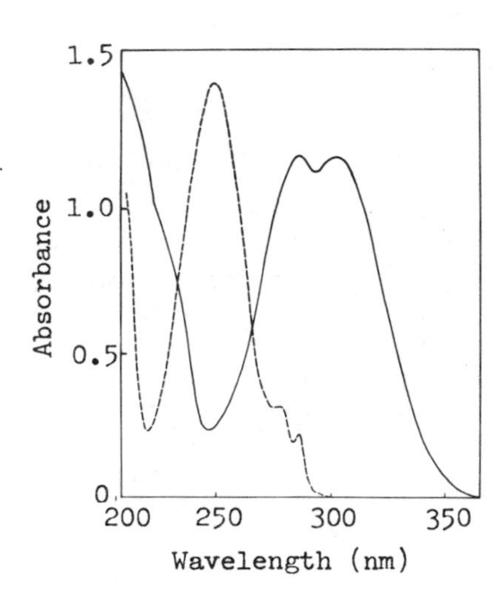

Figure 1. Electronic spectra of 5-isopropyl-1-phenyl-Δ²-1,2,3-triazoline (——) and phenylazide (– – –) in methanol

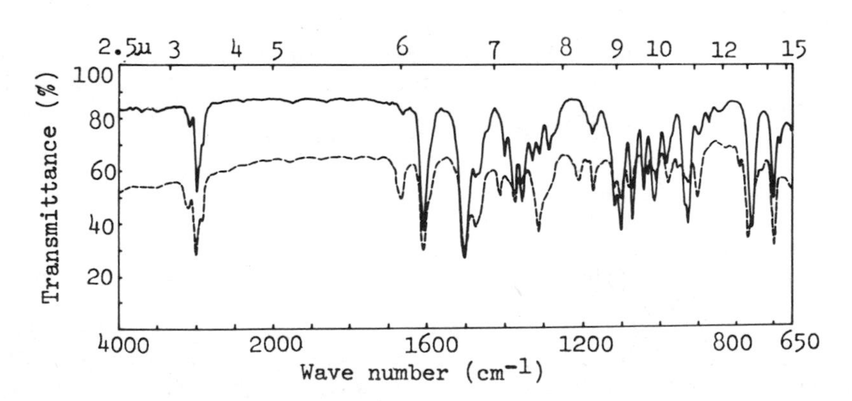

Figure 2. IR spectrum of 5-isopropyl-1-phenyl-Δ²-1,2,3-triazoline and its change by UV irradiation (liquid film method using NaCl cell): unirradiated (——) and irradiated (– – –)

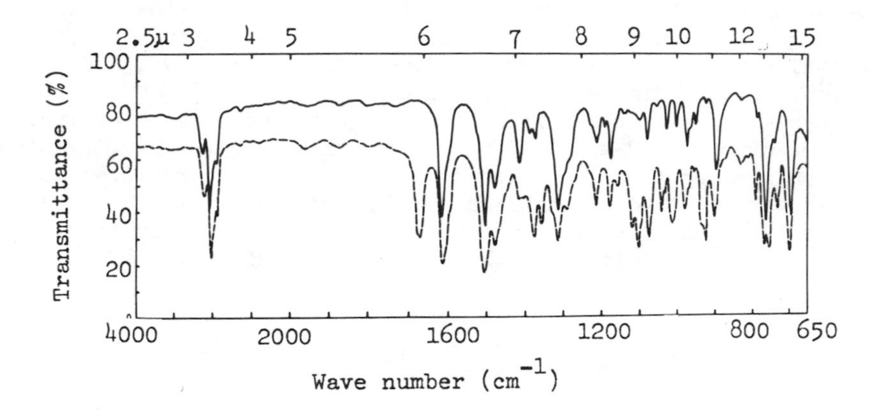

Figure 3. IR spectra of 1-phenyl-2-isopropylaziridine (——) and 3-methyl-2-butylideneaniline (– – –)

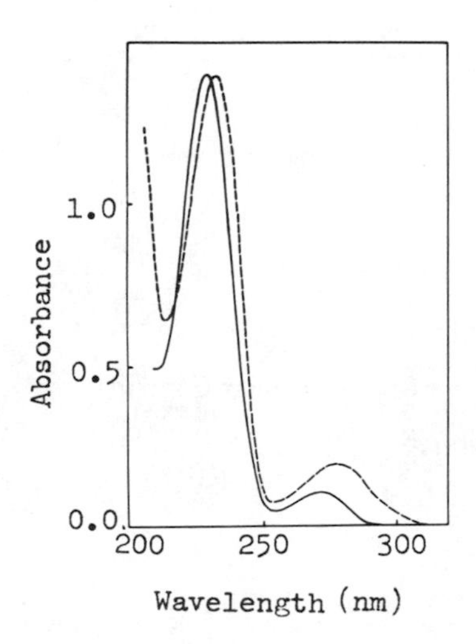

Figure 4. Electronic spectra of 1-phenyl-2-isopropylaziridine (——) and aniline (– – –) in methanol

couple of days. The IR spectrum of the crystal shows absorption peaks at 3450cm-1 and 1689cm-1, which are due to ν(NH) and ν(C=C) respectively. The mass spectrum gives peaks at m/e=322 and 161. The former is equal to the molecular weight of the dimerized species of 3-methyl-2-butylideneaniline[5], and the latter, to the molecular weight of the fraction which is formed by the symmetric cleavage of the dimerized species. The elemental analysis of the crystal agrees with those of the dimer(Found; C:81.89%, H:9.51%, N:8.65%, Calcd; C:81.92%, H:9.39%, N:8.69%).

From these results, the crystaline compound was determined to be a dimerized 3-methyl-2-butylideneaniline[6]. The formation of such a dimer from the residue suggests that the residual matter is 3-methyl-2-butylideneaniline. This presumption is supported by the fact that a similar product can be obtained from the reaction of aliphatic nitrene with unsaturated olefines.

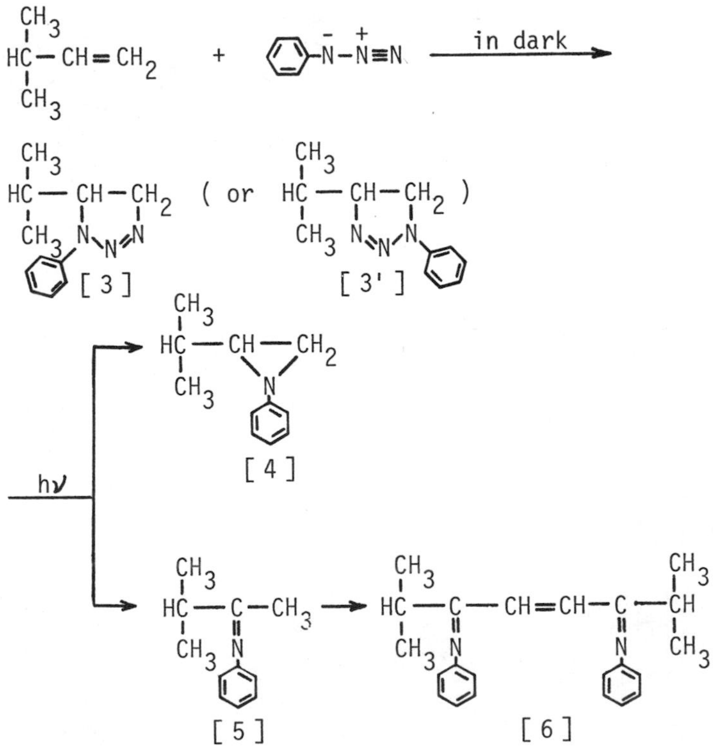

$$(CH_3)_2CH{-}CH{=}CH_2 + C_6H_5{-}\overset{-}{N}{-}\overset{+}{N}{\equiv}N \xrightarrow{\text{in dark}} [3] \text{ (or } [3'] \text{)}$$

[3] [3']

$$\xrightarrow{h\nu} [4]$$

[4]

$$\xrightarrow{h\nu} (CH_3)_2CH{-}C({=}NC_6H_5){-}CH_3 \; [5] \longrightarrow (CH_3)_2CH{-}C({=}NC_6H_5){-}CH{=}CH{-}C({=}NC_6H_5){-}CH(CH_3)_2 \; [6]$$

[5] [6]

The reaction of aziridine on silicagel

The benzene solution of 1-phenyl-2-propylaziridine[4] was developed in the column chromatography using active silicagel as an adsorbent, then separated into the adsorption and the effusion. The mass spectrum of the adsorption shows a peak at m/e=322 which is eual to the molecular weight of 3-methyl-2-butylideneaniline dimer[6] described in the above section. However, both NMR and IR spectra differ from those of the 3-methyl-2-butylideneaniline dimer. The IR spectrum of the adsorption gives no absorption peak due to ν(C=C), while it gives the peak due to ν(NH) at $3500cm^{-1}$. From these results, the compound having structure[7] is proposed for the adsoption.

The IR spectrum of the effusion gives a strong absorption peak at $1740cm^{-1}$ due to ν(C=O) of carbonyl group. This absorption indicates that the aziridine was oxidized on the silicagel surface. Because of such a fact, silicagel can not be used for the analysis of the reaction products of aromatic nitrene with unsaturated olefines. The present experiment, therefore, was repeated by replacing silicagel with alumina.

[4] —Silicagel→ $(CH_3)_2CH$—CH(NH—C_6H_5)—CH_2—CH_2—CH(NH—C_6H_5)—$CH(CH_3)_2$

[7]

Formation of aziridine by direct reaction of phenylnitrene with 3-methyl-1-butene

The products obtained by reacting phenylnitrene with 3-methyl-1-butene in benzene solution as has been described in the experimental section was analysed by gas chromatography. It gave five peaks from A' to E' as shown in Figure 5. Comparing the retention times of these peaks with those which were obtained by the method described in the previous section or with the compounds obtained by known methods, peaks A', B' and D' correspond to those of aniline, aziridine and azobenzene, respectively.

When phenylazide and 3-methyl-1-butene are mixed in benzene and kept standing for a couple of days, triazoline is formed and it gives aziridine by thermal or photodecomposition. However, when the mixture is irradiated as soon as they are mixed, the aziridine which is found in the solution is the one formed, not by the decomposition of the triazoline but, by direct photochemical reaction of phenylnitrene with 3-methyl-1-butene.

Figure 6 shows the gas chromatogram of the reaction products obtained by the ultraviolet irradiation to phenylazide dissolved in 3-methyl-1-butene. It has four peaks from a to d. Peaks a and c agree with those of aniline and aziridine in their retention times, respectively. In this case, the peak due to azobenzene did not appear unlike the case of the reaction in benzene solution mentioned above. The absence of azobenzene suggests that the reaction of phenylnitrene with 3-methyl-1-butene is very rapid since the rate of azobenzene formation by coupling of phenylnitrene is almost diffusion controlled(11).

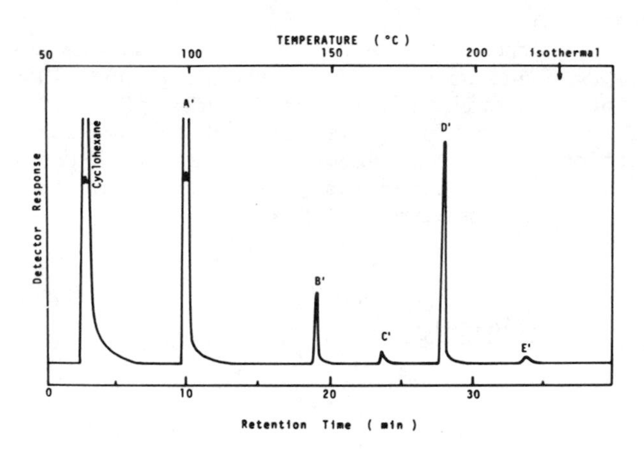

Figure 5. Programmed temperature gas chromatogram of photoreaction products of phenylazide with 3-methyl-1-butene in benzene

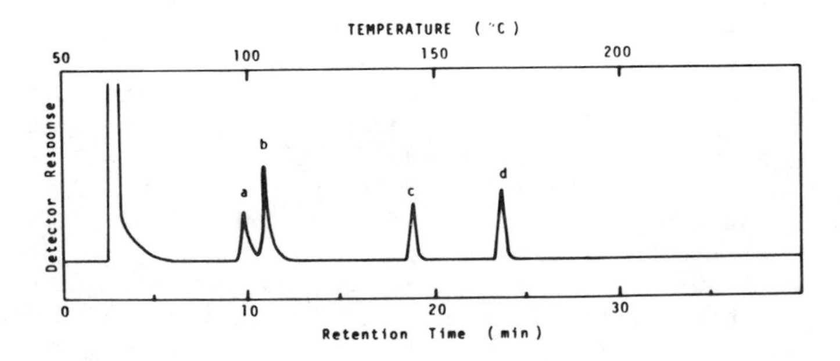

Figure 6. Programmed temperature gas chromatogram of photodecomposition products of phenylazide in 3-methyl-1-butene

Reaction of phenylnitrene with cyclohexene

Phenylazide dissolved in cyclohexene was irradiated with ultraviolet radiation for 39 hours. The unreacted cyclohexene was removed by distillation. The residue was separated by column chromatography using alumina(activity IV) as the adsorbent and n-hexane/diethylether(1:1) as the developer.

The structure of each compound isolated by the above method was determined by comparing its spectra with those of corresponding compound synthesized by the known method. Furthermore, the retention time of the gas chromatogram was compared with those of the known compounds. The following products were obtained from direct reaction of phenylnitrene with cyclohexene; aniline[8](determined by electronic spectrum, gas chromatography), 7-phenyl-7-azabicyclo[4,1,0]heptane[9](determined by electronic, IR, Mass and NMR spectrum), N-(3-cyclohexenyl)aniline[10](determined by electronic, IR, Mass and NMR spectrum) and 3,3'-bicyclohexenyl[11] (determined by IR, Mass and NMR spectrum).

NH_2 N- H N-

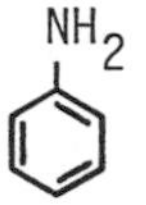

[8] 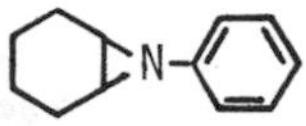 [9] 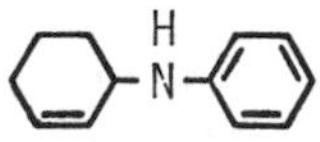 [10] 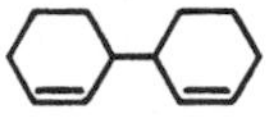 [11]

Quantitative determination of the reaction products of phenylnitrene with cyclohexene

By a method similar to that described in the last section phenylazide in cyclohexene was irradiated with ultraviolet radiation and unreacted cyclohexene was distilled off with evaporation. The residue was extracted with n-hexane. The extract was separated into several products by gas and liquid chromatography. The gas chromatogram and the liquid chromatogram are shown in Figures 7 and 8, which give five peaks from A to E, and four peaks from A' to D', respectively in addition to the peak due to the solvent. Peaks A and A' were determined to be aniline by their retention times. Peaks B and C' are due to 3,3'-bicyclohexenyl. Peaks C and D' are those of aziridine[9] and the product which was formed by the insertion of phenylnitrene to C-H bond of cyclohexene.

The retention time of aziridine and the insertion product [10] are different when each of them is injected separately. However, their peaks are not separated when the mixture is injected. Therefore, peak B' is considered as the overlap of peaks due to aziridine and the insertion product. The structure of the product corresponding to peak E could not be determined since its yield was very low as is shown by the low intensity of the peak in Figure 7.

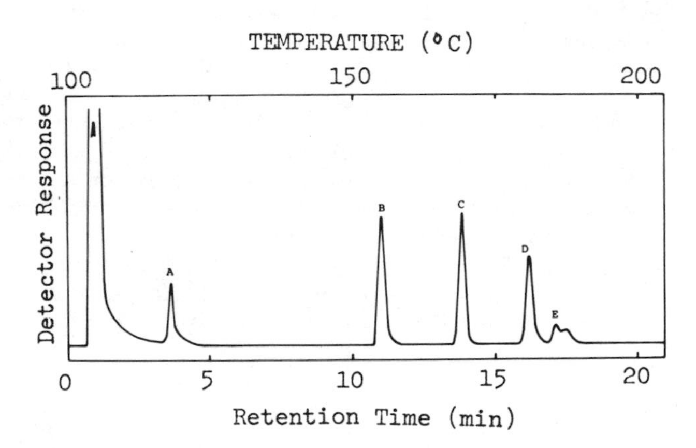

Figure 7. Programmed temperature gas chromatogram of photodecomposition products of phenylazide in cyclohexene

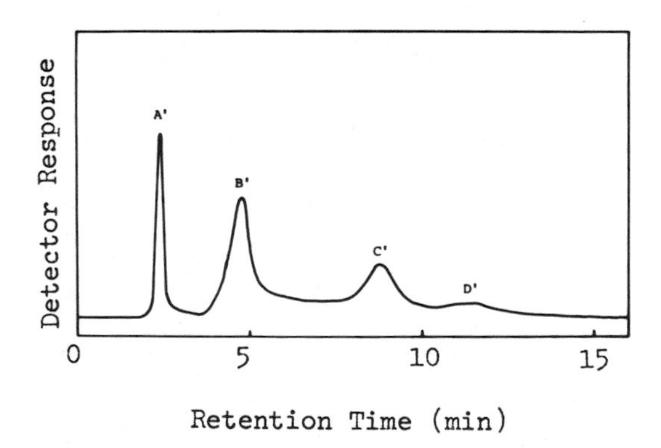

Figure 8. Liquid chromatogram of photodecomposition products of phenylazide in cyclohexene

The yields of each product were quantitatively determined with gas chromatography as illustrated in Figure 7, which are summerized in the first column of Table 2. The yield of aziridine is predominant. It is noticed that the yield of 3,3'-bicyclohexenyl is larger than that which is expected from reactions, (1)-(6) which are stoichiometrically written.

$$R\dot{N}\cdot + C_6H_{10} \longrightarrow R\dot{N}H + \dot{C}_6H_9 \quad (1)$$

$$R\dot{N}H + C_6H_{10} \longrightarrow RNH_2 + \dot{C}_6H_9 \quad (2)$$

$$2\ \dot{C}_6H_9 \longrightarrow C_6H_9{-}C_6H_9 \quad (3)$$

$$R\dot{N}\cdot + C_6H_{10} \longrightarrow RNH{-}C_6H_9 \quad (4)$$

$$R\dot{N}\cdot + C_6H_{10} \longrightarrow C_6H_{10}N{-}R \quad (5)$$

$$R\dot{N}H + \dot{C}_6H_9 \longrightarrow RNH{-}C_6H_9 \quad (6)$$

$$R\dot{N}H + C_6H_{10} \longrightarrow RNH{-}C_6H_9 + 1/2\,H_2 \quad (7)$$

Such mechanism, however, requires equivalent yields for aniline and 3,3'-bicyclohexenyl. The results of the present study (Table 2) give less yield for aniline than expected from eqs.(1)-(6).

Lwowski et al.(12) have studied the reaction product of ethylcarbethoxynitrene with cyclohexene. Their results also do not agree with these equations because the yield of 3,3'-bicyclohexenyl is larger than that of amine compound. This discrepancy suggests that reaction(7) may also occur in addition to reactions from eq.(1)-eq.(6).

Phenylazide and a triplet sensitizer were dissolved in cyclohexene and irradiated with ultraviolet radiation of wavelengths longer than 300nm so that only the sensitizer is excited. The photolysis of phenylazide is occures by the energy transfer from the excited triplet sensitizer. The yields of products are listed in the second and third column of Table 2. Where either benzophenone and acetophenone is added as the sensitizer, the yields of aniline and 3,3'-bicyclohexenyl are increased whereas the yields of 7-phenyl-7-azabicyclo[4,1,0]heptane and N-(3-cyclohexenyl)aniline are decreased, compared with situations employing the direct photolysis of phenylazide. The direct photolysis of phenylazide may give both singlet and triplet nitrenes. Since the hydrogen abstraction of triplet nitrene is well known(13), the results in Table 2 suggest that aniline and 3,3'-bicyclohexenyl were produced by hydrogen abstraction of phenylnitrene. The decrease in the yields of 7-phenyl-7-azabicyclo[4,1,0]heptane and N-(3-cyclohexenyl)aniline in the sensitized photolysis shows that a hydrogen abstraction reaction of triplet nitrene occures prior to the insertion reaction to the C-H bond or ethylenically unsaturated group.

Table 2 Relative ratio and yield(%) of driect and sensitized photodecomposition products of phenylazide in cyclohexene

Products	Direct photolysis		Sensitized photolysis			
			Benzophenone		Acetophenone	
	Ratio	%	Ratio	%	Ratio	%
Aniline	0.4	13	3.1	19	6.5	31
3,3'-Bicyclohexenyl	0.9	30	10.0	60	11.8	57
7-Phenyl-7-azabicyclo[4,1,0] heptane	1.0	32	1.0	6	1.0	5
N-(3-cyclohexenyl) -aniline	0.7	21	2.4	15	1.2	6
unknown	0.1	4	0.1	-	0.1	-

Reaction of 4,4'-dinitrenodiphenyl with 1,2-polybutadiene in matrix

1,2-Polybutadiene(JSR PB-1000, Mn;1000) film containing 0.025mol/l of 4,4'-diazidodiphenyl was prepared. The electronic spectrum of this film is similar to that of 4,4'-diazidodiphenyl in wavelengths longer than 250nm since the polymer is transparent in this wavelength region. Furthermore, its electronic spectrum showed that triazoline was not formed between azide and ethylenic groups of the polymer.

The film was exposed to ultraviolet radiation in a vacuum cell of 10^{-5} toll using a Pyrex filter. Figure 9 shows the change in the IR spectrum of the film due to irradiation with ultraviolet radiation. The peaks at $2150cm^{-1}$ and $1300cm^{-1}$ are due to ν_{as}(N=N=N) and ν_s(N=N=N), respectively. These peaks are decreased in their intensities, showing the decomposition of the azido group. Further, the following absorptions originated from 1,2-polybutadiene were remarkably decreased in their intensities; 3100, $3020cm^{-1}$ ($\nu(=CH_2)$ of alkene group), $1850cm^{-1}$(δ(CH), out of plane, double tone), $1650cm^{-1}$(ν(C=C)), $1410cm^{-1}$(δ(CH), in plane), $990cm^{-1}$, $960cm^{-1}$, $905cm^{-1}$(δ(CH), out of plane). These changes in the peak intensities show that the ethylenic group of 1,2-polybutadiene was reacted with 4,4'-dinitrenodiphenyl.

Absorption peaks due to alkane groups at $2990cm^{-1}$ (ν(CH)), $1450cm^{-1}$(δ(CH), in-plane of methylenic group of the main chain), and $1350cm^{-1}$(δ(CH), in-plane of tertiary hydrogen of the main chain) were also decreased, but the extent of the decrease is not large compared with those due to the alkene groups.

The irradiated film was washed with benzene and the components which are easily soluble in benzene were isolated. Then, the film was treated with the Soxhlet extraction method using benzene. Figure 10 shows the IR spectra of the film after 50, 150 and 200 hours of Soxhlet extraction. The absorption peak at $2150cm^{-1}$ decreases with increase of the extraction time. This fact means that undecomposed azide was removed from the film by Soxhlet extraction. The peak at $1500cm^{-1}$ is considered as the overlap of absorptions due to the benzene ring of undecomposed azide and that which was bonded to the polymer via the reaction of nitrene. The plots of the ratio of the peak intensities for the azide group($2150cm^{-1}$) and for the benzene ring($1500cm^{-1}$) against the extraction time showed that the ratio increases with increase of extraction time, suggesting the presence of the benzene ring which is not removed from the film by extraction, and therefore chemically bonded to the polymer.

In order to determine if the diminish of ethylenic double bond of the polymer by the attack of nitrene occurs , the ratio of IR absorbances due to ν(C=C) of ethylenic double bond($3100cm^{-1}$) and due to ν(CH) of alkane($2990cm^{-1}$) has been determined for the unirradiated film, irradiated film, benzene extract and Soxhlet extract(Table 3).

With the decomposition of the azido group, the double bonds

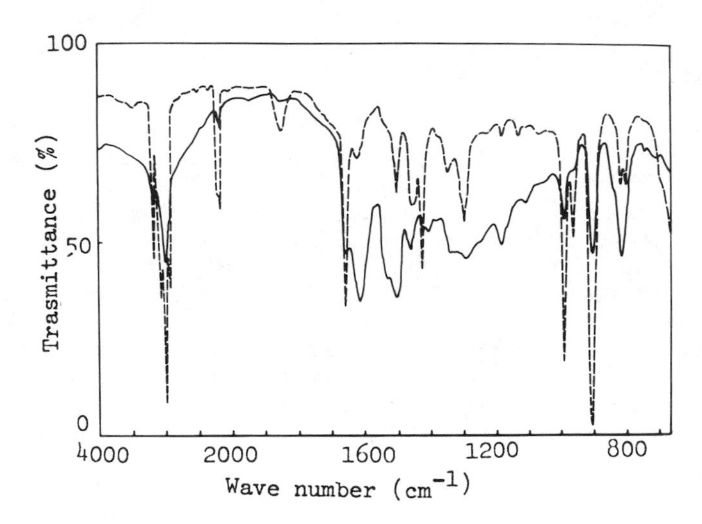

Figure 9. Change in IR spectrum of 1,2-polybutadiene film containing 4,4′-diazidodiphenyl with UV irradiation in vacuum, before (– – –) and after (——) irradiation

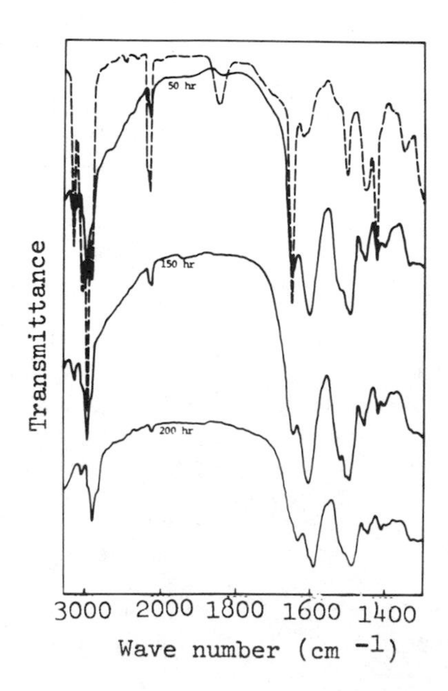

Figure 10. Change in IR spectrum of UV-irradiated 1,2-polybutadiene film containing 4,4′-diazidodiphenyl with 50, 150, and 200 hr Soxhlet extraction by benzene

were decreased, showing the reaction of nitrene with the double bonds. The values for the irradiated film and the benzene extract do not show a remarkable decrease of the double bond. This may be because the large amount of the unreacted polymer is contained in the extract. The Soxhlet extract showed the evident decrease of the double bond. The Soxhlet extract seems to consist of cross-linked polymer which was extracted due to the low molecurar weight of the parent polymer.

Table 3 The ratio of IR absorbances for ν(C=C) of double bond and ν(CH) of alkane

	D(C=C)/D(CH)
Unirradiated film	0.42
Irradiated film	0.31
Benzene extract	0.28
Soxhlet extract	0.07

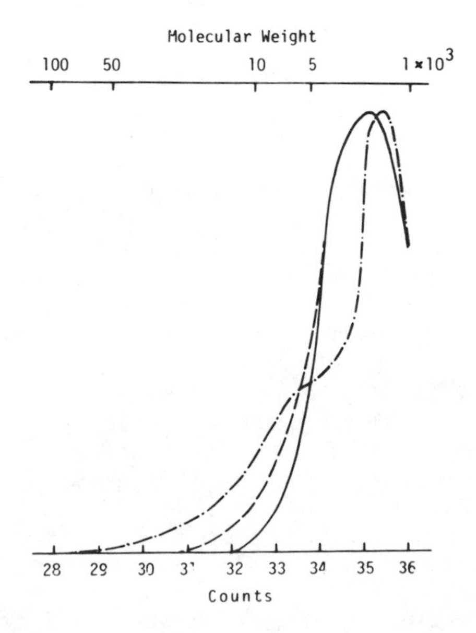

Figure 11. GPC curves of 1,2-polybutadiene (——), benzene extracted component (– – –), and Soxhlet extracted component (– · – ·)

Figure 11 shows GPC curves for 1,2-polybutadiene, the benzene extract and the Soxhlet extract. It is seen that the molecular weight of 1,2-polybutadiene is about 1000 with a narrow molecular weight distribution. The GPC curve of the benzene extract is almost similar to that of 1,2-polybutadiene, but shows a slight increase in the higher components. This fact means that the benzene extract contains a small amount of crosslinked polymer in addition to the unreacted polymer. On the other hand, the GPC curve of the Soxhlet extract differs from that of 1,2-polybutadiene in its peak position and the shape, showing the existence of a component of higher molecular weight.

It has been found from analysis by thin layer chromatography that the benzene extract contains a considerable amount of 4,4'-diaminodiphenyl in addition to the unreacted polymer. The presence of 4,4'-diaminodiphenyl means that 4,4'-dinitrenodiphenyl abstracted hydrogen from 1,2-polybutadiene. With this abstraction unpaired electrons may be generated within the polymer molecule and it may be crosslinked by the recombination of these unpaired electrons. The crosslinking of the polymer by such a mechanism will be possible with not only bisazides but also monoazides. This means that crosslinking by such a way contributes to the photocuring of the polymer, but the efficiency is not very high since the value of the gel fractions of the polymers sensitized with monoazide is usually lower than those with bisazides.

Results for the reaction of 4,4'-dinitrenodiphenyl with 1,2-polybutadiene are summerized as follows;

(1) ethylenic double bond of 1,2-polybutadiene is evidently decreased,
(2) alkane C-H group, especially tertiary C-H group is decreased,
(3) phenyl group bonded with 1,2-polybutadiene is present,
(4) a considerable amount of 4,4'-diaminodiphenyl is formed.

Based on these results and the reaction mode of phenylazide with unsaturated olefine monomers as the model compound of the polymer, mechanisms [I] - [III] are proposed for the crosslinking of 1,2-polybutadiene by bisazide.

Mechanisms [I] and [II] may occur through the recombination of unpaired electrons which are formed by the hydrogen abstraction of nitrenes, and the insertion reaction of nitrenes to C-H bonds, respectively. However, it has not been revealed in this study which C-H bond is attacked by the nitrene.

Mechanism [III] represents crosslinking due to aziridine ring formation. This mechanism is supported by the decrease of ethylenic double bond of 1,2-polybutadiene and the fact that a large amount of aziridine compound is formed in the reaction of phenylnitrene with unsaturated olefine monomers, although the direct observation of it in 1,2-polybutadiene film matrix has not been accomplished in the present study.

Consequently, the photo-crosslinking of 1,2-polybutadiene by bisazide might be due to the contribution of all of the mechanisms [I] - [III].

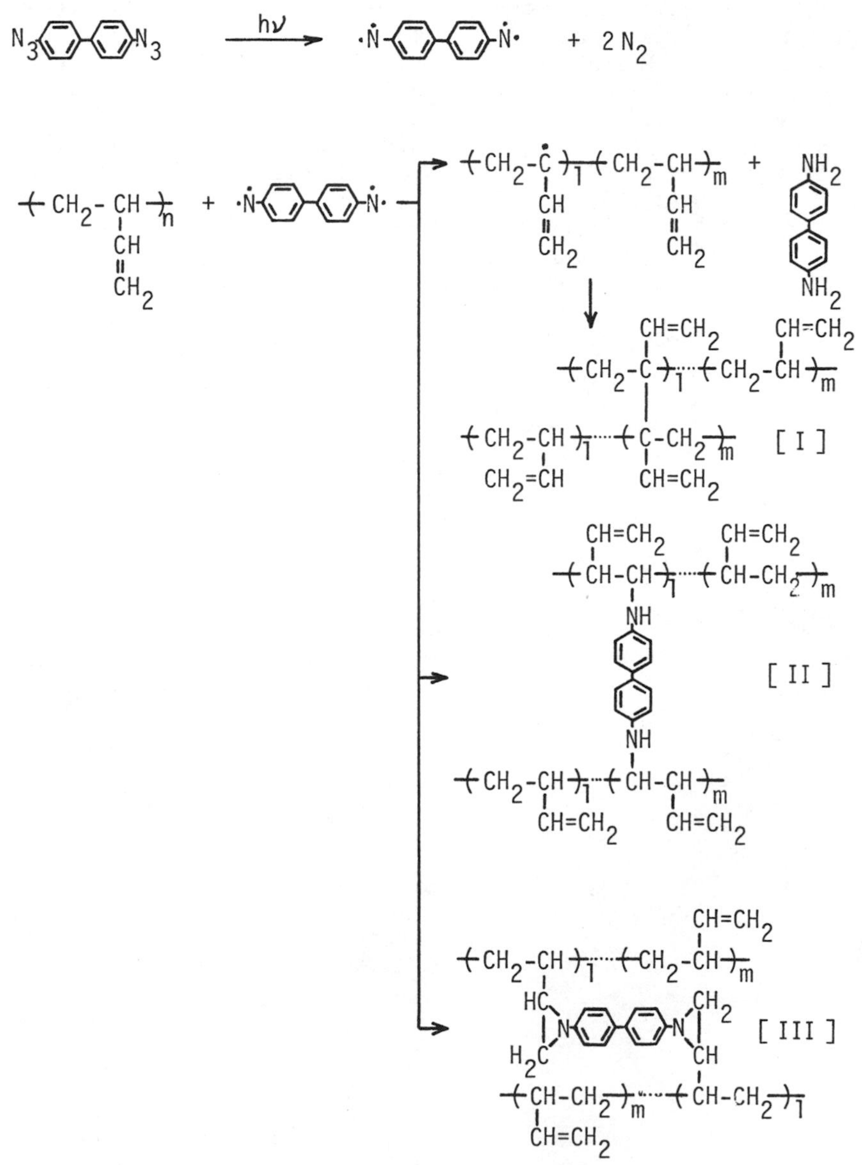

N_3–⟨biphenyl⟩–N_3 $\xrightarrow{h\nu}$ ·N̈–⟨biphenyl⟩–N̈· + 2 N_2

$\left(CH_2 - CH \right)_n$ (with side group $CH = CH_2$) + ·N̈–⟨biphenyl⟩–N̈· →

$\left(CH_2 - \dot{C} \right)_l \left(CH_2 - CH \right)_m$ (side groups $CH = CH_2$) + NH_2–⟨biphenyl⟩–NH_2

↓

$\left(CH_2 - C \right)_l \cdots \left(CH_2 - CH \right)_m$ (side groups $CH = CH_2$)

$\left(CH_2 - CH \right)_l \cdots \left(C - CH_2 \right)_m$ (side groups $CH_2 = CH$, $CH = CH_2$) [I]

$\left(CH - CH \right)_l \cdots \left(CH - CH_2 \right)_m$ (side groups $CH = CH_2$)

NH–⟨biphenyl⟩–NH [II]

$\left(CH_2 - CH \right)_l \cdots \left(CH - CH \right)_m$ (side groups $CH = CH_2$)

$\left(CH_2 - CH \right)_l \cdots \left(CH_2 - CH \right)_m$ (side group $CH = CH_2$)

HC, H_2C – N–⟨biphenyl⟩–N – CH_2, CH [III]

$\left(CH - CH_2 \right)_m \cdots \left(CH - CH_2 \right)_l$ (side group $CH = CH_2$)

References

1) USP 2,940,853.
2) USP 2,852,379.
3) For example, H.P.Thomas, Proceeding of the Microelectronics Seminar, Interface '74, (1974)p.89.
4) Y.Harita, M.Ichikawa, K.Harada, T.Tsunoda, Technical Papers Photopolymer Conference, Soc.Plastics Engineers,Inc.,Oct.13 (1976)p.84.
5) S.Shimizu, G.R.Bird, J.Electrochem.Soc.,124(9),1394(1977).
6) A.Reiser, H.M.Wagner, R.Marley, G.Bowes, Trans.Faraday Soc., 63, 2404(1967).
7) W.L.Hunter, P.N.Crabtree, Photo.Sci.Eng.,13,271(1969).
8) L.S.Efros, T.A.Yurre, Polymer Sci.,USSR, 12(10),2505(1970).
9) R.Scheiner, Tetrahedron, 24,2757(1968). J.Org.Chem.,30, 7(1965). J.Amer.Chem.Soc.,87, 306(1969).
10) H.T.Hoffman, J.Amer.Chem.Soc.,73, 3028(1951). J.B.Patrick, J.Amer.Chem.Soc.,86, 1889(1964).
11) A.Reiser, F.W.Willets, G.C.Terry, V.Williams, R.Marley, Trans.Faraday Soc.,64(12), 3265(1969).
12) W.Lwowski, J.Amer.Chem.Soc.,87, 1947(1965).
13) For example, "The Chemistry of the Azido Group",Ed.,S.Patai, Interscience Publishers (1971)p.463.

RECEIVED July 12, 1979.

14

Modification of Electrical Properties in Poly-*N*-Vinylcarbazole by UV Light—Thermally Stimulated Current

H. KITAYAMA, T. FUJIMOTO, M. YOKOYAMA, and H. MIKAWA

Department of Applied Chemistry, Faculty of Engineering, Osaka University, Yamadakami, Suita, Osaka, Japan, 565

Thermally stimulated hole current in poly-N-vinylcarbazole shows distinct maximum at around 5°C and another large current above 100°C. This 5°C maximum is due to 0.56 eV hole traps of 7×10^{15} cm^{-3} density. Photoconductivity in poly-N-vinylcarbazole increases appreciably when irradiated with UV-light in air at room temperature and this increase accompanies the formation of 0.56 eV hole traps. The nature of this traps has been discussed.

Poly-N-vinylcarbazole is one of the very important organic photoconductors now in use in electrophotography. For practical use, a little amount of sensitizer is usually added. Carrier generation increases usually due to the electron accepting property of the sensitizer in its ground state or in its excited state. The detailed mechanism of the carrier generation is the subject to be reported in another symposium of the present meeting (1).

Another important problem of the photoconductivity of poly-N-vinylcarbazole is the carrier transport. With the so-called time of flight method this problem is well investigated. A very important progress has been made recently on the theoretical background of the analysis of the transport problem also (2). Although it has been made clear that multi-trapping process is determining the carrier transport in the polymer, not only the nature of the trap but also the trap depth and the trap population are not definitely known even in pure poly-N-vinylcarbazole. The present report concerns with this problem.

So-called thermally stimulated current is the most direct method to investigate the nature of the traps, the current being due to the carriers comming out thermally from the traps. Although the thermally stimulated current of poly-N-vinylcarbazole has already been investigated by Pai (3) and Patora (4), the results are somewhat different. We investigated this problem with many

0-8412-0540-X/80/47-121-205$05.00/0

different samples by different groups of members and obtained quite reproducible results as summarized at the top. Although many approximate methods are reported (5) (6) for the analysis of the thermally stimulated current curve, we obtained the values of trap depth by numerical analysis of the theoretical equation without any approximation. Thus establishing the experimental method of obtaining reproducible results and the method of evaluating the parameters, the films modified by UV-light in air were also investigated and confirmed that trap population increased with the same trap depth of 0.56 eV as of the virgin sample.

EXPERIMENTAL

Poly-N-vinylcarbazole

N-vinylcarbazole with 0.1 mol% of AIBN was dissolved in benzene, degassed and polymerized at 80°C for 15 hrs. The polymer was purified by reprecipitation 3 times from THF-methanol followed by extraction with methanol.

Sample Cell Assembly and Cryostat

Onto the surface of a copper plate, the polymer film

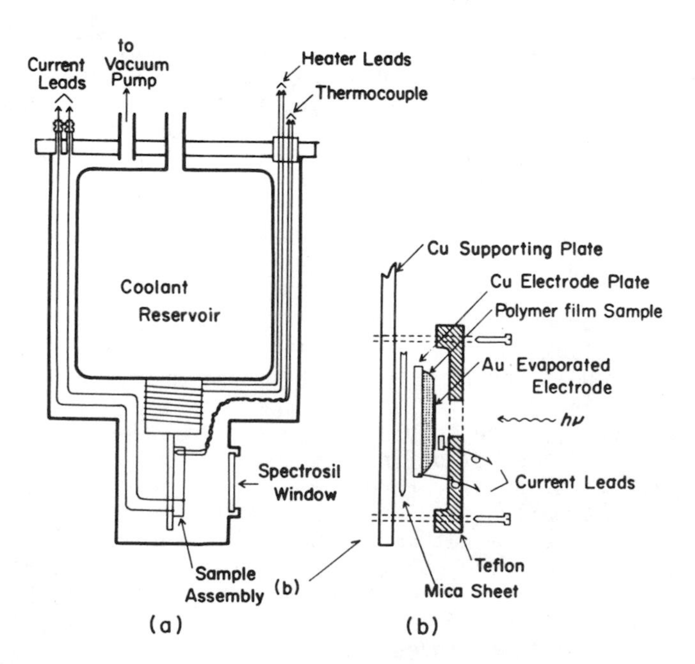

Figure 1. Cryostat (a) and cell assembly (b)

was cast from 5% benzene solution. Gold was evaporated as electrode. To obtain reproducible results, it is of absolute importance to keep the polymer in good contact with the copper base plate over the whole temperature range of the measurement from -150～100°C. This was achieved by etching the copper plate with $FeCl_3$ solution prior to the casting of the film. Raising rate of the temperature was kept constant typically at 3.3°C/min.

Annealing and cleaning up of the eventually existing residual trapped carriers; filling up of the traps with UV-irradiation; and the measurement of thermally stimulated current

First, the whole assembly was evacuated and cooled down to -150°C. Electric field was applied to the film and the temperature was elevated at the rate of 3.3°C/min. Carriers eventually comming out from the traps are collected by the applied field and the observed current due to these carriers and also due to relaxation of internal stress of the polymer film if any was measured. Typical example of this current is shown by curve (i) of Fig. 2. As shown typically by curve (ii) of the same figure, when this sample was cooled down again to -150°C and the same measurement was repeated, no thermally stimulated current was observed till 100°C. This result shows that the heating cycle of curve (i) cleans up the trapped carriers and also the annealing of the internal stress if any is completed. Thus cleaning up the film, the film was again cooled down to -150°C and irradiated by the total light of the 500W Hg lamp for definite time in order to fill up the traps. The energy distribution of the light is shown in Fig. 3. With this low temperature irradiation, photocarriers are generated and captured by traps. After filling up the traps in this way, electric field was applied and the temperature was elevated and the carriers coming out from the traps were collected by the applied field and measured by a vibrating-reed electrometer.

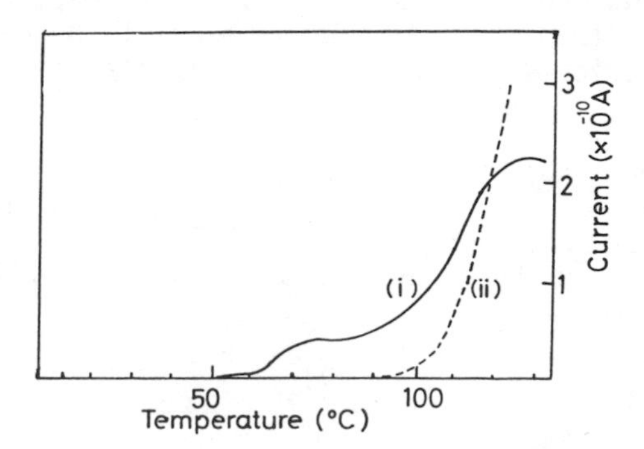

Figure 2. Thermally stimulated current of poly-N-vinylcarbazole film: (i) The current observed for the first time without any pretreatment; (ii) the same after this first measurement (collecting voltage 30V (Au+); heating rate 3.3°C/min)

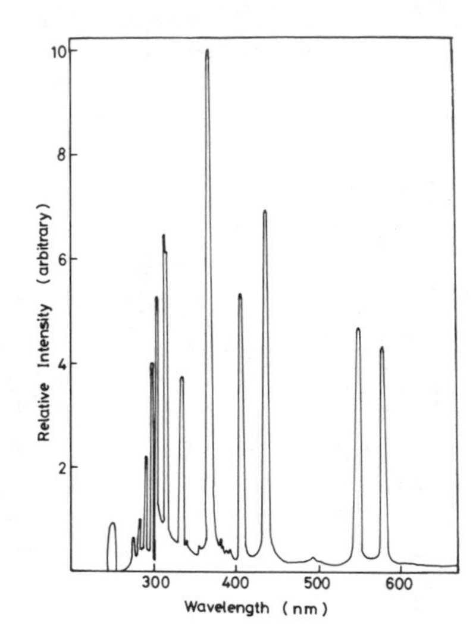

Figure 3. Intensity distribution of 500W Hg lamp

Fig. 4 (a) and (b) explain the principle. This is the thermally stimulated current to be used for the analysis of the trap depth and its population. In so far as the film has no experience of being heated above 110°C, the same film assembly could be used repeatedly many times with satisfactory reproducible results.

Measurement of the thermally stimulated current of the poly-N-vinylcarbazole film irradiated prior to the measurement with UV-light in air at room temperature.

Prior to the measurement of the thermally stimulated current, the film was irradiated at room temperature with the total UV-light from the same 500W Hg lamp. The whole was evacuated and the film was cooled down to -150°C, irradiated with Hg lamp to fill up the traps and the measurement of thermally stimulated current was performed in the same way as stated above.

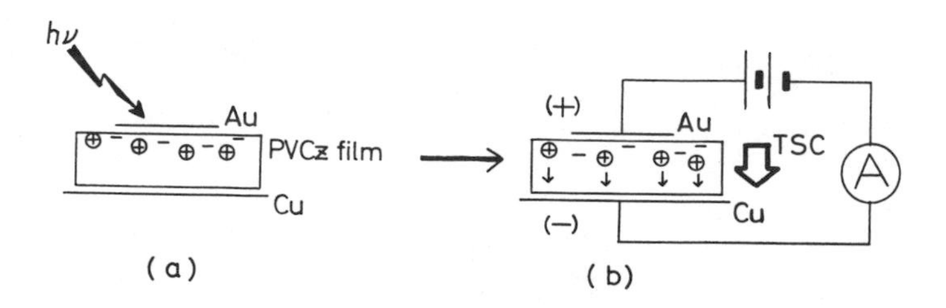

Figure 4. Principle of the measurement of thermally stimulated current: (a) when the sample is irradiated at −150°C photocarriers are generated and the traps are filled with the carriers; (b) measurement of the thermally stimulated current

RESULTS

Fig. 5 shows the thermally stimulated current of poly-N-vinylcarbazole measured with the above mentioned method. The curves show clearly single peak at about 5°C and this peak increases with the increase of the time of UV-irradiation at low temperature. This means that the amount of the trapped carriers increase with the increase of the trapping irradiation. When the polarity of the collecting field is reversed, i.e. in contrary to Fig. 4, the gold electrode is negatively biased, the current decreases appreciably. This shows that the detrapping of hole carriers are predominant in poly-N-vinylcarbazole.

As stated already in the introductory section, main feature of the carrier transport in the polymer is known to be the multi-trapping process of the hole carrier. So, the fundamental equation due to Haering and Adams (6) of the fast retrapping limit (equation (i)) will be used to analyze the results:

$$I(T)=I_0\exp[-\Delta E/kT - \alpha/b\int_{T_0}^{T}\exp(-\Delta E/kT)dT], \quad \alpha=N_v/N_t\tau \qquad (i)$$

On differentiating equation (i),

$$\exp[\Delta E/kT_m] = \alpha kT_m^2/b\Delta E \qquad (ii)$$

where, $I(T)$ is the thermally stimulated current at T, N_v effective state density of the valence band, N_t effective state density of the traps, b the heating rate °K/min., τ the recombination life time, T_m the temperature of maxium $I(T)$ and ΔE is the trap depth. As T_m and b are known, one can calculate α by equation (ii) for some trial value of ΔE. Using this α, b and the trial value of ΔE, one can calculate equation (i) and compare the result with the experiment. Thus, with the iteration of this procedure, we can find ΔE value which fits the experiment best. Fig. 6 shows a typical result.

In this way, the trap depth of the poly-N-vinylcarbazole was calculated as ΔE =0.56 eV and from the area of the 5°C peak the trap density was estimated to be of the order of 7×10^{15} cm^{-3}.

As to be noticed in Fig. 5, the 5°C peak shows considerable tailing to the lower temperature region. This means the presence of some amount of shallower trapps in the polymer. As shown by Fig. 7, the thermally stimulated current from these traps becomes somewhat clearer with slower heating rate. However, the density of these traps is too small to analyze. Pai and Patora show also the presence of such kind of traps (3), (4).

As shown in Figs. 2,5 and 7, the curves show large thermally stimulated current in higher temperature region also. It is difficult, however, to observe any maximum in this current and is impossible to analyze the trap parameters.

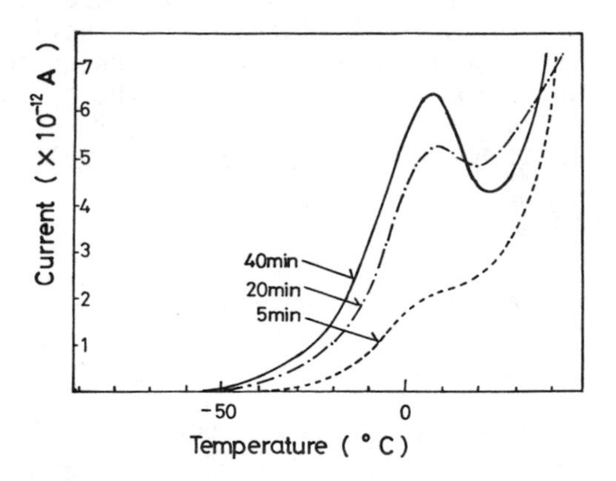

*Figure 5. Thermally stimulated current of poly-*N*-vinylcarbazole showing clear peak at 5°C. Peaks increase with the increase of UV illumination of the total light of the Hg lamp of Figure 3, at −150°C in vacuo 5–40 min. Collecting voltage 30V (Au +); heating rate 3.3°C/min. Films were cleaned prior to the measurement.*

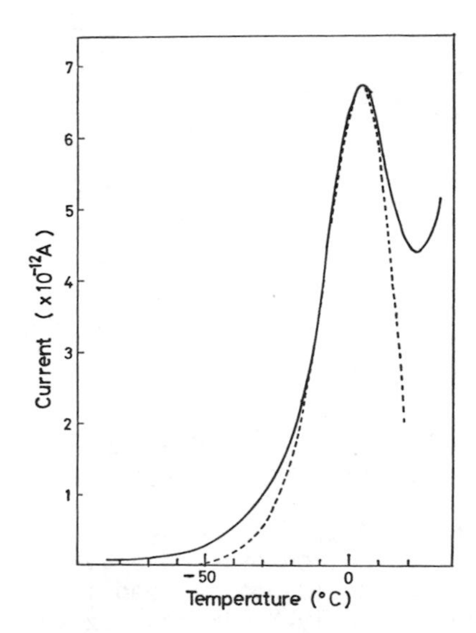

Figure 6. Typical experimental curve of the thermally stimulated current in poly-N-vinylcarbazole near 5°C peak (——) and the calculated theoretical curve with the trap depth $\Delta E = 0.56$ *eV (- - -)*

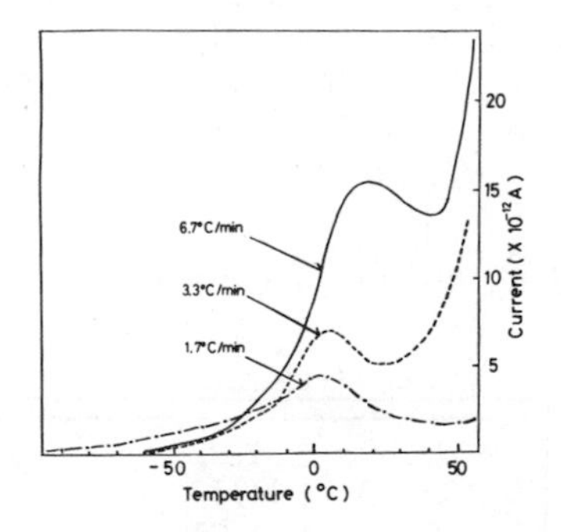

Figure 7. Thermally stimulated current in poly-N-vinylcarbazole with different heating rates (1.7–6.7°C/min; collecting voltage 30V (Au +)). Films were cleaned first and UV illuminated for 20 min at −150°C in vacuo with the total light of the Hg lamp of Figure 3.

As shown in Fig. 8, when the film is irradiated at room temperature in air by UV-light, thermally stimulated current at 5°C increases considerably. This result shows clearly the increase of the density of the same 0.56 eV traps.

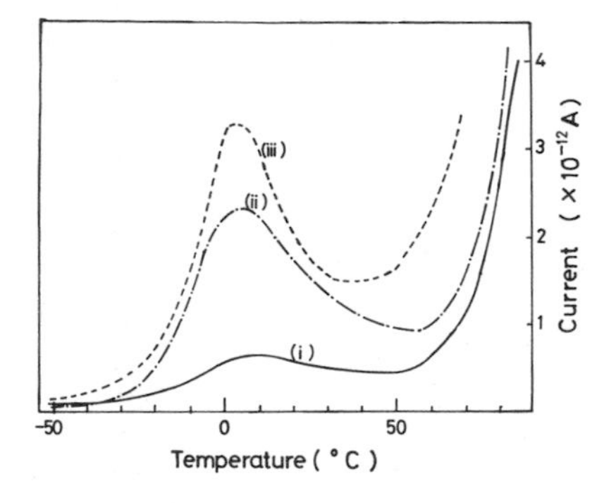

Figure 8. Thermally stimulated current in photooxidized poly-N-vinylcarbazole film. Films are (i) not illuminated; (ii) illuminated for 60 min; (iii) illuminated for 2 min with the 330-nm UV light in air at room temperature. In all cases, these photooxidized films were cleaned by heating to 100°C prior to the measurement, cooled to −150°C, and illuminated with the total light of the Hg lamp for 20 min in vacuo.

DISCUSSIONS

<u>On the methods of analysis of the thermally stimulated current; and the values of the trap depth in poly-N-vinyl-carbazole</u>

Various methods are proposed (5, a—g), (6). Among these, the methods in which monomolecular recombination or bimolecular recombination of the carriers are assumed could not be used in our case, because the carrier transport in poly-N-vinylcarbazole is known to be the multi-trapping process of the hole carrier. The values of the trap depth ΔE of 5°C peak by these several methods are summarized in Tab. 1. Values are widely scattered and it seemed that this is due to the approximations involved in the method of analysis. Our value is calculated by the

Table 1
Values of trap depth ΔE corresponding to the 5°C peak

Analysis by the method of	ΔE (eV)
Garlick-Gibson	0.39 ± 0.02
Haering-Adams	0.40 ± 0.12
Bube	0.54 ± 0.01
Luschik	0.32 ± 0.14
Halperin-Braner	0.68 ± 0.19
This work	0.56

With Booth's method, $\Delta E = 0 \sim 1.1$ eV;
Keating's method could not be applied, since γ value was not in $0.75 \sim 0.9$.

fundamental theoretical equation due to Haering and Adams in the limit of fast retrapping without any approximation in the calculation and is believed to be the best one now available.

According to Gill (7), the activation energy required for hole hopping depended on electric field and this energy was 0.65 eV when the collecting voltage was 30V. The average fluctuation amplitude of the energy levels of hopping sites in Seki's model (8) was 0.57 eV. The activation energy of hole reported by Pai (3) was 0.36 eV which did not depend on electric field and temperature. The values of Mort (9) and Regensburger (10) were also 0.4—0.7 eV. Further, according to the measurenent of the space-charge-limited currents in poly-N-vinylcarbazole film by Kato, Fujimoto and Mikawa, the trap depth was 0.5 eV without depending on the electric field (11). Since the carrier transport in poly-N-vinylcarbazole is known to be the multi-trapping process of the hole carrier, the activation energy in the above reports is thought to be the same with the trap depth which corresponds to the 5°C peak of the thermally stimulated current in the present studies.

On the nature of the traps

As well known, so-called excimer sites exist in poly-N-vinylcarbazole. It is well established that these excimer sites are the efficient traps for the singlet and triplet excitons, which migrate along the polymer chain. The structure of these sites are thought to be a special conformation having a pair of carbazolyl groups arranged parallel each other.

It seems to be natural to suspect if these excimer forming sites were the effective trapping center also for hole carriers. If this were the case, the hole carriers would be trapped in excimer sites as dimer cation radical state having lower energy than the isolated cation radical state (free hole). However, at least for the 0.56 eV trap of the 5°C peak of the thermally stimulated current curve, this possibility is ruled out, because the excimer population in poly-N-vinylcarbazole is so much as one per several hundred carbazolyl grops (12), while that of 0.56 eV trap is so low as 7×10^{15} /cm^3. As evidently shown in several figures in the preceeding sections, poly-N-vinylcarbazole shows besides 5°C peak large thermally stimulated current in higher temperature region. As we have no information on the trap depth and the density of this higher temperature current, there remains a possibility of these traps being due to excimer forming sites in the polymer.

As already stated and shown by Fig. 8, when the film is irradiated at room temperature in air by UV-light prior to the measurement of thermally stimulated current, the current peak at 5°C increased considerably, i.e. the population of the 0.56 eV traps increased by UV-irradiation in air. The photocarriers generated in poly-N-vinylcarbazole with the excitation in its lowest $\pi-\pi^*$ absorption region are mainly holes and the electrons are deeply trapped and immobilized by the electron accepting impurities. Photo-oxidation product of poly-N-vinylcarbazole was suggested by Pfister and Williams (13) as the electron accepting impurity and they suggested the formation of the compounds having structures like

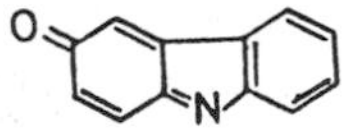

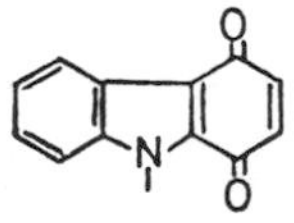

Recently, Itaya, Okamoto and Kusabayashi (14) made an estimation that the concentration of the electron-accepting photo-oxidation product of about 10^{-3} mol/mol monomer unit is necessary to interpret the high yield of the photocarrier generation of ca. 0.1 of the photo-oxidized poly-N-vinylcarbazole film. The fact that the UV-irradiation of the film in air at room temperature increase the 0.56 eV trap density shows close connection between photo-oxidation product and this trap. The amount of 0.56 eV trap is , however, too small when compared with the estimated amount of the electron accepting photo-oxidation product. Moreover, from the organic chemical view, it seems to be difficult to conceive the electron accepting moiety as hole trapping structure. Some kind of molecules such as for example water might be formed by photo-oxidation and function as 0.56 eV trap.

Literature Cited

(1) cf. Paper by M. Yokoyama, Y. Endo, A. Matsubara and H. Mikawa contributed to the Symposium 26-e, Photo- and Radiation Chemistry in Polymer Science, ACS/CSJ Chemical Congress, Hawaii, April 1—6, 1979. M. Yokoyama, Y. Endo and H. Mikawa, Bull. Chem. Soc. Japan, 49, 1538 (1976).
(2) H. Scher and E. W. Montroll, Phys. Rev., B., 12, 2455 (1975).
(3) D. M. Pai, J. Chem. Phys., 52, 2285 (1970).
(4) J. Patora, J. Piotrowski, K. Kryszewski and A. Szymanski, Polym. Lett., 10, 23 (1972).
(5) a) L. J. Grossweiner, J. Appl. Phys., 24, 1306 (1953);
b) R. H. Bube, J. Chem. Phys., 23, 18 (1955); R. H. Bube, Photoconductivity of Solids, John Wiley and Sous Inc., New York, 1960, p. 46-55, 292-296;
c) K. H. Nicholas and J. Woods, J. Appl. Phys., 15, 783 (1964);
d) P. N. Keating, Proc. Phys. Soc., 78, 1408 (1961);
e) A. H. Booth, Can. J. Chem., 32, 214 (1954);
f) A. Halperin and A. A. Braner, Phys. Rev., 117, 408 (1960);
g) G. F. J. Garlick and A. F. Gibson, Proc. Roy. Soc. (London), A60, 574 (1948).

(6) R. H. Haering and E. N. Adams, Phys. Rev., 117, 451 (1960).
(7) W. D. Gill, J. Appl. Phys., 43, 5033 (1972).
(8) H. Seki, Amorphous and Liquid Semiconductors, ed. by J.Stuke and W. Brenig, Taylor and Francis, 1974, p. 1015.
(9) J. Mort, Phys. Rev. B, 5, 3329 (1972).
(10) P. J. Regensburger, Photochem. Photobiol., 8, 429 (1968).
(11) K. Kato, T. Fujimoto and H. Mikawa, Chem. Lett., 63 (1975).
(12) W. Klöpffer, J. Chem. Phys., 50, 2337 (1969); K. Okamoto, A. Yano, S. Kusabayashi and H. Mikawa, Bull. Chem. Soc. Japan 47, 749 (1974).
(13) G. Pfister and D. J. Williams, J. Chem. Phys., 61, 2416 (1974).
(14) A. Itaya, K. Okamoto and S. Kusabayashi, Bull. Chem. Soc. Japan 52, in press (1979).

RECEIVED July 12, 1979.

15

A Novel Modification of Polymer Surfaces by Photografting

SHIGEO TAZUKE and TAKAO MATOBA—Research Laboratory of Resources Utilization, Tokyo Institute of Technology, 4259 Nagatsuta, Midori-ku, Yokohama, Japan

HITOSHI KIMURA—DIT Laboratory, Kansai Paint Co., Ltd., Hiratsuka, Kanagawa, Japan

TAKESHI OKADA—Ibaraki Electrical Communication Laboratory, Nippon Telegraph and Telephone Public Corporation, Tokai-Mura, Ibaraki, Japan

When any polymer is to be used as film, plate, fiber, or molded material, the surface properties are as important as the bulk properties. In comparison with the large number of works devoted to the development of new polymers, relatively minor efforts have been directed to the modification of polymer surface. In particular, owing to the difficulties of studying chemical and physical properties of polymer surface, few articles have been published on the correlation between the condition of surface treatments and the imparted surface properties.

The known techniques of surface treatments(1) are i) corona discharge, ii) surface degradation or oxidation by oxidizing agents such as chromic acid and others, iii) plasma treatment and/or plasma polymerization, iv) graft polymerization, and v) coating. Among these, i) and v) are most widely used processes because of their excellent workability and low cost of operation whereas duability of treated surface is often unsatisfactory. Since i) is essentially the oxidation of polymer surface, the imparted functional groups on polymer surface are limited to oxygen containing polar groups and the effects are not sufficient for many purposes. Furthermore, the effects of corona discharge treatment gradually dissipate during storage. The process v) can provide various surface properties whereas the coated layer is not strong enough unless chemically bound to the base polymer. These processes would be positioned as low grade treatments for general purposes. The wet process ii) is less important than i) owing to its poor workability. Use of plasma is a promising technology to enable extremely thin layer surface modification. A drawback is to operate under reduced pressure. Graft polymerization initiated by high energy irradiation, photo-irradiation, or catalytic processes has been well documented. However, the aim of graft polymerization has not been oriented to surface modification. Most of published results deal with very high graft yield and consequently the bulk properties of the base polymer are altered by grafting. In principle, graft polymerization is an attractive method to impart a variety of surface properties on the condition

0-8412-0540-X/80/47-121-217$06.25/0

that a very thin graft layer can be produced

Among these grafting processes, photochemical reactions are best suited for surface grafting. The reasons are as follows. i)Photochemically produced triplet states of carbonyl compounds abstract hydrogen atoms from almost all polymers so that graft polymerization can be initiated. ii)High concentration of active species can be locally produced at the interface between the base polymer and the reacting solution containing sensitizer when photoirradiation is applied through the base polymer film. This condition could not be achieved either by thermal or radiation chemical initiation. Selective excitaiton is a big merit of photografting. Energy absorption of ionizing radiation is determined by the total number of electrons, but not by functional groups. iii)In comparison with radiation grafting, photochemical processes are much simpler with respect to engineering. In addition, the cost of energy source is cheaper for photochemistry.

In spite of these expected advantages of photografting as a surface modification process, most of studies have been directed to achieve high graft yield and high efficiency(2 - 13). An exception is the gas phase photografting under high vacuum, which enabled the use of various monomers inert to ordinary solution polymerization(14). We thus tried to develop a workable surface photografting technique. The basic concepts are as follows. i)interactions between base polymer and reacting solution is controlled so as to produce very thin graft layer. ii)Benzophenone(BP) or other triplet sensitizers are in solution but not in the base polymer so that any commercial polymers can be treated. iii)When the base polymer is in a form of transparent film, photoirradiation is applied through the base polymer so that the highest energy absorption and consequently the highest active spacise concentration are expected on the solid-liquid interphase. iv)Acrylamide and other water soluble monomers are used to impart hydrophilicity to the base polymer surface.

The prototype of the present surface photografting is the surface photoreaction of maleic anhydride(MAH) onto poly(butadiene) film(15). Although fair improvement of surface wettability was achieved, photoaddition of MAH cannot be applied to other polymers having no ethylenic double bonds. The present process is applicable to almost all polymers except for poly(tetrafluoroethylene) and its analogues.

Experimental

The experimental procedures were essentially the same as reported (Figure 1, (16, 17)). Reagents other than those described in (16, 17) were as follows. Acrylic acid(Wako Pure Chemicals) was distilled once under reduced pressure under a nitrogen stream. Solvents(acetonitrile, n-hexane, and benzene) were purified by accepted procedures.

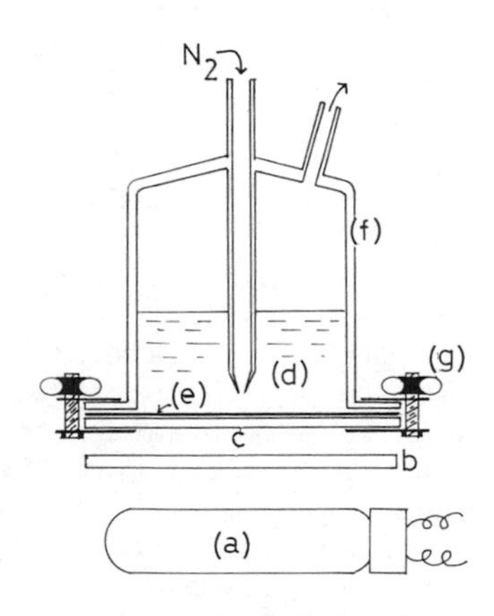

Figure 1. Surface photografting apparatus: (a) 100W high-pressure Hg lamp; (b) interference filter; (c) borosilicate glass plate; (d) reacting solution; (e) polymer film; (f) glass vessel; (g) clamp screw(17)

A Hitachi EPI-S2 infrared spectrometer, a contact angle meter (Kyowa Kagaku Co. Type CA-P) and an ESCA spectrometer(Kokusai Denki Co.) were used for relevant measurements.

Results and Discussion

Surface Photografting onto Various Polymers. The results of surface photografting onto various polymers are shown in Table 1 (16). The graft yield is so small that the weight increase of treated polymer film cannot be detected gravimetrically. However, the carbonyl absorption around 1600-1700cm^{-1} and improvement of wettability clearly indicate the formation of distinct graft layer. Furthermore, as the durability tests in Table 2(16) manifest that the graft layer is strongly adhered to the base polymer. Any solvent can hardly remove the graft layer. Weathering resistance seems to depend much on the nature of base polymer. polypropylene(OPP) film is known for its poor weathering resistance and the loss of graft layer upon weathering test will be a result of base polymer degradation. Exposure test of surface grafted polyethylene(LDPE) film is now under way. Good antifogging property is sustained after one year out-door exposure under natural conditions. The accelerated exposure test predicts at least two years service life for the antifogging LDPE film grafted with acrylamide (AM).

Another extreme case is surface grafting on to plasticized poly(vinyl chloride) film which is widely used for green house and covering. Although the grafting is easy, the graft layer is physically unstable and does not seem to stay on the surface. Since T_g of plasticized PVC is lower than room temperature, polymer molecules would migrate together with the graft layer. The excellent wettability right after surface grafting is lost during outdoor exposure test within few weeks. Also, the surface graft layer is lost by heating the sample in an oven above 50°C. The stability of graft layer is thus obviously related to the base polymer properties.

Unstability of surface graft layer was recently discussed by Hoffman(18) for radiation-grafted hydrogels consisting of poly(dimethylsiloxane), poly(ester-urethane) and polyethylene grafted with 2-hydroxyethyl methacrylate(HEMA), AM and ethyl methacrylate. The surface stability during dehydration treatment depends strongly on the base polymer properties. When HEMA or AM is grafted onto the silicone rubber, the graft polymers locate on the surface under hydrated condition whereas they migrate under the surface during dehydration. Once the graft polymers are barried in the base polymer, it is not easy to restore a hydrophilic surface by re-hydration. On the other hand, polyethylene has a rigid surface and the graft layer stays on the surface regardless of the monomer used or the de-hydration treatment.

Except for soft PVC, we have not found confirmative evidence indicating unstability of graft layer. Nevertheless, the contact

Table 1. Examples of Surface Photografting (16)

polymer film	monomer	solvent	light source	reaction time(hr)	IR absorbance of >C=O[a]	contact angle of water: untreated film(degree)	grafted film(degree)
poly(vinyl chloride)	acrylamide(4.5M)	methanol	I	1	0.6 - 0.7[b]	92	60
poly(vinylidene chloride)			I	1	0.20	75	57
cellulose triacetate			I	1	c	55	60
1,2-polybutadiene	acrylamide(2.0M)	acetone	I	1	d	94	57
low density polyethylene			I	1	0.7 - 0.8	90	45
oriented polypropylene			I	1	0.67	101	40
	N-vinylpyrrolidone (2.0M)		II	4	0.20		not determined.
	methacrylic acid(2.0M)		II	3	0.11		

a 1660 cm^{-1} for acrylamide and N-vinylpyrrolidone, 1700 cm^{-1} for methacrylic acid.
b Plasticizers and additives were extracted after grafting.
c IR spectroscopy was not possible due to the strong background absorption.
d Excessively grafted, absorbance > 1.

Light source. I: a 200-W medium-pressure Hg lamp 2X; irradiation through a glass plate (λ>300 nm). II: a 100-W high-pressure Hg lamp; monochromatic irradiation at 366 nm, 20 W hr/m^2.

Photografting procedure. The monomer solution containing 0.2M of benzophenone was placed in contact with polymer film. The reaction system was deoxygenated by bubbling nitrogen and then irradiated through the polymer film at room temperature. The irradiated film was washed with water and the solvent successively and then immersed in water overnight, giving a transparent film with smooth surface.

Journal of Polymer Science, Polymer Letters Edition

Table 2 Durability Test of Grafted Polymer Surface.
Sample: Polypropylene Film Photografted with Acrylamide (16)

	IR absorbance of amide initial	IR absorbance of amide after test	change in IR absorbance(%)
immersion in acetone, 24hr	0.863	0.848	-1.7
immersion in DMF, 24hr	0.787	0.848	+7.8[a]
boiling in water, 15hr	0.646	0.590	-8.7
abrasion resistance (dropping sand methode) 100g	0.775	0.766	-1.2
400g	0.775	0.731	-5.7
1,500g	0.775	0.663	-14.5
weathering test (sunshine weather-o-meter[c]) 24hr	0.769	0.686	-10.8
65hr	0.769	0.507	-34.1
145hr	0.769	0.317	-58.8

[a]Due to absorbed DMF. [b]ASTM D-969. [c]ASTM G 23

Journal of Polymer Science, Polymer Letters Edition

angle measurement is not always highly reproducible by several degrees, indicating that the surface structure is sensitive to the history of samples, temperature, humidity and others. Water content on surface may be an important factor determining wettability. For further discussions, advancing and receding contact angles should be determined.

When a fixed monomer is used, the grafting rate and the properties of grafted surface depend very much on the kind of solvent. The requirements for appropriate solvents are as follows. 1)The solvent is to be non-solvent of base polymer. In order to obtain a thin graft layer, the solvent must not swell the base polymer excessively. Slight but definitive interactions are however, necessary to provide reaction sites for grafting. This control is delicate and will be discussed later. ii)Although the growing graft chain is not necessarily soluble in the solvent, good solvent-growing chain interactions are advantageous to assist the propagation of graft chain outside the base polymer surface. iii)The solvent has to be inert to the triplet excited state of sensitizer. This is particularly important for base polymers which are not much susceptible to radical attack. OPP and LDPE could not be surface grafted from methanol solution whereas PVC and poly(vinylidene chloride) could.

Choice of sensitizer is related to the base polymer properties. Besides triplet sensitizers, dyes(4, 5), metal salts(11, 12), and radical initiators(7, 11, 12) have been used in photografting. These sensitizers other than triplet sensitizers are, however, not capable of initiating surface photografting onto polyolefins. Although benzoin isopropyl ether has been used for photografting of polypropylene, the reaction conditions seem to be in favor of deep grafting(12).

Among the polymers in Table 1, OPP and LDPE are least active polymers. The low reactivity is not only attributable to the chemical structures having no functional groups but also stemmed from high crystallinity in the case of OPP which would prevent the penetration of reacting solution into the base polymer. Comparing OPP film with polypropylene plate which is not oriented, the later is more reactive. For surface photografting onto unreactive OPP film, the reaction under unaerobic condition is absolutely necessary using BP as a sensitizer. Internal(deep) grafting using swelling solvents, however, seems to be less susceptible to oxygen. In the following sections, detailed studies on surface photografting onto OPP film are described.

Time-Conversion Profile of a Surface Photografting System, OPP - AM - BP in acetone. As discussed above, there are many complicated factors controlling the mode of surface photografting. At first, concentration effects of sensitizer(BP) and monomer(AM) were studied. The energy input/graft yield curves are shown in Figures 2 and 3. Being a radical process, this graft polymerization is retarded by oxygen. Besides the radical scavenging action of oxygen, deactivation of BP^{*3} by oxygen(Eq.(1)) has to be taken into account. The present photografting system is much more sensitive to air than photografting to PVC or cellulose triacetate which are easily attacked by radicals, indicating oxygen effects in the initiation process.

$$BP^{*3} + O_2^3 \rightarrow BP + O_2^{*1} \qquad (1)$$

The S shaped curves in Figures 2 and 3 are explained as follows. The initial slow reaction is considered as an induction period or an acceleration period. A possibility is inhibition by oxygen, particularly that in the base polymer. Although nitrogen bubbling will be sufficient to remove oxygen from solution, it is more difficult to remove oxygen in polymer film. Prolonged nitrogen bubbling does not shorten the induction period. When the reaction system is intermittently irradiated and air is once allowed to enter the reaction vessel and is again removed during the dark period, graft polymerization does not proceed even if the total irradiation time exceeds the induction period observed in the continuously irradiated system. This result indicates oxygen to be responsible for the induction period, at least in part. The autoacceleration phenomena could also be brought about by the

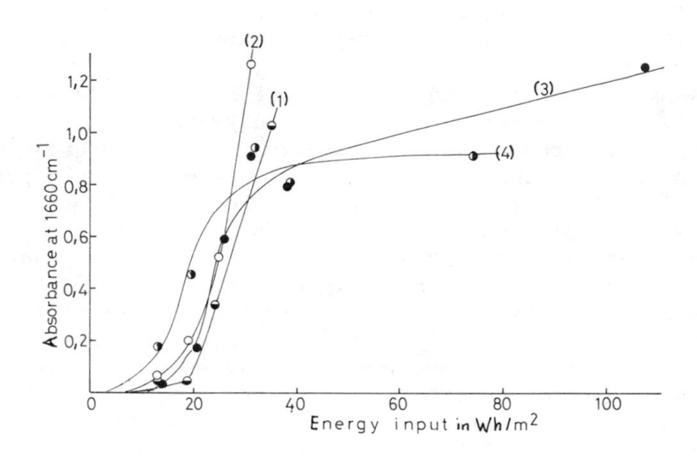

Die Makromolekulare Chemie

Figure 2. Effect of [BP] on grafting: [AM] = 2.00M in acetone; [BP] = 0.05M (1), 0.10M (2), 0.20M (3), 0.40M (4)(17)

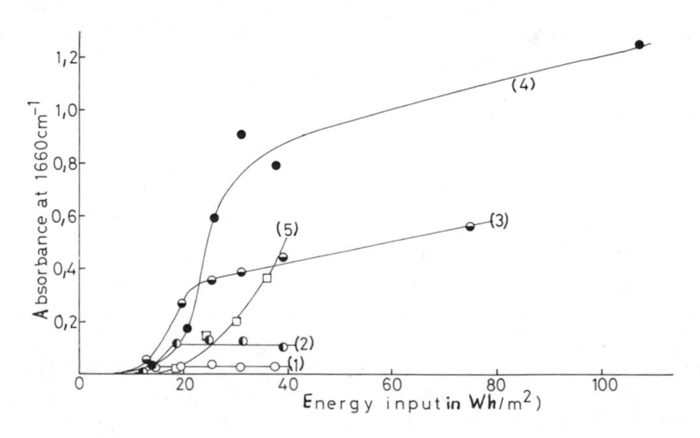

Die Makromolekulare Chemie

Figure 3. Effect of [AM] on grafting: [BP] = 0.20M in acetone; [AM] = 0.50M (1), 1.00M (2), 1.50M (3), 2.00M (4), for (5): [AM] = 1.00M, [BP] = 0.05M (17)

gradual change in the surface properties with the progression of graft polymerization. The increased wettability of the OPP film by the reaction mixture will enhance the rate of further grafting. When OPP film which was once photografted to the absorbance of ca. 0.1 is subject to re-grafting, the second grafting is faster than the first. Consequently, the initial surface properties of the OPP film may be of great importance. Since OPP is easily oxidized to give carbonyl and/or hydroxyl groups, the ease of surface photografting would reflect the history of the OPP film such as the conditions of casting film and the storage period. As shown in Figure 5(A), the surface of the present OPP film before grafting is slightly oxidized. Such impurity sites may participate in radical generation.

After the induction period, the energy input/graft yield curve rises sharply and then levels off depending upon the reaction conditions. The final stage is explainable as a result of consumption of active sites on polymer surface. Under the condition of grafting from acetone solution, the thickness of graft layer is estimated to be in the order of $10^{-1}\mu$ assuming complete coverage of surface by polyAM and the molar extinction coefficient of amide carbonyl at $1660cm^{-1}$ to be $10^2 \sim 10^3$. Furthermore, grafting is considered to be initiated only by the direct attack by BP^{*3} generated at the solid-liquid boundary region to the tertiary hydrogens of OPP film. Consequently, when either active hydrogens on OPP film or BP have been consumed, the graft yield will level off. In a somewhat different system, dead end graft polymerization was observed(8). When nylon-6 film doped with BP from vapor phase onto which methacrylic acid was absorbed was irradiated, BP was rapidly consumed and the rate of grafting decreased with irradiation period. In the present system, there is abundant amount of BP in solution and BP will be quickly supplied to the reaction site. A separate experiment showed that uptake of BP by OPP film from acetone solution is quick enough although the amount is small (see later section). We therefore consider the leveling off phenomena to be attributable to the consumption of reactive sites on OPP film.

Along the discussion developed above, the shapes of time-conversion curves in Figures 2 and 3 are qualitatively interpreted. In Figure 2 where [AM] is constant and [BP] is variable, the induction period decreases with increasing [BP]. On the other hand high graft yields are obtained when [BP] is reduced. With increasing [BP], the rate of production of initiating active species increases and consequently the kinetic chain length decreases. Under such conditions, the active sites in the polymer surface will be rapidly consumed and short polyAM chains would be densely grafted. It results in the saturation phenomenon of graft yield as shown by curves 3 and 4 of Figure 2. Curves 1 and 2 of Figure 2 would represent long polyAM chains which are thinly populated on the polymer surface.

Figure 3 shows the dependence of graft polymerization on

[AM] at constant [BP]. As anticipated from the constant rate of active spacies production, the length of induction period is nearly identical for all runs whereas the saturation level of graft yield increases with [AM]. Thus, concentration of AM and BP should be balanced to achieve efficient surface graft polymerization.

In the present discussions, we neglect the effects of homopolymer formation. HomopolyAM deposited on OPP surface in the course of grafting may prevent diffusion of AM to the grafting site, which may slow down the rate of grafting. It is however difficult to explain by this reasoning the change in time-conversion profile as a function of [AM]/[BP] ratio. Since our aim is surface modification, the amount of homopolymer was not determined.

Also we have not considered the possibility of grafting on grafted polyAM chains. If this could happen efficiently, the graft layer should have grown much thicker. Relevant to this problem, grafting onto poly(ethylene-g-maleimide) was not observed in photoinduced graft polymerization of maleimide to polyethylene (10).

Effect of Sensitizers. Figure 4 shows the effect of sensitizers on graft polymerization. Methyl 2-benzoylbenzoate is as efficient as BP, except for a slightly longer induction period, whereas 4-bromobenzophenone is totally inefficient. Homopolymerization alone was induced with the latter sensitizer. The triplet energy(ET) of benzophenone derivatives lies between 65 and 70 Kcal/mol in general(19) and the transition is n,π^{*3} for both BP and 4-bromobenzophenone. They are equally capable of abstracting a hydrogen atom from 2-propanol with a quantum yield of 1(20). We have no immediate explanation for the inefficiency of 4-bromobenzophenone.

Fluorenone has a lower initiation efficiency than BP although its absorbance is higher(BP: ε_{313}=140 and ε_{366}=50-70 $mol^{-1}cm^{-1}$; fluorenone: ε_{313}=1300 and ε_{366}=200 $mol^{-1}cm^{-1}$)(19). A recent theory (21) of hydrogen abstraction by the n,π^{*3} state of carbonyl compounds treated as a radiationless transition claims that the rates are dependent on the electronic energy, the vibrational frequencies, the reduced mass of the oscillators, the C-H bond strength of the hydrogen donor, and the bond distances. In cases where a charge transfer mechanism is involved, the carbonyl compound reduction potential and the hydrogen donor ionization energy are also rate determining factors. Excellent agreements between observed and calculated rate constants were reported(21, 22) for a variety of hydrogen abstraction reactions. In the present case where the hydrogen donor is OPP and contribution of charge transfer interaction is unlikely, an increase in E_T of sensitizer is expected to enhance the rate of hydrogen abstraction. In this context, the low efficiency of fluorenone having much lower E_T(53 kcal/mol (23)) than BP derivatives is under-

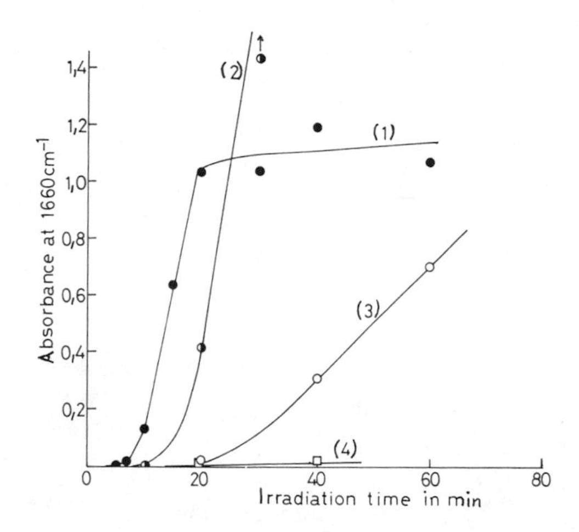

Die Makromolekulare Chemie

Figure 4. Photografting with various sensitizers: [AM] = 2.00M, [sensitizer] = 0.20M in acetone; light source: two 200W medium-pressure Hg lamps with a borosilicate glass filter; sensitizer: BP(1), methyl 2-benzoylbenzoate(2), 9-fluorenone(3), 4-bromobenzophenone(4) (17)

standable.

In the absence of sensitizer, neither graft nor homopolymerization was induced by irradiation at 366nm. However, irradiation by a low pressure mercury lamp through a quartz plate(the glass plate(c) in Figure 1 was replaced by a quartz plate.) induces slow surface grafting sensitized by acetone.

Combination of BP with 2-propanol or amines induces homopolymerization alone. The rate constants of BP^{*3} - isopropylamine and triethylamine are $2.95\ 10^8$ and $2.42\ 10^9 M^{-1}s^{-1}$, respectively(22) whereas that of BP^{*3} - isooctane as a model of OPP is $1.0\ 10^5 M^{-1} s^{-1}$ (24). Also hydrogen abstraction from 2-propanol($k=1.0\ 10^6\ M^{-1} s^{-1}$) (25) is much more efficient than that from aliphatic hydrocarbons. Even methanol is more reactive($k=2.8\ 10^5 M^{-1}s^{-1}$)(25) than OPP towards BP^{*3}. The aforementioned results and the finding that surface grafting does not occur in methanol are well interpreted by the following elementary reactions.

$$BP \xrightarrow{h\nu} BP^{*1} \longrightarrow BP^{*3} \quad (2)$$

$$BP^{*3} + RH \text{ or } AM \longrightarrow \longrightarrow \xrightarrow{AM} \longrightarrow P_H\cdot \quad (3)$$

$$BP^{*3} + \text{-}(CH_2\underset{CH_3}{CH})_n\text{-} \longrightarrow \underset{\text{I}}{Ph_2\underset{OH}{C}\cdot} + \underset{\text{II}}{\text{-}(CH_2\underset{CH_3}{\dot{C}})_n\text{-}} \quad (4)$$

$$\text{I} + AM \longrightarrow P_H\cdot \quad (5)$$

$$\text{II} + AM \longrightarrow P_G\cdot \quad (6)$$

$$P_H\cdot + P_H\cdot \text{ or I} \longrightarrow \text{homopolyAM} \quad (7)$$

$$\left.\begin{array}{l} P_H\cdot + \text{II} \\ P_G\cdot + P_H\cdot\text{, I, or II} \end{array}\right\} \text{grafted polyAM} \quad (8)$$

RH: solvent or impurities with active hydrogen atoms
Suffixes "H" and "G" represent homo and graft propagating chains, respectively.

Dependence of Surface Properties on Graft Yield. Properties of grafted surface are functions of the amount and distribution of graft polymer. We will first discuss the surface property - graft yield correlation while the reaction system is fixed to the OPP-AM-BP-acetone combination. The IR absorbance is a measure for the total amount of polyAM in the grafted film whereas surface density of polyAM may not run parallel with total graft yield. We determined the surface density of polyAM by ESCA measurements. Although there is a limitation of accuracy if one intends to achieve an absolute surface analysis, ESCA is very useful for surface analysis on a relative scale. Examples of

ESCA spectra before and after surface grafting are shown in Figure 5. Taking the C_{1S} peak intensity as the standard, the relative amounts of $C_{1S(C=O)}$, N_{1S}, and O_{1S} were calculated and plotted against the IR absorbance in Figure 6. Since the C_{1S} intensity is attributed to both OPP and the main chain of polyAM, the relative intensities of $C_{1S(C=O)}$, N_{1S}, and O_{1S} should represent the coverage of the film surface by polyAM. It is apparent that the ESCA intensities of polyAM increase sharply during the early stage of grafting but soon level off whereas the IR absorbance increases continuously.

As a measure of surface properties, we chose contact angle data. The plots of contact angle against IR absorbance shown in Figure 7 indicate that the values of contact angle level off at IR absorbance of 0.2, which agrees with the leveling off point of ESCA intensities. These results suggest that the coverage of film surface by polyAM is completed at an IR absorbance of 0.2. Further grafting increases the thickness of grafted layer whereas the surface properties remain unchanged beyond this threshold graft yield. For the purpose of surface modification, heavy grafting is unnecessary indeed!

The surface properties as studied by contact angle measurement are not affected much by the composition of reacting solution so far the solvent is the same. Wettability of all samples plotted in Figures 2 and 3 depends merely on the total amount of graft polymer and the contact angle - IR absorbance plots fall on the same line as shown in Figure 7.

Solvent Effects on Grafting Rate and Surface Properties. Apart from the chemical effects of solvent as hydrogen donor, being heterogeneous systems, an important role of solvent is to provide appropriate reaction sites on the balance of solvent interactions with base polymer and growing graft chain. For the study of solvent effects, AM is not an adequate monomer since it is scarcely soluble in non-polar solvents. Instead, acrylic acid(AA) was employed and photografting was conducted as shown in Figure 8.

Solvent effects on grafting are rather drastic. Being a non-polar polymer, OPP film has higher affinity to n-hexane or benzene than to acetonitrile. Acetone is in the middle. As a matter of consequence, the polymer surface will be slightly swollen in contact with these non-polar solvents and deep grafting is expected. This indicates the volume of base polymer available for graft polmerization to be lager and the rate tends to be faster. On the contrary, acetonitrile cannot provide grafting site since the solvent is too incompatible with OPP film. Acetone shows an intermediate behavior. The results suggest the possibility of controlling the thickness of graft layer by controlling solvent - base polymer compatibility.

A general trend of the shape of time - conversion plots is disappearance of induction period when non-polar solvents are

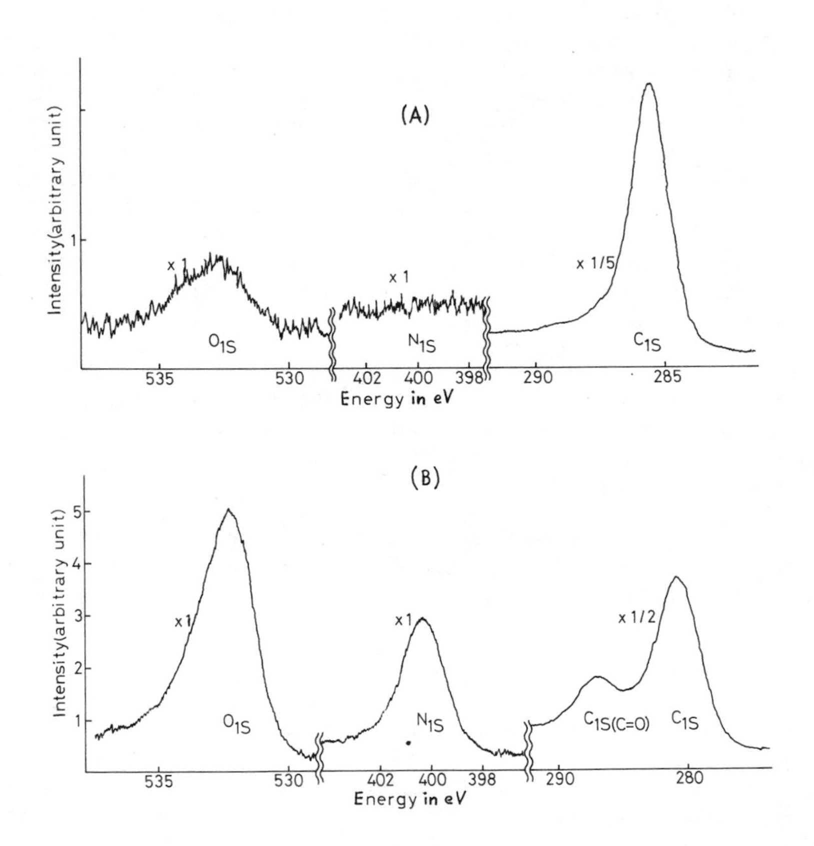

Die Makromolekulare Chemie

Figure 5. ESCA measurements: (A) untreated OPP film; (B) OPP film after surface grafting with AM (17)

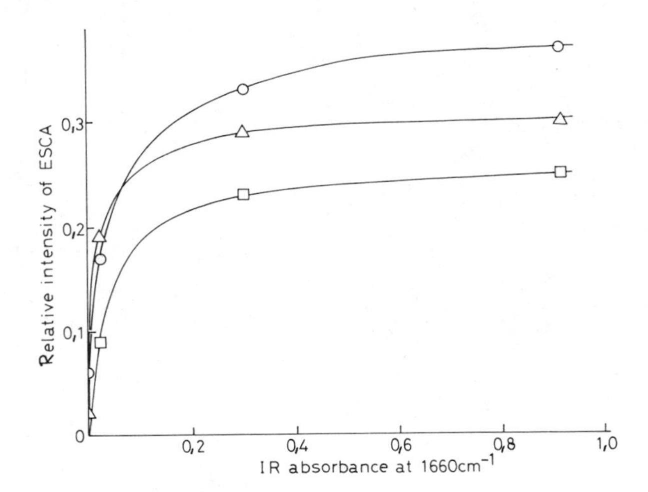

Die Makromolekulare Chemie

Figure 6. Plots of ESCA intensity vs. IR absorbance. The ESCA intensities are expressed by the peak height of the specific atoms relative to C_{1S} intensity. Sensitivities of C_{1S}, O_{1S}, and N_{1S} are assumed to be 0.25, 0.60, and 0.40, respectively (and standardized). (○) $C_{1S(C=O)}$; (□) N_{1S}; (△) O_{1S} (17)

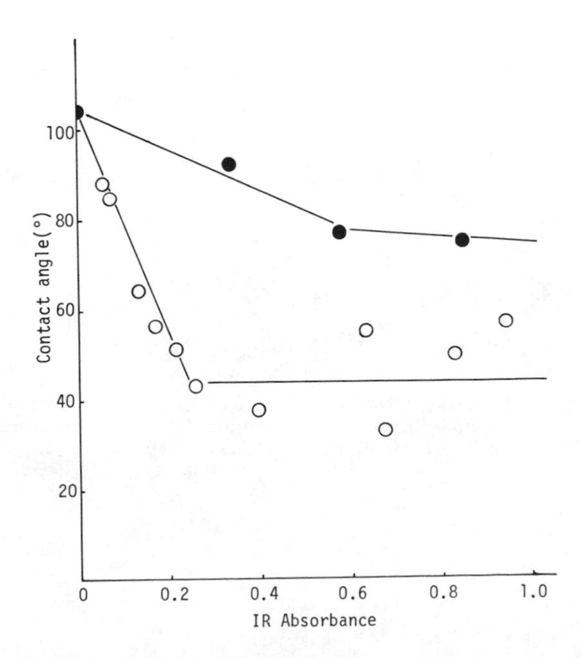

Figure 7. Plots of contact angle of OPP film to water vs. IR absorbance: [AM] = 2.00M, [BP] = 0.20M. Solvent for grafting: (○) acetone, (●) acetone/n-hexane-(7/3).

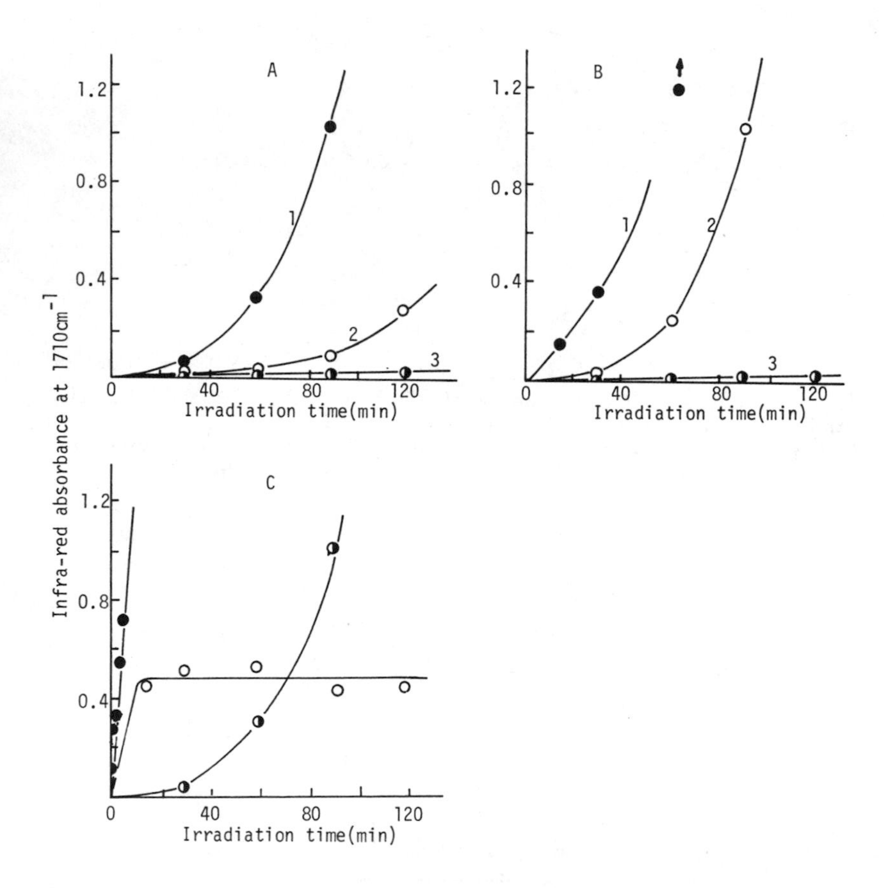

*Figure 8. Solvent effects on grafting of AA onto OPP film: [AA] = 2.00*M*, [BP] = 0.20*M*. (A) (1) acetone; (2) acetone/acetonitrile(3/1); (3) acetonitrile; (B) (1) benzene; (2) benzene/acetonitrile(1/1); (3) acetonitrile; (C) (1)* n*-hexane; (2)* n*-hexane/acetone(1/1); (3) acetone.*

used. As the rate of BP uptake by OPP film as shown in Figure 9 manifests, BP is rapidly transferred to OPP film from n-hexane solution, but not from acetonitrile solution, reflecting the degree of solvent - base polymer interaction. The high initial grafting rate in n-hexane will then be attributed to abundant BP in the solution - OPP boundary.

From the comparison between Figure 8 (a) and (c), the induction period in (a) could not be attributed to oxygen effect alone. Probably in the case of (c), the rate of grafting would decrease with increasing graft yield if the reaction could be followed up to a higher conversion, since n-hexane - OPP interaction must be stronger than n-hexane - poly(propylene-g-AA) interaction. The behavior of n-hexane/acetone mixed solvent systems is a reasonable consequence of balancing two opposing factors. One is to reduce the thickness of graft layer with decreasing solvent - OPP interaction when acetone content is increased. Another is to increase solvent - grafted OPP interaction with increasing acetone content. Consequently, if a part of OPP film is convered with polyAA, subsequent grafting is facilitated in acetone rich systems but suppressed in n-hexane rich systems.

The same trends of rate dependence on solvent polarity are observed for AM systems as shown in Figures 10 and 11. Other relevant results are given in Table 3.

Table 3. Solvent Effects on Photografting to OPP Film.

solvent	sensitizer	IR absorbance	contact angle(°)
acetone	BP	1.00	38 ± 12
acetone	BIPE	0	99.0
THF	BP	0.70	76.5
THF	BIPE	0.42	102.3

[AM] = 2.00M, [sensitizer] = 0.20M, 90 min irradition.
BIPE: benzoin isopropyl ether

These peculiar solvent effects are often found in heterogeneous graft polymerization. An example is radiation or photo-induced grafting of styrene on to polyethylene and cellulose in which the maximum graft yield has been reported at a certain monomer/ solvent ratio depending upon the natures of solvent and the method of initiation(12). This phenomenon was explained by assuming Trommsdorf effect. If the solvent does not dissolve polystyryl growing chain, the explanation may be acceptable. However, when a good solvent for polystyrene is used, other factors such as change in solution - base polymer, solution - grafted base polymer, and solution - growing graft chain interaction are to be taken into account as functions of monomer/solvent mixing ratio. Analysis of graft yield - reaction time profile will provide key information.

The reaction in THF requires comments. In THF, the reaction may not be true grafting. THF is known to be very susceptible to hydrogen abstraction and furthermore, benzoin isoproply ether

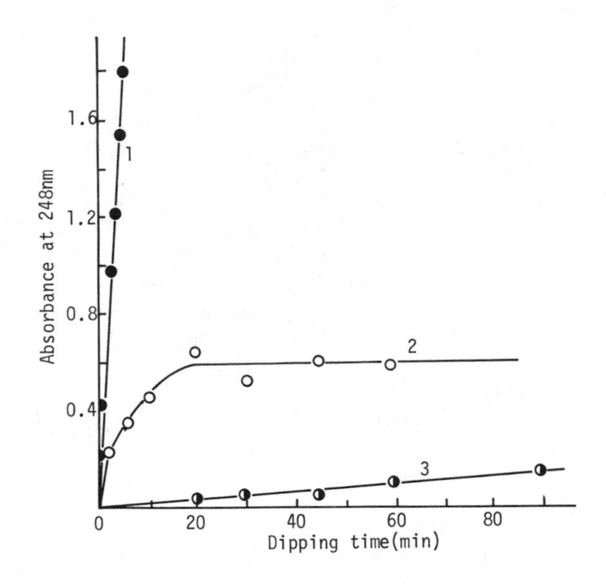

Figure 9. Transfer of BP from various solvent systems to OPP film: [BP] = 0.20M in (1) n-*hexane, (2)* n-*hexane/acetone(1/1), (3) acetone*

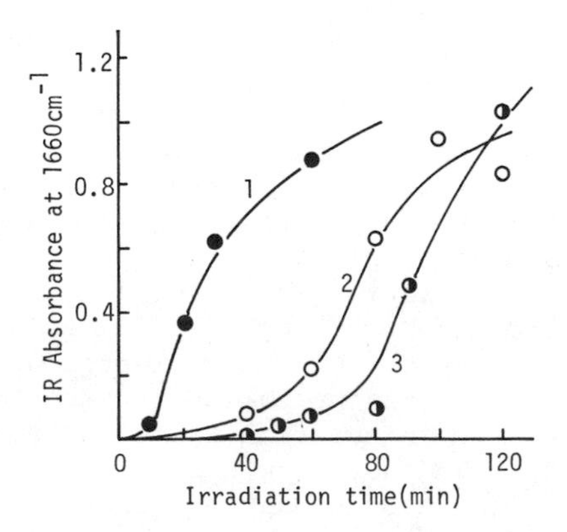

Figure 10. Solvent effects on photografting with AM: [AM] = 2.00M, [BP] = 0.20M in (1) n-*hexane/acetone(1/1), (2) acetone, (3) acetone/acetonitrile(3/1).*

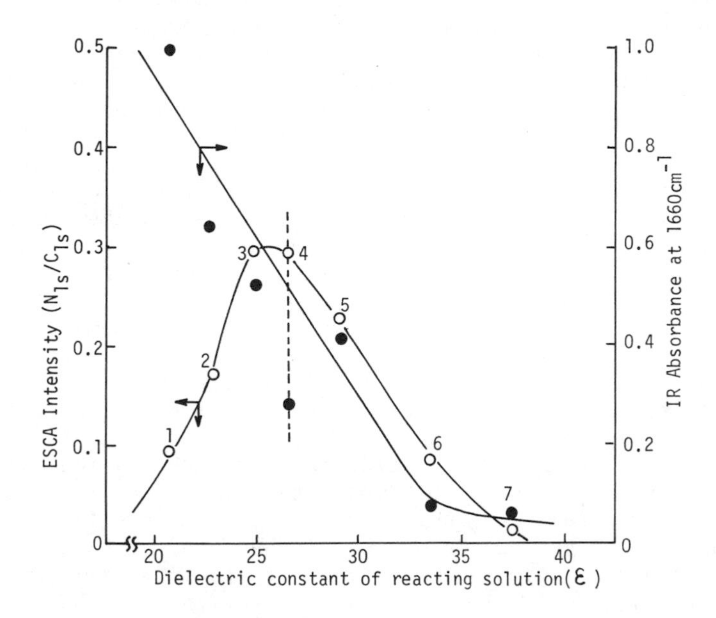

Figure 11. Plots of graft yield and surface polyAM concentration vs. dielectric constant of reacting solution: [AM] = 2.00M, [BP] = 0.20M, irradiation for 90 min. Solvent compositions: (1) acetone alone; (2) acetone/acetonitrile (8.6/1.4); (3) acetone/acetonitrile(3/1); (4) acetone/H_2O(9/1); (5) acetone/acetonitrile(1/1); (6) acetone/acetonitrile(1/3); (7) acetonitrile alone.

which does not initiate surface photografting in acetone could induce apparent photografting in THF. Although polyAM cannot be removed by refluxing the seemingly grafted OPP film in boiling water, there is a possibility of inducing internal homopolymerization of AM rather than true graft polymerization. In fact, the wettability to water is not much improved by photografting in the THF containing systems as shown in Table 3.

Effects of solvent on the properties of grafted surface are striking. In spite of the high graft yield when non-polar solvents are used, the wettability to water is poor and the ESCA spectrum of N_{1S} is weak or not observed at all. The discrepancy between total yield of grafting and surface concentration of polyAM determined by ESCA is confirmative evidence for the change in the graft layer location depending upon the solvent used. When the reacting solution is non-polar, the graft layer is deeper and consequently, the surface concectration of polyAM is low. When the reacting solution is very polar and the affinity to OPP is reduced, graft polymerization proceeds with a great difficulty while a distinct polyAM layer is formed on the surface. As a result, when OPP film is subject to surface grafting with a fixed time of irradiation, the maximum surface concentration of polyAM is obtained for a moderately polar reacting solution as shown in Figure 11. This is a reasonable consequence as revealing the balance between the depth and rate of photografting.

The ESCA intensity, however, does not exactly correspond to wettability. The contact angles of samples 1-8 in Figure 11 are plotted in Figure 12. Interestingly, the contact angles are related to the total graft yields but not with the relevant ESCA intensities. This is indicative of the difference in the depth of polyAM layer detected by ESCA and that responsible for imparting wettability. Depth of 50Å or less is analysed by ESCA whereas a deeper layer of polyAM seems to participate in water - polymer interaction.

Another factor affecting surface grafting is the solubility of growing polymer chain to reacting solution. Acetone, THF, and acetonitrile are all non-solvent for polyAM. The addition of a small amount of water reduces the solvent - OPP film interaction resulting in a diminished yield of grafting while the enhanced solubility of polyAM brings about a higher surface concentration of the polymer by drawing out the growing polymer chain into solution(Figure 11. sample 4).

All discussions on solvent effects are summarized in Table 4 and suggested graft polymer structures are sketched.

We could not conclude at the moment whether the solvent dependent surface properties are to be explained only by the difference in the depth of graft layer. Another possibility is the change in polar group orientation in graft layer as suggested by Hoffman(18). This arguement will be settled by direct determination of the thickness of graft layer prepared under various conditions. Clarification of the surface layer thickness - surface property correlation is our immediate future concern. How thin can the graft layer be to exhibit the intrinsic surface properties of grafted chain? Or, we may question the other way around. How do the properties of base polymer influence the grafted surface when the graft layer is extremely thin? The answer will probably depend on the kind of surface functionality we are looking at. The establishment of a relation between various surface properties and surface thickness would be a key step in giving an inside look into the physical chemistry of polymer surfaces.

Oxygen Effect. To avoid inhibitory effect of oxygen is very important if a large scale practical application is aimed at. For internal graft polymerization or grafting on to polymers reactive to radical attack, inhibition by oxygen is not a serious problem. However, surface photografting of polyolefins in an open system is very inefficient. The results in Figure 13 indicate that oxygen diffused through or in OPP film is an efficient retarder.

A theory of radiationless transition(26) claims that the rate of energy transfer is inversely related to the energy difference between sensitizer triplet state and energy acceptor ground state. In the case of energy transfer from triplet state aromatic hydrocarbons to oxygen, the rates are below the diffusion controlled value when E_T is higher than ca. 47 Kcal/mol(27). Use of high energy triplet sensitizers will be a strategy to reduce oxygen inhibition as well as to increase the rate constant of hydrogen abstration.

Practical Applications. The facile conversion of hydrophobic polymer surface to hydrophilic by the present procedure suggest a variety of practical applications in the fields of printing, coating, packaging, compositing, biomedical, and agricultural materials. The expected useful properties imparted by surface photografting are adhesiveness, printability, paintability, dyability, anti-fogging property, anti-static property, anti-staining property, bio-compatibility, and so on.

At the moment, the required time or irradiation ranges from 10^1 to 10^2 seconds depending upon the kind of base polymer, the monomer, and the out-put and spectrum of the light source. This reaction time is still too long for the on-line application to comodity polymer films. However, unoriented amorphous polymers available as plates or molded materials are much more susceptible to the present surface modification process. An example of applying the present process to the undercoating for painting on polystyrene plates is given in Table 5. Further applied researches are now underway.

Table 4. Summary of Solvent Effects on Surface Photografting.
Base Film: OPP, Monomer: AM, Sensitizer: BP, Irradiation at 366nm.

Solvents (S)	S - OPP interaction	BP[*3] + S reaction	Surface concentration of polyAM	Rate of grafting	Proposed structure of grafted OPP film
acetone - water	small	no	high - medium	slow - medium	PolyAM, OPP
acetone - CH_3CN	moderate	no	high - medium	fast	
THF - CH_3CN	strong	yes	low	slow - medium	

(These classifications are very qualitative and depend on the solvent composition.)

Table 5. Improved paintability of Polystyrene Plate Treated by Surface Photografting Method.

	Top Coating with Nitrocellulose Lacquer		
	Gloss	Impact Resistance ASTM G14-72	Tape Hatch Test ASTM D3002-71
Without under-coating	poor	<5cm	0/100
photografting for 5min with acrylic acid in air	excellent	>50cm	100/100

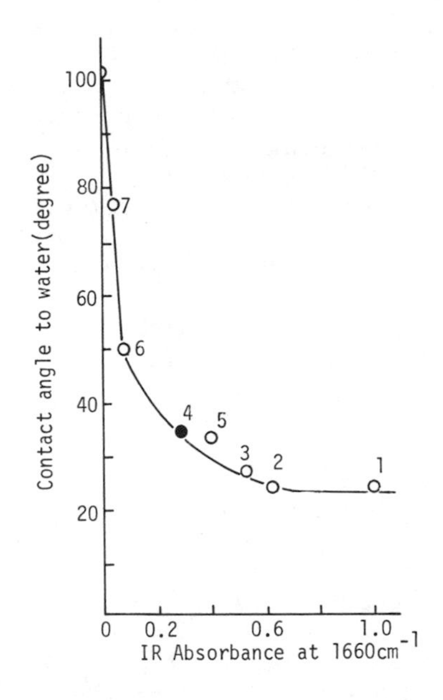

Figure 12. Plots of contact angle of samples in Figure 11 to water vs. IR absorbance

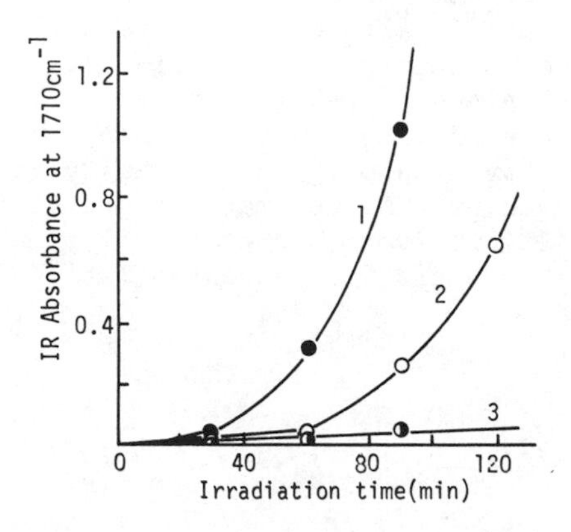

Figure 13. Effects of oxygen on photografting to OPP film: [AM] = 2.00M, [BP] = 0.20M in acetone. The standard procedure as indicated in Figure 1 was applied. However, a spacer was inserted between c and e of the apparatus in Figure 1 so that the irradiating side of the OPP film was replaced with nitrogen (1) or oxygen (3). The reacting solution (d) was always deaerated by bubbling with nitrogen. For (2), the apparatus in Figure 1 was used without modification.

Conclusion

We have developed a novel method of polymer surface modification. Analysis of various factors influencing the thickness of graft layer leads to the conclusion that solvent interactions with base polymer, grafted base polymer, and propagating graft chain are determining factors for the rate and thickness of grafing. Reduction in solvent - base polymer interaction is a necessary condition to confine the graft layer on the base polymer surface whereas it is a negative factor for grafting rate. Consequently, the use of solvent systems having balanced properties is an essential requirement for thin layer surface photografting.

The present procedure is certainly applicable to convert hydrophilic surfaces to hydrophobic, or to impart more sophisticated functionalities other than polar property to various polymer materials. From the view point of practical polymer processing, in particular, of chemically inert polyolefins or highly crystalline polymers, inhibitory effect of oxygen is a difficulty to be overcome.

Abstract: Surface photografting method was shown to be an excellent procedure to improve the surface properties of various polymers(Table 1). Particularly it is useful to impart wettability to hydrophobic polymer surface using hydrophilic monomers such as acrylamide and acrylic acid and triplet sensitizers such as benzophenone and its derivatives. Detailed studies on photografting onto polypropylene under nitrogen atmosphere indicated that the reaction was initiated by hydrogen abstraction by the excited sensitizer and the rate of surface grafting as well as the structure of grafted surface depended very much on solvent used. When the solvent-base polymer interaction was too strong, the grafted surface was not sufficiently hydrophilic although the rate was fast. This was interpreted as due to deep grafting(Table 4). The surface structure as studied by ESCA and wettability were discussed as functions of reaction conditions. It was then concluded that the use of solvent having appropriate affinity to the base polymer was essential to compromise the rate of grafting and the degree of surface modification. Possible practical applications were discussed and an example was presented(Table 5).

Literature Cited

1) Kobunshi(High Polym. Jpn.), 1977, 26, 759.
Special issue for "Polymers and Their Surface Treatments".
Yamakawa, S., J. Adh. Soc, Japan, 1977, 13, 211, Shinbo, M., J.Japan Soc. Colour Material, 1975, 48,517.
2) Oster, G.; Shibata,O. J. Polym. Sci., 1957, 26, 233.
3) Oster, G.; Oster, G. K. and Moroson, H., J. Polym. Sci., 1959, 34,671.

4) Geacintov, N.; Stannett, V.; Abrahamson, E. W., Makromol. Chem., 1960, 36, 52.
5) Geacintov, N.; Stannett, V.; Abrahamson, E. W.; Hermans, J. J., J. Appl. Polym. Sci., 1960, 3, 54.
6) Reine, A. H.; Arthur, J. C. Jr., Text. Res. J., 1972, 42, 155.
7) Cooper, W.; Vaughan, G.; Miller, S.; Fielden, M., J. Polym. Sci., 1959, 34, 651.
8) Howard, G. J.; Kim, S. R.; Peters, R. H., J. Soc. Dyer Color., 1969, 85, 468.
9) Hayakawa, K.; Kawase, K.; Yamakita, H., J. Polym. Sci., A-1, 1970, 8, 1227.
10) Hayakawa, K.; Kawase, K.; Yamakita, H., J. Polym. Sci., Polym. Chem. Ed., 1974, 12, 2603.
11) Davis, N. P.; Garnett, J. L.; Urguhart, R. G., J. Polym. Sci., Polym. Symp., 1976, 55, 287.
12) Ang, C. H.; Davis, N. P.; Garnett, J. L.; Yen, N. T., Radiat. Phys. Chem., 1977, 9, 831.
13) Schindler, A.; Gratzl, M.; Platt, K. L., J. Polym. Sci., Polym. Chem. Ed., 1977, 15, 1541.
14) Wright, A. N., Nature, 1967, 215, 953.
15) Tazuke, S.; Kimura, H., J. Polym. Sci., Polym. Chem. Ed., 1977, 15, 2707.
16) Tazuke, S.; Kimura, H., J. Polym. Sci., Polym. Lett. Ed., 1978, 16, 497.
17) Tazuke, S.; Kimura, H., Makromol. Chem., 1978, 179, 2603.
18) Ratner, B. C.; Weathersby, P. K.; Hoffman,; A. Kelly,; M. A. Sharpen, L. H. J. Appl. Polym. Sci., 1978, 22, 643.
19) Murov, S. L., "Handbook of Photochemistry", Marcel Dekker, New York, 1973, pp3.
20) Lamola, A. A.; Turro, N. J., "Energy Transfer and Organic Photochemistey", Interscience, New York, 1969, p220.
21) Formosinho, S. J., J. Chem. Soc., Faraday Trans. II, 72, 1313 1976.
22) Abbott G. D.; Phillips, D., Mol. Photochem., 1977, 8, 289.
23) ref. 20, p201.
24) Giering, L.; Berger, M.; Steel, C., J. Am. Chem. Soc., 1974, 96, 953.
25) Scaiano, J. C., J. Photochem., 1973/4, 2, 81.
26) Koizumi, M.; Kato, S.; Mataga, N.; Matsuura, T.; Usui, Y., "Photosensitized Reactions", Memorial publication for late Profesor Koizumi, M., Kagaku Dojin, 1977, pp251.
27) Gijzeman, O. L.; Kaufman, F.; Porter, G., J. Chem. Soc., Faraday Tans. II, 1973, 69, 708.

RECEIVED July 12, 1979.

16

Acid Effects in the Radiation Grafting of Monomers to Polymers, Particularly Polyethylene

JOHN L. GARNETT and NGUYEN T. YEN

Department of Chemistry, The University of New South Wales, Kensington, N.S.W., 2033, Australia

Radiation grafting is an extremely valuable one-step method for directly modifying the properties of polymers (1,2). The technique has been used with a wide variety of naturally occurring macromolecules such as wool (3) and cellulose (4) and also with many synthetic polymers, particularly the polyolefins (2,5,6). Both the pre-irradiation and simultaneous irradiation procedures have been utilised for this purpose, however the latter technique is the more flexible and more extensively investigated. In the simultaneous procedure, the monomer, usually in the presence of solvent, is simultaneously irradiated in contact with the backbone polymer. The predominant variables which influence the grafting yield include radiation dose and dose-rate, concentration of monomer in solvent and the structures of both monomer and backbone polymer.

The inclusion of mineral acid in the grafting solution has recently been shown to increase the radiation copolymerisation yield, particularly when styrene is grafted to trunk polymers like wool (3) and cellulose (4) i.e. polymers which readily swell in polar solvents such as methanol. This acid effect is important since for many copolymerisation reactions, relatively low radiation doses are required to yield finite graft. The process is particularly valuable for monomers and/or polymers that are either radiation sensitive or require high doses of radiation to achieve the required graft.

A theory for this acid effect has been developed essentially from the wool and cellulose work (3,4). Recently, in a brief communication, we reported analogous acid enhancement effects in the radiation grafting of monomers such as styrene in methanol to non-polar synthetic backbone polymers like polypropylene and polyethylene (5). In the present work, detailed studies of this acid enhancement effect are discussed for the radiation grafting of styrene in various solvents to polyethylene. The results are fundamentally important since most of the experiments reported here have been performed in solvents such as the low molecular weight alcohols which, unlike cellulose and wool systems, do not swell polyethylene.

0-8412-0540-X/80/47-121-243$05.00/0

However the acid effect is just as significant with the polyolefins as with the naturally occurring macromolecules.

Experimental

The following grafting technique was a modification of that previously used for analogous experiments with wool (3), cellulose (4) and polypropylene (5). Styrene (ex-Monsanto, Aust. Pty. Ltd.) was purified by column chromatography on alumina, a procedure that has previously been satisfactory for radiation copolymerisation (3,4). High purity methanol (acetone free, ACS reagent, code 1212) was obtained from Allied Chemicals whilst the remaining alcohols, dimethyl formamide, dimethyl sulfoxide, acetone and dioxan were AR grade and used without further purification. These materials were satisfactory in earlier grafting reactions to cellulose (4).

The copolymerisation experiments were performed in pyrex tubes, solvent being added first followed by acid or a concentrated solution of acid in solvent, then monomer to make up a total volume of 20 ml. Polyethylene (ex-Union Carbide) strips (37.5 x 25 x 0.01 mm) were then fully immersed in the monomer solutions, only homogeneous solutions being used, since with acid additives there is a limit of acid concentration before phase separation occurs. Irradiation of the lightly stoppered tubes containing the reagents was carried out in cobalt-60 or spent fuel element facilities at the Australian Atomic Energy Commission. After irradiation, the grafted polymer film was quickly removed from the monomer solution, soaked for two days in benzene to remove homopolymer, benzene being changed several times during the process. The film was then washed again in methanol and dried in a ventilated oven at 60° to constant weight. The grafting yield was the percent increase in weight of the original film.

Homopolymerisation was determined by the following modification of the Kline procedure (7). The grafting solution from the irradiation was poured into methanol (200 ml) to precipitate homopolymer and the sample tube rinsed with methanol (50 ml). Homopolymer which adhered to the polymer film and to the tube was dissolved carefully in dioxan (20 ml) and the dioxan solution added to the methanol in a beaker, together with the benzene washings from the extraction of the original film. The beaker was heated on a steam bath with frequent stirring until all polystyrene coagulated, the mixture cooled, filtered through a tared sintered glass crucible, washed three times with methanol (100 ml) and the crucible dried to constant weight at 60°. The percentage homopolymer was calculated from the weight of homopolymer divided by the weight of monomer in solution. The grafting efficiency was ratio of graft to graft plus homopolymer.

Results

The data in Table. I show that the low molecular weight alcohols are useful solvents for radiation grafting styrene to polyethylene. These results are thus consistent with earlier reports by Odian and coworkers (8) who only used methanol and also preliminary experiments by ourselves (2,5). Isomeric structure of the alcohol is important in determining solvent to be used for optimisation of grafting, (Table II).

In addition, in the higher molecular weight normal alcohols, copolymerisation reaches a maximum with n-heptanol and then decreases (Table III). With all of these alcohols (Tables I-III), there is a Trommsdorff peak observed at 30-40% monomer concentration. Addition of sulfuric acid to the grafting solutions leads to an enhancement in copolymerisation at most monomer concentrations, especially at the gel peak. Nitric, sulfuric and perchloric are the most effective mineral acids for increasing grafting yields (Table IV), nitric being the most efficient. With perchloric acid, the Trommsdorff effect shifts to 50-60% monomer concentration. Organic acids are not as effective as inorganic analogues in increasing grafting yields.

Homopolymerisation which accompanies copolymerisation increases at higher styrene concentrations, thus the grafting efficiency decreases with increasing styrene concentration (Table V). These results are similar to analogous data for polypropylene (5,6). Inclusion of mineral acid increases homopolymerisation, however not to the same degree as it enhances copolymerisation, thus, overall, grafting efficiency is significantly improved in the presence of acid.

Efficient copolymerisation can also be achieved in solvents other than the alcohols (Table VI). Thus the order of effectiveness for the present copolymerisation of these additional solvents is DMSO>DMF>dioxan>acetone>>chloroform>hexane. Acid enhancement is also observed in the first of these four solvents (Table VI). Characteristically (5), acid increases the intensity of the Trommsdorff peak if it is already present in the system (dioxan) or alternatively induces the formation of the gel peak if it is not present in the solutions prior to acid addition (DMSO).

Radiation dose rate significantly affects the copolymerisation yield (Table VII). As the dose rate is increased from 10,000 rad/hr to 74,000 rad/hr there is a gradual decrease in grafting yield at all monomer concentrations studied for copolymerisation of styrene in DMF. Addition of acid to these solutions leads to significant increases in grafting at all three dose rates studied (Table VII). The effect of acid is more marked when methanol is used as solvent (Figure 1). In the absence of acid, the grafting yield in methanol decreases with increasing dose rate over the range of 117,000 - 546,000 rad/hr, consistent with the DMF data in Table VII. However, if acid is added to the methanolic solutions, the copolymerisation yield remains virtually constant at the Trommsdorff peak for all dose rates studied. More importantly, the percentage increase in grafting yield at the gel peak is very much higher at the highest dose rate in the presence of acid. As

TABLE I. Acid Effect in Radiation Grafting to Polyethylene using Styrene in Low Molecular Weight Alcohols.[a]

Styrene (%v/v)	Graft % in							
	Methanol		Ethanol		n-Propanol		iso-Propanol	
	0	0.2M[b]	0	0.2M[b]	0	0.2M[b]	0	0.2M[b]
10	5	18	10	11	14	19	16	15
20	57	63	35	44	44	63	53	59
30	75	130	91	115	92	121	73	101
40	79	100	86	101	108	105	72	90
50	68	75	74	79	76	84	69	78
60	60	83	64	87	64	69	63	74
70	56	66	58	60	56	83	58	66
80	52	-	53	-	48	-	48	-
90	46	-	46	-	43	-	42	-

[a]Total dose 0.2 x 10^6 rad at 0.040 x 10^6 rad/hr.
[b]H_2SO_4

TABLE II. Acid Effect in Radiation Grafting to Polyethylene using Styrene in the Isomeric Butanols.[a]

Styrene (%v/v)	Graft % in					
	n-Butanol		iso-Butanol		tert-Butanol	
	0	0.2M[b]	0	0.2M[b]	0	0.2M[b]
10	7	10	25	31	22	25
20	29	38	81	106	59	60
30	78	89	104	125	78	96
40	69	100	89	113	75	84
50	73	82	74	78	62	73
60	63	64	62	79	50	79
70	55	60	54	80	44	69
80	53	-	43	-	34	-
90	41	-	39	-	41	-

[a]Total dose 0.2 x 10^6 rad at 0.040 x 10^6 rad/hr.
[b]H_2SO_4

TABLE III. Acid Effect in Radiation Grafting to Polyethylene using Styrene in the Higher Molecular Weight Alcohols.[a]

Styrene (%v/v)	Graft % in									
	n-Pentanol		n-Hexanol		n-Heptanol		n-Octanol		n-Decanol	
	0	0.2M[b]	0	0.2M[b]	0	0.2M[b]	0	0.2M[b]	0	0.2M[b]
10	8	11	13	12	12	17	12	16	-	-
20	39	41	43	51	71	82	38	47	-	-
30	70	100	84	116	119	163	85	116	45	32
40	87	116	91	130	125	170	107	130	49	55
50	77	87	76	83	101	142	91	112	-	-
60	64	78	64	74	68	123	74	128	-	-
70	53	78	51	92	68	119	61	123	-	-
80	57	-	43	-	43	-	51	-	-	-
90	40	-	56	-	40	-	41	-	-	-

[a]Total dose 0.2 x 10^6 rad at 0.040 x 10^6 rad/hr.
[b]H_2SO_4.

TABLE IV. Effect of Acid Structure in Grafting of Styrene in Methanol to Polyethylene.[a]

Styrene (%v/v)	No Acid	0.1N CH_3COOH	0.1N HCOOH	0.1N HCl	0.1N $HClO_4$	0.1N H_2SO_4	0.1N HNO_3
20	17	25	26	31	52	36	55
30	34	48	47	65	87	94	110
40	52	46	46	53	81	118	127
50	42	37	32	36	104	108	109
60	34	28	26	28	100	98	100

[a]Total dose 0.2 x 10^6 rad at 0.080 x 10^6 rad/hr.

TABLE V. Effect of Acid on Homopolymer Formation in the Radiation Grafting of Styrene in Methanol to Polyethylene.[a]

Styrene (%v/v)	Graft (%)		Homopolymer (%)		Grafting Effic. (%)	
	0	0.1M[b]	0	0.1M[b]	0	0.1M[b]
20	13	44	0.6	1.3	30	45
40	80	109	1.3	1.5	40	45
60	65	75	1.4	1.7	26	27
70	-	73	-	1.7	-	22
80	52	-	1.6	-	16	-

[a]Total dose 0.2 x 10^6 rad at a dose rate of 0.032 x 10^6 rad/hr.
[b]H_2SO_4.

TABLE VI. Acid Effect in Radiation Grafting using Styrene in DMF, DMSO, Acetone, Dioxan, Chloroform and Hexane.[a]

Styrene (%v/v)	Graft % in									
	DMSO		DMF		Dioxan		Acetone		Chloroform	Hexane
	0	0.2M[b]	0	0.2M[b]	0	0.2M[b]	0	0.2M[b]	0	0
10	11	24	5	16	1	7	1	4	0	0
20	14	41	29	44	9	16	6	18	0	0
30	18	60	48	73	35	35	12	30	0	0
40	33	83	57	89	29	75	20	55	2	0
50	41	115	61	91	50	93	28	50	4	2
60	82	104	71	104	55	89	39	44	7	4
70	85	101	64	105	44	61	36	40	10	8
80	87	-	63	-	59	-	43	-	16	10
90	82	-	51	-	41	-	33	-	29	-

[a]Total dose 0.2 x 10^6 rad at 0.040 x 10^6 rad/hr.
[b]H_2SO_4.

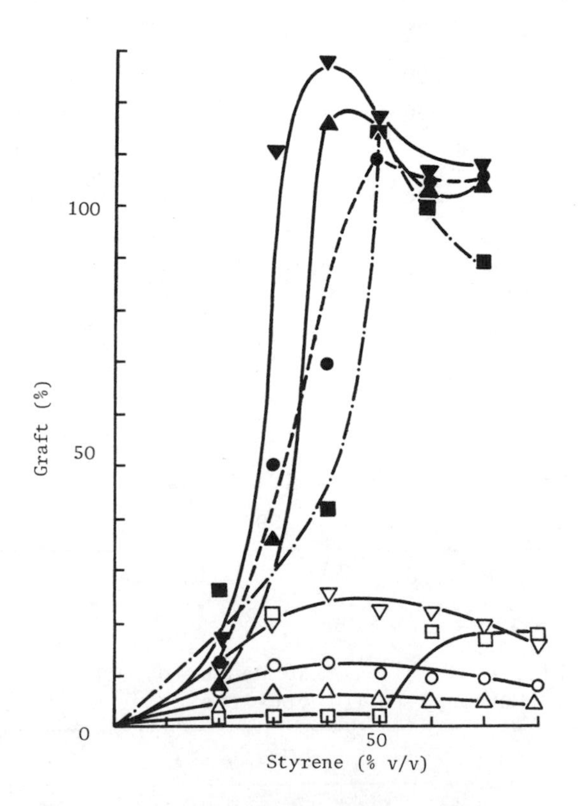

Figure 1. Grafting of styrene in methanol to polyethylene at various dose rates; total dose 0.2 × 10^6 rads. (▽) 0.117 × 10^6 rad/hr, without H_2SO_4; (▼) with 0.1N H_2SO_4; (○) 0.196 × 10^6 rad/hr, without H_2SO_4; (●) with 0.1N H_2SO_4; (△) 0.279 × 10^6 rad/hr, without H_2SO_4; (▲) with 0.1N H_2SO_4; (□) 0.546 × 10^6 rad/hr, without H_2SO_4; (■) with 0.1N H_2SO_4.

TABLE VII. Effect of Dose Rate in the Presence of Acid on the Grafting of Styrene in DMF to Polyethylene.[a]

Styrene (%v/v)	Graft (%) at					
	0	0.1M[c]	0	0.1M[c]	0	0.1M[c]
10			20	20		
20	158	208	63	72		
30	217	286	107	129		
40	255	336	146	156	64	84
45	-	-	-	-	66	94
50	174	378	147	168	81	103
55	-	-	-	-	85	111
60	287	388	163	173	91	123
65	-	-	-	-	92	122
70	282	371	147	168	95	128
75	-	340	-	165	95	128
80	266	-	138	-	96	-
90	257	-	124	-	91	-
100	153	-	59	-	-	-

[a]Total dose 0.5×10^6 rad. [b]Dose rate in rad $\times 10^3$/hr.
[c]H_2SO_4.

TABLE VIII. Effect of Acid in the Radiation Grafting of Styrene in Methanol to Polyethylene using Open Tubes.[a]

Styrene (%v/v)	Graft (%)	
	Without H_2SO_4	With H_2SO_4 (0.1M)
20	5.0	6.1
40	31.0	54.6
60	59.4	51.2
80	51.9	48.8

[a]In contrast to data in previous tables, these runs were performed in unstoppered, open tubes. Polyethylene (37.5 x 25 x 0.004 mm) was used to a dose of 0.17×10^6 rad at 0.038×10^6 rad/hr.

the acidity is increased, the grafting yield also increases at constant dose and dose rate for copolymerisation in DMF (Figure 2). If irradiations are performed in unstoppered (i.e. open) tubes, acid enhancement is still observed around the Trommsdorff peak, although the overall graft is reduced in this region (Table VIII).

Discussion

The parameters which predominantly influence the acid effect in radiation grafting of styrene monomer to polyethylene film are the structure of solvent, the concentration of monomer and the dose rate. Because these three variables are inter-related, it is difficult to predict, *a priori*, the conditions required to yield an optimum in grafting. In this respect the type of solvent used is particularly important.

Structure of Solvent. In previous radiation grafting studies of styrene to polyethylene, Odian et al. (8) and Silverman and coworkers (9) have shown that methanol is a suitable solvent for this reaction. The present data (Tables I-III) confirm these earlier results and also show that other low molecular weight alcohols up to n-octanol are efficient solvents for this process. This observation is important since analogous experiments with copolymerisation to wool (3) and cellulose (4) indicate that only those alcohols which wet and swell the trunk polymer are suitable for grafting, i.e. alcohols up to propanol whereas butanol is not effective. By contrast, for grafting to polyethylene, especially with styrene, swelling and wetting effects of solvents are not necessary since this trunk polymer is already extensively swollen by the monomer. Both Odian (8) and Silverman (9) groups attribute the accelerating affect of methanol in polyethylene grafting to physical changes in the system caused by methanol. In the present discussion, the radiation chemistry of the trunk polymer, monomer and solvent will also be shown to be significant, especially in determining the conditions for peak grafting.

The Trommsdorff Effect. In the copolymerisation reactions using the solvents in Tables I-III, there is a particular monomer concentration at which the grafting reaches a maximum, e.g. 30% for methanol. From their styrene grafting studies in methanol, Odian and coworkers (8) attribute the peak to a gel type effect. They suggested that it occurred because the particular solvent was a precipitant of polystyrene. As the grafted polystyrene chains are precipitated, they become immobilised. Further collisions with the precipitated polystyrene are thus inhibited, the termination rate is reduced while there is no reduction in initiation rate. Rate of grafting is thus increased. This situation is further accentuated by the fact that the termination rate in the grafting of undiluted styrene is already very hindered because of the high viscosity of the grafting medium. This hindrance is

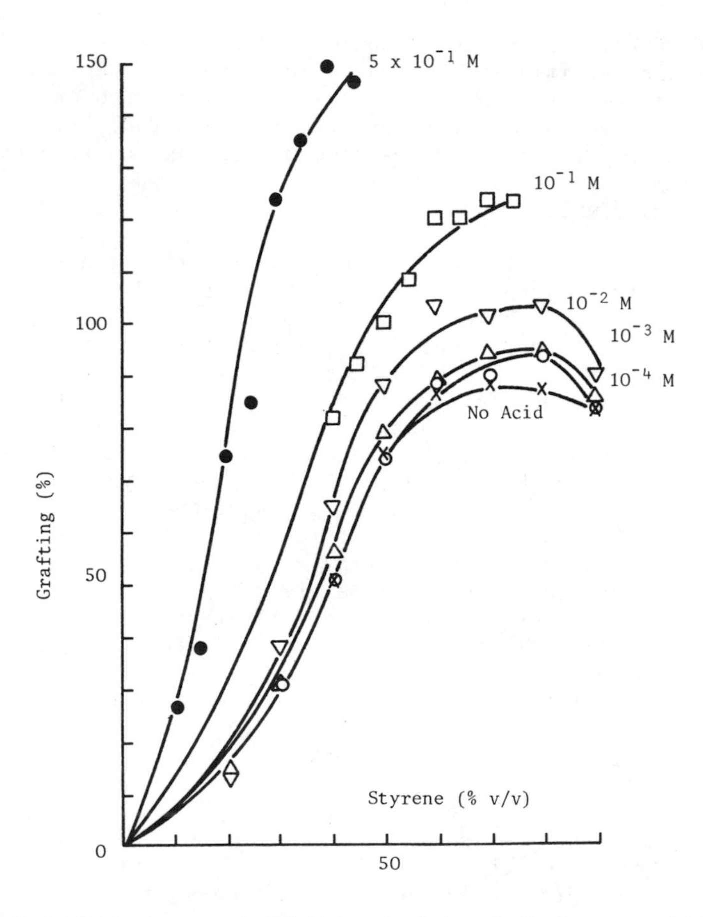

Figure 2. Grafting of styrene in DMF to polyethylene in the presence of H_2SO_4 at various concentrations; total dose 0.5×10^6 rad at dose rate 0.080×10^6 rad/hr

further increased in the presence of methanol, due to the precipitating effect. Thus the Trommsdorff effect which already exists in the system is enhanced by the presence of methanol. This mechanism for the accelerating effect of methanol in grafting styrene to polyethylene is also applicable to all of the alcohols up to n-octanol (Tables I-III), due to their similar precipitating effect on polystyrene. Thus, Chapiro (1) observed that in the radiation polymerisation of styrene in different alcohols, the polymer is precipitated at high dilution (90 mole percent methanol, 85 mole percent n-propanol, 80 mole percent n-butanol or 60 mole percent n-octanol, all these being 70 percent alcohol by volume). In the present work these concentrations of alcohols were also found to give maximum grafting yields.

Machi, Kamel and Silverman (9) modified the above mechanism of the Trommsdorff effect. These authors analysed mixtures of styrene and methanol which were absorbed in polyethylene films prior to grafting. They found that the methanol fraction in the mixture was very low (≤4%), and was not enough to precipitate the polystyrene grafted chains. They proposed a new mechanism of the gel effect based on the concentration of occluded styrene in polyethylene and the viscosity of the amorphous region of polyethylene which is swollen by styrene. In terms of this mechanism, methanol reduces the concentration of occluded styrene in polyethylene, hence the swelling of polyethylene decreases and the viscosity of the grafting medium increases. At low styrene concentrations, the initiation and propagation rates are low and increase with increasing styrene concentration, while the termination rate is also low because of the high viscosity of the medium. At high styrene concentrations, the propagation rate tends to increase, but the termination rate also increases due to the decreased viscosity of the swollen polymer. There is thus an optimum styrene concentration at which the grafting rate is a maximum.

Both Odian and Silverman models satisfactorily explain most of the observed results in all solvents (Tables I-III, VI) used in the present study, however there are some exceptions especially when solvents other than the alcohols are used (Table VI). Thus the Odian mechanism is not consistent with the DMF data nor can the Silverman model account for the acetone results. In addition, in further preliminary studies with grafting of styrene to polyethylene (10) in solvents other than those reported here both Odian and Silverman mechanisms are deficient. The problem is that possible contributions from the radiation chemistry of the components in the grafting reaction need to be considered in formulating a complete mechanism for the overall process.

Radiolysis Effects. Radicals formed in solvent (SH) and trunk polymers (PH) are important in the grafting of monomers (MH) with gamma radiation. With polymers such as polyethylene, grafting sites are formed by direct bond rupture (Equation 1). Additional sites are also

$$PH \rightsquigarrow P^{\cdot} + H^{\cdot} \qquad (1)$$

formed by hydrogen abstraction reactions involving radiolysis fragments ($S^{\cdot}$, $H^{\cdot}$) of the solvent (Equations 2 and 3). Thus G(H)

$$PH + H^{\cdot} \longrightarrow P^{\cdot} + H_2 \qquad (2)$$

$$PH + S^{\cdot} \longrightarrow P^{\cdot} + HS \qquad (3)$$

values of solvents are important in radiation grafting (5,6). In this concept monomers such as styrene are considered to be radical scavengers similar to the mode of action of benzene in the radiolysis of benzene-methanol solutions (2,5,11), scavenging occurring on either the aromatic ring of styrene by predominantly addition reactions (Equation 4) or by radical attack on the double bond of the side chain. Thus for the grafting of styrene to polyethylene,

$$C_6H_5CH{=}CH_2 + R \longrightarrow \text{(I)} \qquad (4)$$

$R = H^{\cdot}, S^{\cdot}$

(I): cyclohexadienyl radical bearing CH=CH₂, with H and R on the same ring carbon

a mechanism for the accelerating effect of methanol can be proposed based on the relative numbers of styrene molecules and methanol radicals. At low concentrations of styrene, the monomer will essentially scavenge methanol radicals while at high concentrations of styrene, the monomer will scavenge other styrene radicals. Homopolymerisation is thus preferred to grafting in both regions. In the middle range of styrene concentrations, a compromise is attained where there is sufficient styrene to scavenge all excess methanol radicals not involved in activation of the trunk polymer, yet an excess of monomer remains for grafting by the charge-transfer mechanism proposed by Dilli and Garnett (12) originally for copolymerisation to cellulose (4) and subsequently extended to wool (3), polyolefin (2,5) and PVC (13) systems. The data in Table V are consistent with this interpretation.

An additional contributing factor to the mechanism of the present grafting reaction is the role of radiolytically produced hydrogen atoms. In the radiolysis of binary mixtures of aromatic and aliphatic compounds such as styrene-methanol, the concentration of aromatic strongly influences the $G(H_2)$ obtained from the methanol. In the most extensively studied binary mixtures of benzene-methanol (11) and pyridine-methanol (10), it is found that the yield of H atoms is important in determining product yields and types. Small additions (5%) of benzene and pyridine significantly reduce $G(H_2)$ from the methanol by scavenging H atoms. Above 5% additive, $G(H_2)$ is reduced further, but at a slower rate. These data for benzene-methanol and pyridine-methanol can be extrapolated to the styrene-methanol grafting solution of the present polyethylene grafting system. Thus although G(H) is reduced by the presence of scavenger styrene, there remains a sufficient concentration of H atoms compared with other radicals to make a significant contribution to the accelerating effect of methanol by abstracting hydrogen from the backbone polymer to give additional grafting sites (Equation 2). Thus the enhancement of the initiation rate by methanol decreases with increasing styrene concentration. Simultaneously, the propagation rate increases with increasing styrene (14). Between these two extremes, there is a concentration of styrene where grafting is maximised.

A similar explanation of accelerative effects observed in grafting with other solvents such as the higher straight chain alcohols, DMF, DMSO, acetone, chloroform and cyclohexane (Table VI) has been advanced (6). It appears that there is a relationship between G(H) value of solvent and the extent to which the solvent participates in accelerated grafting. The radiolysis pathway thus contributes, but not exclusively, to the mechanism of the overall copolymerisation reaction.

The Acid Effect. The possible mechanistic role of hydrogen atoms in the current radiation grafting work becomes even more significant when acid is used as an additive to enhance the copolymerisation. At the concentrations utilised, acid should not affect essentially the physical properties of the system such as precipitation of the polystyrene grafted chains or the swelling of the polyethylene. Instead the acid effect may be attributed to the radiation chemical properties of the system. Thus Baxendale and Mellows (15) showed that the addition of acid to methanol increased $G(H_2)$ considerably. The precursors of this additional hydrogen were considered to be H atoms from thermalised electron capture reactions, typified in Equation 5.

$$CH_3OH_2^+ + e \longrightarrow CH_3OH^* + H \qquad (5)$$

Acid enhancement in the radiation grafting of styrene in methanol to cellulose (4), wool (3) and in preliminary work with the polyolefins (5,6) has been proposed as being predominantly due to such reactions.

Further work (10) with acid effects in the radiolysis of binary mixtures such as benzene-methanol and pyridine-methanol indicates that the acid phenomenon is more complicated than the simple H atom model originally developed (4). These more recent experiments (10) show that whilst increased hydrogen atom yields in the presence of acid enhance the overall grafting yield, other mechanisms also contribute to this acid effect. Thus the acid stability of intermediate radicals (I-III) and also analogous species involving the trunk polymer are important. With radicals (I-III), at low styrene concentrations in methanol, these intermediates (MR·) will predominantly react with other available

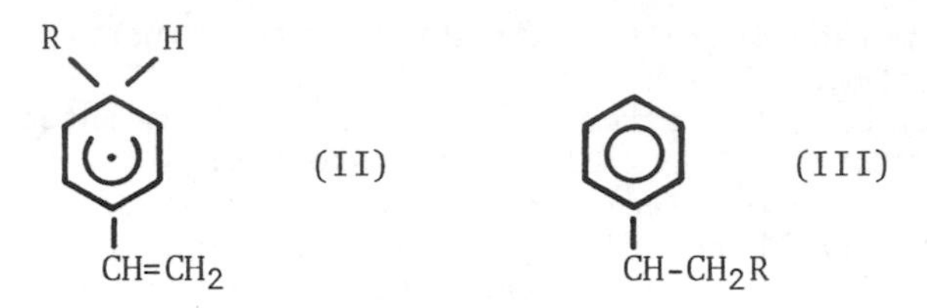

solvent radicals by addition or disproportionation reactions. On a statistical basis, the probability of MR· reacting with another styrene molecule in this concentration region is relatively low, hence homopolymerisation is relatively low. Grafting is also relatively low in this region as is the grafting efficiency (Table 5). Again on a statistical basis, at relatively high styrene concentrations, MR· will react predominantly with other available styrene molecules to yield homopolymer, consistent with the observed increasing homopolymer formation, decreasing graft and grafting efficiency with increasing styrene concentration (Table V).

Between these two styrene concentrations, there exists an intermediate region where the probability of MR· abstracting a hydrogen atom from the adjacent polymer chain to give a polymer radical (Equation 6) is high and grafting can thus be induced in the cage (Equation 7) as discussed previously (16). Copolymerisation yields and efficiency thus attain a maximum in the medium range of styrene concentrations. This mechanism, involving

$$PH + MR\cdot \longrightarrow P\cdot + MRH \qquad (6)$$

$$P\cdot + MRH \longrightarrow [P\cdot \rightarrow MRH] \longrightarrow PMRH\cdot \qquad (7)$$

styrene-solvent radicals (MR·) is consistent with the observed yields of the predominant scavenging products found in the radiolysis of benzene-methanol and pyridine-methanol mixtures (10,11). In the former system, cyclohexadiene-methanol and anisole are the relevant products whilst in the latter mixture, pyridylmethanol and methylpyridine are the corresponding products. In both systems, these products display maximum yields at 20-30% aromatic in methanol, i.e. close to the region where the Trommsdorff peak appears for the present grafting process.

For the purpose of this discussion, styrene and pyridine may be considered to be structurally analogous since the aromatic ring of styrene is complemented by the electron rich $-CH=CH_2$ group whilst pyridine has the nitrogen lone pair. Acid effects observed in the pyridine-methanol system may thus be extrapolated to styrene-methanol. In separate studies, it has been shown that inclusion of acid increased the quantities of scavenging products, pyridyl methanol and methylpyridine. In addition, one scavenging product, pyridylethanol, disappeared and was replaced by a number of unidentified products when acid was included in the radiolysis mixture. The mechanism whereby acid increases the yield of scavenging products has not been completely clarified (10) especially the role of the anion, however a plausible interpretation is that acid enhances the disproportionation reaction of the aromatic-methanol radicals (Equations 8 and 9) to give higher yields and a wider variety of scavenging products due to the greater migration of electrons from the ring to a more positive environment, thus facilitating the rupture of the C-H bond in the disproportionation process.

It is now proposed that acid acts in an analogous manner in the styrene-methanol grafting system. No data for the scavenging role of styrene in this reaction is yet available although the data from preliminary results are consistent with this concept (10). It is thus proposed that acid affects the stability and reactivity of the intermediate charge-transfer complex formed in

$$C_5H_5N + \cdot CH_2OH \longrightarrow C_5H_5N(H)(CH_2OH)\cdot \xrightarrow[H^+]{R'\cdot} R'H + C_5H_4N(CH_2OH) \quad (8)$$

$$C_5H_5N + \cdot CH_3 \longrightarrow C_5H_5N(H)(CH_3)\cdot \xrightarrow[H^+]{R'\cdot} R'H + C_5H_4N(CH_3) \quad (9)$$

the grafting reaction. This may occur by affecting the reactivity of the styrene-methanol radicals (MR·) or acid may facilitate the hydrogen abstraction reaction from an adjacent polymer molecule by MR· to give more grafting sites (Equation 10). Acid can also simultaneously catalyse more efficient reaction between MR· radicals and monomer to give increased homopolymer yields consistent with observation (Table V), especially at low monomer

$$PH + MR\cdot \xrightarrow{H^+} P\cdot + MRH \quad (10)$$

concentrations. At high monomer concentrations, the addition of acid leads to marginal increases in homopolymer yield, presumably because homopolymer yield is already appreciable without additive.

It is significant that grafting yields are enhanced to the same degree by each of the three strongest acids studied (Table IV). This indicates that a mechanism common to all acids prevails in the copolymerisation. The suggestion that acid facilitates hydrogen abstraction from the trunk polymer is consistent with Silverman's observation (9) concerning the small amounts of methanol which were found in swollen polymer. Thus once the polyethylene is swollen by the monomer, a dynamic equilibrium can be established such that intermediates MR· can diffuse to trunk polymer sites and graft.

The fact that acid enhances grafting also indicates the possibility that ionic processes may also contribute to the present grafting mechanism. In this context, acid may be considered to be a catalyst for the cationic process especially since ionising radiation is the initiator for the reaction and both free radicals and ions are known to be species formed from interaction between molecules and such radiation. However, the ionic mechanism for grafting is favoured by anhydrous conditions, thus, in the present system, acid enhancement via the ionic pathway would not appear to be a predominant process.

Acid Effect in Grafting With Isomeric Alcohols. With the isomeric propanols and butanols, some unexpected results were obtained both in the presence and absence of acid (Table II). When n-propanol was substituted by iso-propanol, the grafting yield around the Trommsdorff peak decreased (Table I), but with the corresponding two butanols (Table II), the reverse was observed. This apparent anomaly requires explanation. Branching increases the polarity and molecular size of the alcohol. Higher polarity should improve the precipitating effect of the solvent on the polystyrene chains while the higher molecular size should exert an opposite effect. From n-propanol to isopropanol, the molecule is more bulky (density from 0.803 to 0.785) and the increase in molecular size is more important than the increase in polarity, hence the grafting yield should decrease. From n-butanol to isobutanol, the increase in molecular size is less important (density from 0.808 to 0.802) than the increase in polarity, hence the grafting should increase. However from isobutanol to t-butanol the molecular size increases more significantly (density from 0.802 to 0.789), hence the grafting yield should decrease as observed. In the presence of acid similar trends were observed and the yields were enhanced, as expected.

Grafting in Unstoppered Tubes. The final interesting feature of the present data is the fact that the acid effect is still present in the styrene-methanol grafting system in the region of the Trommsdorff peak even if the tubes are unstoppered (i.e. open) during the irradiation (Table VIII). When compared with the corresponding runs in stoppered vessels the magnitude of the copolymerisation in the gel region is reduced for both acid containing and neutral monomer solutions, this result presumably reflecting the degree to which oxygen retards grafting strongly in this region. This conclusion is consistent with the role of hydrogen atoms, as precursors, contributing to the gel peak (4,5,16), since oxygen would readily scavenge such species. Oxygen could likewise interfere with the stability of the intermediates in the charge-transfer mechanism (Species I-III). The other important aspect of the present acid enhancement observed for grafting in open tubes is that, in this respect, the styrene copolymerisation system is different to the corresponding grafting reaction involving methyl methacrylate where no acid effect was reported for grafting in open tubes (17). In grafting under these conditions with methyl methacrylate, homopolymerisation is much more severe than with styrene (5,6), thus there are mechanistic differences, yet to be resolved, between the two monomer systems. The role of acid in the present polyethylene work is confirmed by analogous studies with polypropylene (6) also preliminary experiments recently reported by Chappas and Silverman (18), our own studies with monomers other than styrene and methyl methacrylate, e.g. acrylonitrile, ethyl acrylate, acrylic acid, acrylamide, etc. (3,6,10), also the work of Hoffman and Ratner (19) with acrylamide.

Advantages of Acid Effect in Synthesis of Copolymers. When polar solvents other than the alcohols are used for grafting (Table VI), strong acid enhancement effects are still observed. If the dose rate is altered at a fixed dose, for a particular solvent (Table VII, Figure 1), addition of acid remains effective in yielding higher grafts. In methanol, the percentage increase in graft increases with increasing dose rate particularly in the region of the Trommsdorff effect. This result is of significant, beneficial, preparative importance, since irradiation at the highest dose rate in the presence of acid would shorten considerably the time of source exposure required to achieve a particular percentage graft. As the acidity is increased for a particular solvent, copolymerisation increases (Figure 2), the limiting property of the grafting system under these conditions being phase separation which occurs at a certain acidity, (e.g. 45% styrene in DMF at 0.5 M H_2SO_4 in Figure 2).

Conclusions

Overall, addition of acid in the present polyethylene system leads to an enhancement in graft over a wide range of monomer conditions. Generally the predominant effect occurs near the gel peak. Thus acid can (i) induce a Trommsdorff effect in a particular system or (ii) enhance the intensity of the gel peak if it is already present. These polyethylene results are similar to the corresponding data obtained with wool (3), cellulose (4), polypropylene (5) and PVC (13). Analogous acid enhancement is also found in grafting monomers other than styrene (3,6,10,19). The acid effect thus appears to be a general phenomenon in radiation grafting.

Acknowledgements

The authors thank the Australian Institute of Nuclear Science and Engineering and the Australian Research Grants Committee for the support of this research.

Abstract

Inclusion of mineral acid in monomer solutions is shown to enhance grafting to polyethylene under certain gamma irradiation conditions. The predominant variables affecting the optimisation of the acid effect in these reactions are considered. These include type of solvent, acid structure, competing homopolymer formation and radiation dose-rate. A mechanism for the acid effect is proposed. The advantages of its use in preparative copolymerisation reactions are discussed. The present polyethylene data are compared with analogous acid enhancement observed in radiation grafting to wool, cellulose, polypropylene and PVC. The data show that the present acid effect appears to be a general phenomenon in radiation copolymerisation reactions.

Literature Cited

1. Chapiro, A., "Radiation Chemistry of Polymeric Systems"; Interscience: New York, 1962.

2. Garnett, J.L., Proc. 2nd Inter. Meeting Radiation Processing, Miami Beach, 1978, in press.

3. Garnett, J.L.; Leeder, J.D., "Textile and Paper Chemistry and Technology", ACS Symp. Series 49; Arthur, J.C. Jr., Ed.; American Chemical Society: Washington, D.C., 1977; p. 197.

4. Garnett, J.L., "Cellulose Chemistry and Technology", ACS Symp. Series 48; Arthur, J.C. Jr., Ed.; American Chemical Society: Washington, D.C. 1977; p.334.

5. Garnett, J.L.; Yen, N.T. Polymer Letters, 1974, 12, 225.

6. Garnett, J.L.; Yen, N.T. Aust. J. Chem., in press.

7. Kline, G.M., "Analytical Chemistry of Polymers"; Part 1, 3rd Edn.; Interscience: New York, 1966.

8. Odian, G.; Acker, T.; Sobel, M. J. Appl. Polymer Sci., 1963, 7, 245.

9. Machi, S.; Kamel, I.; Silverman, J. J. Polymer Sci. Part A-I, 1970, 8, 3329.

10. Fletcher, G.; Garnett, J.L. unpublished work.

11. Ekstrom, A.; Garnett, J.L. J. Phys. Chem., 1966, 70, 324.

12. Dilli, S.; Garnett, J.L. J. Appl. Polymer Sci., 1967, 11, 859.

13. Barker, H.; Garnett, J.L.; Levot, R.; Long, M.A. J. Macromol. Sci.-Chem., 1978, A12(2), 261.

14. Odian, G.; Sobel, M.; Rossi, A.; Klein, R. J. Polymer Sci., 1961, 55, 663.

15. Baxendale, J.H.; Mellows, F.W. J. Am. Chem. Soc., 1962, 83, 4720.

16. Dilli, S.; Garnett, J.L. J. Polymer Sci. A-I, 1966, 4, 2323.

17. Pinkerton, D.M.; Stacewicz, R.H. Polymer Letters, 1976, 14, 287.

18. Chappas, W.J.; Silverman, J. Proc. 2nd Inter.Meeting Radiation Processing, Miami Beach, 1978, in press.

19. Hoffman, A.S.; Ratner, B.D. Polymer Preprints, 1979, 20, 423.

RECEIVED July 12, 1979.

Synthesis and Properties of Photopolymer Printing Plates for a Printing Master Plate by Modification of Polyvinyl Alcohol

HISASHI NAKANE, TOSHIMI AOYAMA, HIROSHI TAKANASHI, BONPEI KATO, and HIROYUKI TOHDA

Tokyo Ohka Kogyo Company, Limited, 150 Nakamaruko, Nakahara-ku, Kawasaki, Japan

1. Introduction

A number of photopolymer printing plates are already known. Their basic structures are to combine one of the general purpose resins such as cellulose (1), polyamide (2), polyester, poly urethane (3), polyvinyl alcohol (4), synthetic rubber (5) and the like with photopolymerizing vinyl monomer, photopolymerization initiator and so on. Any one of the plates of such structures can be used as a press plate, but they can not be used as an original plate for duplicate plate owing to their insufficient hardness, toughness and the similar negative properties. Therefore, metal plates have been conventionally used. We have performed the research and development of a photopolymer printing plate for a master plate with a new basic structure by combining an oligomer of urea structure having a polyvinyl base with polyvinyl alcohol, photopolymerization initiator and other ingredients. The result shows that the newly developed plate (6)* is so good that it has replaced metal plates and has been stably used at leading newspaper companies in Japan where several millions of newspapers are daily issued.

2. Experiment

2-1: Synthesis of Oligomer

The reactive oligomer which is an important ingredient in the present research can be obtained by the polycondensation of an alkylol derivative of urea or alkalation alkylol derivative with N-alkylolacrylamide in the presence of acid or its salt.

(*) This plate is now commercially available under a trade name of RIGILON MASTER of which the pantent, applied by Tokyo Ohka Kogyo Co., Ltd., is pending.

0-8412-0540-X/80/47-121-263$05.00/0

Example of Synthesis
Methylhydroquinone, 0.025 part by weight, is dissolved in 10 parts by weight of water. Added to this is 74 parts by weight of dimethylolurea dimethylether, 202 parts by weight of N-methylol-acrylamide and 2 parts by weight of ammonium chloride. The mixture is agitated for 2 hours with heating at 80°C. Thus, is obtained a transparent and viscous condensation polymer which was then added into 1,000 parts by weight of acetone and the sediments removed by filtration. The filtered solution was then distilled for removal of acetone and viscous oligomer condensate was obtained.

2-1-2: Structure of Oligomer
The oligomer is soluble in water, alcohol, acetone and the like, and it shows a viscosity of about 95cp at 25°C, measured with a B type rotational viscometer.
As shown in the infrared absorbance spectrum of the condensation polymer (See Fig. 1), bands characteristic of absorbance of the urea resin at 3,400 cm^{-1}, 3,000 cm^{-1}, 1,680 cm^{-1}, 1,540 cm^{-1}, 1,380 cm^{-1}, 1,100 cm^{-1}, 1,020 cm^{-1}, 780 cm^{-1}, and characteristic absorbance of acrylamidemethyl base at 800 cm^{-1} are present; judging from this data, it is considered that N-methylacrylamide is connected with the end group of urea resin main chains and the imino group. The nuclear magnetic resonance (See Fig. 2) spectrum of the oligomer shows a resonance value of 6.48 ppm based on CH_2= and a resonance value of 5.74 ppm based on -CH=. The structure of this condensate is believed to be that shown in Fig. 3.

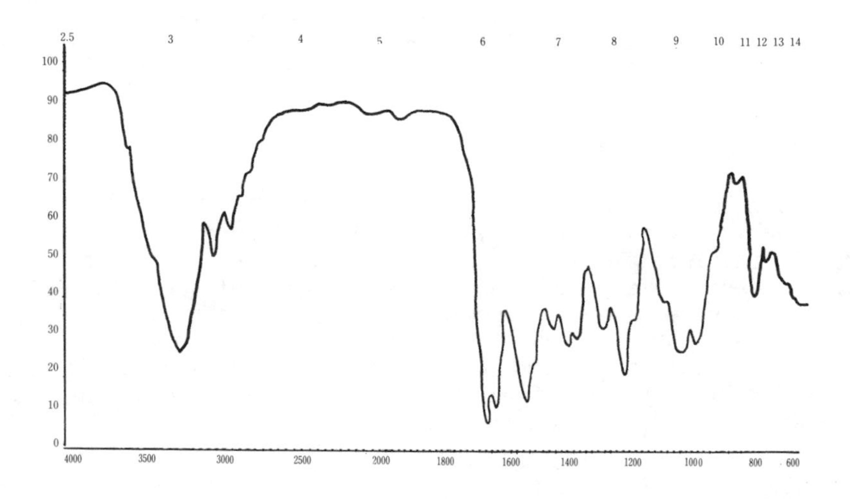

Figure 1. IR absorbance spectrum of condensate

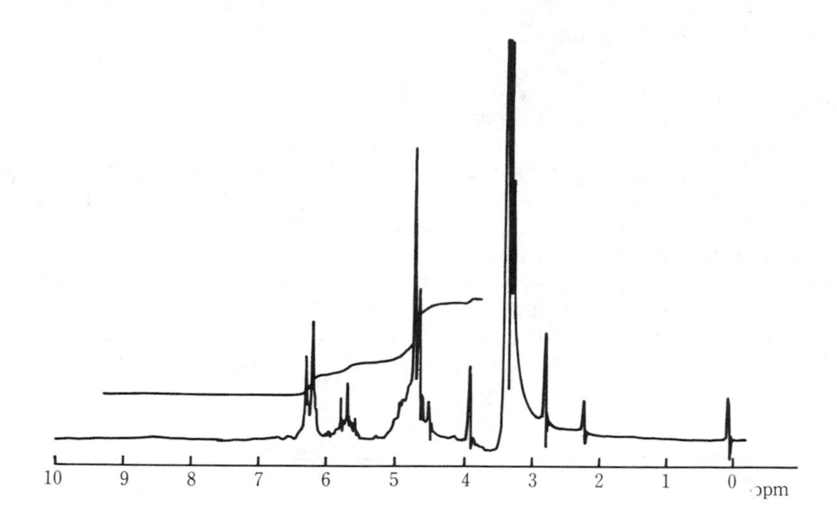

Figure 2. NMR of condensate

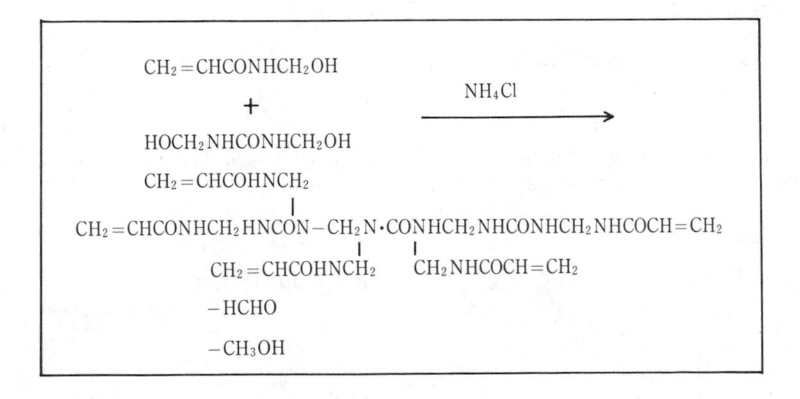

Figure 3. Concluded structure of condensate

2-2: Manufacture of Photopolymer Plate used as Master Plate for Duplicate Plate

2-2-1: Composition

Partially saponified (71-74%) polyvinyl alcohol (degree of polymerization: 500) .... 100 parts by weight

Condensation polymer prepared under paragraph 2-1 .... 95 parts by weight

Benzoin isopropyl ether .... 2 parts by weight·

Methylhydroquinone .... 0.02 part by weight

2-2-2: Manufacture of Plate in Sheet Form

The product prepared under paragraph 2-2-1 above, is mixed with 160 parts by weight of water, 40 parts by weight of methanol, as solvent, and dissolved completely in the water bath of 90°C under agitation in a flask. The solution thus obtained is spread over a glass plate, left on standing over a night for drying, and stripped off the glass plate as a photopolymer layer of 0.7 mm. thickness. This photopolymer layer is pasted on a steel backing sheet on which adhesive has already been coated, and at the same time polyethylene terephthalate film as a protective cover sheet is laminated to make a photopolymer printing plate.
The structure is shown in Fig. 4.

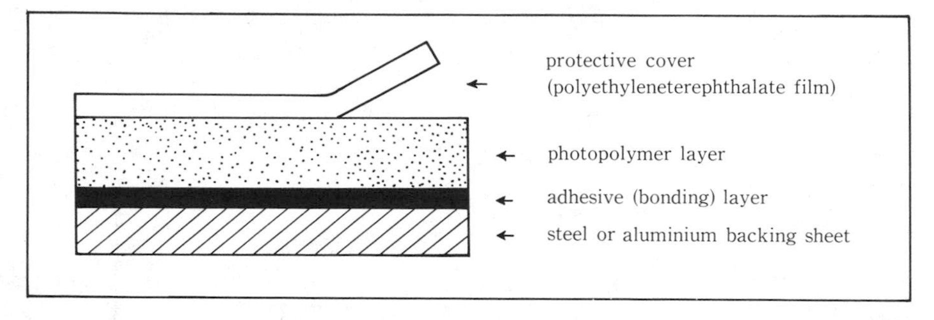

Figure 4. Structure of photopolymer printing plate for duplicate plate. (*) The plate of the present development (hereafter referred to as the Photopolymer Plate).*

2-2-3: Spectral Sensitivity of the Photopolymer Plate

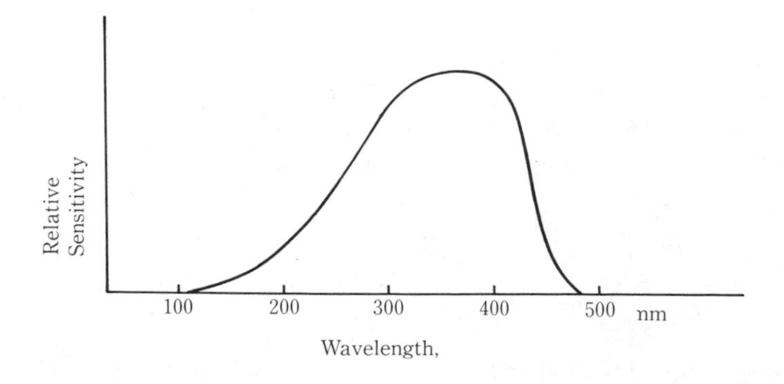

Figure 5. Spectral sensitivity of the Photopolymer Plate

As shown in Fig. 5, chemical lamp or ultra high pressure mercury lamp is considered to be a suitable light source for the Photopolymer Plate.

3. Plate-Making Process

3-1: Exposure

After the protective cover sheet is peeled off from the Photopolymer Plate, a negative film is placed on the plate in a complete contact with the plate under vacuum. The negative film used here, should have a density of below 0.08 in the light transmitting area and over 2.6-3.0 in the non-light-transmitting area.

The light source which is generally recommended, in case of a newspaper size of A-2 (420 X 594 mm) for instance, will be:

a) Ultra high pressure mercury lamp
   In case of one(1) 4 Kw output: about 1 minute exposure.

b) Chemical lamp
   In case of 20 X 10 outputs: about 2-3 minute exposure.

After the negative film is taken off the plate, the exposed area may appear to dent a little. After exposure, the plate is processed for washout.

3-2: Washout

The Photopolymer Plate can be washed out with water alone. In the actual operation, the Photopolymer Plate is processed on an Automatic Processor as shown in Fig. 6 & 7.

3-2-1: Washout Solution of Automatic Processor

Washout water is sprayed through 55 nozzles of 1/4 inch diameter under an optimum condition of nozzle pressure of 6 kg/cm^2, recirculated at 200 lit/min., upward spray power to spout water for 80 meters, output of 7.5 kw, and the water temperature of about 45°C. The washout speed under these conditions is about 20 seconds per 0.1 mm depth of relief. (See Fig. 8).

3-2-2: Drying

Water on the plate surface is removed with the sponge rolls for removal of water, and the surface is dried with 90-100°C convectional air, and then further dried with 90-100°C far-infrared heater.

3-2-3: Post Exposure

By applying ultra violet rays to the plate, light-hardening is promoted. When this process is completed the hardness should have been raised to 84-90 degrees of Shore D. (See Fig. 9). When the hardness has reached this level, the plate can be a practical printing master plate for making paper mold and matrix.

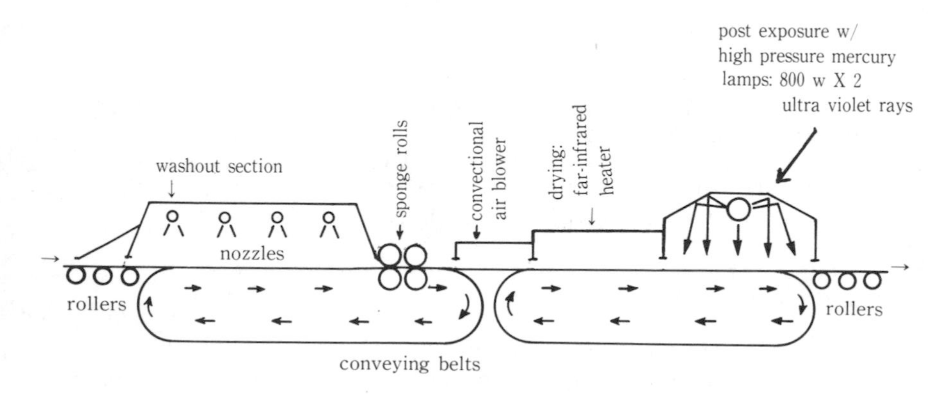

Figure 6. Schematic of automatic processor

Figure 7. Automatic processor

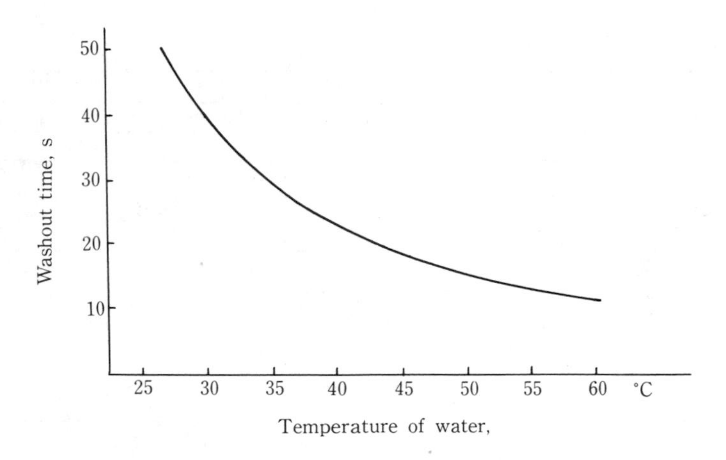

Figure 8. Change of washout time for 0.1-mm relief with temperature of water

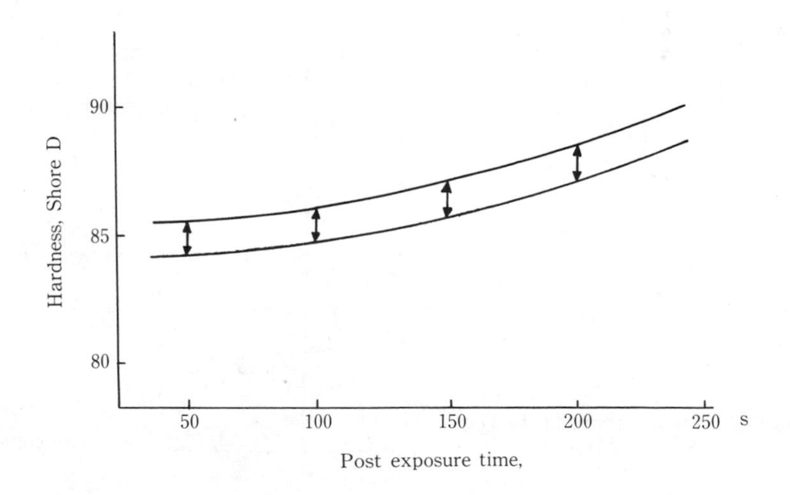

Figure 9. Change of hardness with post exposure time

The aforementioned processes will genera-ly require the time as indicated in the below Table 1.

Order		1st	2nd	3rd	4th
Processes		Exposure	Wash-out	Drying	Post exposure
Time, min.	ea. step	1	2	2	2
	accumulated	1	3	5	7
Conditions		4 kw ultra high pressure mercury lamp	6 kg/cm² temp. of water: 45°C	90−100°C	800 w X 2 high pressure mercury lamps 100°C
			← Automatic Processor →		

2nd, 3rd and 4th processes are performed on Automatic Processor:
Processing example of one newspaper size paper.
Table 1—Processing Time of The Photopolymer Plate

4. On Profile of the Photopolymer Plate for making Paper Mold and Matrix

The shoulder angle of the plate obtained from the Photopolymer Plate has been so designed in the original formula that it will be around 23° (θ) from the time of exposure. Furthermore, the shoulders are flat and smooth. (See Fig. 10)

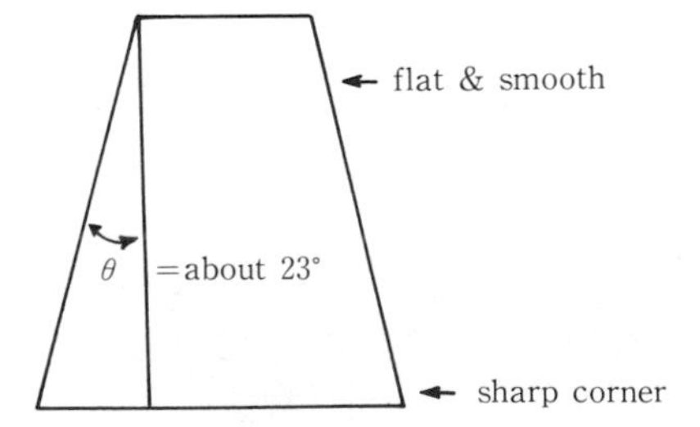

Figure 10. Shoulder shape of the Photopolymer Plate

For this reason, it is possible to obtain even fine pattern areas with a clean reproduceability. The rubber plates obtained with zinc and other metal plates require considerable exacting control of shoulder angles and actually desired adjustments can not be made at will. The reason for this exacting control of shoulder angles may be because the shape of shoulder is dependable upon acid concentration of etching solution, deteriority degree of solution and like factors which are impossible to maintain these factors constant at all times. As shown in Fig. 11, the shoulder angle of the Photopolymer Plate shows a slight change but the change is so small that it can be ignored.

In general, the shoulder angle tends to be low and if it is made to be high it invites a bigger undercut and makes the situation worse. Since the shoulder is shaped as a chemical reaction progresses by action of acid, the shoulder surface can be roughened by inadequate etching caused by small undesirable changes in the processing conditions. That is why the shoulders of rubber plates obtained from matrix are rough and dirty. (See Fig. 12)

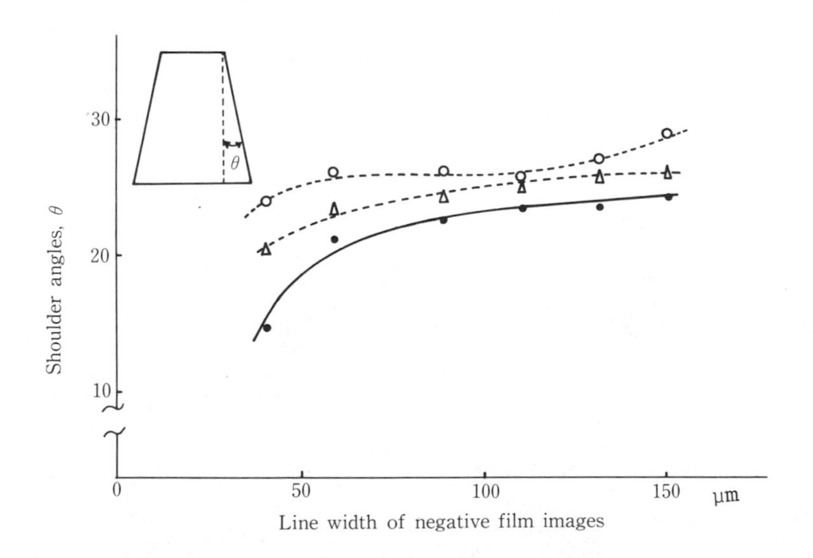

Figure 11. Change in the shoulder angles of image lines by change in the amount of exposure: (○) less exposure (600 mj/cm²); (△) proper exposure (1,000 mj/cm²); (●) over exposure (1,400 mj/cm²) (determined with Actino Integrator made by Oac Seisakusho Co.)

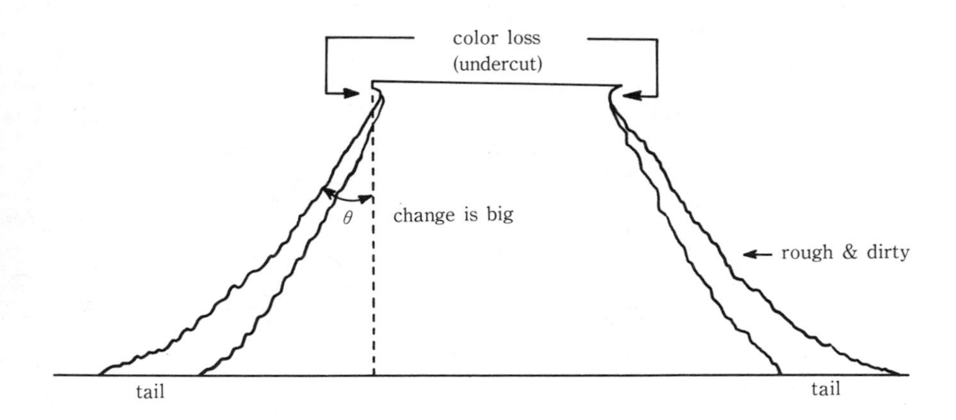

Figure 12. Shoulder shape of Zn powderless-etched plate

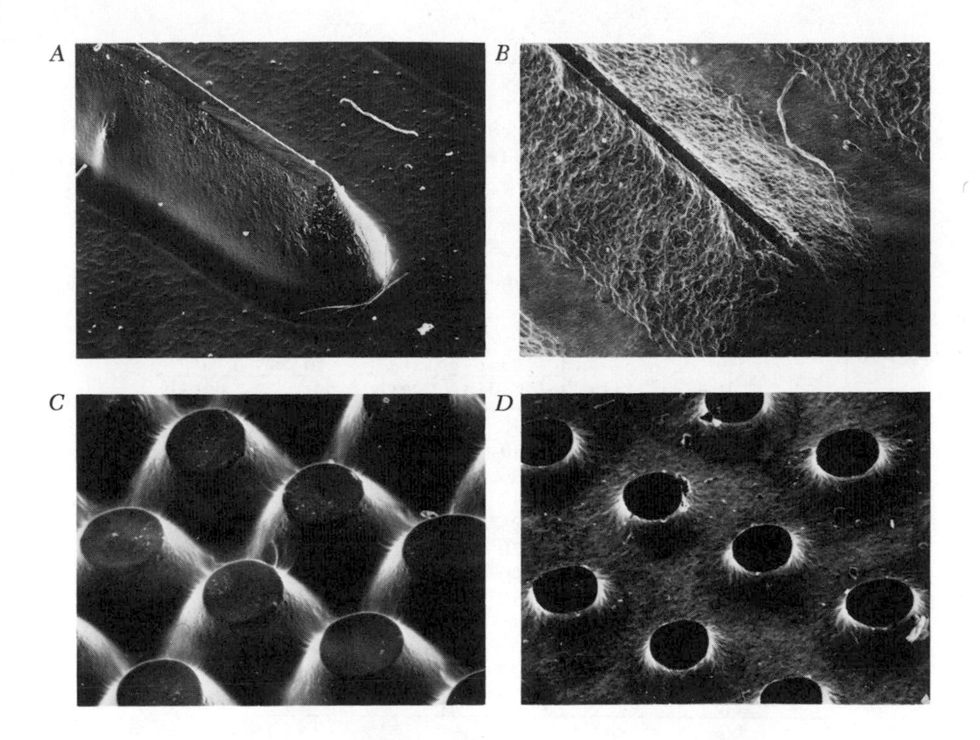

Figure 13. Comparison of reliefs between the Photopolymer Plate and Zn powderless-etched plate: (A) the Photopolymer Plate (100 μm line); (B) Zn powderless-etched plate (100 μm line); (C) plate made from the Photopolymer Plate (picture taken from 45° angle, magnification: 30 ×, showing dot: 25% & 65 lines); (D) Zn plate obtained by powderless etching (picture taken from 45° angle, magnification: 30 ×)

Figure 14. Rolling press

5. On Reproduceability from the Photopolymer Plate to Paper-Mold
The images in a negative film can be reproduced with good fidelity. Presently used zinc plates, because of their suspected low color loss, show a less fidelity compared with photopolymer plates, but their reproduceability are very similar as shown in Table 2.

Table 2 - Comparison of Dot Dimension between Zinc Plate and The Photopolymer Plate

Density	Dot dia. in nega. film	Zinc Plate		The Photo-polymer plate	
		dot dia.	depth	dot dia.	depth
0.10	80 μm	60 μm	-	70 μm	-
0.19	100	90	-	90	-
0.28	140	130	-	130	-
0.37	170	155	-	165	-
0.47	200	185	-	195	-
0.57	230	220	-	230	-
0.67	250	240	-	250	-
0.77	270	250	-	270	82 μm
0.89	265	275	135 μm	245	61
1.00	210	230	129	200	34
1.11	165	170	85	-	-
1.20	-	120	65	-	-

The comparison of reproduceability of diameters of dots in the highlight areas between the Photopolymer Plate and zinc plate from a negative film, showed 90-95% for the Photopolymer Plate and about 75% for a zinc plate because the latter showed color loss. (See Fig. 15)

Comparison of Reproduceability of Halftones between the Photopolymer Plate and Powderless Etched Zinc or Magnesium Plate
As shown in Fig. 16, the Photopolymer Plate will reproduce the halftone almost completely from the negative film. This is in comparison with photopolymer plates, zinc plates or magnesium plates which give much less reproduceability.

Comparison of Reproduceability of Isolated Image Lines between the Photopolymer Plate and Powderless Etched Zinc or Magnesium Plate
As shown in Fig. 17, the Photopolymer Plate will reproduce the isolated image lines almost completely from the negative film. Contrary to the Photopolymer Plate, zinc plate or magnesium plate gives much less reproduceability of only 76-65%.
The Photopolymer Plate produces images by simple dissolution of unexposed areas of the photopolymer resin. However, in case of

metal plates, images are produced by etching through the chemical reaction of acid; therefore, metal plates show that undercut is an unavoidable phenomenon to some extent with powderless etching which is a considerably excellent etching technique.

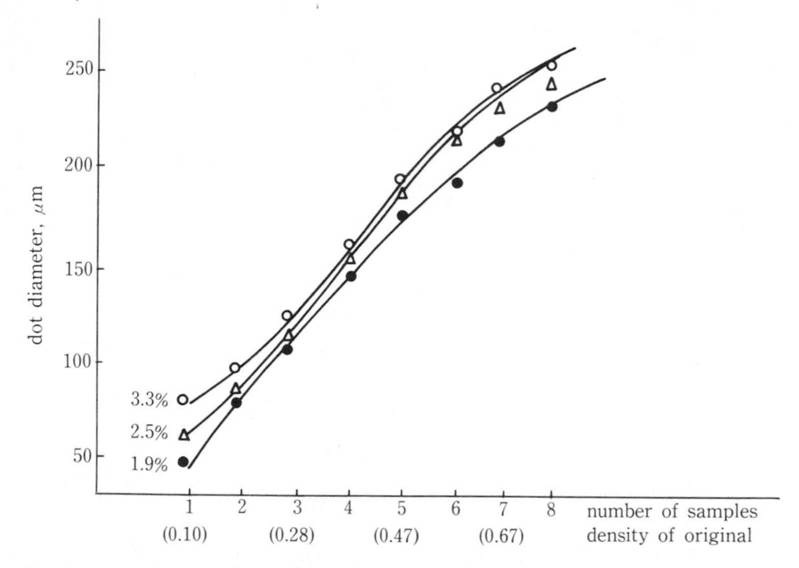

Figure 15. Comparison of dot diameters in the highlight areas of the Zn plate and the Photopolymer Plate: (○) dot diameter of negative; (△) dot diameter of the Photopolymer Plate; (●) dot diameter of Zn plate; (%) halftone % of dot diameter

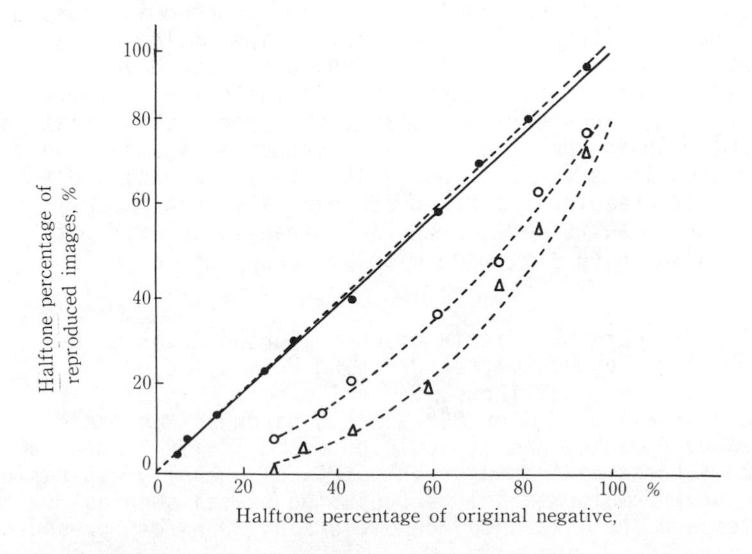

Figure 16. Comparison of reproduceability of halftones between the metal plate and the Photopolymer Plate: (●) the Photopolymer Plate; (△) Zn plate (powderless-etched); (○) Mg plate (powderless-etched)

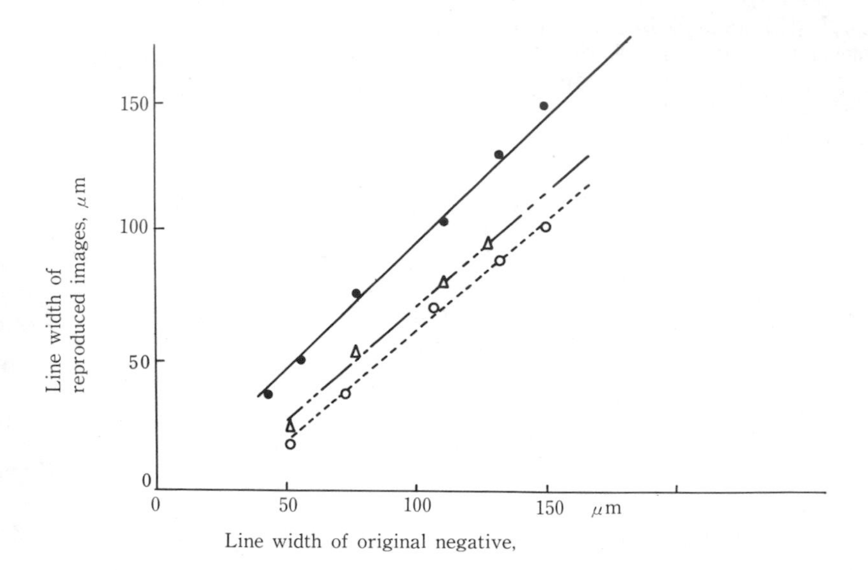

Figure 17. Comparison of reproduceability of isolated image lines between the Photopolymer Plate and metal (Zn & Mg) plates: (●) the Photopolymer Plate; (△) Zn plate (powderless-etched); (○) Mg plate (powderless-etched)

6. On Reproduceability of Paper Mold made from the Photopolymer Plate to Stereo of Lead

As is clear from Table 2, zinc plates gave relatively deep images compared with those of the Photopolymer Plate. However, there is not much difference in the depth between those stereos obtained from zinc plates and the Photopolymer Plate, as shown in Table 3. This means that in reproduction work from a paper mold into a stereo, even if an effort is made to give more depth beyond necessity, it is not actually reproduced in the stereo. The Photopolymer Plate can show a satisfactory reproduceability if it has 30-40 μm depth in the shadow area. Further evaluations were made on isolated lines (in case of 60-150 μm line width) and depth in reverse area for their reproduceability onto paper surface, and the results were more stable than those with metal plates. Stereos can be also made with polypropylene as well as with lead.

7. On Making Thermoformed Matrix of Phenol Group Resin to be used for Thermoformed Polymer Printing Plates including Rubber Plates and the like.....

The Photopolymer Plate of over 85° Shore D hardness can stand temperatures over 160°C, and it could be successfully used as a master plate for making thermoformed matrix of phenol group resin to be used for thermoformed polymer printing plates such as rubber plates and the like under conditions of 30 kg/cm^2 pressure and temperature of 150°C for 8 minutes. In this case, it is more efficient to use a quick-drying spray of graphite group together as a releasing agent. The matrixes thus obtained are widely used for reproducing flexographic rubber printing plates.

Table 3- Comparison of Reproduceability of Dot Diameters in Stereos made from Zinc Plate and the Photopolymer Plate

Density	Dot dia. in negative film	Stereo made from			
		zinc plate		Photopolymer Plate	
		dot dia.	depth	dot dia.	depth
0.10	80 μm	50 μm	-	-	-
0.19	100	70	-	80	-
0.28	140	100	-	100	-
0.37	170	120	-	150	-
0.47	200	150	-	170	-
0.57	230	200	-	195	-
0.67	250	205	-	220	-
0.77	270	230	-	245	41 μm
0.89	265	270	47 μm	220	36
1.00	210	230	34	160	25
1.11	165	145	20	-	-

Figure 18. Thermoformed matrix of phenol groups by the Photopolymer Plate

8. Waste Solution

No toxic substance is contained in the waste water of the Photopolymer Plate washout solution, and it is safe.

9. Influence of the Photopolymer Plate to the health
Acute toxic tests with mouses show that the exposed Photopolymer Plate has no toxidity at all and even the unexposed Photopolymer Plate has only a very weak toxidity; therefore, there are no bad influences to the health and the Photopolymer Plate can be handled safely.

Figure 19. Thermoformed rubber printing plate and matrix by the Photopolymer Plate

10. Conclusion
The Photopolymer Plate, a water developable photopolymer relief printing master plate made by modification of polyvinyl alcohol with urea group oligomer having a functional polyvinyl base, for making paper mold and matrix for printing master plates, has characteristics which are very close to those of the conventionally etched metal relief printing plates and has some superior points to them.
Furthermore, the Photopolymer Plate does not require handling of dangerous chemicals which must be handled with care and which create pollution problems, and it can be safely processed with water alone.
Because of these advantages, the Photopolymer Plate has been accepted by most of the leading Japanese newspaper companies and it is attracting more people in different field of the printing industry.

11. Gratitude
We would like to express our sincere appreciation to the engineers of the major newspaper companies in Japan for their cooperation in presenting their valuable information and data on the evaluation of the Photopolymer Plate.

References

1. L. Plamberk Jr., U.S.P. 2,760,863 (1956)
   " U.S.P. 2,791,504 (1957)
2. Jpn, Patent Kokai Tokkyo Koho 73-22343
   " 77-39287
   " 77-40862
   " 74-27522
   " 76-39846
   " 71-9284
   " 71-26125
   M. Hasegawa, U.S.P. 3,890,150 (1975)
3. Jpn, Patent Kokai Tokkyo Koho 76-58102
4. Y. Takimoto, U.S.P. 3,801,328 (1974)
5. H. Toda, U.S.P. 4,045,231 (1979)
6. Jpn, Kokoku Tokkyo Koho 79-3790

RECEIVED July 12, 1979.

18

Sensitized Photodegradation of Polymethylmethacrylate

MINORU TSUDA and SETSUKO OIKAWA—Laboratory of Physical Chemistry, Pharmaceutical Sciences, Chiba University, Chiba 260, Japan

YOHICHI NAKAMURA, HIDEO NAGATA, AKIRA YOKOTA, and HISASHI NAKANE—Tokyo Ohka Kogyo Co., Ltd., Samukawa, Kanagawa 253-01, Japan

TOSHIRO TSUMORI and YASUAKI NAKANE—Sony Corporation, Semiconductor Development Division, Atsugi, Kanagawa 243, Japan

Poly(methylmethacrylate), PMMA, is a well-known degradable polymer in the radiation chemistry of macromolecule (1). Hatzkis reported that PMMA is an excellent resist material usable in the microfabrication technology for manufacturing the microelectronic devices where X-rays and electron beams are used as radiation sources (2).

There is an empirical rule proposed by Millar, Wall and Charlesby which is useful for the prediction of the radiation effect of polymers, i.e., crosslinkable or degradable (1); where vinyl polymers in which there are two side chains attached to a single carbon (namely,$-CH_2-CR_1R_2-$) degrade while those with a single side chain or no side chain (namely, $-CH_2-CR_1H-$ or $-CH_2-CH_2-$) crosslinks. However, considerable exceptions are known to this rule (1). Recently, the authors demonstrated theoretically that there is a fundamental difference between degradable and crosslinkable polymers, which is due to the chemical or electronic structures of polymers themselves (3). The radiation effects of polymers are governed by the reaction in the excited states without exceptions. Millar-Wall-Charlesby's empirical rule is exact only when the shape of the adiabatic potential curve in the ground state reflects those in the excited states. Typical examples illustrating this for degradable and crosslinkable polymers are shown in Fig. 1 and Fig. 2, respectively (4). We find that the activation energies of the main chain cleavage reactions in the excited states are very small in the former and quite large in the latter. On the other hand, it is easily shown by the same procedure that the C-H bond is very weak in the excited states and a polymer radical which crosslinks (4) will be formed especially readily in the latter case.

If the newly proposed theory is applicable in the case of PMMA, the polymer should be degraded by the irradiation of UV light because PMMA has a weak absorption band near the 230 nm wavelength region (Fig. 5). Lin demonstrated that PMMA is degradable under the irradiation of Deep UV (200∿350 nm light) giving a high-quality resist image (5). Since the deep UV lithography

0-8412-0540-X/80/47-121-281$05.00/0

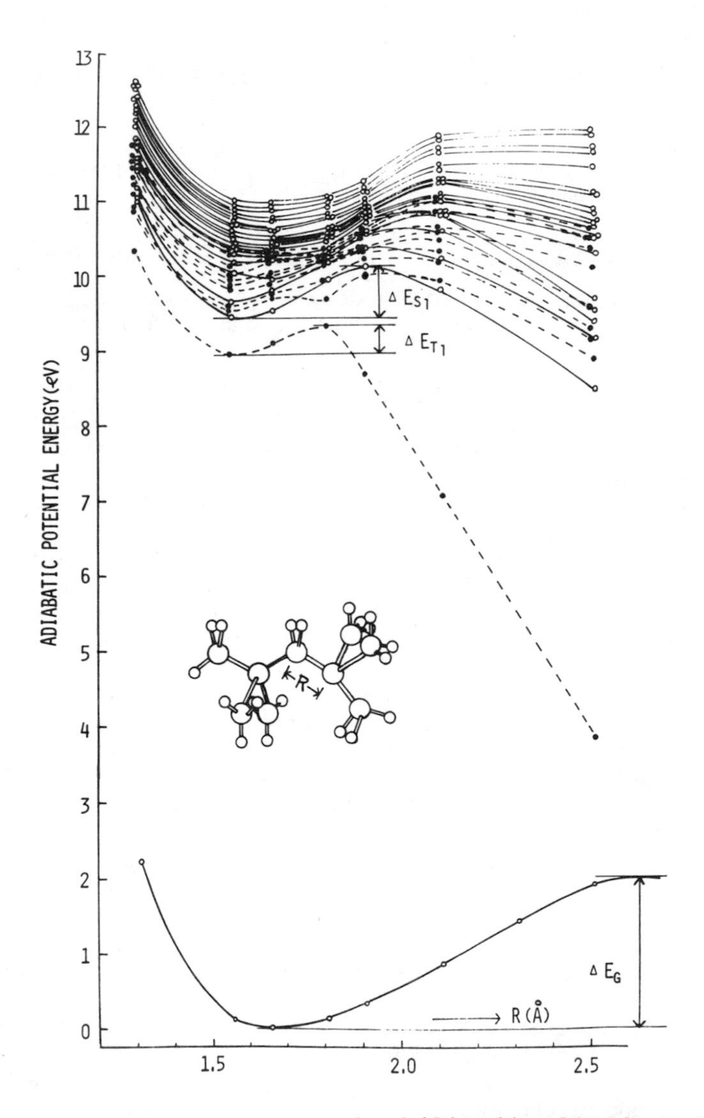

Journal of Polymer Science, Polymer Chemistry Edition

Figure 1. Adiabatic potential curves in the main chain scission of a model compound of poly(isobutylene): 2,2-, 4,4-tetramethylpentane (4). ΔE_{S_1}*(=0.61eV),* ΔE_{T_1}*(=0.35eV), and* ΔE_G*(=2.05eV) are the activation energies of the main chain scission in the lowest singlet excited state* (S_1)*, the lowest triplet state* (T_1)*, and the ground state, respectively.*

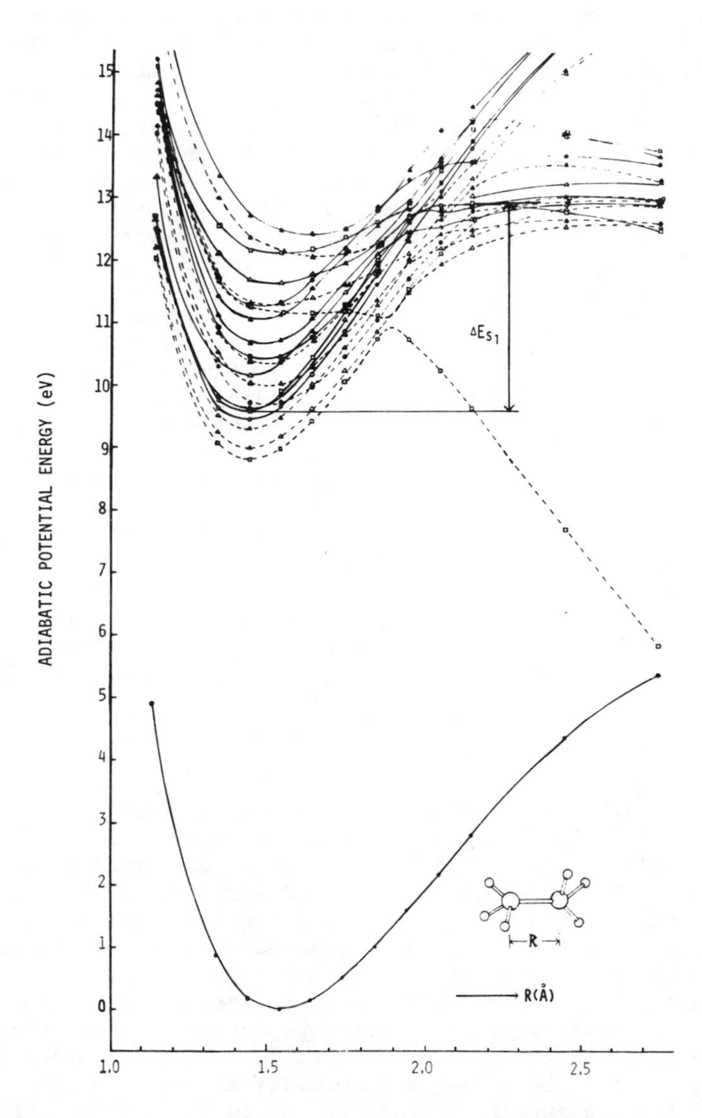

Journal of Polymer Science, Polymer Chemistry Edition

Figure 2. Adiabatic potential curves in the main chain scission of a model compound of polyethylene: ethane (●) A1g; (○) A1u; (■) A2g; (□) A2u; (▲) Eg; (△) Eu; (——) singlet; (– – –) triplet (4)

is expected to be a near-future technique for the production of the microelectronic devices (6), the development of polymer materials which have excellent properties as a Deep UV resist is an important problem. Although PMMA gives a high-quality resist image, its disadvantage from the practical viewpoint is its very low sensitivity to Deep UV light (the slow reaction rate of the photodegradation) and its weakness as a resistive material in the plasma etching which is used in the microfabrication process. Generally speaking, it is not easy to find a new polymeric material which gives a high-quality image. So, the accerelation of the reaction rate of the photodegradation (the phenomenon is called sensitization) of PMMA is preferable for the present purpose. A successful example of sensitization is found in the case of poly(vinyl cinnamate) (7).

We recently found that some aromatic compounds accerelate the degradation reaction of PMMA under the Deep UV light irradiation. The sensitivity of the sensitized PMMA is 4 times larger than that of unsensitized PMMA, when the sensitized material was used as a positive type Deep UV resist for the formation of microimage by the practical aligner which loads a Xe-Hg short arc lamp newly developed for the Deep UV lithography (8). We confirmed that the main chain degradation of PMMA is accerelated by the senistization. The measurement of the spectral dependence of the absolute quantum yield spectra of PMMA sensitized by some of the novel sensitizers revealed that the quantum yield itself increases even in the wavelength region where PMMA itself absorbs light and has its intrinsic sensitivity. This paper presents these new findings from the experimental point of view.

EXPERIMENTAL

Measurement of Sensitivity

PMMA in ethyleneglycol monoethylether monoacetate (7 wt%) was coated in 0.5 μm thickness by the spinning method on a thermally oxidized silicon wafer. The coated polymer was dried at 80°C for 30 min (prebaking), and then irradiated stepwise at the regular time intervals by a ultra high-pressure Hg short arc lamp (100 W) at the distance of 11 cm. The irradiated polymer coating was immersed in the developer (mixture of ethylacetate 1 volume and isoamylacetate 9 volumes) at 25°C for 1 min (development) and linsed. The relative sensitivity of polymer was defined as the reciprocal of the least irradiation time required for the dissolution of the polymer where more than 90 % of the original thickness was maintained at the non-irradiated part of the coating. (See Formula (3. a)) The time interval was ajusted in order to obtain the same significant figures. The sensitized polymer solution was prepared by adding the sensitizer (10 wt% of the polymer) to the PMMA ethyleneglycol monoethylether monoacetate solution. Similar measurements were also made using a Xe-Hg short arc lamp (500 W) for Deep UV lithography.

Mesurement of Spectral Sensitivity

The polymer coating prepared as described above was irradiated stepwise at the regular intervals by the monochromatic light obtained by the monochrometer mounting the 200 nm-blaze grating and 5 kW Xe short arc lamp (9) (Fig. 3). Light emitted from the Xe lamp was introduced to the spectroscope using a reflective grating. The spectrogram of the sensitivity of the monochromatic light is obtained when the resist coating was irradiated at the sample holder. The relative sensitivities over a wide range of the wavelength are obtained at once by this method as shown in Fig. 4. The intensity of the monochromatic light was measured using a vacuum thermocouple with a quartz window (10). The spectral sensitivity was corrected by using the intensity data shown in Fig. 6.

Measurement of Quantum Yield

The number of photons, incident to 1 cm^2 surface of the polymer coating, of 2537 Å light from the low pressure Hg long arc lamp was measured by a chemical actinometer (11), where a solution of $K_3Fe(C_2O_4)\cdot 3H_2O$ in a cell with a quartz window was irradiated in place of a polymer coating sample. The least number of photons required for the dissolution of the PMMA coating when immersed in the developer was $1.457\cdot 10^{-4}$ einstein/cm^2.

It was found from the absorption spectrum that 1.1 % of the incident photons were absorbed at 2537 Å by a PMMA film of 0.5 μm thickness (Fig. 5). The molecular weight distribution and the average molecular weight of the coated polymer which was irradiated for the least irradiation time required for the dissolution of polymer coating in the developer were measured by gel-permeation chromatography (Fig. 7).

From noting the changes of the molecular weight distributions and the average molecular weight before and after UV light irradiation, we tentatively postulate the reaction mechanism as follows:

$$\left[-\underset{\underset{\underset{CH_3}{|}}{\underset{O}{|}}}{\underset{C=O}{\overset{CH_3}{\overset{|}{C}}}}-CH_2-\right]_{m+n} \xrightarrow{h\nu} \sim\underset{\underset{\underset{CH_3}{|}}{\underset{O}{|}}}{\underset{C=O}{\overset{CH_3}{\overset{|}{C}}}}\bullet \;+\; \bullet CH_2-\underset{\underset{\underset{CH_3}{|}}{\underset{O}{|}}}{\underset{C=O}{\overset{CH_3}{\overset{|}{C}}}}\sim \longrightarrow \left[-\underset{\underset{\underset{CH_3}{|}}{\underset{O}{|}}}{\underset{C=O}{\overset{CH_3}{\overset{|}{C}}}}-CH_2-\right]_{m} + \left[-\underset{\underset{\underset{CH_3}{|}}{\underset{O}{|}}}{\underset{C=O}{\overset{CH_3}{\overset{|}{C}}}}-CH_2-\right]_{n}$$

Using the molecular weight change and the number of photons required for the change, the absolute quantum yield at 2537 Å of PMMA was obtained where the value of 0.9 was used as the density of PMMA film. Once the absolute quantum yield at 2537 Å was obtained, the absolute quantum yields over all the spectral range can be calculated from the relative data of the spectral sensiti-

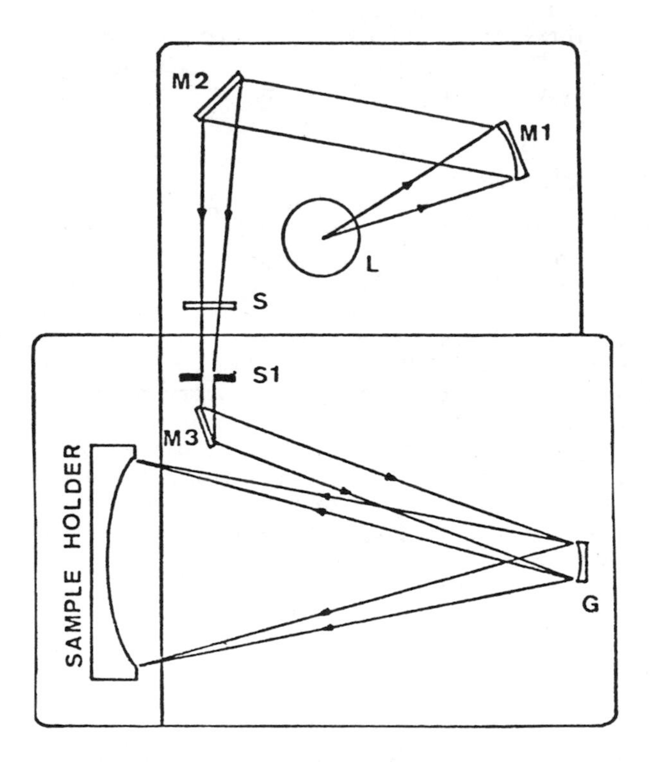

Figure 3. Optical system of instrument of spectral sensitivity measurement: (L) Xe arc lamp; (M1) concave mirror; (M2, M3) mirror; (S) shutter; (S1) slit; (G) concave reflective grating

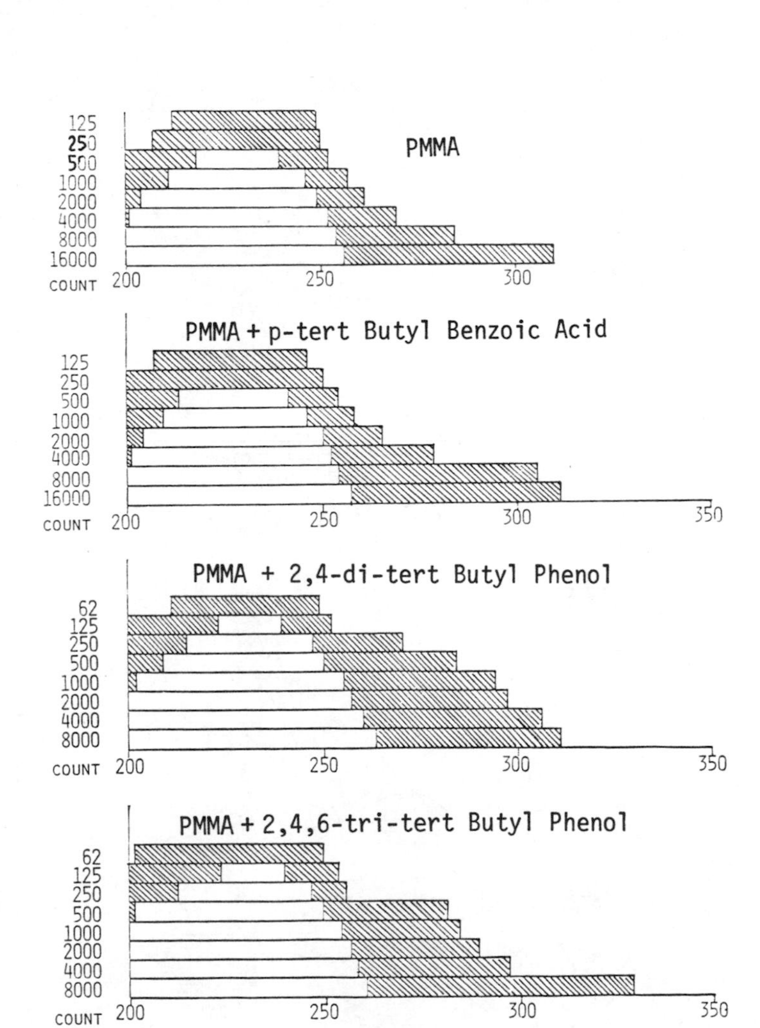

Figure 4. Spectral sensitivity data of PMMA and sensitized PMMA obtained by the equipment shown in Figure 3: (A) PMMA; (B) PMMA + p-tert-*butyl benzoic acid; (C) PMMA + 2,4-di-*tert-*butylphenol; (D) PMMA + 2,4,6-tri-*tert-*butylphenol*

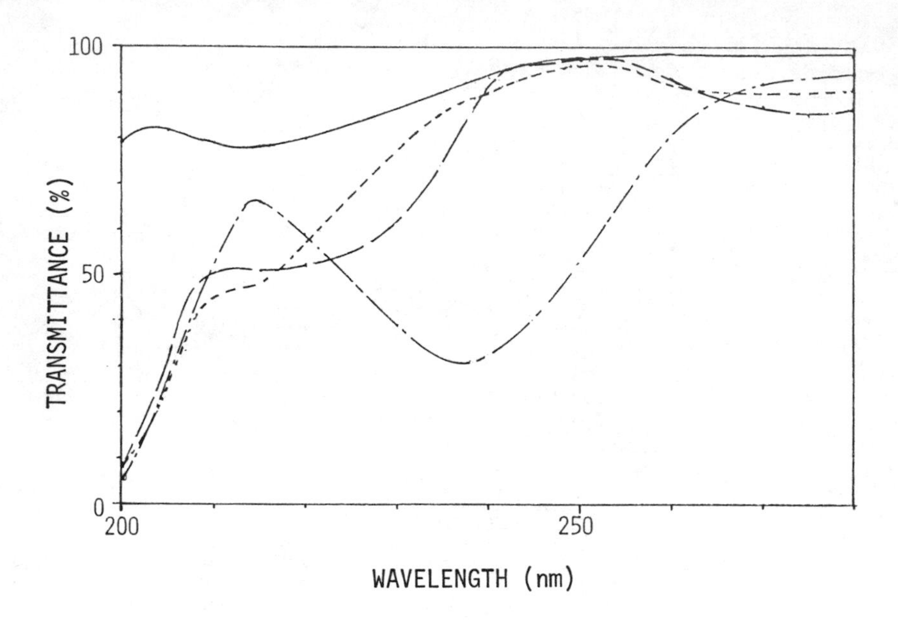

*Figure 5. Spectral transmittance of resist film: (——) PMMA; (- - -) PMMA + 2,4,6-tri-*tert*-butylphenol; (— —) PMMA + 2,4-di-*tert*-butylphenol; (- · -) PMMA +* p-*tert*-*butyl benzoic acid*

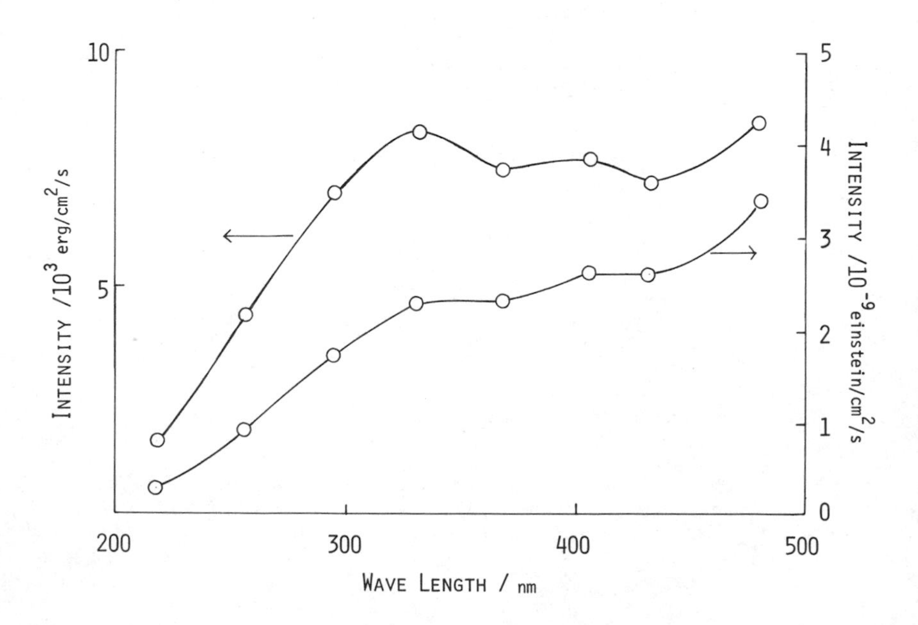

Figure 6. Emission spectrum of Xe lamp

vity as will be shown later (Fig. 8). For the sensitized polymer coating, the same method is applicable. Variables are the photon absorption data in the % scale which depends upon the absorption spectrum of the sensitizer, and the least irradiation time required for dissolution of polymer coating in the developer. The absorption spectra of sensitizers were measured in PMMA film of 0.5 μm thickness and was always 0.5 μm as measured by TALYSTEP (Tayler-Hobson, England). The density of the film was assumed to be 0.9 in all cases.

RESULTS AND DISCUSSIONS

Definitions of Sensitivity and Quantum Yield

Sensitivity, S, of the thin resists film is defined as the number, N, of a specific reaction occured in one polymer molecule, which forms the film, by one photon irradiated on a unit surface (cm^2). In this case quantum yield, ϕ, is defined as the number, N, of a specific reaction occured in one polymer molecule by one photon absorbed per one polymer molecule; namely, ϕ is the number N of a specific reaction caused by one photon absorbed:

$$S = dN\ (\text{reaction/molecule})\ /\ dE_{irrad.}(\text{photon/cm}^2) \qquad (1.a)$$

$$\phi = dN\ (\text{reaction/molecule})\ /\ dE_{absorb.}(\text{photon/molecule}) = dN\ /\ dE_{absorb.}\ (\text{reaction/photon}) \qquad (1.b)$$

$$dE_{irrad.} = d[I_o t] \qquad (2.a)$$

$$dE_{absorb.} = (1/n)\ d[I_o(1-\exp(-kc))t] \qquad (2.b)$$

where I_o is the number of photons incident on the 1 cm^2 surface of the resist film during a unit time at a specific wavelength, k the absorption coefficient per one chromophore at the same wavelength and c the number of chromophore in the 1 cm^2 of the resist film of a definite thickness. The value (1-exp(-kc)) is the absorption of the resist film in the transmittance scale (%), n is the number of polymer molecule in the 1 cm^2 resist film of a definite thickness and t is the irradiation time.

A definite number of main chain cleavages should occur in the soluble part of the film when a developer is specified and the least amount of photons required for the dissolution of the film in the developer used. If the definite number is K, the following formulae are obtained;

$$S = \frac{\int_o^K dN}{I_o \int_o^{\Delta t} dt} = \frac{K}{I_o \Delta t} \qquad (3.a)$$

$$\phi = \frac{\int_0^K dN}{I_0(1-\exp(-kc))/n \int_0^{\Delta t} dt} = \frac{K}{\Delta t \ I_0(1-\exp(-kc))/n} \tag{3.b}$$

where Δt is the least irradiation time required for the dissolution of the irradiated part of the resist film in the developer. From (3.a) and (3.b) we obtain the relationship between the sensitivity S and the quantum yield ϕ; namely,

$$S = \frac{\phi \, (1-\exp(-kc))}{n} \tag{4}$$

Using formula (4), we obtain the absolute spectral sensitivities of various resist films (Fig. 9).

Relative Sensitivity of Sensitized PMMA

The relative sensitivity of the thin resist film is defined as the reciprocal of the least irradiation time required for the dissolution of the polyemr coating when the light source, the irradiation conditions, the conditions of the development and the developer are specified; because in this case I_0 and K of the formula (3.a) become constant.

Some examples of the sensitivities of the sensitized PMMA coatings are shown in Table 1, where 2,4,6-tri-tert-butyl phenol is most effective.

Table 1. Relative Sensitivity Using Ultra-High Pressure Hg Lamp as a Light Source

SENSITIZER	R.S.
(PMMA)	100
p-tert-butyl benzoic Acid	200
p-tert-Butyl Benzene	180
p-tert-Octyl Phenol	200
tert-Butyl Hydroquinone	120
2,5-Di-tert-Butyl Hydroquinone	220
2,6-Di-tert-Butyl p-Cresol	131
2,4-Di-tert Butyl Phenol	250
2,4,6-Tri-tert-Butyl Phenol	330
4,4'-Thio-bis(6-tert-Butyl-3-methyl Phenol)	150
2,2'-Methylene-bis(4-methyl-6-tert-Butyl Phenol)	131
2,2'-Methylene-bis(4-ethyl-6-tert-Butyl Phenol)	133

R.S.: Relative Sensitivity (PMMA=100)

Main Chain Degradation of PMMA

In order to prove that the sensitizer is effective in the main chain degradation of PMMA, the change in the molecular weight distribution of PMMA film before and after irradiation was measured by GPC using tetrahydrofuran as a elution developer and caliblated with a standard sample of polystylene. The caliblation method is known to give a reliable result when the elution developer is tetrahydrofuran (12). The least dosage of radiation required for the dissolution of the sensitized film in the developer was used in this experiment. The sensitizer used was 2,4,6-tri-tert- butyl phenol. The results appear in Fig. 7. It is clear from the results of Table 1 and Fig. 7 that the main chain cleavage reaction is accerelated by addition of sensitizers.

The feature of the profile of the curves in Fig. 7 is that the initial wider molecular weight distribution range converges to that of a smaller range by the irradiation. These moleuclar weight distribution changes are given in two ways in Fig. 7; i.e., the linear scale and the exponential scale in the molecular weight (abscissa). The linear scale has such a simple physical meaning that the area under the curves are the total weight of the sample. The exponential scale is not as easily interpreted although this scale is often used. It is interesting that the exponential distribution curves reflect the commonly used distribution measure, $\overline{Mw}/\overline{Mn}$. The values of $\overline{Mw}/\overline{Mn}$ are approximately equal before and after irradiation (Table 2), reflecting that these two curves are similar. The values of Table 2 were calculated from the data of Fig. 7.

Spectral Dependence of the Absolute Quantum Yield of the Main Chain Cleavage Reaction

The raw data of the spectral sensitivities of the sensitized and unsensitized PMMA, which were obtained using the equipment shown in Fig. 3, appear in Fig. 4. These data were corrected at first by the intensity data shown in Fig. 6; and then the spectral dependence of the quantum yield was obtained by the following procedure: in equation of (3.a), $I_0\Delta t$ is the number of the photon quanta(per 1 cm^2 of the film) irradiated at the ridge line of the mountain in Fig. 4, because the least amount of the photon quanta required for the dissolution of PMMA film was irradiated at the ridge line; $(1-\exp(-kc))$ is the absorption of the film in a definite thickness (0.5 μm in this experiment) in the transmittance scale (%); n is the number of polymer molecule containing in the 1 cm^2 film at the definite thickness. These values are all measurable. The value K is the number of the main chain fission occurring in one polymer molecule by the irradiation of the photon quanta of $I_0\Delta t$. The value of K is easily calculated to be about 4 from the data in Table 2. The value of $I_0\Delta t$ was precisely determined at 2537 Å using a chemical actinometer.

Using this data the spectral dependence of the absolute quantum yield of the main chain fission of PMMA was calculated follow-

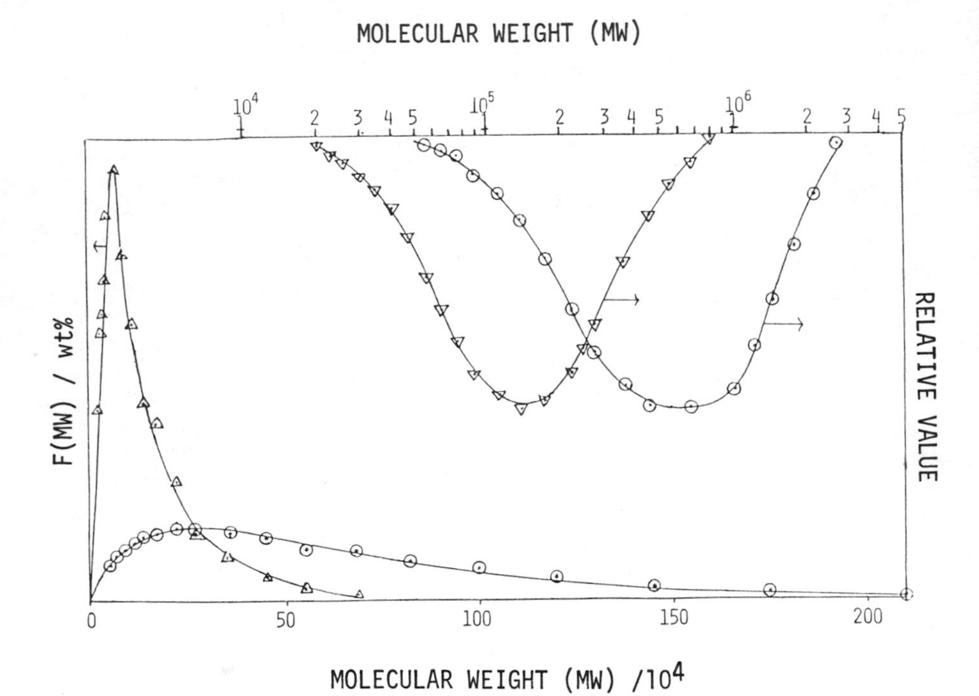

Figure 7. Molecular weight distribution of PMMA. F*(mol wt) is normalized in such a way that* $\int$ F*(mol wt)*d*(mol wt) = 1: (⊙) before exposure; (△) after exposure*

Table 2. Molecular Weight Change by UV-Irradiation*

Irradiation	$\overline{Mw}$	$\overline{Mn}$	$\overline{Mw}/\overline{Mn}$
before	685,000	341,000	2.01
after	160,000	91,400	1.74

* The least irradiation required for the dissolution of polymer coating in the developer.

ing the formula (3.b), and the results are shown in Fig. 8. The quantum yield of the sensitized PMMA increases over wide range of wavelength when we compared it with that of PMMA itself.

Absolute Spectral Sensitivity

The absolute sensitivity was defined as the number of main chain scissions occurring in one photon molecule when one photon was irradiated on a unit surface (1 cm^2) of the film. If the spectral dependence of the quantum yield has been obtained, the absolute sensitivity is calculated following the formula (4). The results obtained are shown in Fig. 9. The practical sensitivity will be the integral of the product of the absolute intensity of the irradiated light and the absolute sensitivity in Fig. 9 over all the wavelength of the spectra.

APPLICATION TO DEEP UV MICROLITHOGRAPHY

Experiments in this section were carried out using practical equipment, because the evaluation of the PMMA resist should have practical meaning. For the production test of the micro-patterns a Canon PLA-520F aligner for Deep UV lithography loading a Xe-Hg short arc lamp was used (8), and for plasma etching test a Tokyo Ohka OAPM-300 plasma etching machine was used; both operated under practical conditions.

Relative Sensitivity of the Sensitized PMMA

Relative sensitivities of PMMA sensitized by various sensitizers were measured using practical conditions. The results are shown in Table 3.

Formation of Micro Patterns

The electron microscope images of the fine patterns obtained by PMMA itself and PMMA sensitized by 2,4,6-tri-tert-butyl phenol on the silicon wafer appear in Figs. 10 and 11, respectively. The slit width of the fine patterns is 0.5, 1.0 and 1.5 μm in (a), and the magnifying image of the 0.5 μm area is shown in (b). The thickness of the coating is 0.5 μm.

Resistivity to the Plasma Etching

The comparison of the resistive property in the plasma etching process between PMMA itself and PMMA sensitized by 2,4,6-tri-tert-butyl phenol is shown in Fig. 12. Following the etching time, the thickness of the PMMA coating becomes thinner. The rate of the decreasing of the film thickness is proportional to the etching time in the former case, but it becomes very slow in the case of the sensitized PMMA. Therefore, the sensitized PMMA is a superior resist than PMMA itself in both properties of the sensitivity and the resistivity. This fact is true in the cases of other sensitizers.

The reason for the increase of the resistivity of the sensi-

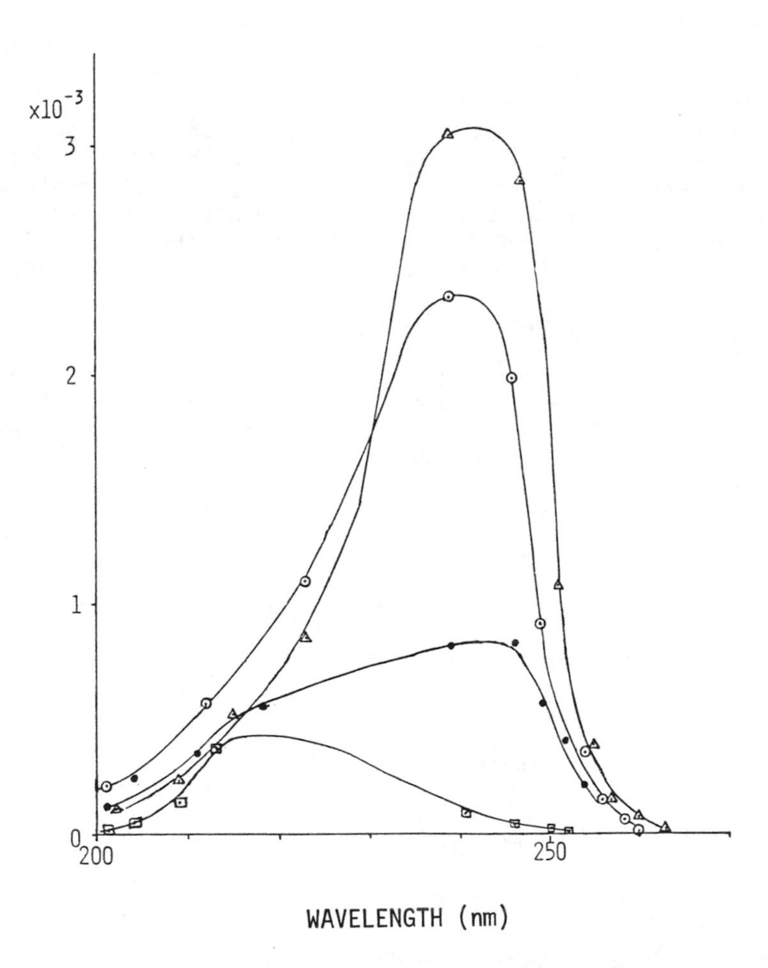

Figure 8. Spectral dependence of quantum yield: (●) PMMA; (⊡) PMMA + p-tert-*butyl benzoic acid; (△) PMMA + 2,4-di-*tert-*butylphenol; (⊙) PMMA + 2,4,6-tri-*tert-*butylphenol*

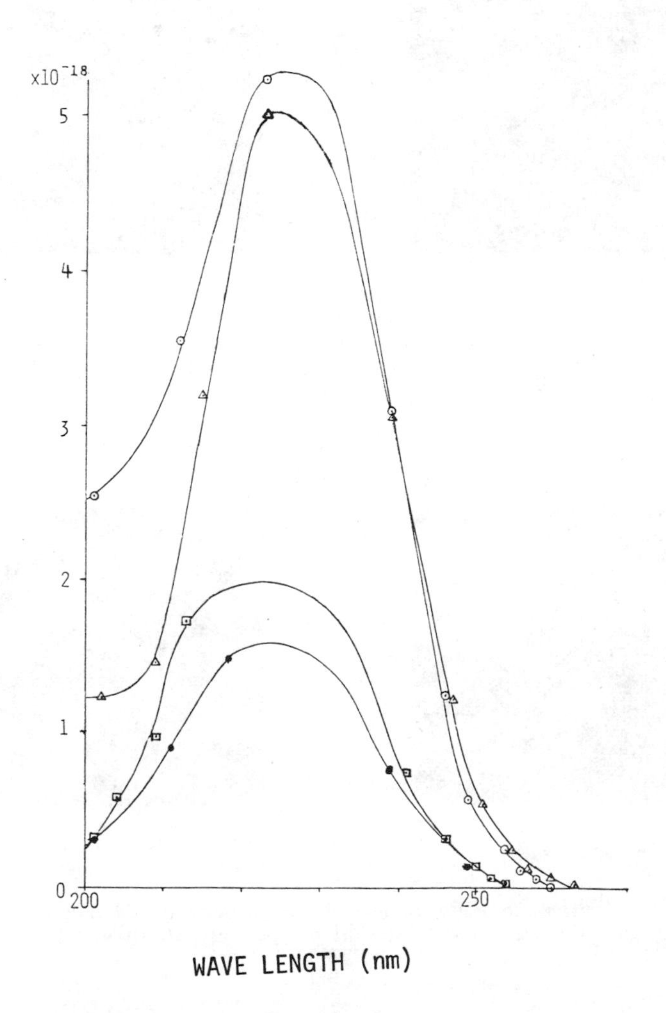

Figure 9. Spectral sensitivities of PMMA and sensitized PMMA: (●) PMMA; (⊡) PMMA + p-tert-*butyl benzoic acid; (△) PMMA + 2,4-di*-tert-*butylphenol; (⊙) PMMA + 2,4,6-tri*-tert-*butylphenol*

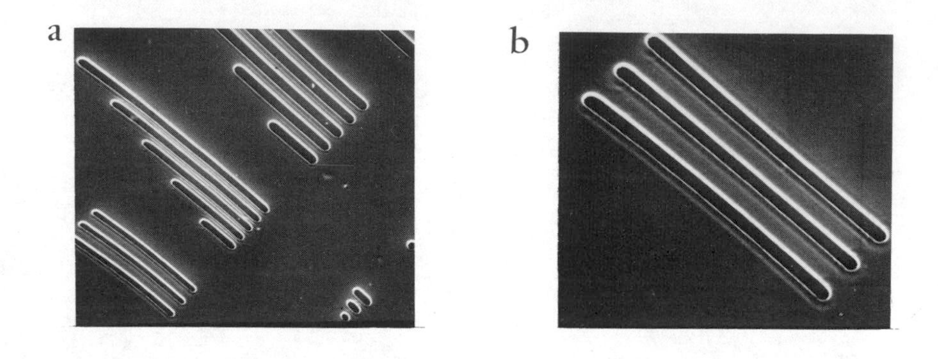

Figure 10. Electron microscope images of the fine patterns obtained by PMMA (window opening pattern) (a) × 1000; (b) × 2500

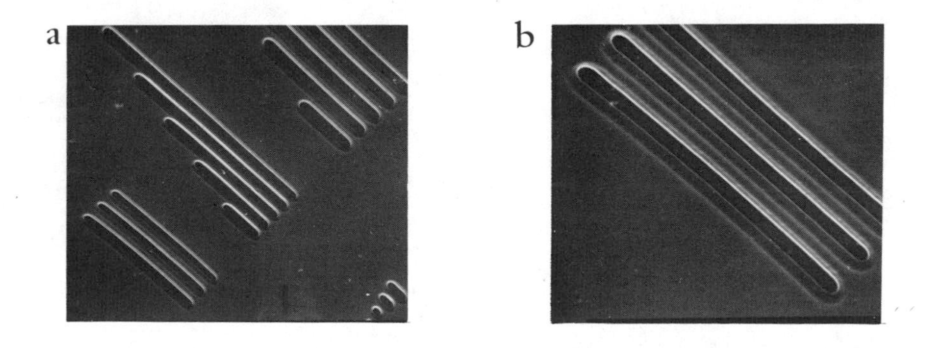

*Figure 11. Electron microscope images of the fine patterns obtained by 2,4,6-tri-*tert*-butylphenol-sensitized PMMA (window opening pattern) (a) × 1000; (b) × 2500*

Table 3. Relative Sensitivity of Sensitized PMMA Using Practical Conditions

SENSITIZER	
(PMMA)	100
p-tert-Butyl Benzoic Acid	120
2,4-di-tert-Butyl Phenol	200
2,4,6-tri-tert-Butyl Phenol	400

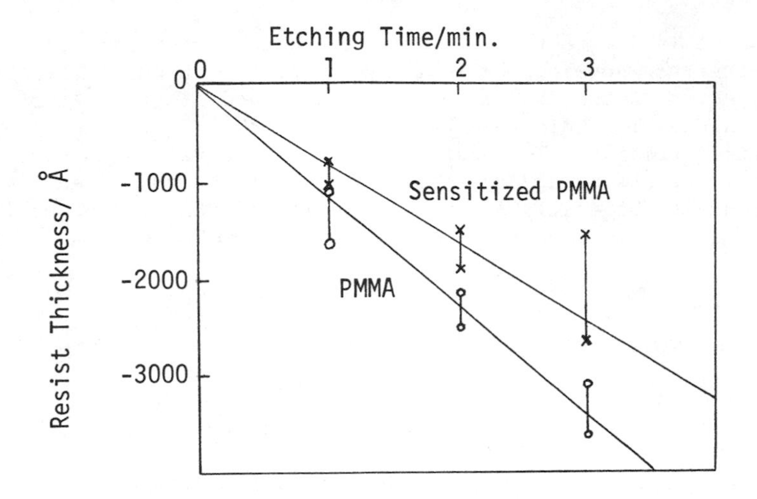

Figure 12. Resistive property of sensitized PMMA: decrease of resist thickness by plasma (Apparatus: OAPA-300; RF: 200 W; vac: 0.55 torr; gas: CF_4 95% and O_2 5%; flow rate: 1.2 L/min)

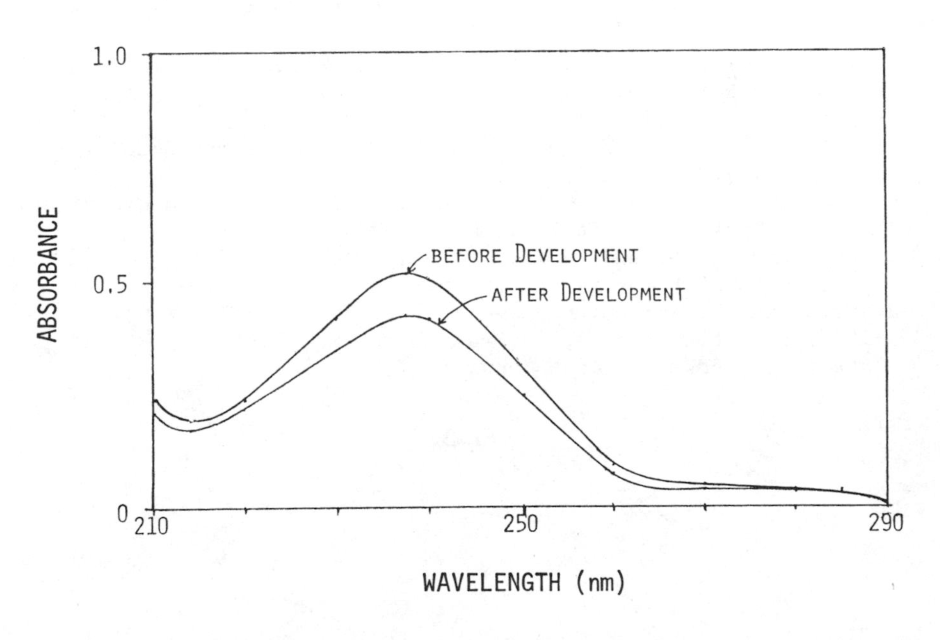

Figure 13. Sensitizer remaining in thin resist film after development (PMMA + p-tert-*butyl benzoic acid)*

tized PMMA is due to the fact that the added sensitizer remains even after the development in the thin film (0.5 μm). The absorption spectra of the resist coating before and after development shown in Fig. 13 indicates that more than 80 % of the unchanged sensitizer remains. This data means that the resist coating scarecely swells in the development process and that fine micro-patterns will be available by the sensitized PMMA resist newly developed in this research.

LITERATURE CITED

1. Charlesby, A., Atomic Radiation and Polymers, Pergamon, Oxford (1960).
2. Hatzkis, M.,J. Electrochem. Soc., (1969) 116, 1033.
3. Tsuda, M., Oikawa, S. and Suzuki, A., Polymer Eng. Sci., (1977) 17, 390.
4. Tsuda, M. and Oikawa, S., J. Polymer Sci., Polymer Chem. Ed., (1979), in press.
5. Lin, B. J., IBM J. Res. Develop., May 213 (1976); J. Vac. Sci. Tech., (1976) No. 6, 1317.
6. Tsuda, M., Oikawa, S., Nakamura, Y., Nagata, H., Yokota, A. and Nakane, H., Photo. Sci. Eng., (1979) in press.
7. Tsuda, M., Bull. Chem. Soc. Japan, (1969) 42, 905, and literatures cited therein.
8. Nakane,Y., Tsumori, T., Mifune, T., Kira, K., Momose, K. and Kanoh, I., The 25th Spring Meeting of Japan Society of Applied Physics, 30aM2 (1978).
9. Tanaka, H., Tsuda, M. and Nakanishi H., J. Polymer Sci., (A-1), (1972) 10, 1729
10. Tsuda, M., Oikawa, S. and Miyake,R., Nippon Shashin Gakkai-shi (J. Soc. Photo. Sci. Tech. Japan) (1972) 35, 90.
11. Hatchard, C. G. and Parker, C. A., Proc. Roy. Soc. (London), (1956) A234, 518.
12. Grubisic, Z., Rempp, P. and Bonoit, J., J. Polyemr Sci., (1967) B5, 753; Hattori, S., Hamashima, M., Nakahara, H. and Kamata, T., Kobunshi Ronbunshu (1977) 34, 503.

RECEIVED July 12, 1979.

19

The Effect of Photolysis on the Biodegradation of Some Step-Growth Polymers

SAMUEL J. HUANG, CATHERINE BYRNE, and JOSEPH A. PAVLISKO

Department of Chemistry and Institute of Materials, University of Connecticut, Storrs, CT 06268

Earlier studies on the biodegradation of polymers were initiated by the desire to avoid degradation, thereby obtaining long lasting materials. As the disposal of bio-inert synthetic polymer wastes became increasing difficult, the emphasis in studies on the biodegradation of polymers shifted to the design and synthesis of biodegradable polymers. Since many low molecular weight organic materials can be utilized by soil microorganisms, one of the approaches to biodegradable polymers has been the preparation of photodegradable materials that will give fragments that are biodegradable on photolysis. (1,2,3) One of the most important applications of biodegradable polymers is the use of these materials as agricultural mulches and controlled release agricultural chemical formulation. For use in agriculture, one cannot depend on photodegradation as the primary degradation process since the available amount of sunlight can vary considerably, and, furthermore, soil might cover the materials, thereby prevent light from reaching them. Biodegradable polymers without the need of photolysis for degradation must therefore be developed also. The optimal might be a polymer that is both biodegradable and photodegradable. We report here our recent results on the effects of photolysis on the biodegradation of step-growth polymers.

Experimental

Synthesis. Poly(ethylene sebacamide) (A), poly(1,2-propylene sebacamide) (B), poly(2-hydroxy-1,3-propylene sebacamide) (E), poly(2-hydroxy-1,3-propylene sebacamide-co-2propylene sebacamide) (F), poly(benzylethylene sebacamide) (H), and poly(piperazinyl sebacamide) (J) were prepared by interfacial polymerization using carbon tetrachloride and water as solvents. (4)

Poly(1,1-dimethylethylene sebacamide) (C) was prepared by solution polymerization of 1,1-dimethylethylenediamine and sebacoyl chloride in chloroform using triethylamine as base.

0-8412-0540-X/80/47-121-299$05.00/0

Poly(dodecamethylene D-tartrate) (G) was prepared by melt polymerization of 1,12-dodecanediol and tartaric acid with p-toluenesulfonic acid as catalyst. (5)

Polyureas D and I were prepared in solution with 1,6-diisocyanatohexane and the corresponding diamine using bicyclo-[2,2,2]-diazaoctane as catalyst. (6)

Photolysis. Polymer films suspended in water were irradiated with a water-cooled 450 w medium pressure Hanovia Hg lamp through quartz filter at a distance of 2.5 inches in the presence of air at r. t. for 24 hr. The intrinsic viscosities of the polymer samples before and after photolysis were compared. Weight losses of the sample were also measured.

Biodegradation. The abilities of the polymer samples to support the growth of the fungus *Aspergillus niger* as the only carbon nutrient source were compared before and after photolysis. American Society of Testing for Materials recommended procedure and rating of growth were used to report the results in Tables 1 & 2: 0 = no visible growth; 1 = 10% surface covered; 2 = 10-30% surface covered; 3 = 30-60% surface covered; 4 = 60-100% surface covered. Incubation period at r. t. was four weeks. Details of the biodegradation procedure were reported earlier. (4,6)

Results and Discussion

Since the primary mechanism for biological systems to degrade macromolecules is hydrolysis followed by oxidation we have directed our research efforts firstly to the degradable polymers containing hydrolyzable groups such as amide, ester, urea, and urethane. We reasoned that synthetic polymers containing structural and stereochemical features that are similar to that of proteins might be degradable by common proteases and esterases. Since many proteases are specific in cleaving peptide linkages adjacent to substituent groups, we decided to prepare substituted polymers, anticipating that the introduction of the substituents would make the polymers more degradable. Methylated, benzylated, and hydroxylated polymers were prepared and their biodegradations have been studied. Compared to the unsubstituted polymers we found the substituted polymers more susceptible to attack by enzymes and microorganisms. (4,6)

Although high molecular weight polyamides such as nylon-6, nylon-6,6, and nylon-12 were found to resist microbial (7,8) and enzyme attack (9) low molecular weight cyclic and linear oligomers of ε-aminocaproic acid were found to be utilized by certain bacteria isolated from the effuent water of a nylon-6 plant. (10,11, 12) Photolysis of polymers might result in fragmentation of the polymer chains thus leading to improved biodegradabilities. On the other hand, if photocrosslinking occurs one can expect a decrease in biodegradability.

Photolysis of the unsubstituted poly(ethylene sebacamide) (A), methylated poly(1,2-propylene sebacamide) (B), and poly(1,1-dimethylethylene sebacamide) (C) resulted in mostly chain fragmentation as indicated by the decreases in intrinsic viscosities of the polymer samples, Table 1. The same decrease in intrinsic viscosity was also observed for polyurea D. Polymer A and D remained bio-inert under the testing condition whereas the abilities for polymers B and C to support the growth of Apergillus niger were improved by photolysis.

Table 1. Photolysis and Biodegradation of Polymers that Undergo Primary Photofragmentation.

Polymer		[η]		A. niger growth[a]	
		Before	After	Before	After
$-NHCH_2CH_2NHCO(CH_2)_8CO-$	A	0.36	0.24	0	0
$-NHCH(CH_3)CH_2NHCO(CH_2)_8CO-$	B	0.57	0.51	1	2
$-NHC(CH_3)_2CH_2NHCO(CH_2)_8CO-$	C	0.06	0.05	1	2
$-NHCH(CH_3)CH_2NHCONH(CH_2)_6NHCO-$	D	1.00	0.80	0	0

[a] ASTM rating, see experimental.

Photolysis of hydroxylated polymers E, F, G, benzylated polymers H and I, and poly(piperazinyl sebacamide) J all resulted in crosslinking as the polymer samples became insoluble after photolysis. Interestingly enough, however, with the exception of polymer H, all polymer samples were found to be better carbon nutrients for the fungus Aspergillus niger after photolysis. Small amounts of weight losses (up to 4%) were also observed after photolysis. Since the samples were suspended in water during photolysis these results suggested that although photocrosslinking was the primary process photofragmentation also occurred to a small extent to give water soluble fragments. Photofragmentation also produced materials that were better

Table 2. Photolysis and Biodegradation of Polymers that Undergo Primary Photocrosslinking.

Polymer		A. niger growth[a] Before	After
$-NHCH_2CH(OH)CH_2NHCO(CH_2)_8CO-$	E	3	4
$-(NHCH_2CH(OH)CH_2NHCO(CH_2)_8CO)_x-$ $-(NHCH(CH_3)CH_2NHCO(CH_2)_8CO)_y-$	F		
x/y = 1/3		1	4
x/y = 1/1		0	4
x/y = 3/1		0	4
D- $-O(CH_2)_{12}OOCCH(OH)-CH(OH)CO-$	G	2	4
D,L- $-NHCH(CH_2Ph)CH_2NHCO(CH_2)_8CO-$	H	0	0
L- $-NHCH(CH_2Ph)COOCH_2CH_2OOCNH(CH_2)_6NHCO-$	I	1	1
$-N\langle piperazine \rangle NCO(CH_2)_8CO-$	J	0	1

[a]ASTM rating, see experimental.

nutrients for the fungus than the original polymers, as indicated by the substantial increases of fungal growth, Table 2.

There are several mechanisms by which carbonyl-containing polymers can react photochemically. The most important two are the Norrish Type I and Norrish Type II processes, both of which might result in chain fragmentation. However, the recently reported studies favored Norrish Type II process as the more important of the two for photodegradation of polymers. (13,14,15,16) It has been suggested that γ-hydrogen abstraction by the excited carbonyl group to give a short-lived biradical is responsible for the eventual fragmentation to olefin and carbonyl end groups.

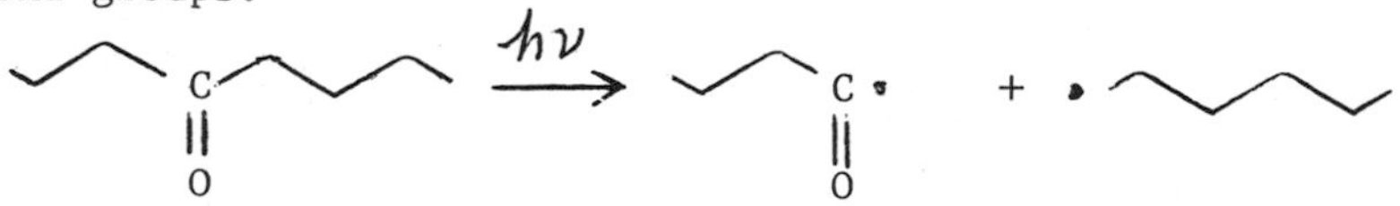

C=O —hν→ C•=O + •

Norrish Type I

R = H

R = Ph

Crosslinking

* Excited carbonyl

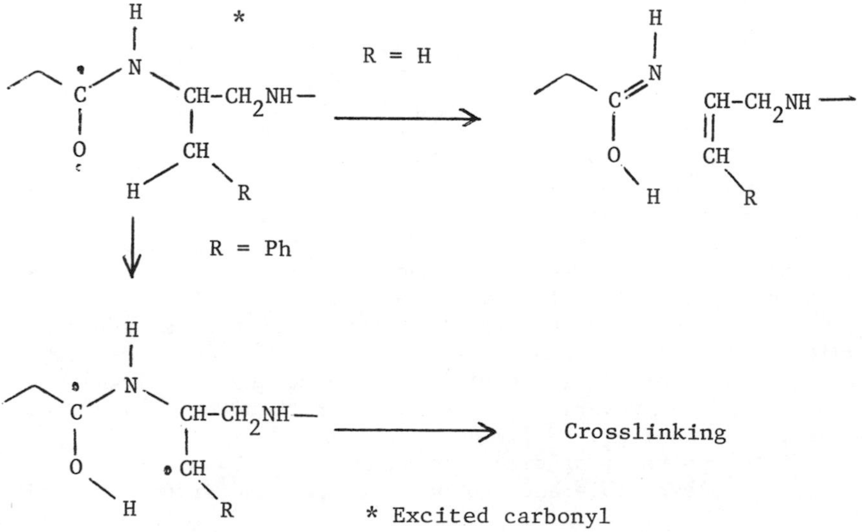

Norrish Type II

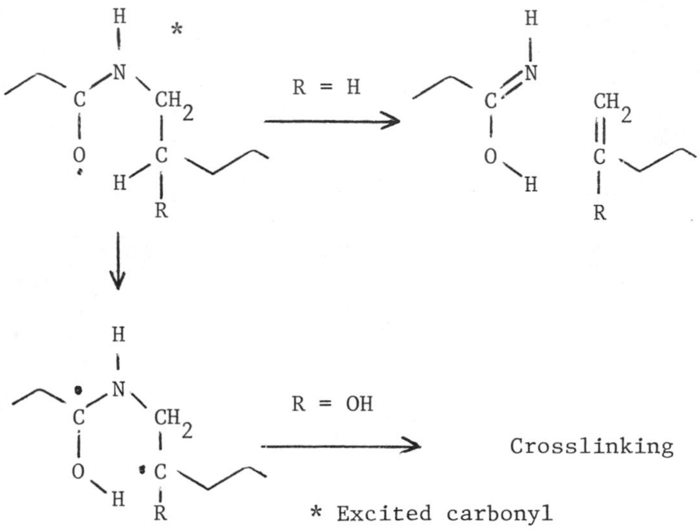

R = H

R = OH

Crosslinking

* Excited carbonyl

Norrish Type II

Our results suggest that in the cases of benzylated and hydroxylated polymers the biradical intermediates are stabilized by the phenyl and hydroxy group, resulting in a longer life for the intermediates. As a result, crosslinking becomes the favorable process. In the cases of the unsubstituted and the methylated polymers the short-lived biradicals lack such stabilization and the unimolcular fragmentation becomes the favorable process.

Since in most cases the abilities of biodegradable polymers to support fungal growth were found to be improved by photolysis it can be concluded that photolysis produced small fragments. These are better nutrients for the fungi. Of the few reports available on the effect of molecular weight on the biodegradability of polymers it is generally accepted that the lower molecular weight materials degrade faster than their higher molecular weight analogs. (8,17)

Of the polymers studied the methylated polyamides B and C are both biodegradable and photodegradable. Their use in agriculture are being studied.

Acknowledgement

We thank the National Science Foundation (Grants DMR 75-16912 and DMR 78-13402) and the University of Connecticut Research Foundation for financial supports.

Literature cited

1. Guillet, J. E., "Polymers and Ecological Problems", J. E. Guillet, Ed., Plenum Press, New York, 1973, pp. 1-26.
2. Scott, G., ref. 1, pp. 27-44.
3. Baum, B., and White, R. A., ref. 1, pp. 45-60.
4. Huang, S. J., Bitritto, M., Leong, K. W., Pavlisko, J., Roby, M., and Knox, J. R., Advances in Chemistry Series, No. 169, D. L. Allara and W. L. Hawkins, Eds., Am. Chem. Soc., 1978, pp. 205-214.
5. Bitritto, M., Ph. D. Dissertation, Univ. of Connecticut 1975.
6. Huang, S. J., Bansleben, D. A., and Knox, J. R., J. Appl. Polym. Sci., 1979, 23, 429.
7. Rodriquez, F., Chem. Technol., 1971, 409.
8. Potts, J. E., Clendenning, R. A., Ackart, W. B., and Niegisch, W. D., ref. 1, pp. 61-79.
9. Bell, J. P., Huang, S. J., and Knox, J. R., U.S. NTIS, Ad-A Rep. No.009577, 1974; No. 029935, 1975.
10. Fukumura, T., Plant Cell Physiol., 1966, 7, 93.
11. Fukumura, T., J. Biochem., 1966, 59, 537.
12. Kinoshita, S., Kageyama, S., Iba, K., Yamada, Y., and Okada, H., Agr. Biol. Chem., 1975, 39, 1219.
13. Pload, P. I., and Guillet, J. E., Macromolecules 1972, 5, 405.
14. Ranby, B., and Rabek, J. F., "Photodegradation, Photooxidation, and Photostabilization of Polymers", John Wiley and Sons, New York, 1976, pp. 48-49.
15. Sugita, K., Kilp, T., and Guillet, J. E., J. Polym. Sci. Polym. Chem. Ed., 1976, 14, 1901.
16. Wagner, P. J. Acc. Chem. Res., 1971, 4, 168.
17. Fields, R. D., and Rodriquez, F., "Proc. 3rd. International Biodegradation Symposium", J. M. Sharpley and A. M. Kaplan, Eds., Applied Science Publishers, London, 1976, pp. 775-784.

RECEIVED July 30, 1979.

20

Acceleration of Radiation-Induced Cross-Linking of Polyethylene by Chlorotrifluoroethylene

T. KAGIYA, T. WADA, N. YOKOYAMA, and H. ONO

Department of Hydrocarbon Chemistry, Kyoto University, Kyoto, Japan

Polyethylene is well known to be cross-linked by ionizing radiation through the recombination of polyethylene radicals produced by the irradiation (1). The efficiency of the reaction is not so high and the G-value has been reported to be 1-5 (2, 3).

In the previous papers, we reported that the radiation-induced cross-linking of polyethylene was accelerated by acetylene (4) and by the mixtures of acetylene and some fluorine-containing monomers (5).

The present paper is a continuation of these studies. We found that the cross-linking of polyethylene was promoted by the presence of chlorotrifluoroethylene (CTFE) and the unsaturated groups contained in polyethylene decreased rapidly with the proceeding of the cross-linking reaction. The cross-linkings of buatdiene- and isoprene-grafted polyethylenes in the presence of CTFE were carried out in order to elucidate the role of unsaturated groups in the cross-linking reaction. In addition the cross-linking of polyethylene was also studied in the presence of the mixture of CTFE and butadiene. On the basis of these results, the acceleration mechanism of the cross-linking of polyethylene by CTFE was discussed.

Experimental

Material. Films (thickness; 20μ) of low density polyethylene (LDPE) (Sumikathene:number average molecular weight; 2.0×10^4, density; 0.92) and high density one (HDPE) (Hizex: 1.4×10^5, 0.96) were used in the experiment. Polyethylene films which contained high concentration of unsaturated groups were prepared without gel formation by the radiation-induced grafting of butadiene and isoprene in gas phase onto LDPE film at the irradiation dose lower than 1.5 Mrad in the same method as reported previously (6). The concentrations of the total unsaturated group ($[U]_0$), trans-vinylene ($[Ut]_0$), vinyl ($[Uv]_0$) and vinylidene ($[Ud]_0$) in polyethylene used are shown in Table I.

0-8412-0540-X/80/47-121-307$05.00/0

Table I
Concentration of unsaturated groups in polyethylenes

Polyethylene	Number of unsaturated group per 1000 carbons $[U]_0$	$[Ut]_0$	$[Uv]_0$	$[Ud]_0$
HDPE	0.18	0.00	0.11	0.07
LDPE	0.41	0.06	0.21	0.14
BG-LDPE-1[a)]	1.32	0.93	0.35	0.04
BG-LDPE-2[a)]	4.32	3.19	1.00	0.03
BG-LDPE-3[a)]	9.25	6.76	2.40	0.09
IG-LDPE[b)]	1.91	0.04	0.93	0.98

a) Butadiene-grafted low density polyethylene
b) Isoprene-grafted low density polyethylene

CTFE was used as supplied from Daikin Kogyo Co. Ltd. (purity; 99.5% up). Commercially available butadiene was used as obtained.

Procedure. The radiation-induced cross-linking was carried out as follows. About 0.1 g of polyethylene film was placed in a glass ampoule of 30 mm diameter and 200 mm long. Gaseous CTFE and the mixture of CTFE/butadiene was introduced into the ampoule under the gas pressure of 1 atm. after evacuation of the ampoule. The ampoule was irradiated by γ-ray with a cobalt-60 at the dose rate of 0.05Mrad/hr at room temperature.

After the irradiation, the surface of the film was wiped with a soft cloth wetted with acetone, dried under vacuum at 40°C for 10 hours and weighed. The degree of grafting (Dg) was determined by the equation (1).

$$Dg = \frac{W_g - W_0}{W_0} \times 100 \quad (\%) \tag{1}$$

Where, W_0 and W_g are the weight of the sample films before and after the irradiation, respectively.

The polyethylene film thus obtained was packed in a 100 mesh stainless steel basket, extracted with hot p-xylene in a Soxhlet extractor for 48 hours, washed with acetone for 4 hours in the same type extractor, dried in vacuum for 20 hours at 40°C and weighed. The gel fraction (Gf) of the polymer was detrmined by the equation (2) from the weight change by the extraction.

$$Gf = \frac{W_a}{W_b} \times 100 \quad (\%) \qquad (2)$$

Where W_b and W_a are the weight of the samples before and after the extraction, respectively.

The IR-spectrum of the polymer was measured with a Nippon Bunko Model DS-403G infrared spectrometer. The concentration of the unsaturated groups in the polymer (number of unsaturated group per 1000 carbons) was determined from the absorbances at 966 cm^{-1} (trans-vinylene), 910 (vinyl) and 890 (vinylidene) in the IR-spectrum by the method reported by Cernia et al. (7).

The composition of the co-graft polymer obtained by the irradiation in CTFE/butadiene mixture was determined from the over all degree of grafting by weight measureement and the degree of grafting of CTFE (Dg(CTFE)) by IR-spectrum. It was recognized from the IR-spectrum of CTFE-grafted polyethylene that the value of Dg(CTFE) was proportional to the absorbance (D_{1215}) at 1215 cm^{-1} (assigned to C-F bond) as expreesed by the equation (3).

$$Dg(CTFE) = 91.6 D_{1215} \quad (wt\%) \qquad (3)$$

Results and Discussion

Radiation-Induced Cross-Linking in the Presence of Some Fluorine-Containing Monomers. The results of the radiation-induced cross-linking of low density polyethylene in the presence of various fluorine-containing monomers at a dose of 2.5 Mrad are summarized in Table II. Both the formation of gel and the decrease in the amount of unsaturated groups contained in polyethylene were found in the irradiation in the presence of these monomers. The highest values of the gel fraction and the degree of grafting were obtained in the irradiation in the presence of CTFE. In addition, the unsaturated groups decreased in the irradiation in CTFE more markedly than in the other monomers. These results lead to the consideration that the cross-linking of polyethylene in the presence of these monomers is resulted by grafting with the consumption of the unsaturated groups contained in polyethylene.

Role of Unsaturated Group in the Cross-Linking in the Presence of CTFE. In order to make clear the role of unsaturated group contained in polyethylene, the radiation-induced cross-linkings of the polyethylenes containing various concentrat-ions of the unsaturated groups were carried out in the presence of CTFE.

Figures 1-3 show the changes in the gel fraction of irradiated polymer, the degree of grafting of CTFE and the total concentration of the unsaturated groups ([U]) with the irradiation

Table II
Radiation-induced cross-linking of LDPE in the presence of various fluorine-containing monomers at a dose of 2.5 Mrad

Monomer	Dg (%)	Gf (%)	No. of unsatd. groups per 1000 carbons [U]	[Ut]	[Uv]	[Ud]
(Vacuum)	-	0	0.37	0.12	0.15	0.10
TFE[a]	10.0	14.9	0.19	0.05	0.09	0.05
CTFE	23.5	50.0	0.01	0.01	0.00	0.00
HFP[b]	1.4	11.8	0.30	0.07	0.14	0.09
VdF[c]	0.2	4.7	0.31	0.05	0.17	0.09
VF[d]	0.6	6.5	0.31	0.04	0.16	0.11
(Non-irradiated LDPE)			0.41	0.06	0.21	0.14

a) Tetrafluoroethylene
b) Hexafluoropropylene
c) Vinylidene fluoride
d) Vinyl fluoride

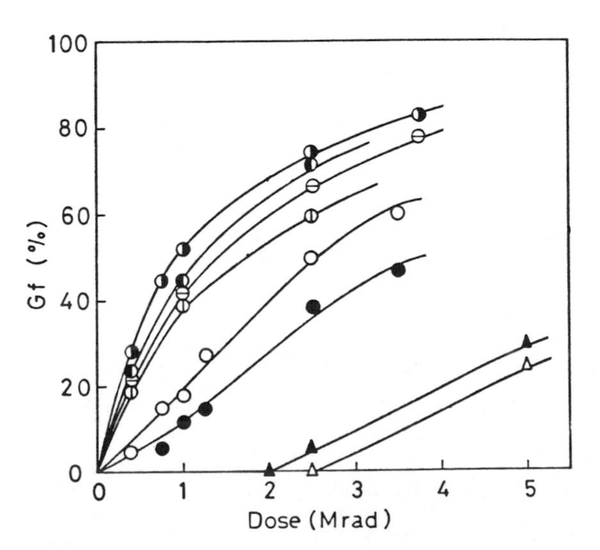

Figure 1. Plot of Gf vs. dose in the irradiation of various polyethylenes in the presence of CTFE ((●) HDPE; (○) LDPE; (⦶) BG-LDPE-1; (◐) BG-LDPE-2; (◑) BG-LDPE-3; (⊖) IG-LDPE) and in vacuum ((△) LDPE; (▲) HDPE)

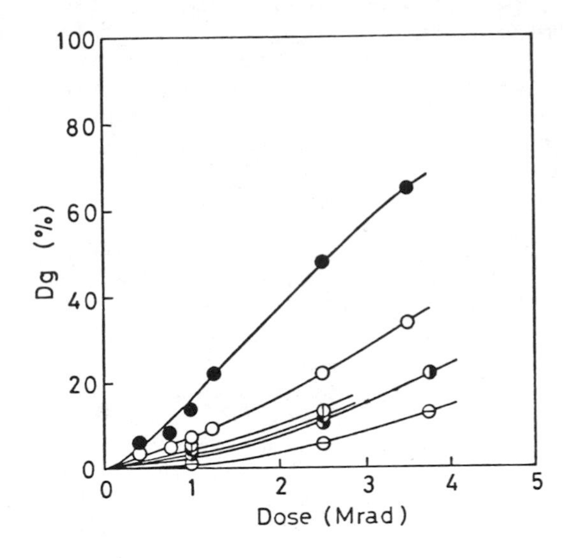

Figure 2. Plot of Dg vs. dose in the irradiation of various polyethylenes in the presence of CTFE (symbols are the same as in Figure 1)

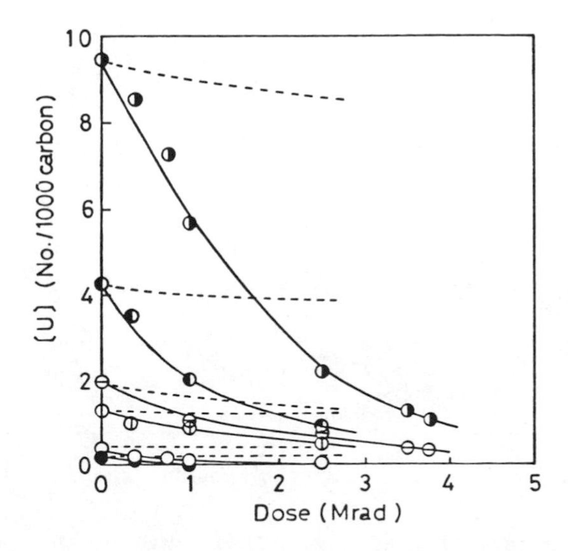

Figure 3. Plot of [U] vs. dose in the irradiation of various polyethylenes in the presence of CTFE (symbols are the same as in Figure 1) and in vacuum (– – –)

dose in the irradiation in the presence of CTFE. As found in Figure 1, the gel formation was extremely accelerated by CTFE and the rates of gel formation of butadiene- and isoprene-grafted polyethylenes were higher than those of the original ones.

On the other hand, the rates of grafting of CTFE onto butadiene- and isoprene-grafted polyethylenes were lower than those onto original ones (Figure 2). As shown in Figure 3, the rate of consumption of the unsaturated groups, as well as the gel fraction, was remarkably accelerated by CTFE.

The average rates of gel formation, grafting of CTFE and consumption of the unsaturated groups in the irradiation in the presence of CTFE were summarized in Table III. The order of the rate of gel formation did not agree with that in the rate of grafting, but those in the initial total concentration and the rate of consumption of the unsaturated groups. Figure 4 shows the plots of the rates of gel formation and grafting against the initial total concentration of the unsaturated groups contained in the polyethylenes. The rate of gel formation increased with the increase in the concentration of the unsaturated groups, while the grafting rate onto the polyethylene decreased. Charlesby-Pinner's plot (8) did not give a straight line for the data of gel formation in the presence of CTFE.

These results lead to the consideration that the cross-linking of polyethylene in the presence of CTFE is not

Table III
Radiation-induced cross-linking of various polyethylenes in the presence of CTFE

Polyethylene	$[U]_0$	$\bar{R}$(gel)[a] (%/hr)	$\bar{R}$(graft)[b] (%/hr)	$\bar{R}$(unsat)[c] (No./1000C/hr)
HDPE	0.18	0.66	0.56	0.005
LDPE	0.41	0.98	0.42	0.009
BG-LDPE-1	1.32	1.43	0.25	0.020
BG-LDPE-2	4.22	2.50	0.22	0.110
BG-LDPE-3	9.25	2.78	0.21	0.167
IG-LDPE	1.91	1.85	0.15	0.037

[a] Average rate of gel formation in the range of the gel fraction 0-50%.

[b] Average rate of grafting in the range of the degree of grafting 0-10%

[c] Average rate of consumption of total unsaturated group of consumption 0-50%

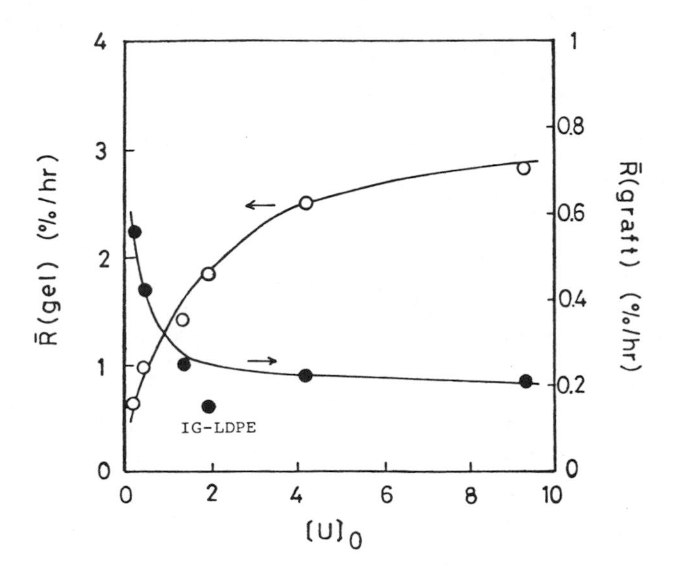

Figure 4. Plots of R̄(gel) (○) and R̄(graft) (●) vs. $[U]_0$ in the irradiation of various polyethylenes in the presence of CTFE

caused by the recombination of radicals, such as polyethylene radical and propagating graft chain radical of CTFE, but the addition reaction of the unsaturated groups to the propagating graft chain rdaical.

Figure 5 shows the relationship between the gel fraction and the total consumption of the unsaturated groups ($-\Delta[U]$ expressed by equation (4)) in the irradiation of butadiene- and isoprene-grafted polyethylenes in the presence of CTFE.

$$-\Delta[U] = [U] - [U]_0 \quad (4)$$

The gel fraction increased with the consumption of the unsaturated groups. The efficiency of gel formation by the total consumption of the unsaturated groups, that is, the slope of the curves decreased with the increase in the initial total concentration of the unsaturated groups. The lower efficiency in the polyethylene with higher concentration of the unsaturated groups may be caused by the more frequent intramolecular cross-linking reaction in the dienes-grafted polyethylenes.

On the basis of the results described above, the chain mechanism for the reactions of cross-linking and grafting in the irradiation in the presence of CTFE can be presented by the equations (5)-(8-3).

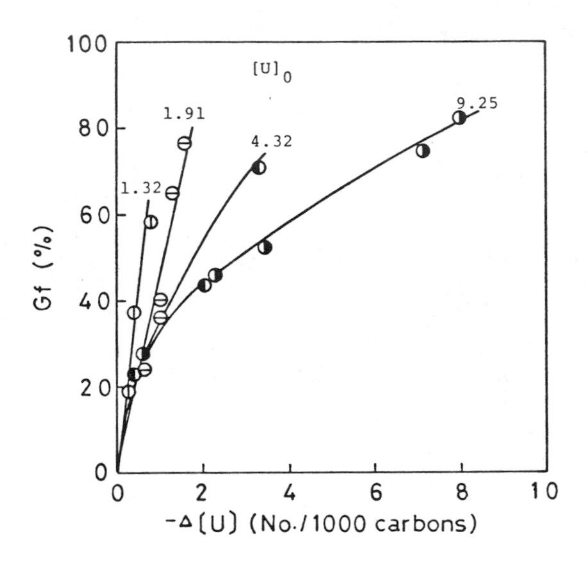

Figure 5. Plot of Gf vs. $-\Delta[U]$ in the irradiation of various polyethylenes in the presence of CTFE (symbols are the same as in Figure 1)

Initiation;

$$\text{HC}\cdot + CF_2{=}CFCl \longrightarrow \text{HC}{-}CF_2{-}\dot{C}FCl \quad (5)$$

Propagation;

$$\text{HC}{\sim}CF_2{-}\dot{C}FCl + CF_2{=}CFCl \longrightarrow \text{HC}{\sim}CF_2{-}CFCl{-}CF_2{-}\dot{C}FCl \quad (6)$$

Cross-linking;

$$\text{HC}{\sim}CF_2{-}\dot{C}FCl + \text{CH}{=}\text{CH} \longrightarrow \text{HC}{\sim}CF_2{-}CFCl{-}\text{CH}{-}\cdot\text{CH} \quad \text{(I)} \quad (7\text{-}1)$$

$$+ CH_2{=}CH{-}CH \longrightarrow \text{HC}{\sim}CF_2{-}CFCl{-}CH_2{-}\dot{C}H{-}CH \quad \text{(II)} \quad (7\text{-}2)$$

$$+ CH_2{=}CR{-}CH \longrightarrow \text{HC}{\sim}CF_2{-}CFCl{-}CH_2{-}\dot{C}R{-}CH \quad \text{(III)} \quad (7\text{-}3)$$

Re-initiation;

$$(I) + CF_2{=}CFCl \rightarrow HC\!\sim\!CF_2\text{-}CFCl\text{-}CH(\text{-}CH\text{-}CF_2\text{-}\dot{C}FCl) \quad (8\text{-}1)$$

$$(II) + CF_2{=}CFCl \rightarrow HC\!\sim\!CF_2\text{-}CFCl\text{-}CH_2\text{-}CH(CF_2\text{-}\dot{C}FCl)\text{-}CH\!\sim \quad (8\text{-}2)$$

$$(III) + CF_2{=}CFCl \rightarrow HC\!\sim\!CF_2\text{-}CFCl\text{-}CH_2\text{-}CR(CF_2\text{-}\dot{C}FCl)\text{-}CH\!\sim \quad (8\text{-}3)$$

(R = alkyl group or polymer chain)

As discussed kinetically in detail (9), the reactivity of each unsaturated group in the cross-linking reactions (7-1)-(7-3) was in the order, vinylidene > vinyl > trans-vinylene.

In order to discuss the results described above, the activation energies of the model reactions (7-1)-(7-3) were estimated by author's method (10). The activation energies were calculated as 1.5 kcal/mole for (7-1), 0.5 kcal/mole for (7-2) and 0.3 kcal/mole for (7-3), respectively. The order in the reactivity of each unsaturated group in the cross-linking reactions (7-1)-(7-3) can be well explained by these theoretical considerations. The activation energies of the reactions (7-1)-(7-3) were much smaller than those (> 4 kcal/mole) of the addition reactions of the unsaturated groups to the propagating chain radicals produced from the conventional vinyl monomers, such as styrene, methyl methacrylate, methyl acrylate, vinyl acetate and vinyl chloride. Therefore, the promotion of gel formation of the polyethylenes containing the unsaturated groups by CTFE can be ascribed to the high reactivity of CTFE propagating radical in the addition reaction to the unsaturated groups.

As has been shown in Figure 4, the grafting rate decreased with the increase in the unsaturated groups contained in the polyethylenes. This result may indicate that the rate constants of the re-initiations (8-1)-(8-3) are considerably smaller than that of propagation (6).

Since BG-LDPEs contain mainly trans-vinylene group and vinyl one, the reactions (7-1), (7-2), (8-1) and (8-2) take place in the irradiation of these polymers. While, the reactions (7-2), (7-3), (8-2) and (8-3) take place in the irradiation of IG-LDPE containing vinyl and vinylidene groups. The lower rate of grafting onto IG-LDPE may be caused by the lower reactivity of tertiary carbon radical (III) in the re-initiation (8-3) than those of secondary radicals (I) and (II) in the re-initiation (8-1) and (8-2).

Radiation-Induced Cross-Linking in the Presence of CTFE/Butadiene Mixture. On the basis of the results mentioned in the previous section, it is concluded that in the irradiation of polyethylenes in the presence of CTFE the polyethylenes are mainly cross-linked through the addition reaction of the unsaturated groups contained in the main and the side chains of the polymers to the propagating graft chain radical of CTFE. Therefore, the radiation-induced cross-linking of polyethylene is expected to be accelerated by the presence of the mixture of CTFE and a diene monomer effectively than the presence of pure CTFE.

The radiation-induced cross-linking of polyethylenes in the presence of CTFE/butadiene mixture with various compositions are shown in Figure 6. The overall degree of co-grafting decreased gradually with the increase in the mole fraction of butadiene in the mixture, while the gel fraction of the polymer was increased rapidly by the addition of a small amount of butadiene to CTFE and then decreased with the increase in butadiene mole fraction in the mixture. The maximum of the gel fraction was found at about 0.1-0.2 of butadiene mole fraction in the mixture.

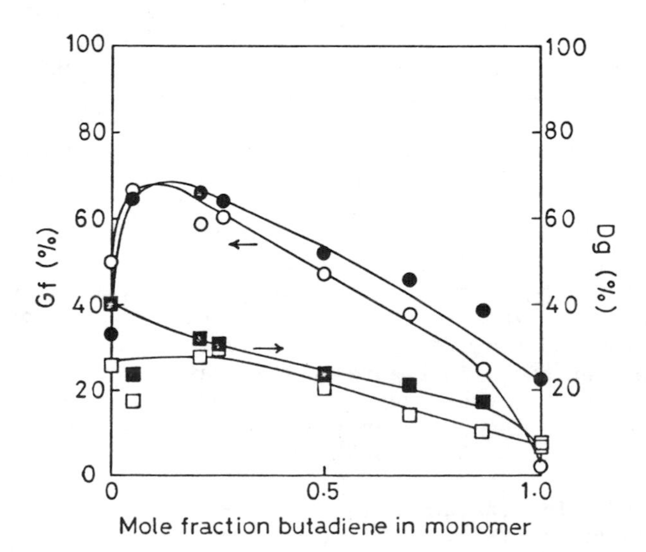

Figure 6. Plots of Gf and Dg vs. butadiene mole fraction in the irradiation of LDPE (○, □) in the monomer mixture and HDPE (●, ■) in the CTFE/butadiene mixture

The composition curve in the co-graft polymerization of CTFE and butadiene is shown in Figure 7. From the result, the monomer reactivity ratios were obtained as r_{CTFE}=0.10±0.06 and $r_{Butadiene}$=16±3 from the Finemann-Ross plot. Since the product of $r_{CTFE} \times r_{Butadiene}$ is nearly equal to one, a random co-grafting takes place in the reaction system. The high reactivity of butadiene in the co-grafting results in the remarkable acceleration of the gel formation by the addition of a small amount of butadiene to CTFE (Figure 6).

The total concentration of unsaturated groups ($[U]_a$) observed in the polymer, which was obtained by the irradiation in the presence of CTFE/butadiene mixture, was lower than the concentration ($[U]_b$) calculated based on the assumption that butadiene introduces equimolar unsaturated groups into the side chain of the polymer. As shown in Figure 8, a relationship was obtained between the gel fraction and the consumption of the unsaturated groups ($[U]_b-[U]_a$) independent of the composition of the mixture. From this figure, it is found that the gel fraction increased with the increase in the value $[U]_b-[U]_a$.

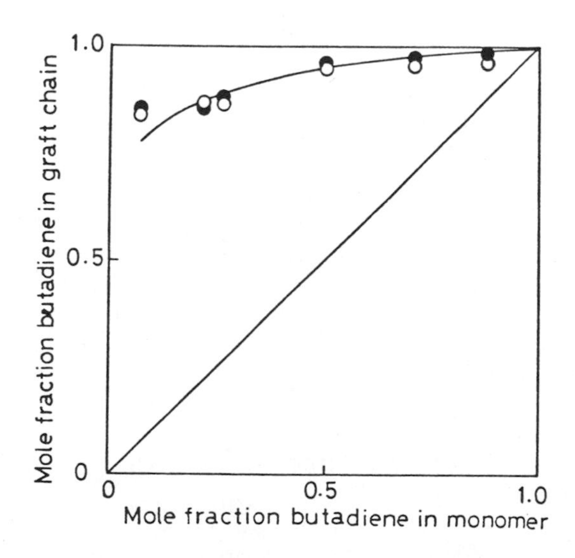

Figure 7. Composition curve for the co-graft polymerization of CTFE/butadiene onto HDPE (●) and LDPE (○)

From these results, it is concluded that the acceleration of cross-linking of polyethylene by the presence of CTFE/butadiene mixture is due to the addition reaction (9) of the propagating

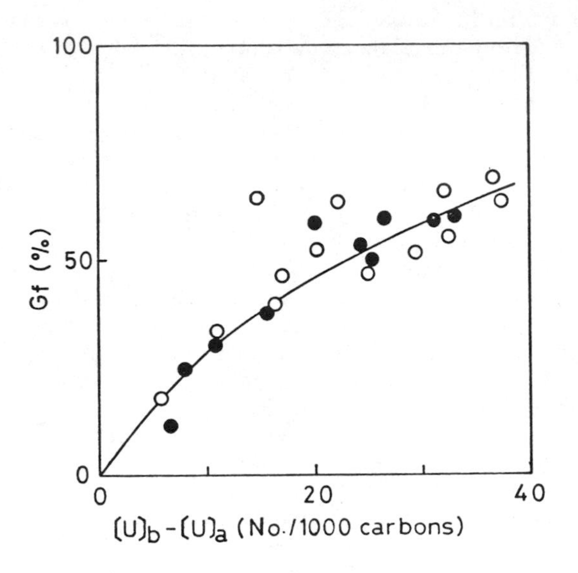

Figure 8. Plot of Gf vs. $[U]_b - [U]_a$ in the irradiation of HDPE (●) and LDPE (○) in the presence of CTFE/butadiene mixture

graft chain radical of CTFE to the unsaturated groups incorporated by the grafting of butadiene.

$$\text{HC}\sim\text{CF}_2\text{-CFCl} + \text{HC}\sim\text{CH}_2\text{-CH=CH-CH}_2\sim\text{CH}_2\text{-CH}(\text{-CH=CH}_2)\sim \longrightarrow$$

$$\text{HC}\sim\text{CF}_2\text{-CFCl-CH}(\text{-CH}_2\text{-}\sim\text{CH}\sim)\text{-CH-CH}_2\sim \qquad (9)$$

$$\text{or HC}\sim\text{CF}_2\text{-CFCl-CH}_2\text{-CH-CH-CH}_2\sim\text{CH}$$

Literature Cited

1. Charlesby, A., Proc. Roy. Soc., Ser. A, 1952, 215, 187
2. Brandrup, J., Immergut, E. H., Ed., "Polymer Handbook" Interscience Publishers, New York, 1966;p v-26
3. Chapiro, A., "Radiation Chemistry in Polymer Systems" Interscience Publishers, New York, 1962; p418
4. Mitsui, H., Hosoi, F. and Kagiya, T., Polymer J., 1974, 6, 20
5. Mitsui, H., Hosoi, F. and Kagiya, T., Polymer J., 1972, 3, 108
6. Wada, T. and Kagiya, T., Bull. Inst. Chem. Res. Kyoto Univ., 1978, 56, 27
7. Cernia, E., Mancini, C. and Mantaudo, G., J. Polym. Sci., 1963, B1, 371
8. Charlesby, A. and Pinner, S. H., Proc. Royal Soc. (London), 1959, A249, 367
9. Wada, T., Yokoyama, N. and Kagiya, T., Polymer, in press.
10. Kagiya, T. and Sumida, Y., Polymer J., 1970, 1, 137

RECEIVED July 12, 1979.

21

Effect of Discharge Frequency on the Plasma Polymerization of Ethane

S. MORITA and S. ISHIBASHI

Department of Electrical Engineering, Meijo University, Tenpaku-ku, Nagoya 468, Japan

M. SHEN and A. T. BELL

Department of Chemical Engineering, University of California, Berkeley, CA 94720

Plasma polymerized films have potential applications as dielectrics, protective coatings (1,2), and as shown most recently, as coatings on the fabrication of pellet target for the use of laser nuclear fusion (3). The plasma polymerization method has several distinctive features compared to other methods (1,2). For example, many kinds of monomers may be polymerized by this method and it is possible to make a coating on porous or smooth substrates. Also, the properties and morphology of the polymer may be controlled relatively easily by selecting the discharge conditions such as pressure, monomer flow rate, discharge frequency, discharge power level and so on. However, it is at present not possible to predict apriori the desired film properties and the morphology because of the very complex polymerization mechanisms, principally, due to the lack of sufficient experimental data on the relation between discharge phenomena and polymerization mechanism.

Most of plasma polymerizations have been carried out in the frequency range from 50 Hz to 13.56 MHz with using the capacitively coupled discharge system (4,16). For the inductively coupled discharge system, a frequency of 13.56 MHz was mostly used as a discharge frequency (17,18). In this paper, the discussion will be concentrated on the discharge in the capacitively coupled discharge system.

The polymerization mechanisms have been extensively studied in a wide discharge frequency region (4-16). But these studies are confined to the rather limited region of pressure, flow rate, discharge frequency and power level (19). However, there are some studies on the effect of the discharge frequency. Taniguchi (8) measured the growth rate in the discharge frequency range from 100 Hz to 100 KHz. Brown (13,14) studied at several discharge frequencies from 3.14 MHz to 14 MHz. Carcano (4) discussed the effects of discharge frequency on the plasma polymerization, but he measured only the flashover voltage of styrene vapour as a function of dis-

0-8412-0540-X/80/47-121-321$05.00/0

charge frequency from 50 Hz to 6 MHz and the actual experiment on the plasma polymerization was done at the frequency of 2 KHz.

Most plasma polymerizations were discussed by analyzing the growth rate as a function of a discharge current or a discharge power level. The discharge current was used mostly for a parameter in the low discharge frequency region (4-9,12,13), but the discharge power was used in the high discharge frequency region (10,11, 14-18). The relations between the growth rate and a discharge current or a discharge power level were variable for each experiment. When the growth rates were plotted against the discharge current or the power level, the curves were usually linear or convex (4-9,11-15,18), but concave curves were also obtained (10, 17).

In this study, the plasma polymerization and the discharge phenomena were studied as a function of the discharge frequency. In order to support the speculation of the polymerization mechanism, infrared spectra and the dielectric properties were also measured for the samples formed in the same discharge frequency range.

Experimentals

A schematic diagram of apparatus is shown in Fig. 1. For the discharge power supply, three amplifiers were used. One covered the frequency range from 40 Hz to 20 KHz in a continuous manner. It consisted of an oscilator (IEC, F33), an amplifier (Borgen, MT-125) and a set-up transformer. In order to control the discharge current, variable resistance was inserted in series between the set-up transformer and the reactor. Next one covered the range from 20 KHz to 10 MHz. It consisted of an oscilator (HP, 651A), two amplifiers (ENI,240L) and a set-up transformer. Two amplifiers were used in parallel run. Last one was available only at 13.56 MHz and it consisted of a generator (IPC,PM104B) and an impedance matching circuit.

A tubular type reactor was used in this experiment as shown in Fig. 2. Two copper electrodes with an area of 100 cm^2 each were set in parallel with a gap of 3.5 cm. In order to diminish the eddy current in the gas flow, Teflon inserts were set in a tubular reactor as shown in Fig. 2. The bottom electrode was cooled by circulating water. The temperature of water was almost 19 °C throughout this experiment.

The capacitance of a parallel electrode system was about 3 pF in the air. The impedance of the glow discharge was estimated to be the order of 10 KΩ in the low frequency region and the order of 1 KΩ in the high frequency region from preliminary experiment. The impedance of 3 pF at 500 KHz is about 100 KΩ. Therefore, matching circuit was used for these discharge systems at the higher frequency than 500 KHz.

Gaseous monomer of ethane was purchased from the Matheson Gas Co.. The plasma polymerized ethane (PPE) was deposited on aluminum foil set on the discharge electrode throughout this work and the

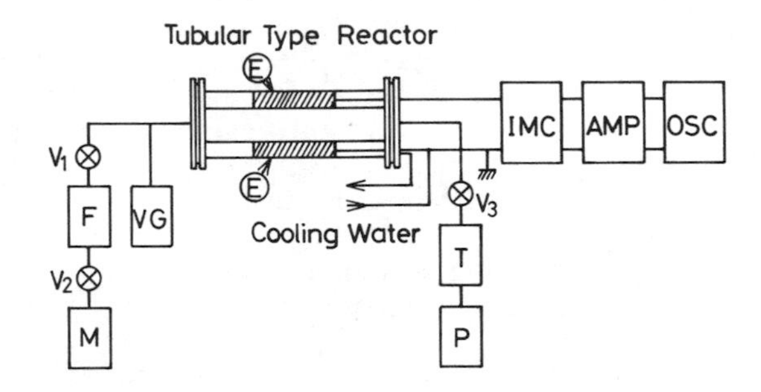

Figure 1. Diagram of apparatus: (M) monomer reservoir; (F) flow meter; (VG) vacuum gage (mercury manometer); (E) electrode; (T) liquid nitrogen trap; (P) mechanical pump; (V_1) needle valve; (V_2) stop valve; (V_3) pressure control valve; (OSC) discharge frequency oscillator; (AMP) amplifier; (IMC) impedance matching circuit

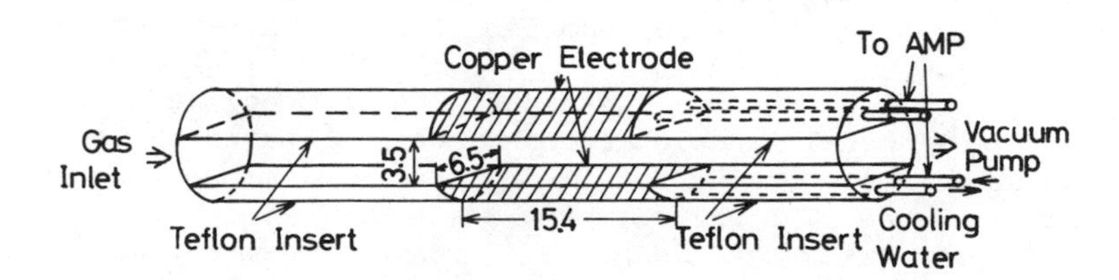

Figure 2. Tubular type reactor (numbers show length of parts in cm)

polymer growth rate was obtained by weighing. From the preliminary experimental results, the discharge condition was selected as a pressure of 0.5 torr, a flow rate of 20 cm^3STP/min and a power level of 10 watts. Under this condition, almost complete film was obtained in the discharge frequency range from 50 Hz to 13.56 MHz.

The discharge voltage and current were measured with a VTVM (HP,410B) by using voltage divider and series resistance in circuit, which were carefully designed at the high frequency in order to minimize the effect of stray capacitance. However, the discharge voltage and current were not measured at the higher frequency than 2 MHz because of the frequency limit of measuring circuit. The wave shapes of voltage and current and the phase shift between them were observed by an osciloscope. The wave shapes were not sine waves, especially in the low discharge frequency region, but the phase shift was negligibly small. In order to confirm the value of discharge current measured by a VTVM, a thermocouple type current meter was used in the low discharge frequency region, too. For the same discharge current measured by the two measuring methods, the growth rate of the polymerization was larger for the experiment by using a thermocouple type current meter than that by using a VTVM. But the difference between them was within about 10 % error of that by a thermocouple type current meter in our experimental condition. Therefore, the VTVM was used throughout this experiment instead of a thermocouple type current meter.

The PPE films for the infrared spectra measurement were formed on NaCl or KCl crystals at three different discharge frequencies. For the discharge frequencies of 1 KHz and 100 KHz, the two films were polymerized at 0.5 torr, 20 cm^3STP/min and 10 watts. Another one was prepared at 13.56 MHz, 2.0 torr, 20 cm^3STP/min and 50 watts. Infrared spectra were recorded with a Perkin-Elmer Model 137 Spectrometer.

For the dielectric loss measurement by a bridge (Ando Co. TR-10C),metal-PPE-metal sandwich specimens were prepared on the silicon dioxide substrate (Corning 7059). Evaporated aluminum was used as a metal electrode. The PPE film for the use of dielectric measurement was formed with the discharge electrode whose surface area was 26 cm^2 and the remainder of the electrode was covered by the Teflon plate. Two kinds of samples were prepared for this experiment. One of them was formed at 5 KHz, 0.5 torr, 20 cm^3STP/min and 5 watts. The other was formed at 13.56 MHz, 0.5 torr, 40 cm^3STP/min and 25 watts.

Results

The effects of discharge frequency on the polymer growth rate were studied at a pressure of 0.5 torr, a flow rate of 20 cm^3STP/min and a power level of 10 watts. All other parameters remained constant. The growth rates were plotted in the frequency range from 50 Hz to 13.56 MHz as shown in Fig. 3(a). The frequency dependence on growth rate may be divided into three regions of dis-

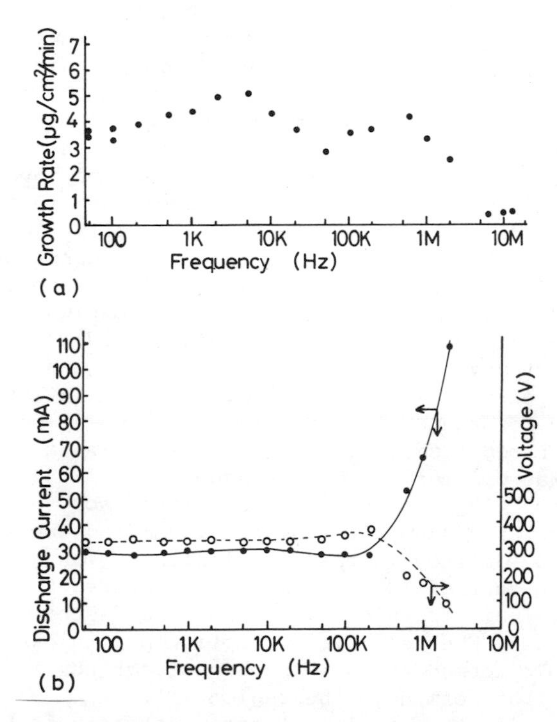

Figure 3. Effect of discharge frequency on the growth rate of plasma-polymerized ethane and the discharge voltage and current of ethane at 0.5 torr, 20 cm³ STP/min, and 10 W at room temperature: (a) growth rate; (b) discharge voltage and current

charge frequency. In the frequency range from 50 Hz to 50 KHz (Region I), the deposition rate increased with increasing frequency up to approximately 5 KHz and decreased from 5 KHz to 50 KHz. The deposition rates were 3.5 $\mu g/cm^2/min$ and 5.0 $\mu g/cm^2/min$ at 50 Hz and 5 KHz respectively. In the frequency range from 50 KHz to 6 MHz (Region II), the deposition rate increased with increasing frequency up to approximately 600 KHz and decreased from this frequency to 6 MHz. The deposition rates were 3.0 $\mu g/cm^2/min$ and 4.2 $\mu g/cm^2/min$ at 50 KHz and 600 KHz respectively. In the high frequency region above 6 MHz (Region III), the deposition rate was almost constant. The deposition rate was 0.4-0.5 $\mu g/cm^2/min$ and this value is almost one order smaller compared to those in region I and II.

The effects of discharge frequency on the discharge voltage and current were measured simultaneously when the growth rate was obtained, and the results are shown in Fig. 3(b). The discharge voltage was almost constant in the frequency range from 50 Hz to 200 KHz, but decreased with increasing frequency over 200 KHz. On the other hand, the discharge current increased in the same region at a constant discharge power level of 10 watts.

The form of deposited polymer usually depends on the discharge condition (16,21-23). In region I and II, most deposited polymers were cracked or powder like films whose colour was yellow or dark brown, but continuous films were also obtained in the small discharge current and low pressure region. In region III, a transparent or a yellow film was obtained in the relatively wide discharge parameter range.

The infrared absorption spectra are shown in Fig. 4 for the film which was polymerized at a frequency of 13.56 MHz and whose thickness was 2.02 μm. The relatively large absorptions of peaks were observed at 3.4 μm, 6.9 μm and 7.3 μm and the weak absorption band was also admitted in the wave length region over 8 μm in comparision with that of KCl crystal. The infrared absorptions at photon of 10.6 μm were plotted against the discharge frequency as shown in Fig. 5. The relative absorption per thickness was larger for the sample formed in the discharge frequency in the region I and II than that formed in the region III.

The experimental results of $\tan\delta$ are shown in Fig. 6 and 7, for the samples formed at 5 KHz and 13.56 MHz, respectively. For both samples, a relatively large loss peak was observed at about -30 °C for 1 KHz. The difference in dielectric loss between the two samples was admitted at the higher temperature region than 60 °C. For the sample formed at 5 KHz, the dielectric loss was almost constant in the high temperature region. However, for the sample formed at 13.56 MHz it increased with increasing temperature in the same temperature region.

Discussions

Discharge Phenomena in the Organic Gas. In discussing the discharge phenomena, the difference of the definition between the

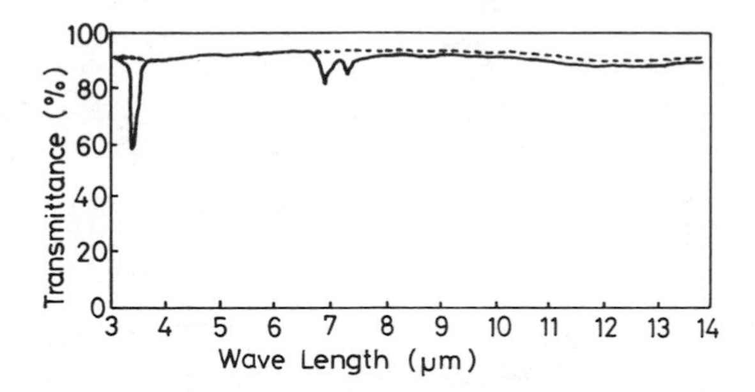

Figure 4. IR spectrum for plasma-polymerized ethane at 13.56 MHz, 2.0 torr, 20 cm³STP/min, and 50 W: (——) plasma-polymerized ethane on KCl (2.02 μm); (– – –) KCl crystal

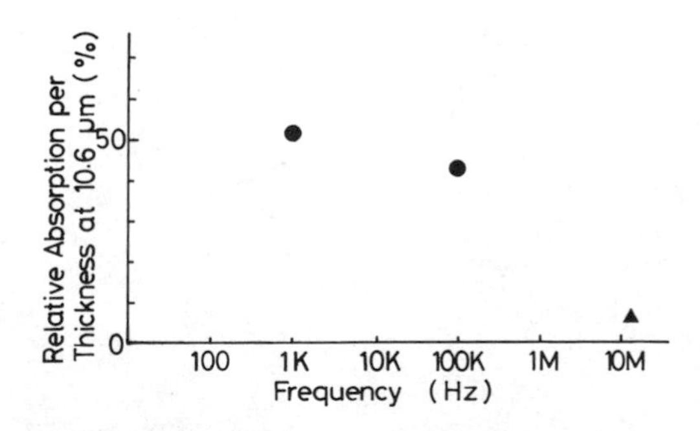

Figure 5. IR absorption at photon of 10.6 μm vs. the discharge frequency used for film formation: (●) polymerized at 0.5 torr, 20 cm³STP/min, and 10 W; (▲) polymerized at 2.0 torr, 20 cm³STP/min, and 10 W

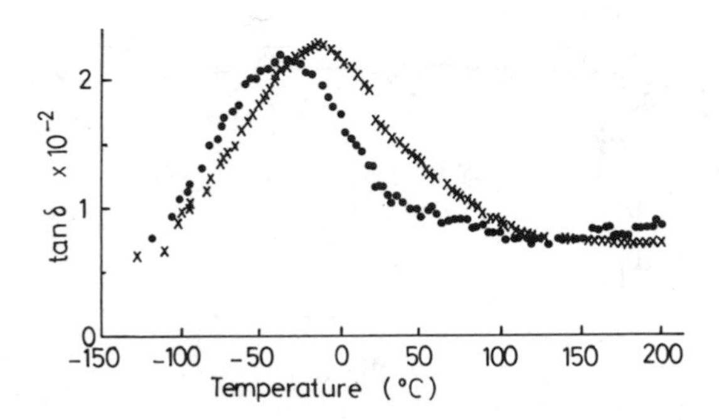

Figure 6. Plot of tan δ vs. temperature for plasma-polymerized ethane formed at 0.5 torr, 20 cm^3STP/min, 5 W, and KHz. Film thickness was 820 Å. (●) Measured at 1 KHz and (×) measured at 10 KHz.

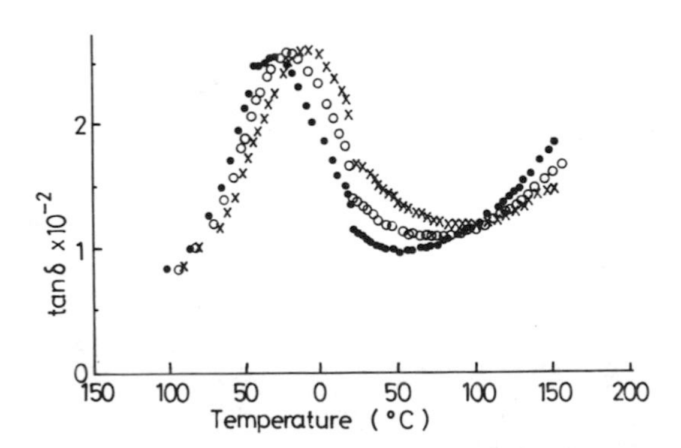

Figure 7. Plot of tan δ vs. temperature for plasma-polymerized ethane formed at 0.5 torr, 40 cm^3STP/min, 25 W, and 13.56 MHz. Film thickness was 9460 Å. (●) Measured at 1 KHz; (○) measured at 3 KHz; and (×) measured at 10 KHz.

flashover voltage (4) and the discharge voltage must be delineated. The flashover voltage is defined as the maximum voltage at which the discharge initiated and the discharge voltage is as that when the steady state discharge is obtained. Until now, most discussions were concentrated on the phenomena of flashover voltage. Numerous experimental results were presented in these discussions. Takeda (24) and Muehe (25) summarized the effects of discharge frequency on the flashover voltage for air and inorganic gases. However, there is only limited literature data on the discharge voltage in the wide discharge frequency range, especially for the discharge phenomena in organic gases.

From our experimental results, it was shown that the discharge voltage was almost constant in the frequency range from 50 Hz to 200 KHz. This means that the discharge is maintained by the same ionization mechanism in this frequency range. Above 200 KHz, the discharge voltage decreased with increasing discharge frequency. The implication is that here a different discharge phenomenon is operative.

In the low discharge frequency region, the discharge phenomena in air and inorganic gases are basically the same as in dc discharge as long as the ion and/or electron inertia are not effective in the charge transportation. The electron will be injected from the cathode by the positive ion bombardment and/or the photon absorption. This effect is known as γ effect (24). There is a significant difference in the discharge phenomena between ethane and air. In the case of ethane, the discharge electrode will be coated by the insulating polymer as soon as the discharge is initiated. As long as the film is thiner than 100 A, the electron may be injected by the tunnel effect from the cathode and the continuous discharge will be obtained. But, if the film becomes thicker than 100 A, the discharge will be stopped because the electron injection will be forbidden. If the higher voltage is applied at that time, the electrical breakdown of the film will occur in order to continue the discharge. The above hypothesis was confirmed experimentally. In the low discharge frequency region, a stable discharge was observed for a certain duration after the discharge initiated, but the electrical breakdown of the film occurs after a certain period of the discharge. However, in a limited discharge condition, the continuous discharge without a breakdown of the film was obtained also. The limited maximum discharge current is a function of discharge frequency and the breakdown strength of the film. The limited current increases with increasing discharge frequency and the breakdown strength of the film (26).

Therefore, it must be supposed that the continuous electron emission from the surface of insulating film on the cathode will be existed by the ion bombardment. The work function for the electron emission from the surface of insulating film may be larger than that from the metal surface. Thus the cathode fall in front of the insulating film must be larger than that of metal surface. At a high discharge frequency where the inertia of an electron is effective for the charge transportation between the electrodes, the electron emission from the cathode is not enough in order to sustain the discharge current. Thus, the electron multiplication in gas phase becomes significant. For this discharge, the cathode fall is not necessary because the discharge will continue without electron emission from the cathode. Therefore, the discharge voltage becomes small compared to that in the low discharge frequency range. The effect of insulating film on the electrode needs not be considered after the discharge initiation.

Polymerization Mechanism in Region I. The effect of discharge frequency on the polymer growth rate showes different features compared that of discharge voltage. The polymer growth rate mainly depends on the number of active particles like as ions, electrons anf free radicals, while the discharge voltage principally is a function of the energy of charged particles. The effect of discharge frequency on the deposition rate was devided into three regions of I, II and III. From the consideration of the discharge phenomena, it may be inferred that three regions result from the behavior of charged particles. In the case of dc discharge in organic gas, the polymer was deposited mostly on the cathode (4). The film growth on the anode was negligibly small. This fact suggests that the polymerization is caused on the positive ion bombardment to the cathode (4,26).

In region I, polymerization may be initiated by ion bombardment and the deposition rate may be propotional to the number of ions (27). In this study, the discharge current was almost constant in the discharge frequency region from 50 Hz to 200 KHz. If the effective discharge current is I and the shape of discharge current wave is sinusoidal, the mean current which will flow to either electrode may be calculated by the following equation:

$$\frac{1}{T}\int_0^{\frac{T}{2}} \sqrt{2}\ I \sin\omega t\ dt = \frac{\sqrt{2}}{\pi} I \qquad (1)$$

This equation is not a function of the discharge frequency. However, the actual growth rate was a function of the frequency. In order to reconcile this discrepancy, an additional mechanism must be introduced. In the dc discharge, the cathode fall is formed in front of the cathode. In the cathode fall region, electrons are multiplied in the high electric field and positive ions are formed simultaneously. As long as a continuous dc discharge is sustained, the distribution of charges in the cathode fall is constant. However, in an ac discharge, the positive ions in the cathode fall will disappeard in each half cycle of the discharge frequency. The disappeared positive ions might be recombined with electrons in gas phase or diffused to the electrodes and the wall of reactor vessel, where ions was recombined with electrons. These positive ions, which formed positive space charge in front of the cathode, may also contribute to the polymerization on the discharge electrode (27).

This contribution will be calculated. Assuming that the electric field in the cathode fall is almost a linear function of the distance from the cathode, following equation can be obtained.

$$E = \frac{E_0}{d}(d-x) \qquad (2)$$

where E = the electric field in cathode fall region
E_0 = the maximum electric field near the cathode
d = the width of the cathode fall region
x = the distance from the cathode.

From Poisson's equation in one dimention, the charge distribution is defined by

$$\frac{\partial E}{\partial x} = -\frac{E_0}{d} \equiv -\frac{en_{+0}}{\varepsilon_0} = \text{constant} \qquad (3)$$

where e = the electric charges of ion
n_{+0} = the net ion density.

Then,

$$\int_0^d E\,dx = \frac{E_0 d}{2} \equiv V_C \qquad (4)$$

where V_C = the cathode fall which is almost same as the applied voltage between the electrodes.

The number of ions in the cathode fall region may be calculated by following equation:

$$N_{+0} = n_{+0}d = \frac{\varepsilon_0 E_0}{ed}\,d = \frac{2\varepsilon_0 V_C}{ed} \qquad (5).$$

Part of the space charge will arrive on the discharge electrode and contribute to the polymerization. The contribution of the space charge to the polymerization will become

$$\Delta f N_{+0} = 2\Delta\frac{\varepsilon_0 V_C}{ed}\,f \qquad (6)$$

where Δ is a contribution ratio of space charge on the cathode. It is a function of pressure, density of space charge, dimention of the reactor and so on. The calculated value is a linear function of the discharge frequency. Actual growth rate is proportional to the summation of the $\sqrt{2I/\pi e}$ and $\Delta f N_{+0}$.

At higher frequencies, all of the ions formed in gas phase cannot arrive at the cathode within a half cycle because of the inertial effect of ions. The growth rate begins to decrease at the frequency where the inertial effect of ions is effective. The number of ions which will arrive at the cathode in a half cycle is calculated by

$$N_C = \overline{n}_+ \overline{v}_+ \frac{1}{2f} \quad (7)$$

where $\overline{n}_+$ = the mean charge density of ion
$\overline{v}_+$ = the mean drift velocity of ion
N_C = the number of ions which will arrive to the cathode in a half cycle.

N_C is a function of reverse of frequency. In the low frequency region, most ions will arrive to the cathode in a half cycle of the discharge frequency. The low frequency was defined as follows,

$$\overline{v}_+ \frac{1}{2f} \geqq \Lambda \quad \text{or} \quad f \leqq \frac{\overline{v}_+}{2\Lambda} \quad (8)$$

where $\Lambda = \pi L$ and L is a gap length between the electrodes. At the higher frequency than $f = \overline{v}_+/2\Lambda$, some parts of ions formed in a half cycle cannot arrive at the cathode and most ions leave as space charge in gas phase.

From the above calculations, the discharge frequency effect on the growth rate of the film can be explained qualitatively. From the frequency at the maximum growth rate that was obtained in region I, mean drift velocity of ion is calculated as $\overline{v}_+ = 2f\Lambda = \omega L = 1.1 \times 10^5$ cm/sec. This value of mean drift velocity is not unreasonable compared to those which were observed for air and inorganic gases (28).

Polymerization Mechanism in Region II. In region II, the charge distribution in the discharge and its fluctuation in an alternative electric field are not known exactly. According to Asami (28), the two types of discharge were observed in the frequency range from 30 KHz to 5.08 MHz. The type of discharge depends on the discharge current. One of them was the low frequency type which was same as the glow discharge observed in the low frequency region. Another one was the high frequency type which had positive columns only. The high frequency type glow discharge was observed at low discharge current. The low frequency type glow discharge was observed at large discharge current. The discharge voltage of the low frequency type glow discharge was almost constant in the frequency range of the experiment. But, the discharge voltage of the high frequency type glow discharge decreased with increasing frequency in the same frequency range. On the basis of Asami's experimental results, the type of discharge in region II is presumably the low frequency type.

If the growth rate of the film is depending on the number of ion only, the growth rate will decrease continuously with increasing frequency in region II. But, the actual growth rate increased with increasing frequency from 5 KHz to 600 KHz and decreased from 600 KHz to 6 MHz. This fact needs to be explained by another mechanism of polymerization. In order to sustain the low frequency type discharge, electrons must be injected from the cathode by ion bombardment. Because of the inertial effect of ion, most of the ions will be in the gas phase in the region II. However the cathode will be bombarded by very small part of ions that are leaved near the cathode, in order to sustain the discharge by γ effect. These speculations may be supported by the experiment of the discharge voltage. In this frequency range, the discharge voltage was almost the same as that in region I. On the other hand, electrons can be transferred to the anode within a half cycle of discharge frequency and form the space charge of electron near the anode. In the case of the discharge in air, electrons will be injected in the anode easily because there are not insulating film on the anode. But in the case of organic vapour, the electrodes are coated by insulating film and the electrons cannot be recombined on the anode so that the space charge of electron will be formed near the anode. These electrons will also contribute to the polymerization. At first, the growth rate will increase with increasing frequency as discussed for the ion in region I and it will decrease when the inertial effect of electron is effective. From the frequency at the maximum growth rate in region II, the mean drift velocity of the electron can be calculated as $\overline{v}_e = 2f\Lambda = 6.3 \times 10^6$ cm/sec. This value is comparable to those observed in air and inorganic gases (28).

Polymerization Mechanism in Region III. In region III, all the electrons cannot be transported to the anode in a half cycle of the discharge frequency. A possible charge transportation mechanism is an ambipolar diffusion of ion and electron pairs which will cause polymerization. The diffusion of free radicals may also contribute to the polymerization. In our experiment, the contribution of these two mechanisms cannot be distinguished because the ion and electron pairs behave as neutral gases.

Charge Exchange on the Surface of Discharge Electrode. In general, the polymerization process in plasma may be divided into three processes, i. e. the ionization of monomer, the transportation of active particles and polymerization. In a certain discharge condition, the polymerization was supposed to occur in gas phase and powder like polymers were obtained. In our experimental condition, no powder was obtained. Therefore, the polymerization must be initiated on the substrate.

In the polymerization process, the charge exchange may be supposed to occur on the substrate. The electron emission from a cathode by ion bomberdment was supposed to exist from the discussion of the discharge phenomena in region I and II. For the extraction of electron from the bound electrons in the atom, relatively large energy may be necessary compared to that from the trapped electrons on the surface site on insulating film. Therefore, the electron emission process may be expressed in two ways. One is caused by the energy delivered from the recombination between the bomberded ion and the trapped electron. Another one comes from the recombination between the bomberded positive ion and the trapped negative ion. At the cathode, there is a third process. The energetic ion will transfer the energy to the substrate and form free radicals or an activated site, and the ion which lost the energy will be trapped on the surface. These processes are described by

$$[M^+]_{gas} + 2[e]_{surface} \rightarrow [M^*]_{surface} + [e]_{gas} \quad (9)$$

$$[M^+]_{gas} + [e]_{surface} + [P_n^-]_{surface} \rightarrow [P_{n+1}^*]_{surface} + [e]_{gas} \quad (10)$$

$$[M^+]_{gas} + [P_n]_{surface} \rightarrow [M^+]_{surface} + [P_n^*]_{surface} \quad (11)$$

where $[M^+]$ is a positive ion, [e] is an electron, $[P_n]$ is a polymer and $[P^-]$ is a negative ion and * means the radical or active particle. Actually, a trapped electron and negative ion can not be distinguished.

At the anode, the charge exchange will also occur. Some electrons will recombine with the trapped positive ion, but most electrons may be trapped on the surface. The energetic electron may form the activated site on the surface:

$$[e]_{gas} + [M^+]_{surface} \rightarrow [M^*]_{surface} \quad (12)$$

$$[e]_{gas} + [P]_{surface} \rightarrow [P^*]_{surface} + [e]_{surface} \quad (13)$$

The negative ions will also exist in the gas phase. However, the contribution of the negative ions may be negligibly small, because the mobility of electron is much larger than that of the negative ions.

In region III, the discharge is maintained only by ionization in the gas phase without electron injection from the cathode. Because of the inertial effect of ions and electrons, only small part of charged particles in the gas phase can arrive on the electrode. Therefore, polymerization may be induced principally by diffused free radicals and/or ion-electron pairs:

$$[M^*]_{gas} \rightarrow [M^*]_{surface} \quad (14)$$

$$[M^+]_{gas} + [e]_{gas} \rightarrow [M^*]_{surface} \quad (15).$$

Usually, the radiation of light is observed in the discharge region. By absorbing photons on the electrode, electrons are emitted from the cathode or radicals formed on the subdtrate (30). In these cases, the following equations must be added to the above equations.

$$h\nu + [P]_{surface} \rightarrow [P^*]_{surface} \quad (16)$$

$$h\nu + [P]_{surface} \rightarrow [P^+]_{surface} + [e]_{gas} \quad (17)$$

In spite of same power consumption, the deposition rate in region III is smaller compared to those in regions I and II, but the discharge current is larger in region III than that in regions I and II. This may be caused by the large displacement current in region III, which does not accompany the actual charge tranportation.

Infrared Spectrum. The plasma polymerized organic film shows features distinctive from the conventional polymer. According to ESR measurements (31), the film contains a high concentration of residual free radicals, which showed a relatively long life time. The free radicals were oxidized in air and the oxidization is promoted significantly at elevated temperatures. The film is not soluble in usual solvents and it is more thermally stable than the conventional polymers. These properties are thought to be caused by the highly crosslinked structure of the film (32).

The molecular stracture and morphology of the film may be affected by the polymerization mechanism. The infrared spectra and dielectric properties were measured in order to correlate with the polymerization mechanisms.

Figure 4 illustrates the infrared spectrum for a sample of PPE. The absorptions of the peaks at 3.4, 6.9 and 7.3 μm were assigned to C-H stretch and C-H bending frequencies in CH_2 and CH_3 (33). These absorptions are proportional to the surface density of deposited ethane (16). However, the absorptions at photons near 10 μm are attributable to OH deformations and CO stretchings of alcoholic groups and vibrations of alkyl ketones (22). They also indicate the existence of branches in unsaturated chain (33). These absorptions may reflect the degree of crosslinking and/or the degree of degradation which is proportional to the amount of residual free radicals. The relative absorptions per thickness at 10.6 μm are plotted against the discharge frequency as shown in Fig. 5. The absorptions near 10 μm of the film formed in region III.

Dielectric Loss. Usually, the dielectric loss of plasma polymerized hydrocarbon is larger than that of the conventional polymer by more than one order of magnitude. This difference is supposed to be caused by the oxidation of the film (34). For both samples of PPE, a loss peak appeared at -30 °C at the measurement frequency of 1 KHz. The activation energy of this peak was 0.68 eV as shown in Fig. 8 for both samples. This value was almost same as

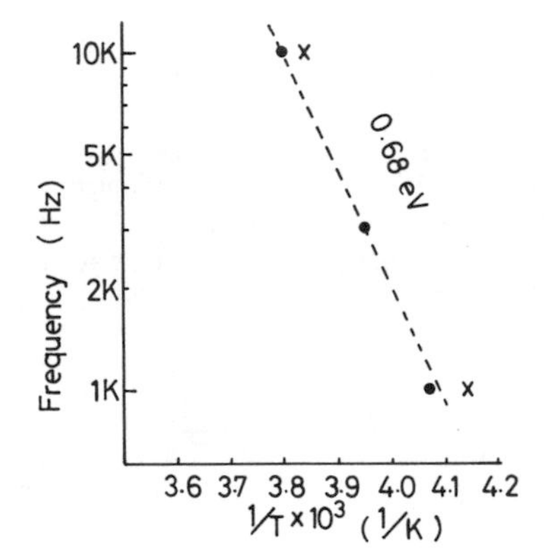

Figure 8. Activation energy plot for plasma-polymerized ethane formed at discharge frequencies of 5KHz (×) and 13.56 MHz (●)

the activation energy of γ_p measured by Tibbitt et. al. (34) within 10 % error. Thus, this peak may be assigned to motions of the carbonyl group. The dielectric loss of the sample formed at 13.56 MHz increased with the increasing of the temperature at the high temperature region from 60 °C to 160 °C. However, this increase was not observed in the sample formed at 5 KHz. The degree of crosslinking may be higher in the sample formed in regions I and II as suggested by the infrared spectra. The segmental movement of the mesh structure and the branch of the polymer are more active in the sample formed in region III. The increase in dielectric loss may suggest the existence of the peak at a higher temperature, or it may be caused by charged particles in the film.

Summarization

The effect of discharge frequency on plasma polymerization was studied in the wide frequency range of 50 Hz to 13.56 MHz. It is hypothecized that the polymerization is characterized by nature of active particles at the discharge conditions that the continuous film is obtained. The active particles are positive ions, electrons and free radicals. The growth rate of the film was almost proportional to the number of active particles arriving on the substrate. The energetic charged particles will promote crosslinking in the film, as well as the formation of active sites on the substrate.

Abstract

The effect of discharge frequency on the kinetics of plasma polymerization of ethane was studied over the frequency range from 50 Hz to 13.56 MHz, in a tubular reactor at a pressure of 0.5 torr, a flow rate of 20 cm^3STP/min and a discharge power level of 10 watts. The discharge voltage and current were also measured at the same time. On the basis of these data, the following polymerization mechanisms are proposed, i. e., ion bombardment in the frequency range from 50 Hz to 50 KHz, electron bombardment from 50 KHz to 6 MHz and free radical and/or ion-electron pair diffusion from 6 MHz to 13.56 MHz. In the region that ion or electron bombardment is the predominant mechanism, most films were cracked, but in the region that the free radical and/or ion-electron pair diffusion are predominant, transparent or colored films were obtained in the relatively wide discharge parameter range. The infrared spectra and dielectric properties of the films formed in the three frequency regions were also measured. The infrared spectrum showed a broad absorption in the wave length region over 8 μm. This absorption depends on the polymerization condition, suggesting the existence of OH and CO as well as unsaturated groups. The dielectric properties are also affected by the condition of polymerization, treatment after film formation, film thickness and so on. In our experimental condition, an appearent loss peak can be observed at a temperature of -30 °C for a measuring frequency of 1 KHz. The activation energy of this loss peak was about 0.68 eV for each sample formed at the different discharge conditions. It is proposed that this loss peak is attributable to γ_p assigned to the carbonyl group.

References

1. Shen M. Ed. "Plasma Chemistry of Polymers" Marcel Dekker Inc.: New York, 1976.
2. Hollahan J. R.; Bell A. T. Ed. "Techniques and Application of Plasma Chemistry" John Wily & Sons: New York, 1974.
3. Johnson W. L.; Letts S. A.; Hatcher C. W.; Lorensen L. E.; Hendricks C. D., ACS Polymer Preprints, 1978, 19,(2), p. 544.
4. Carchano H., J. Chem. Phys., 1974, 61, p. 3634.
5. Williams T.; Hays M. W., Nature, 1966, 209, p. 769.
6. Poll H. U., Z. angew. Phys., 1970, 26, p. 260.
7. Poll H. U.; Arzt M.; Wickleder K. H., Europ. Polymer J., 1976, 12, p. 505.
8. Taniguchi I.; Tsuneto K., ACS Polymer Preprints, 1978, 19,(2), p. 447.
9. Lam D. K.; Baddour R. F.,"Plasma Chemistry of Polymer" Shen M. Ed., Marcel Dekker Inc.: New York, 1976, p. 53.

10. Denaro A. R.; Owens P. A.; Crawshaw A., Europ. Polymer J., 1968 4, p. 93.
11. Denaro A. R.; Owens P. A.; Crawshaw A., Europ. Polymer J., 1969 5, p. 471.
12. Westwood A. R., Europ. Polymer J., 1971, 7, p. 363.
13. Brown K. C., Europ. Polymer J., 1972, 8, p. 117.
14. Brown K. C.; Copsey M. J., Europ. Polymer J., 1972, 8, p. 129.
15. Kobayashi H.; Shen M.; Bell A. T., J. Macromol. Sci.-Chem., 1970, A8, p. 1345.
16. Hiratsuka H.; Akovali G.; Shen M.; Bell A. T., J. Appl. Polymer Sci., 1978, 22, p. 917.
17. Dynes P. J.; Kaelble D. H., "Plasma Chemistry of Polymer" Shen M. Ed., Marcel Dekker Inc.: New York, 1976, p. 167.
18. Yasuda H.; Hirotsu T., J. Appl. Polymer Sci., 1977, 21, p. 3139.
19. Morita S.; Shen M.; Bell A. T., 26th ICPAC at Tokyo, Sep. 1977, 8E211.
20 Niinomi M.; Kobayashi H.; Bell A. T.; Shen M., J. Appl. Phys., 1973, 44, (10), p. 4317.
21. Kobayashi H.; Bell A. T.; Shen M., Macromolecules, 1974, 7, p. 277.
22. Kobayashi H.; Shen M.; Bell A. T., J. Macromol. Sci.-Chem., 1974, A8, (2), p. 373.
23. Tibbitt J. M.; Bell A. T.; Shen M., J. Macromol. Sci.-Chem., 1977, A11, (1), p. 139.
24. Takeda S., Japan. J. IEE, 1951, 71, p. 13.
25. Muehe C. E., Technical Report 380, MIT Lincoln Lab., 1965.
26. Williams T.; Hays M. W., Nature, 1966, 209, (5025), p. 769.
27. Morita S.; Sawa G.; Ieda M., Japan. J. Appl. Phys., 1975, 14, (10), p. 1459.
28. "Discharge Handbook" IEE of Japan: Tokyo, 1974.
29. Asami Y., Japan. J. IEE, 1929, 49, p. 710.
30. Bell A. T., "Plasma Chemistry of Polymers" Shen M. Ed., Marcel Dekker Inc.: New York, 1976, p. 1.
31. Morita S.; Mizutani T.; Ieda M., Japan. J. Appl. Phys., 1971, 10, p. 1275.
32. Bradley A.; Hammes J. P.; J. Electrochem. Soc., 1963, 110, (1), p. 15.
33. Tibbitt J. M.; Shen M.; Bell A. T., J. Macromol. Sci.-Chem., 1976, A10, (8), p. 1623.
34. Tibbitt J. M.; Bell A. T.; Shen M., "Plasma Chemistry of Polymers" Shen Ed., Marcel Dekker Inc.: New York, 1976, p. 151.

RECEIVED July 12, 1979.

NATURAL POLYMERS

22

Chiral Organofunctional Polysiloxanes: Synthesis, Properties, and Applications

ERNST BAYER and HARTMUT FRANK

Institute for Organic Chemistry, University of Tübingen Auf der Morgenstelle 18, D-74 Tübingen, Fed. Rep. Germany

The binding sites of most enzymes and receptors are highly stereoselective in recognition and reaction with optical isomers (1,2), which applies to natural substrates and synthetic drugs as well. The principle of enantiomer selectivity of enzymes and binding sites in general exists by virtue of the difference of free enthalpy in the interaction of two optical antipodes with the active site of an enzyme. As a consequence the active site by itself must be chiral because only formation of a diasteromeric association complex between substrate and active site can result in such an enthalpy difference. The building blocks of enzymes and receptors, the L-amino acid residues, therefore ultimately represent the basis of nature's enantiomer selectivity.

The goal of the current investigation was to achieve further insight into the nature of these interactions and to understand the stereoselectivity of biological systems, partially a result of orientation factors. It has been observed for a number of enzymes that the environment of the active site is relatively

0-8412-0540-X/80/47-121-341$05.00/0

non-polar. This has the effect that any non-directional interactions of the enantiomeric substrate with achiral polar moieties are excluded and a definite relative orientation of receptor and substrate results in respect to their assymetric centers. We attempted to impart to synthetic polymers such enantio-selectivity, which should be particularily interesting for the development of stereoselective catalysts. As the polymeric basis for such a system we chose the polysiloxanes due to their unusually weak intramolecular forces. In addition the synthesis of appropriately functionalized monomers is relatively uncomplicated and the proportions of polar chiral binding sites and apolar environment can freely be chosen by equilibration of the proper amounts of the corresponding homopolysiloxanes. This in turn determines the relative distance between the incividual chiral groups.

Polysiloxanes are known to be chemically and thermally stable. Due to this feature they can be used as stationary liquids in gas chromatography (3,4,5). The same is true for the chiral polysiloxanes described here. Their use as stationary phases in gas chromatography allows the calculation of the differences in enthalpy and entropy for the formation of the diastereomeric association complexes between chiral receptor and two enantiomers from relative retention time over a wide temperature range. Only the minute amounts of the polysiloxanes required for coating of a glas capillary are necessary for such determinations. From these numbers further conclusions are drawn on the stereochemical and environmental properties required for designing systems of high enantio-selectivity in condensed liquid systems.

The novel class of polymeric organo silicones offer some new aspects to phase selectivity in gas-liquid chromatography. Conventional silicones used for this purpose are homopolymers of the polydimethylsiloxane type or copolmyers from dimethylsiloxane and simple organofunctional siloxanes units, e.g. cyanoalkylmethylsiloxane units. Enantioselective silicones can be prepared by first synthesizing appropriately functionalized polysiloxanes, to which a polar chiral moiety can be bound covalently. Chiral groups of high selectivity for amino acids and many other chiral compounds are amino acid derivatives themselves (6,7). Therefore coupling of an amino acid residue to a polysiloxane with organofunctional carboxyl groups provides an adequate way of synthesizing enantio-selective silicones. An important parameter for achieving optimum properties of the silicones is the proper adjustment of the ratio of chiral polar groups and silicone monomer units. If this number is too small, selectivity remains low, but if it is higher than 0.2 the physicochemical properties of the silicone become unfavorable.

The enantio-selectivities are largely dependant upon the structure of the optically active moiety. Highest selectivities for amino acids are exhibited by silicones with amino acid amide groups. The classes of compounds for which high selectivities are observed as well are α-hydroxy carboxylic acids, amino alcohols, amines and glycols (8). Other polysiloxanes containing chiral 1-aryl -1-amino ethane groups (9) exhibit high enantioselectivities for chiral amines and amino alcohols containing aromatic rings.

For a number of adrenergic compounds an interesting parallel of enantio-selectivity of these silicones and the relative adrenergic activities of two antipodes has been observed, suggesting some similarity of interaction of these drugs with their receptor sites and with the chiral groups of the stereoselecitive polysiloxanes. With further insight into the stereochemistry of the diastereomeric complexes on synthetic polymers, generalizations on the nature and structure of biological active sites can be drawn.

Experimental Methods

Synthesis of enantioselective polysiloxanes comprises three main steps: (a) synthesis of appropriately functionalized dichlorosilane monomers, (b) preparation of a fluid copolymer with a specified number of functional groups per weight unit and (c) covalent attachment of a suitable chiral enantio-selective moiety.

The functionalized dichlorosilane monomers are synthesized generally by radical addition of dichloromethylsilane to an unsaturated carboxylic acid ester or an unsaturated nitrile. Catalysts used for this purpose are platinum/charcoal (10,11), or organic peroxides (12), but for laboratory syntheses hexachloroplatinic acid (13,14) proved to be most convenient (scheme 1 a).

scheme 1 a:

$$CH_3SiHCl_2 + CH_2{=}CCH_3COOR \xrightarrow{H_2PtCl_6} CH_3SiCl_2CH_2(CH_3)COOR \qquad I$$

$$CH_3SiHCl_2 + CH_2{=}CH\text{-}CH_2C{\equiv}N \xrightarrow{H_2PtCl_6} CH_3SiCl_2CH_2CH_2CH_2CN \qquad II$$

The compounds I and II represent intermediates for the synthesis of either carboxy functional or amino functional polysiloxanes. To this end I or II is either saponified with sodium hydroxide and concomitantly polycondensated to a poly(carboxylalkyl methyl siloxane) (III) or reduced to an amino alkyl methyl dialkoxy silane with sodium borohydride (15).

scheme 1 b:

$$n\ CH_3SiCl_2CH_2CH(CH_3)COOR \xrightarrow{H_2O(OH^-)} [CH_3SiO_{2/2}CH_2CH(CH_3)COOH]_n \quad \text{IIIa}$$

$$n\ CH_3SiCl_2CH_2CH_2CH_2CN \xrightarrow{H_2O(OH^-)} [CH_3SiO_{2/2}CH_2CH_2CH_2COOH]_n \quad \text{IIIb}$$

scheme 1 c:

$$CH_3SiCl_2CH_2CH_2CH_2CN \xrightarrow{NaBH_4\ (CH_3OH)} CH_3Si(OCH_3)_2CH_2CH_2CH_2CH_2\text{-}NH_2$$

Completeness of the saponification is controlled by IR-spectroscopy monitoring the disappearance of the ester carbonyl band at 1740 cm^{-1} and the simultaneous increase of either the carboxylate absorption at 1580 cm^{-1} or the protonated carboxyl group at 1700 cm^{-1} (figure 1). Another criterion is the disappearance of the ester signal in the proton- or ^{13}C-NMR-spectra.

The analytical control of this step is of special importance: the alcaline saponification is performed at a relatively low pH in order to prevent cleavage of the silicon-carbon bond. The closer the electron-withdrawing carboxyl group is located to the Si-C-bond, the larger is the danger of scission. Therefore, for the ß-silyl carboxylic acid derivatives the pH during saponification should not surpass 10.5; however, at this pH saponification of the methyl ester requires about 1 day, even at 60°C. For the γ-silyl derivatives, the pH of the reaction mixture is not critical. We therefore now exclusively utilize the latter.

The homopolymeric carboxyalkyl silicone is precipitated from the saponification mixture by adjustment of the pH to about 1. After standing over night a clear, viscous silicone is deposited on the bottom of the vessel, The silicone is rinsed acid-free, and heated to 100°C in vacuo to remove last traces of water and low molecular weight substances. Then the resinous silicone is heated to 180°C under nitrogen for approximately 1 hour in order to condense most of the residual silanol groups. The presence of free silanol groups in the silicone is indicated by the presence of downfield shifted satellites of the signals of the two silicon-attached carbons in the ^{13}C-NMR-spectrum (figure 2).

The removal of free silanol groups is important for correct calculation of the ratio of the two monomers in the subsequent coequilibration-step. This step is required in order to generate a copolymer of appropriate viscosity and to separate the functional groups in the polysiloxane by at least five dialkylsiloxy units. The reason for this necessity is discussed below.

Poly(carboxyalkyl methyl siloxane) and octamethyl cyclotetrasiloxane are mixed in a monomer ratio of 1:6, hexamethyl disiloxane is added to bring the polymerization degree of the ensuing silicone to about 60, and 6 volume percent of concentrated sulfuric acid are used as equilibration catalyst.

scheme 2:

$$a\ HOOC(CH_2)_n\ (CH_3)SiO_{2/2} + b\ (CH_3)_2SiO_{2/2} + 2(CH_3)_3 SiO_{1/2} \xrightarrow{H_2SO_4} (CH_3)_3Si\text{-}O\text{-}\left[(CH_3)Si(CH_2)_nCOOH\text{-}O\right]_a - \left[(CH_3)_2SiO\right]_b - Si(CH_3)_3$$

The components are placed in a round bottom flask and shaken vigourosly unitl a homogeneous mixture is obtained. As an increasing size of the organo-residues of the silicone reduces the rate of equilibration considerably (16,17), two weeks are required for a batch of a few gramms depending on the efficiency of the mixing. A solvent at this step is omitted since this may result in a misleading homogeneity at an early stage of equilibration. If a statistical distribution of the functionalized siloxy-units is not achieved the properties of the resulting polysiloxane are unfavorable. After the reaction the mixture is diluted with 20 volume percent of water and shaken for another hour. The two phases are separated by centrifugation, the polysiloxane is diluted with 1 volume of ether and extracted with water until the test for sulfate is negative. The solvent is removed in vacuo.

The final step in this sequence is the coupling of a suitable chiral group to the silicone. In reaction scheme 3 the attachment of an amino acid amide residue is given as an example. The carboxy functional copolymer is dissolved in dichlormethane/dimethylformamide and an amino aid amide is coupled with dicyclohexyl

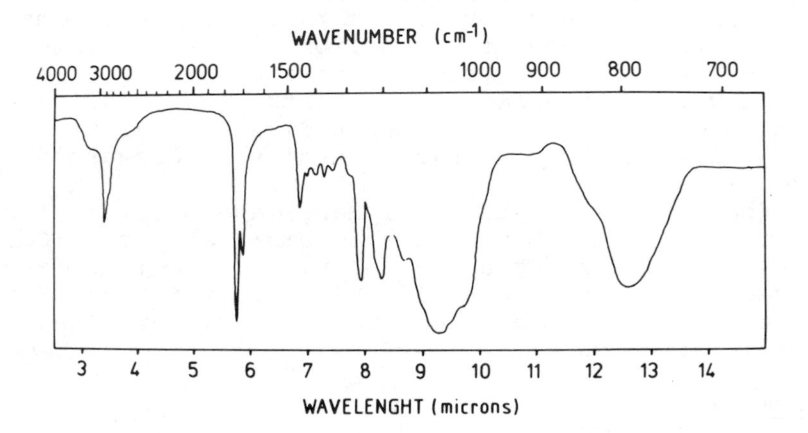

Figure 1. IR spectrum of incompletely saponified poly-2-methoxycarbonylpropyl-methylsiloxane

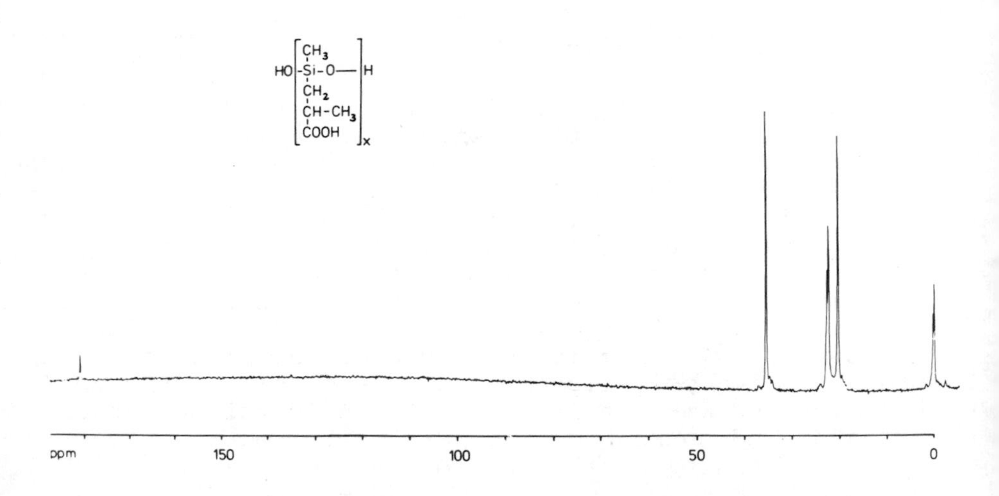

Figure 2. C-13 NMR spectrum of poly-2-carboxypropylmethylsiloxane. Free silanol groups give rise to a downfield satellite of the signals of the two silicon-attached carbons.

carbodiimide in the usual manner (18) with N-methyl morpholine as base.

scheme 3:

$$\left[\begin{array}{l} CH_3 \\ | \\ -Si-O \\ | \\ (CH_2)_n \\ | \\ COOH \end{array}\right]_m + m\ H_2N-\overset{*}{C}H-CONHR' \ (\text{with } R \text{ on } \overset{*}{C}H) \xrightarrow{DCCI} \left[\begin{array}{l} CH_3 \\ | \\ -Si-O- \\ | \\ (CH_2)_n \\ | \\ CO \\ | \\ NH-\overset{*}{C}H-CONHR' \\ \quad\quad\ | \\ \quad\quad\ R \end{array}\right]_m$$

After one hour the organic solution is extracted with 0.5 N hydrochloric acid, then with water and filtered. The solvent is evaporated, the residue redissolved in hexane and the solution is stirred for one hour with 10% acetic acid in water. The solution is filtered again, the aqueous layer is discarded and the organic layer is washed three times with 5% sodium bicarbonate solution.

Finally the solvent is evaporated, the residue redissolved in butanol and applied to a column of LH-20 in the same solvent. The first third of the fractions between dead volume and total bed volume is pooled, the solvent is evaporated, and the silicone is heated to 180°C for one hour in vacuo. A clear silicone of a viscosity of about 50 000 centi-stokes at room temperature is obtained.

Results and Discussion

In scheme 4 the general formula of the silicones described here is shown:

scheme 4:

$$(CH_3)_3Si\text{-}O\text{-}\left[\begin{array}{c} CH_3 \\ | \\ \text{-}Si\text{-}O\text{-} \\ | \\ CH_3 \end{array}\right]_n \text{—} \left[\begin{array}{c} CH_3 \\ | \\ \text{-}Si\text{-}O\text{-} \\ | \\ (CH_2)_x \\ | \\ CO \\ | \\ NH \\ | \\ C^{*} \\ R \diagup \; | \; \diagdown R \\ | \\ R' \end{array}\right]_m \text{-}Si(CH_3)_3$$

The ratio of n to m is between 6 and 7 if the chiral group is an amino acid, but roughly 4 if the chiral moiety is an amine. In figure 3 the IR-spectrum for such a polysiloxane carrying a valine-t-butylamide residue is shown. Of particular diagnostic value is the shape of the absorption band between 9 and 10 μ, a superposition of two bands. A polysiloxane of normal fluidity shows both bands in a roughly equal intensity. The band at 9.8 μ obviously reflects the flexibility of a silicone chain over a longer segment, as it is absent in highly crosslinked polysiloxanes or in linear short-chain silicones. As indicated above coupling of chiral groups to a carboxyfunctional homopolymer does not lead to a suitable silicone. In such a case a material of high rigidity and semi-cristalline consistence is obtained. It does not exhibit any enantio-selectivity and is thermally unstable. Interestingly the IR-spectrum of such a material, as shown in figure 4, reveals a very weak absorption at 9.8 μ, another proof for the low flexibility of the polysiloxane backbone due to formation of a large number of intra- and inter-chain hydrogen-bonds.

We have synthesized several structural analogs of these silicones and examined their enantio-selectivity by gas chromatography. The structure of the connecting group is of little influence on phase selectivity (table 1), but much more important is the structure of the optically active group itself.

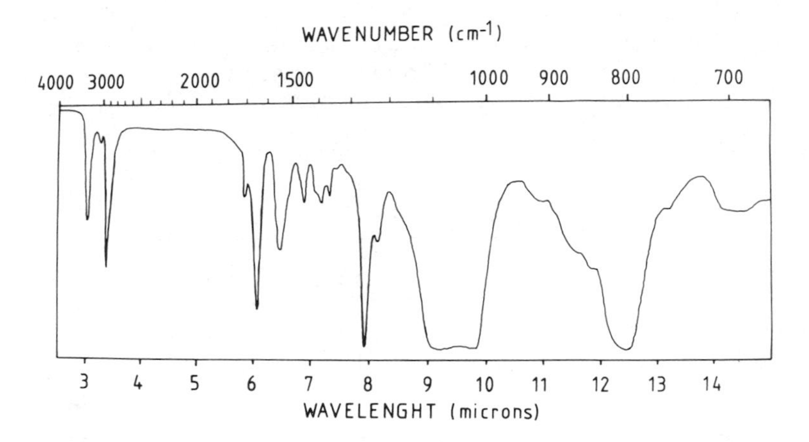

Figure 3. IR spectrum of Chirasil-Val

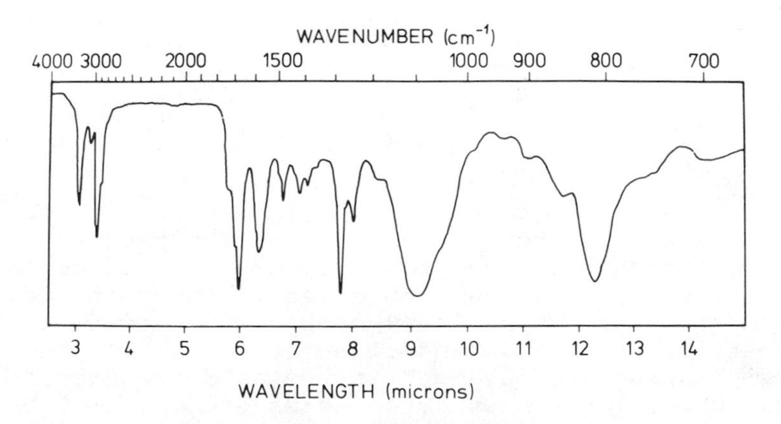

Figure 4. IR spectrum of a chiral polysiloxane prepared without spacing dimethylsiloxy units

Table 1: Gas chromatographic resolution factors of N-pentafluoropropionyl amino acid isopropyl esters on polysiloxanes carrying chiral L-valine t-butylamide groups (Val-NtBu)

Structure	$\alpha_{L/D}$		
	Ala (90°C)	Asp (120°C)	Phe (140°C)
CH_3-Si-$(CH_2)_2$-CO-Val-NtBu	1.156	1.030	1.077
CH_3-Si-$(CH_2)_3$-Co-Val-NtBu	1.159	1.025	1.074
CH_3-Si-CH_2-CH(CH_3)-CO-Val-NtBu	1.158	1.023	1.072

In nearly all cases amino acid residues as enantio-selective groups show highest resolution factors. A large difference in the stability of the diastereomeric association complex seems to arise if the amino acid side chain branches at the ß-carbon adjacent to the assymetric α-carbon as in valine (table 2). Another decisive structural element is the amide residue of the chiral moiety. For most amino acids the valine-t-butyl-amide exhibits highest enantio-selectivities except for the aromatic amino acids. The cyclohexyl-ring obviously brings about a higher selectivity for most phenyl-ring containing compounds except for the 1-phenyl-2-amino ethanol. The leucine silicones generally exhibit lower resolution factors, but the highest for most compounds are found for the n-butylamide derivative.

Table 2: Resolution factors of N(O)-pentafluoropropionyl derivatives of chiral compounds on polysiloxanes carrying different chiral amino acid residues.

Structure of chiral group				CH_3 CH-NH_2 C_6H_5	CH_2NH_2 CH-OH C_6H_5	CH_2-OH CH-OH (C_6H_3(3-OCH_3, 4-OH)
	Ala 90°	Asp 120°	Phe 140°	100°	120°	120°
Val-n-butylamide	1.126	1.020	1.057	1.035	1.017	1.018
Val-cyclohexyl-amide	1.152	1.021	1.079	1.038	1.019	1.040
Val-tert. butylamide	1.159	1.027	1.075	1.022	1.045	1.024
Leu-n-butylamide	1.097	1.019	1.044	1.031		
Leu-cyclohexyl-amide	1.079		1.039	1.026		1.023
Leut-tert.butyl-amide	1.053		1.022	1.016		

The large enthalpy differences of up to 0.5 kcal for amino acids leads into the assumption that a stereoselective autoassociation (20,21) of "receptor" and "substrate" is the decisive aspect for the observed enantioselectivity of these polymers. The structure of the resulting diastereomeric association-complex is depicted in figure 5 and resembles an antiparallel pleated sheet structure known as ß-structure in keratins. This conformation permits formation of the maximum number of hydrogen bonds. The space filling isopropyl residues of the "receptor" and the side chain R of the L-amino acid "substrate" as well as the alkyl residues of the corresponding acyl- and amide-groups fit together in stacked layers and stabilize the structure by van der Waals forces. If the bulky side chain and the hydrogen at the assymetric carbon of the "substrate" exchange their positions, the ensuing complex is destabilized by mutual hindrance of the space filling side chains of "substrate " and "receptor". The significance of the dimethylsiloxane units between two chiral groups becomes apparent from the model in

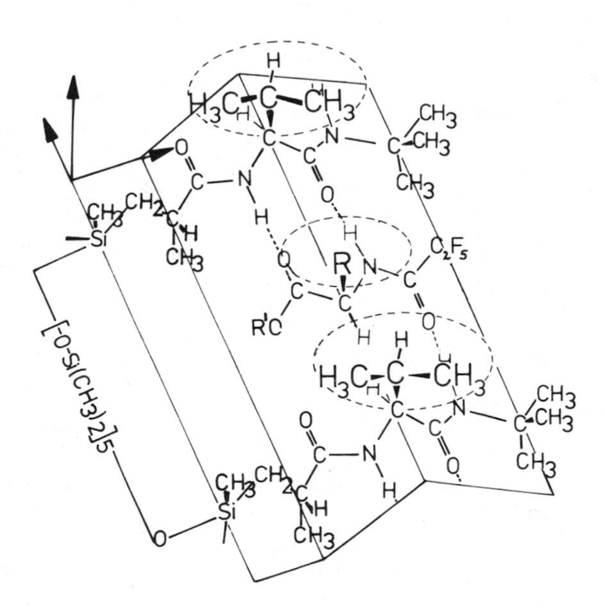

Figure 5. Assumed structure of the diastereomeric association complex

figure 5: they keep the polar valine amide units separated, thus facilitating insertion of a "substrate" and preventing the formation of strong intra-chain hydrogen-bonds. Interestingly, the IR-spectrum of Chirasil-Val (19) exhibits an amide-I band at a wave-number of 1640 cm^{-1}, lower than expected for an unordered amide-I absorption. This phenomenon is reported as a consequence of the formation of a ß-structure (22). In the IR-spectrum of the highly loaded silicone exhibiting no enantio-selectivity the amide-I band occupies a position at 1670 cm^{-1}.

The new polysiloxanes are excellently suited as stationary phases for the gas chromatographic separation of the optical antipodes of different compounds classes over a temperature range from 70° to 240° C.

The thermal stability of these silicones is very high and even at 200° C over months no racemization of the chiral groups can be detected. We attribute this phenomenon to the formation of a specific pleated-sheet structure in an apolar environment, which stabilizes the L-configuration and prevents inversion at the assymmetric carbon.

Due to the high thermal stability these silicones for the first time enable the complete separation of all protein amino acid enantiomers as shown in figure 6. Only glycine, the nonchiral amino acid, in some instances overlaps with one of isoleucine stereomers, but this may be overcome by slightly changing the polarity of the stationary phase. By use of these phases a novel approach to amino acid analysis, called enantiomer-labelling (5), recommends the gas chromatographic procedures as a real alternative to the slower and less sensitive ion exchange procedure (23). The method proves to be especially valuable if larger numbers of serum samples must be processed for clinical studies (24). Other optically active compounds which may be separated into their enantiomers are -hydroxy carboxylic acids, glycols, amino alcohols, amines and atropisomers of the o,o'-substituted biphenyls.

The chiral polysiloxanes open the opportunity of investigating the differences of the stability of "receptor-substrate" complexes for two enantiomers over a wide temperature range through determination of the relative gas chromatographic retention times. An interesting example are the adrenergic drugs of the 1-phenyl-2-amino ethanol and propanol classes. We noted that the differences in enthalpy and entropy for the diastereomeric association with the silicone are well paralleled by the ratios of the maximal adrenergic activities of these compounds in biological systems

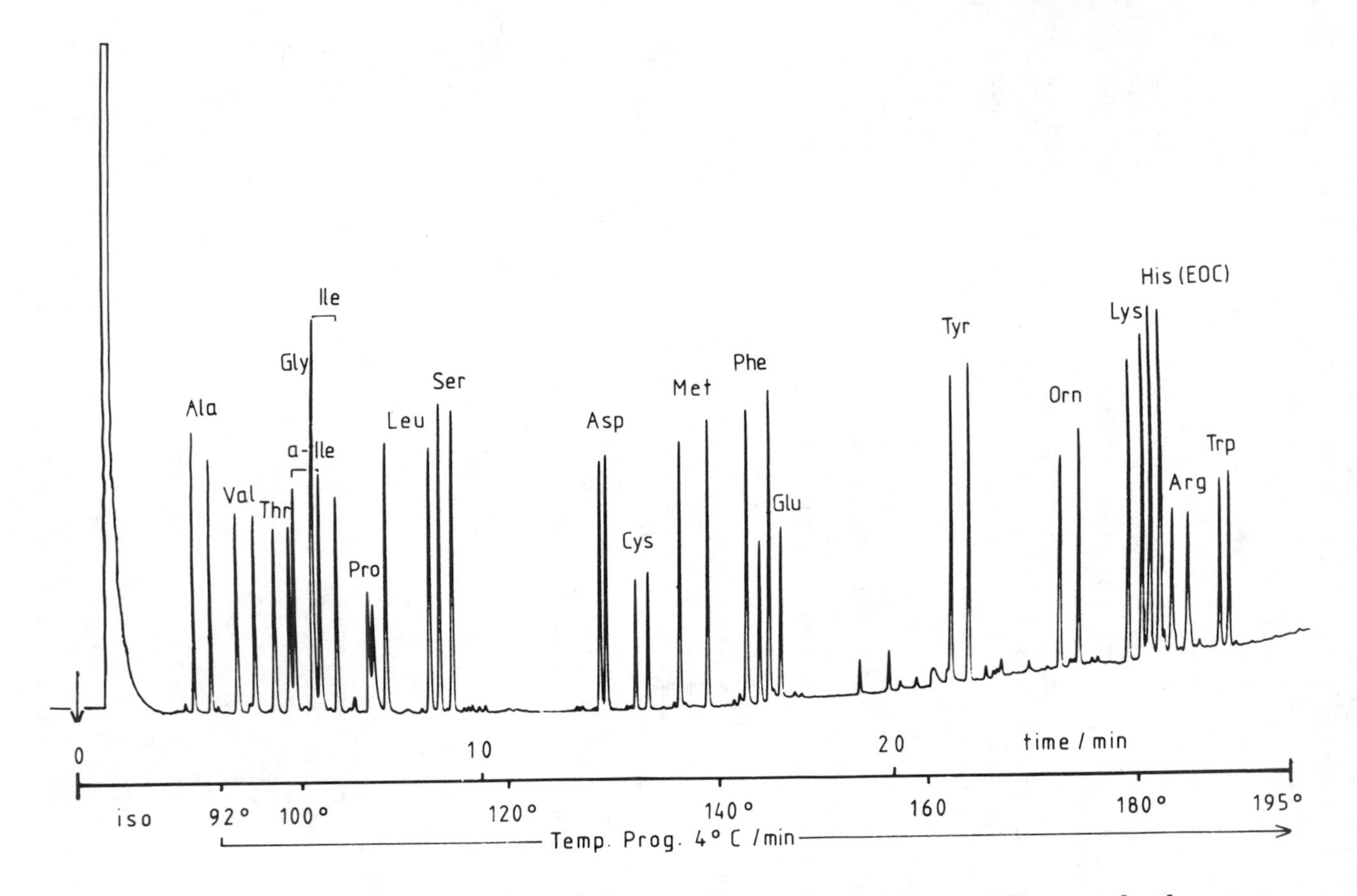

Figure 6. GC separation of the enantiomers of all protein amino acids on a capillary coated with Chirasil-Val (20 m × 0.25 mm)

(numbers of adrenergic activities taken from reference 25) (table 3).

Table 3: Differences in enthalpy (H_o, cal/mole) entropy (S_o, cal/mole.deg) and ratios of adrenergic activities R_{ci}) of the enantiomers of adrenergic drugs.

	H_o	S_o	R_{ci}
Primary Amines			
Norephedrine	- 203	- 0.36	2.3
Metaraminol	- 198	- 0.33	2.0
Octopamine	- 86	- 0.15	1.2
Noradrenaline	- 55	- 0.09	1.1
Secondary Amines			
Pseudoephedrine	- 170	- 0.34	10
Ephedrine	- 92	- 0.18	2.0
Synephrine	- 51	- 0.11	1.8
Adrenaline	- 12	- 0.07	1.2

Higher differences for the interaction of both the biological and the synthetic receptors with the two enantiomers are found for the propanol-derivatives than for the ethanol-derivatives (Norephedrine, Metaraminol vs. Octopamime, Noradenaline; Pseudoephedrine, Ephedrine vs. Synephrine, Adrenaline).

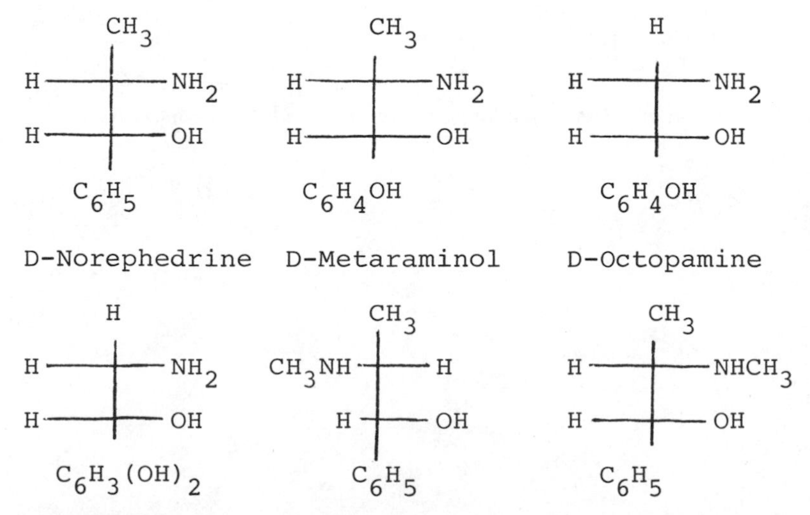

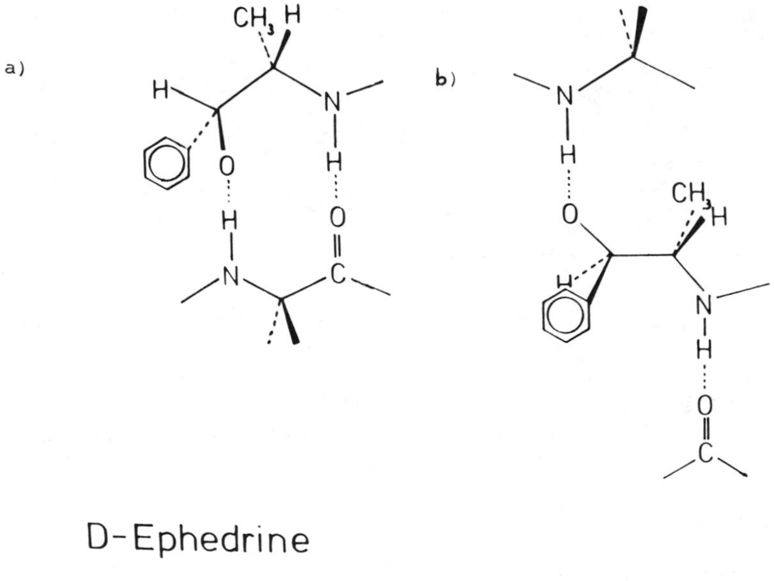

a) b)

D-Ephedrine

Figure 7. Possible structures of adrenergic receptor–substrate complexes (see text)

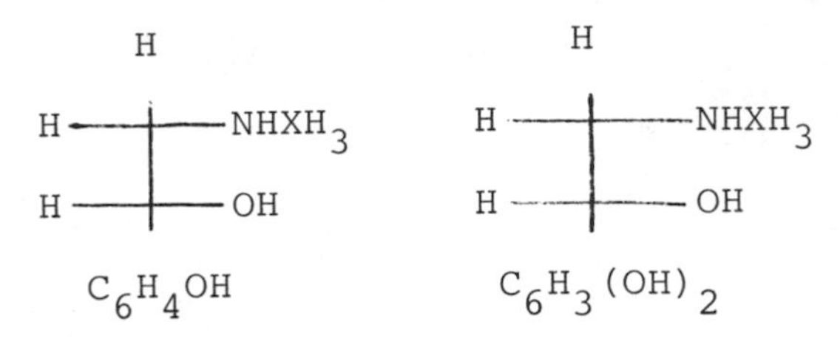

D-Synephrine D-Adrenaline

From calculations of the preferred conformation of the ephedrine and pseudo ephedrine molecules the topography of the adrenergic receptor as a flat surface has been deduced (26). This view, however, does not explain the decisive effect of the additional methyl group; its relative position should not influence the interaction with the proposed receptor to a large extent (Fig. 7a). However, as the differences in activities are especially high for the ephedrine and pseudo ephedrine isomers, we believe that the receptor must have an intercalated structure similar as projected for the "binding sites" of the silicones (Fig. 7b).

Only in this case the relative position of the additional methyl groups can exert such profound differences in the interaction with the receptor, as it is observed for these drugs in the biological system and similarly with the synthetic model. The energy-niveau of the rotamer of D-ephedrine in b) is energetically only slightly higher than of the rotamer in a) (27).

In general, the differences in interaction with a biological receptor are the more pronounced the closer the assymmetric center is located to the area of the molecule interacting with the active site. The same applies to the interactions of enantiomers with chiral groups in the silicones. However, in comparative studies of the resolution factors of a large number of 1-phenyl 2-amino alcohols it became apparant that not only hydrogen-bonding is involved in the stereo-selective association, but van der Waals forces and hydrophobic interactions play also an important role.

REFERENCES

(1) Daniels, T.C., and Jorgensen, E.C., in Wilson, C. O., Gisvold, O., and Doerge, R.F., (Eds.), "Textbook of Organic Medicinal and Pharmaceutical Chemistry", Lippincott, Philadelphia, 1966, pp. 4-64

(2) Albert, A., "Selective Toxicity", Methuen, London 1968

(3) Frank, H., Nicholson, G.J. and Bayer, E., J.Chromatogr. Sci. 15, 174 (1977)
(4) Frank, H., Nicholson, G.J. and Bayer, E., J.Chromatogr. 146, 197 (1978)
(5) Frank, H., Nicholson, G.J., and Bayer, E., J.Chromatogr. 167, 187 (1978)
(6) Gil-Av, E., Feibush, B., and Charles-Sigler, R., in Littlewood, A.B., (Ed.), "Gas Chromatography 1966", Institute of Petroleum, London, 1967, pp. 227-239
(7) Beitler, U., and Feibush , B., J.Chromatogr. 123, 149 (1976)
(8) Frank, H., Nicholson, G.J., and Bayer, E., Angew. Chem. 90, 396 (1978); Angew.Chem.Int.Ed. 17, 363 (1978)
(9) Weinstein, S., Feibush , B., and Gil-Av, E ., J. Chromatogr. 126, 97 (1976)
(10) Wagner, G.H., Union Carbide, A.P. 2 637 738 (17.9.1949)
(11) Goodman, L., Silverstein, R.M., and Benitez, A., J.Amer.Chem.Soc. 79, 3073 (1957)
(12) Gadsby, G.N., Research 3, 338 (1950)
(13) Ryan, J.W., Menzie, G.K., and Speier, J.L., J.Amer.Chem.Soc. 82, 3601 (1960)
(14) Smith, A.G., Ryan, J.W., and Speier, J.L., J.Org.Chem. 27, 2183 (1962)
(15) Niederprüm, H., Horn, E.M., and Simmler, W., Farbenfabriken Bayer, DAS 1 216 873 (14.5.1964)
(16) Sokolov, N.N., Zh.Obshch.Khim. 28, 3354 (1958)
(17) Simmler, W., Makromol.Chem. 57, 12 (1962)
(18) Sheehan, J.C., and Hess, G.P., J.Amer.Chem.Soc. 77, 1067 (1955)
(19) The silicone carrying L-valine t-butylamide residues is referred to as Chirasil-Val; this material is commercially available from Applied Science Laboratories, College Station, Pa., USA
(20) Chung, M.T., Marrand, M., and Neel, J., Biopolymers 15, 2081 (1976)
(21) Chung, M.T., Marrand, M., and Neel, J., Biopolymers 16, 715 (1977)
(22) Toniolo , C., and Palumbo, M., Bipolymers 16, 219 (1977)
(23) Spackman, D.H., Stein, W.H., and Moore, S., Anal.Chem. 30, 1190 (1958)
(24) Frank, H., Rettenmeier, A., Weicker, H., Nicholson, G.J., and Bayer, E., manuscript in preparation
(25) Patil, P.N., La Pidus, J.B., and Tye, A., J.Pharmacol.Exp.Ther. 155, 1 (1967)
(26) Kier, L.B., J.Pharmacol.Exp.Ther., 164, 75 (1968)

RECEIVED July 12, 1979.

23

Synthesis of Poly-L-Lysine Containing Nucleic Acid Bases

Y. INAKI, T. ISHIKAWA, and K. TAYEMOTO

Faculty of Engineering, Osaka University, Yamadakami, Suita, Osaka 565, Japan

Among the naturally occurring polymers with multiple functionality, nucleic acids seem to be the most interesting species. The nucleic acids, that is, DNA and RNA, contain two bases of the purine family, adenine and guanine, and also those of the pyrimidine family, thymine (in the latter case, uracil) and cytosine. The most essential function of the nucleic acid bases is considered to be the formation of base-base pairing through hydrogen bonding between purines and pyrimidines, such as thymine (or uracil) with adenine, and cytosine with guanine, which plays an important role in realizing the replication and transcription of genetic codes for protein synthesis.

uracil

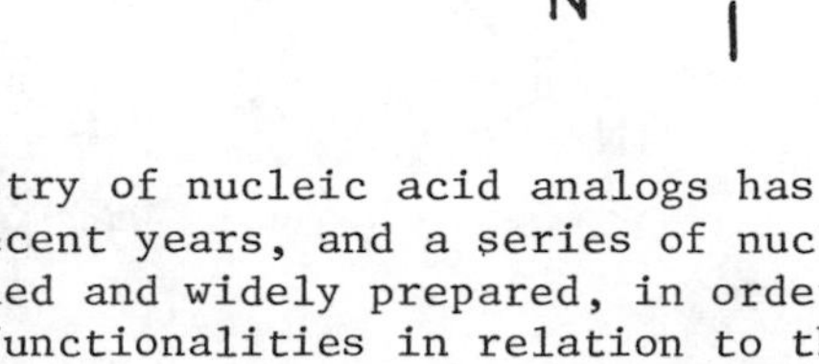

adenine

The chemistry of nucleic acid analogs has received much attention in recent years, and a series of nucleic acid models has been designed and widely prepared, in order to estimate and utilize their functionalities in relation to the specific base-pairing properties (1, 2, 3). These monomers and polymers, particularly those containing purines, pyrimidines, nucleosides, and nucleotides, are not only of interest to the field of heterocyclic organic chemistry, but also to that of biomimetic macromolecular chemistry as synthetic analogs of the nucleic acids.

In the investigations hitherto developed, however, it was

0-8412-0540-X/80/47-121-359$05.00/0

difficult to know the conformation of these polymers in solution, because most of them consist of a vinyl-type backbone and pendant nucleic acid bases. In this respect, a study on nucleic acid analogs having poly-α-amino acid backbone seems also to be attractive. The present article concerns the synthesis of poly-α-amino acids, particularly of ε, N-substituted poly-L-lysine with the nucleic acid bases, as well as the synthesis of amino acids including L-lysine having pendant nucleic acid bases and their polymerization by using the N-carboxyamino acid anhydride method, in order to get detailed informations about the conformational effect of the nature of their backbone structure on the complex formation of the nucleic acid analogs.

Incorporation of nucleic acid bases by polymer modification reactions

The derivatives of poly-L-lysine having pendant nucleic acid bases, that is, adenine, thymine and uracil were prepared as shown in the following scheme:

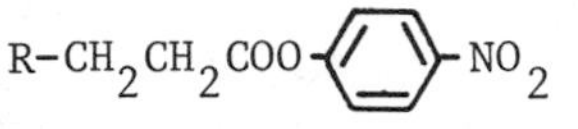

$R-CH_2CH_2COO-C_6H_4-NO_2$ + $-NH-CH(-(CH_2)_4-NH_2\cdot CF_3COOH)-CO-$

R: Adenine (*1*)
Thymine (*2*)
Uracil (*3*)

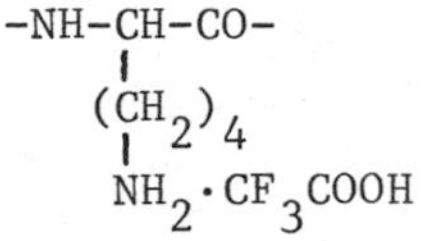

(*4*)

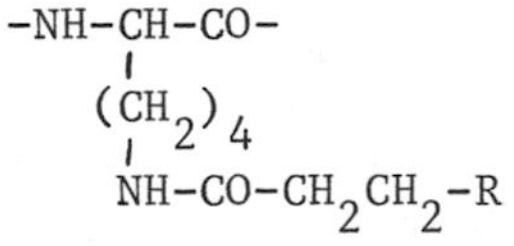

$-NH-CH(-(CH_2)_4-NH-CO-CH_2CH_2-R)-CO-$

R: Adenine (*5*)
Thymine (*6*)
Uracil (*7*)

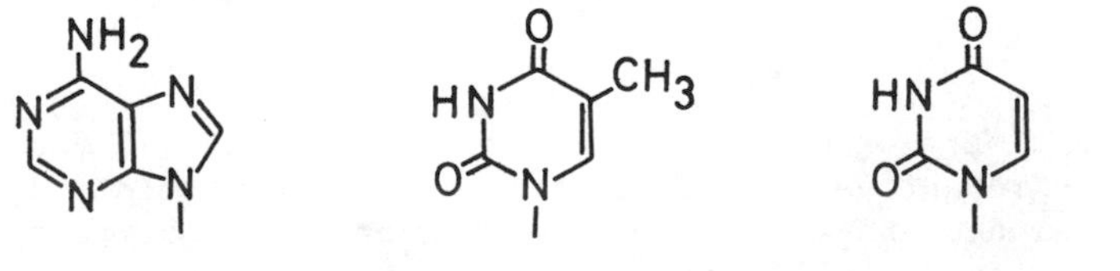

Adenine (Ade) Thymine (Thy) Uracil (Ura)

Carboxyethyl derivatives of the nucleic acid bases were grafted onto poly-L-lysine by using the activated ester method (4). Poly(L-lysine trifluoroacetate)(*4*) was prepared according to the method of Sela (5). The *p*-nitrophenyl esters (*1*, *2*, *3*) were prepared by the reaction of the corresponding carboxyethyl derivatives of the nucleic acid bases with *p*-nitrophenyl trifluoroacetate according to the method of Overberger and Inaki (6), and purified by recrystallization.

The reaction of the *p*-nitrophenyl esters with the polymer (*4*) was studied in dimethyl sulfoxide (DMSO) solution in the presence of triethylamine at 25°C. The poly-L-lysine derivatives obtained have different IR absorption spectra from those of the starting compounds, and have absorptions assigned to the nucleic acid bases. Poly(ε,N-Ade-L-lysine)(*5*) was soluble in DMSO and ethylene glycol, and also in water below pH 3, where it was present as a protonated form. In dimethylformamide (DMF) solution, the poly-L-lysine containing 53 mol % adenine units was soluble, while the polymer containing 74 mol % adenine units was insoluble. Poly(ε,N-Thy-L-lysine)(*6*) and poly(ε, N-Ura-L-lysine)(*7*) was soluble in DMSO, DMF and 6 N-hydrochloric acid.

The contents of the nucleic acid bases in the poly-L-lysine derivatives were determined by UV spectra of the polymers after hydrolysis: The polymers were hydrolyzed in 6 N-hydrochloric acid at 105°C for 24 hr, into lysine dihydrochloride and the carboxyethyl derivatives of the nucleic acid bases. The quantitative calculation was made relative to the standard sample of the carboxyethyl derivative of the nucleic acid bases. The analytical data are listed in Table 1. It was found that the thymine and uracil derivatives was completely substituted to polylysine. Low value in case of adenine base in the polymer may be attributed to the unstability of the activated ester, Ade-PNP (*1*), and may also be explained in terms of the steric interaction among bulky pendant groups of the polymer. When the poly-L-lysine containing about 50 mol % adenine units was again treated with Ade-PNP, the adenine unit content in the polymer increased up to 74 mol % (7).

The spectral data of these polymers are tabulated in Table 1 and 2. From these data, it was concluded that the activated ester of Ade-PNP (*1*) reacted only with ε-amino group of poly-L-lysine, and did not react with amino group of the adenine base. Figure 1, 2 and 3 show their NMR spectra.

In relation to these works, the reaction of *p*-nitrophenyl esters with optically active poly(propyleneimine)(*8*) was studied at 25°C in DMSO solution according to the same procedure described for the case of poly-L-lysine derivatives. The poly(propyleneimine) derivatives thus obtained have different IR and UV absorption spectra from those of the starting compounds, and show absorptions assigned to the nucleic acid bases. However, their contents determined by UV spectroscopy were substantially low as compared with the case of poly-L-lysine derivatives; for (*9*) and (*10*), the base contents were below 30 and 50 %, respectively. The result was explained by a steric hindrance caused by methyl groups on the main chain of poly(propyleneimine):

Table 1. Analytical data of nucleic acid base substituted poly-L-lysines

	Base	mol %[1]	UV[2] $\lambda_{max,nm}$	ε_{max}	$[\alpha]_D$ [3]	$[\eta]$ [4]
(5)	Ade	53	266	12,200	+ 4.8°	0.76[5]
(5)	Ade	74	268	12,400	+ 4.7°	0.34[6]
(6)	Thy	97	273	9,000	+ 1.0°	0.40[5]
(7)	Ura	97	269	8,800	+ 2.7°	0.26[6]

1) From UV spectra of the hydrolyzed samples 2) in DMSO at 25°C. ε value is corrected based on the nucleic acid bases 3) in DMSO (c = 1) at 22°C 4) in DMSO at 25°C 5) [η] = 0.4 in DMSO at 25°C for the original polymer; poly(ε,N-trifluoroacetyl-L-lysine) 6) [η] = 0.3 in DMSO at 25°C for the original polymer; poly(ε,N-trifluoroacetyl-L-lysine)

Table 2. NMR spectra of nucleic acid base substituted poly-L-lysines[1]

Base	mol %[2]	CH_2 c, d, e	CH_2 f	CH b	NH a, g	CH_2 h
(5) Ade	53	1.40, 1.76	3.04	4.15	7.34, 7.52	2.72
(5) Ade	74	1.39, 1.74	3.04	4.08	7.36, 7.58	2.71
(6) Thy	97	1.43, 1.8	3.07	4.15	7.34, 7.5	2.50
(7) Ura	97	1.44, 1.77	3.09	4.21	7.36, 7.5	2.52

CH_2, i	8-H	2-H	6-NH_2	5-CH_3	5-H	6-H	3-NH
4.40	8.03	8.23	5.28				
4.39	8.01	8.22	5.54				
3.89				1.77		7.35	10.43
3.92					5.52	7.52	10.47

1) From UV spectra of the hydrolyzed samples 2) in d_6-DMSO at 150°C, ppm from TMS:

$$-CO-CH^{b}(NH^{a})-CH_2^{c}-CH_2^{d}-CH_2^{e}-CH_2^{f}-NH^{g}-CO-CH_2^{h}-CH_2^{i}-Base$$

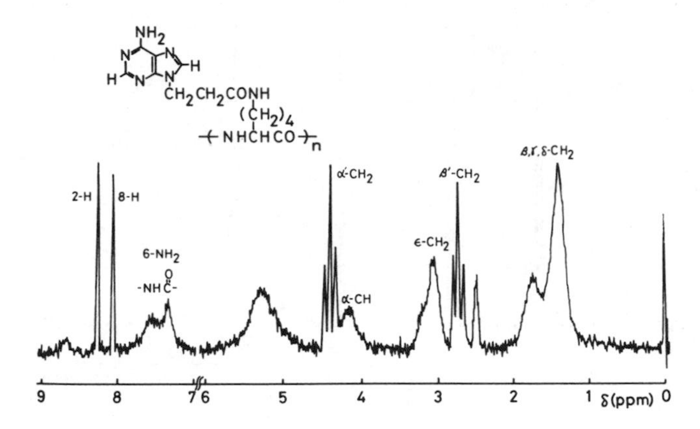

Figure 1. NMR spectrum of poly-L-lysine having pendant adenine moieties

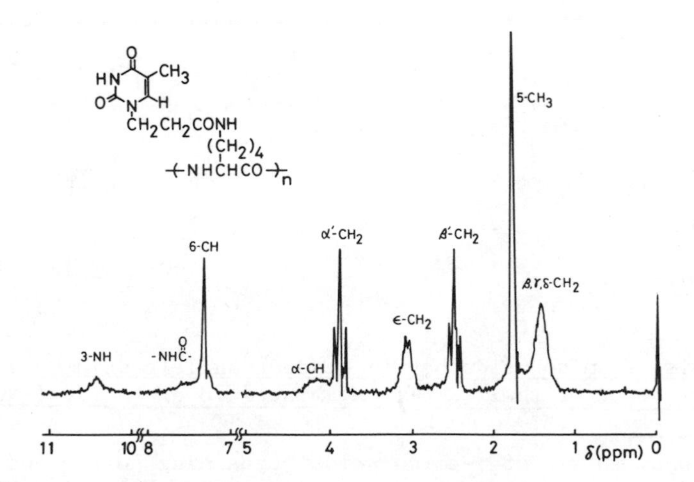

Figure 2. NMR spectrum of poly-L-lysine having pendant thymin moieties

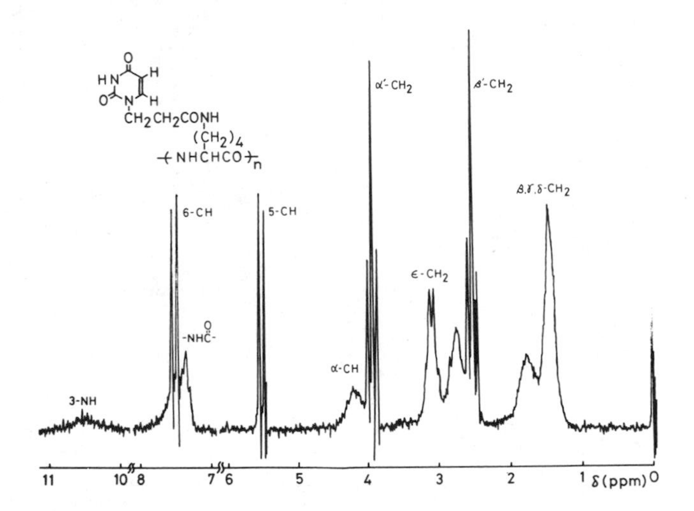

Figure 3. NMR spectrum of poly-L-lysine having pendant uracil moieties

$$-NH-\overset{CH_3}{\underset{*}{CH}}-CH_2- \quad + \quad (1),\ (2) \longrightarrow -N(C(=O)CH_2CH_2-R)-\overset{CH_3}{\underset{*}{CH}}-CH_2-$$

(*8*)

R: Adenine (*9*)
Thymine (*10*)

Synthesis of amino acids having pendant nucleic acid bases and polymerization of their N-carboxyamino acid anhydride (NCA) derivatives

The synthesis of α-amino acids containing purine and pyrimidine side chains was first reported in 1964 (8), with the aim of preparing nucleic acid analogs in which phosphodiester linkages are replaced by peptide linkages. Only one known, naturally occurring compound of this family is L-willardiine, which is an amino acid having uracil moiety as the side group.

D,L-Willardiine was obtained from α,N-tosylasparagine via 2-amino-3-ureidopropanoic acid (8), as the first synthetic example (*11*, R = H). In addition, β-(2,4-dihydroxypyrimidin-5-

yl)alanine (*12*) was prepared by acid hydrolysis of the reaction product from 5-chlorouracil, sodium methoxide, and diethyl acetylaminomalonate in methanol solution (9). Other amino acids containing differently substituted pyrimidine residues were also synthesized (10, 11, 12). The thymine and cytosine analogs of willardiine (*11*, R = CH_3, and *13*, respectively) and D,L-β-(6-aminopurin-9-yl)alanine (*14*) were prepared by treating the corresponding pyrimidines and adenine with bromoacetaldehyde diethyl acetal, followed by hydrolysis to afford the aldehydes, and conversion of these by the Strecker synthesis into D,L-alanine derivatives (13):

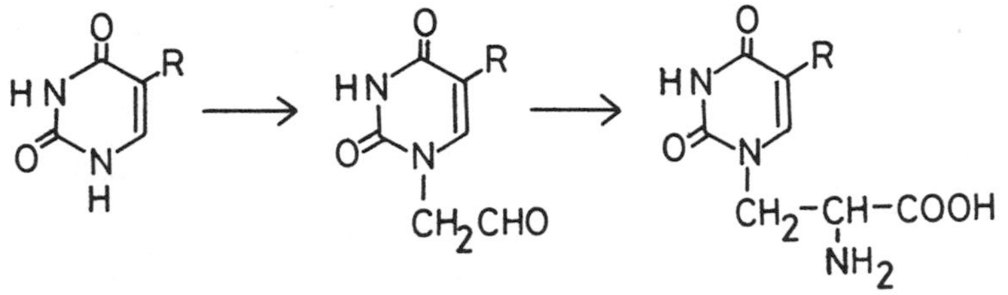

(*11*)

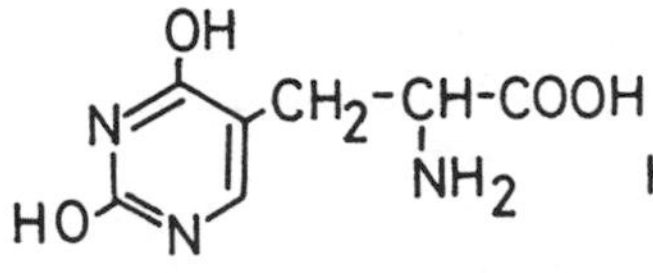

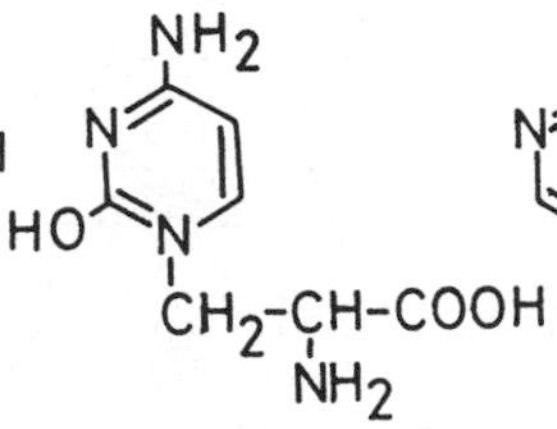

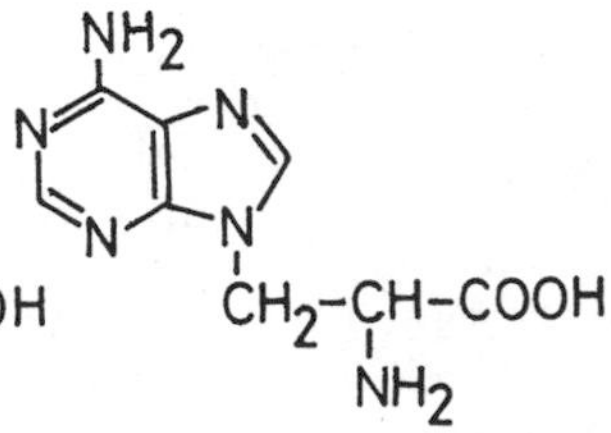

(*12*) (*13*) (*14*)

The preparation of two additional derivatives of amino acids having pendant nucleic acid bases was reported successively (14, 15).

In 1973, the synthesis of amino acids containing pendant theophylline and other nucleic acid bases, and the polymerization of NCA compounds derived therefrom have studied by our group (16).

The polymers obtained, however, were oligomeric ones.

As an extension of our systematic works on these subjects, we found successively a convenient route to a series of new amino acid derivatives having pendant nucleic acid bases: L-lysine deri-

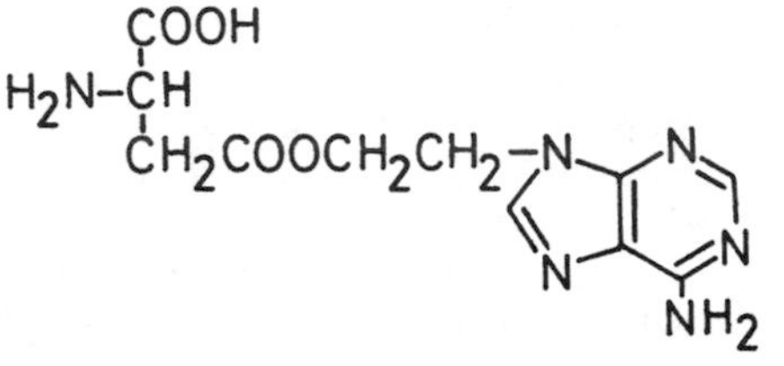
COOH
H_2N-CH
$CH_2COOCH_2CH_2$-N
N
N
N
NH_2

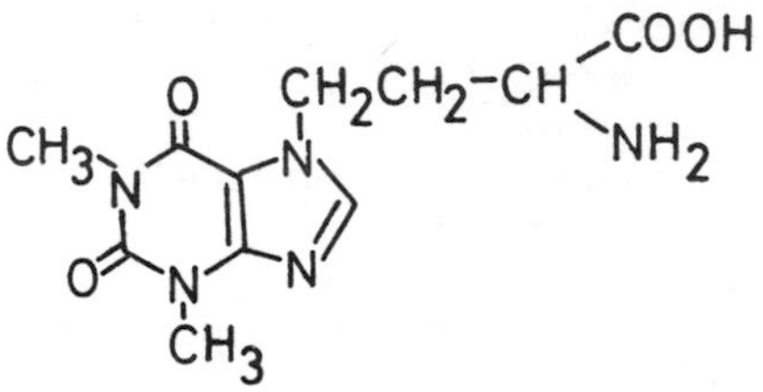
COOH
O
CH_2CH_2-CH
NH_2
CH_3
N
N
N
O
N
CH_3

(*15*) (*16*)

vatives containing the bases were prepared as shown in the following scheme (<u>17</u>):

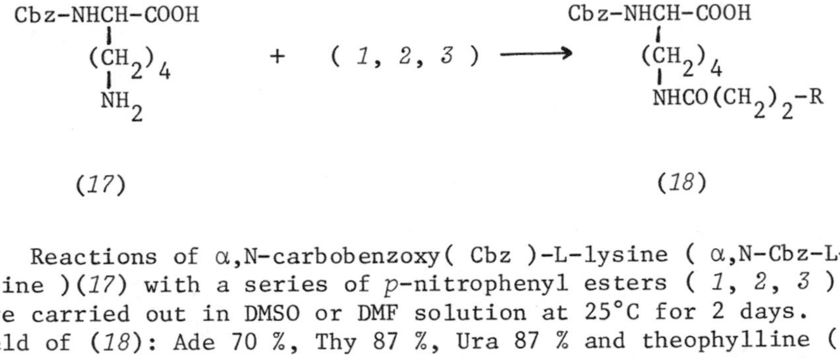
Cbz-NHCH-COOH
$(CH_2)_4$
NH_2
+ (*1, 2, 3*)
Cbz-NHCH-COOH
$(CH_2)_4$
$NHCO(CH_2)_2$-R

(*17*) (*18*)

Reactions of α,N-carbobenzoxy(Cbz)-L-lysine (α,N-Cbz-L-lysine)(*17*) with a series of *p*-nitrophenyl esters (*1, 2, 3*) were carried out in DMSO or DMF solution at 25°C for 2 days. Yield of (*18*): Ade 70 %, Thy 87 %, Ura 87 % and theophylline (The) 69 %. These products were assigned by elemental analysis,

Table 3. Synthesis and properties of nucleic acid base substituted L-lysines

	Base	Solvent[1)]	m.p. (°C)	UV[2)] λ_{max},nm	ε_{max}	$[\alpha]_D$ [3)]
(*18a*)	Ade	DMSO	125-127	263	13,300	- 2.4°
(*18b*)	Thy	DMF	159-161	271	9,500	- 3.2°
(*18c*)	Ura	DMF	169-171	266	10,000	- 2.2°
(*18d*)	The	DMSO	141-143	276	8,400	- 1.0°

1) Solvents used for the reaction 2) In ethanol, at 25°C 3) In DMSO (c = 2), at 21°C

IR, UV and NMR spectra. The result was listed in Table 3. It was also found that the amino group of adenine at 6-position remained unreacted by the attack of the activated ester under the condition used.

In connection with the synthesis of L-lysine derivatives having pendant nucleic acid bases, two types of L-glutamic acid derivatives, that is, those of amide and ester type, were also successfully prepared. The former compound was prepared by the reaction of N-phthalyl-L-glutamic acid anhydride with aminoethyl derivatives of nucleic acid bases, and the phthalyl blocking group of this compound (*19*) was removed by treating with hydrazine to give the aimed L-glutamic acid derivatives (*20*):

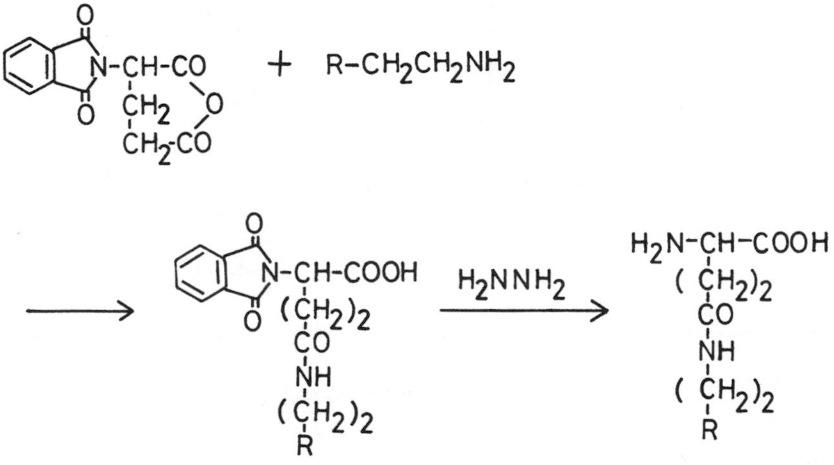

$$\text{Phth}\text{N-CH-CO(O)CO-CH}_2\text{-CH}_2 \;(\text{cyclic anhydride}) + R\text{-}CH_2CH_2NH_2 \longrightarrow \text{PhthN-CH(COOH)-}(CH_2)_2\text{-CO-NH-}(CH_2)_2\text{-R} \xrightarrow{H_2NNH_2} H_2N\text{-CH(COOH)-}(CH_2)_2\text{-CO-NH-}(CH_2)_2\text{-R}$$

(*19*) (*20*)

The latter, ester-type derivatives (*21*) were prepared by the reaction of L-glutamic acid with hydroxyethyl derivatives of nucleic acid bases. The reaction was studied in the presence of *p*-toluenesulfonic acid at 100-110°C in dioxane, and water formed was removed by azeotropic distillation with dioxane (18).

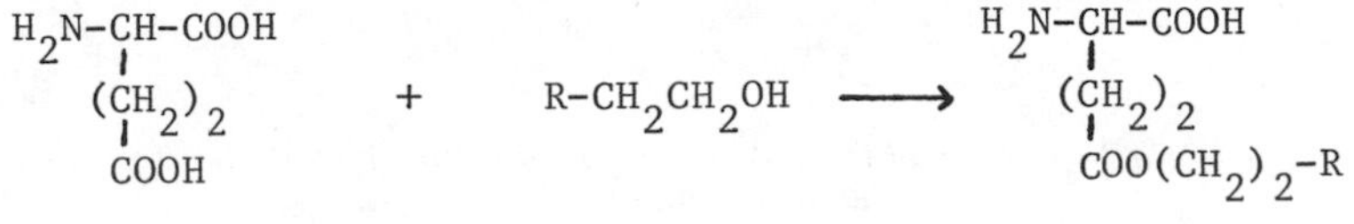

$$H_2N\text{-CH(COOH)-}(CH_2)_2\text{-COOH} + R\text{-}CH_2CH_2OH \longrightarrow H_2N\text{-CH(COOH)-}(CH_2)_2\text{-COO}(CH_2)_2\text{-R}$$

(*21*)

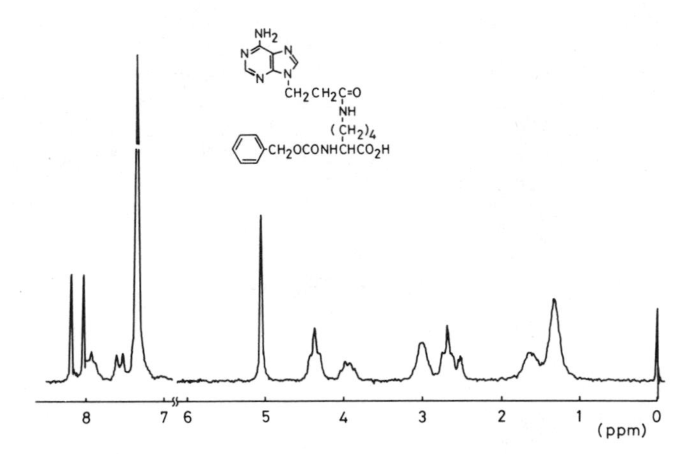

Figure 4. NMR spectrum of α,N-carbobenzoxy-L-lysine having adenine moiety (18)

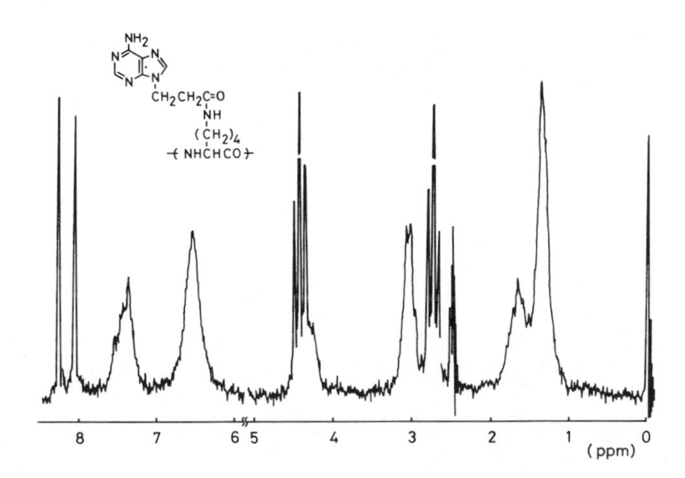

Figure 5. NMR spectrum of the polymer derived from α,N-carbobenzoxy-L-lysine having adenine moieties

These amino acid derivatives containing pendant nucleic acid bases were polymerized by the NCA method. The NCAs of L-lysine derivatives (*22*) were prepared from the corresponding amino acids, for example, (*18*) and thionyl chloride using the Leuchs method, and those of L-glutamic acid derivatives (*23*) were prepared with phosgen in dioxane using the Fuchs method, because of solubility problems. The polymerization of them was carried out using triethylamine as an initiator at room temperature for 2 days. Reprecipitation from DMSO-ethanol gave poly(ε,N-Ade-L-lysine), for example, in 73 % yield. The degree of polymerization was about 15. Figure 4 and 5 show NMR spectra of (*18*) and its polymer, as an example.

$$(18) \xrightarrow{SOCl_2} \underset{(22)}{\begin{matrix} NH\text{-}R \\ | \\ (CH_2)_4 \\ | \\ CH\text{-}CO \\ | \quad \diagdown \\ | \quad\; O \\ | \quad \diagup \\ NH\text{-}CO \end{matrix}} \xrightarrow{N(C_2H_5)_3} \begin{matrix} NH\text{-}R \\ | \\ (CH_2)_4 \\ | \\ \text{-(-NH-CH-CO-)-} \end{matrix}$$

R: Base-$(CH_2)_2$-CO-

$$\begin{matrix} CO\text{-}R \\ | \\ (CH_2)_2 \\ | \\ H_2N\text{-}CH\text{-}COOH \end{matrix} \xrightarrow{COCl_2} \underset{(23)}{\begin{matrix} CO\text{-}R \\ | \\ (CH_2)_2 \\ | \\ CH\text{-}CO \\ | \quad \diagdown \\ | \quad\; O \\ | \quad \diagup \\ NH\text{-}CO \end{matrix}} \xrightarrow{N(C_2H_5)_3} \begin{matrix} CO\text{-}R \\ | \\ (CH_2)_2 \\ | \\ \text{-(-NH-CH-CO-)-} \end{matrix}$$

R:Base-$(CH_2)_2$-NH-

Base-$(CH_2)_2$-O-

Conclusion

A series of new amino acid derivatives having pendant nucleic acid bases was prepared by the reaction of L-lysine and L-glutamic acid with the nucleic acid bases. These amino acids were further polymerized by using the N-carboxyamino acid anhydride (NCA) method. Alternatively, the nucleic acid base substituted poly-L-lysines were also prepared by using polymer reactions which include the reaction of carboxyethyl derivatives of the bases onto poly-L-lysine. Physico-chemical properties of the polymers obtained were given.

Literature Cited

1. Takemoto, K., J.Polymer Sci., Polymer Symposia, 1976, 55, 105.
2. Takemoto, K., "Polymeric Drugs" (Donaruma, L.G. and Vogl, O. Ed.), Academic Press, New York, 1978, p.103.
3. Pitha, J., Polymer, 1977, 18, 425.
4. Anand, N., Murthy, N.S.R., Naider, F. and Goodman, M., Macromolecules, 1971, 4, 564.
5. Sela, M., Arnon, R. and Jacobson, I., Biopolymers, 1963, 1, 517.
6. Overberger, C.G. and Inaki, Y., J.Polymer Sci.Polymer Chem. Ed., in press.
7. Ishikawa, T., Inaki, Y. and Takemoto, K., Polymer Bull., 1978, 1, 85.
8. Shvachkin, Y.P. and Azarova, M.T., Zh.Obshch.Khim., 1964, 34, 407.
9. Shvachkin, Y.P. and Shprunka, I.K., Vestn.Mosk.Univ.Der.II Khim., 1964, 19, 72; Chem.Abst., 1965, 62, 6481.
10. Shvachkin, Y.P. and Syrtsova, L.A., Zh.Obshch.Khim., 1964, 34, 2159.
11. Shvachkin, Y.P. and Krivtsov, G.G., Zh.Obshch.Khim., 1964, 34, 2164.
12. Shvachkin, Y.P. and Azarova, M.T., Zh.Obshch.Khim., 1964, 34, 2167.
13. Doel, M.T., Jones, A.S. and Taylor, N., Tetrahedron Letters, 1969, 2285.
14. Kondo, K., Murata, Y. and Takemoto, K., Techn.Repts.Osaka Univ., 1972, 22, 785.
15. Miyata, M., Kondo, K. and Takemoto, K., Techn.Repts.Osaka Univ., 1973, 23, 339.
16. Takemoto, K., Tahara, H., Yamada, A., Inaki, Y. and Ueda, N., Makromol.Chem., 1973, 169, 327.
17. Ishikawa, T., Inaki, Y. and Takemoto, K., Polymer Bull., 1978, 1, 215.
18. Ishikawa, T., Shigeno, Y., Takahara, T., Inaki, Y., Kondo, K. and Takemoto, K., Nucleic Acids Res.Suppl., 1978, 5, 279.

RECEIVED July 12, 1979.

24

Homogeneous Solution Reactions of Cellulose, Chitin, and Other Polysaccharides

C. L. McCORMICK, D. K. LICHATOWICH, J. A. PELEZO, and K. W. ANDERSON

Department of Polymer Science, University of Southern Mississippi, Hattiesburg, MS 39401

In the past, a lack of suitable organic solvents has prevented facile preparation of a wide range of derivatives from unmodified polysaccharides such as chitin and cellulose. Normally synthetic reactions for modification of these natural polymers have been conducted heterogeneously. In the absence of acceptable solvents, characterization of starting materials is difficult and reaction yields are often low due to unfavorable kinetics. Only in those cases in which the substituted products were soluble, have polymer structures been readily identifiable by instrumental analysis.

In one successful procedure,(1) however, cellulose was dissolved in dimethylsulfoxide in the presence of formaldehyde to yield clear viscous solutions. These methyloylcellulose derivatives were then reacted with anhydrides in the presence of pyridine to form esters.(2) The major advantage of the reported method was that the intermediate methyloylcellulose derivative could be reacted directly without isolation or purification.

Recently, non-aqueous solvent systems for cellulose have been reviewed.(3) Binary systems consisting of a nitrosyl-cation-forming compound like N_2O_4, NOCl, or $NOSO_4H$ with a polar organic liquid successfully dissolve cellulose. Similarily, three component systems of an amine and a polar organic compound with SO_2, $SOCl_2$, or SO_2Cl_2 dissolve cellulose. Unfortunately, the

0-8412-0540-X/80/47-121-371$05.00/0

reactivity of the solvent precludes preparation of a number of simple organic derivatives.

In continuing investigations(4-8) of hydrolyzable polymeric pesticide derivatives, a number of carbamate and ester derivatives of biodegradable polysaccharides have been prepared in our laboratories in organic solvents under homogeneous reaction conditions. The solvent systems utilized in these reactions--N,N-dimethylacetamide with three to eight percent of an inorganic lithium salt--allow facile characterization of both the reactant and derivatized polysaccharide. This product dissolution is especially important for reactions in which a low degree of substitution is desirable.

Lithium salts in organic solvents with high solubility parameters have been previously used to dissolve strongly hydrogen-bonded polyamides(9-11) and polysaccharides(12) for viscosity studies and for preparation of films or fibers. We are not aware of any previous attempts to utilize these solvent systems to prepare ester or carbamate polysaccharide derivatives.(8)

Experimental

A. Isocyanates

Phenyl isocyanate, p-tolyl isocyanate, p-toluenesulfonyl isocyanate, and p-chlorophenyl isocyanate were purchased from Aldrich Chemical Company and were distilled under vacuum prior to use.

4-Isocyanato-6-(1,1-dimethylethyl)-3-(methylthio)-(1,2,4)-triazin-5(4*H*)one(13) was prepared by dropwise addition of 0.46 moles of 4-amino-6-(1,1-dimethylethyl)-3-(methylthio)-(1,2,4)-triazine-5-(4*H*)one in 200 ml of tetrahydrofuran (THF) into a 2 liter, 3-necked flask equipped with stirrer, condenser, and thermometer and containing 2.0 moles of phosgene in 250 ml of THF at -10°C. The reaction was allowed to proceed for one hour with stirring at -5°C to +5°C. The temperature was allowed to rise slowly to 25°C over a one hour period. The excess phosgene and HCl gases were trapped by a sodium hydroxide scrubber solution as the reaction mixture was warmed gently to 40°C. During this process the reaction media became clear. A strong infrared absorbance at 2320 cm^{-1} confirmed the presence of the isocyanate functional group. This isocyanate (101.3 g, 92%) was used without further purification.

B. Acid Chlorides

2,4-Dichlorophenoxyacetyl chloride(13) was prepared by addition of 110.0 g (0.498 moles) of 2,4-dichlorophenoxyacetic acid in 250 ml of benzene and 7 ml of N,N-dimethylformamide (DMF) into a one liter, 3-necked roundbottomed flask equipped with mechanical stirrer, addition funnel, thermometer and condenser connected to a caustic scrubber. After heating to reflux, a solution of

80.0 g (0.676 moles) of thionyl chloride in 100 ml of benzene was added over a five hour period. After removing excess solvent and thionyl chloride, the crude product was distilled yielding 110 g (92%) of the acid chloride (b.p. 124°C @ 120 mm).

2,2-Dichloropropionyl chloride(13) was prepared by charging a two liter, 3-necked, roundbottom flask equipped with mechanical stirrer,thermometer, addition funnel, and condenser connected to a caustic scrubber with 250 g of 2,2-dichlorophenoxyacetic acid in 500 ml of chloroform and 10 ml of DMF. The mixture was refluxed for one hour, followed by the addition of 328 g of thionyl chloride over a period of one hour. The mixture was refluxed for two additional hours. After removal of excess thionyl chloride and solvent, the crude product was vacuum distilled yielding 210 g of the acid chloride (b.p. 75°C @ 20 mm).

C. Carbamate Derivatives of Polysaccharides

Carbamate derivatives (Table 1) of cellulose, chitin, amylose, amylopectin, and dextran were prepared using the isocyanates described in Part A of the Experimental Section. Amylose, amylopectin, dextran, and cellulose were obtained from Polysciences, Inc. and used without further purification. Chitin, obtained from Eastman Kodak, was decalcified and deproteinated by the method reported by Hayes(14) prior to use.

The lithium salt solutions were prepared by adding the chosen salt (LiCl, LiBr, or $LiNO_3$) to the N,N-dimethylacetamide followed by addition of the polysaccharide. Complete dissolution of the polysaccharide required several hours--often with heating and cooling cycles.

In a typical experiment the isocyanate (0.006 moles) was reacted with 1.5 g of the polysaccharide in 150 ml of a 5% LiCl/ N,N-dimethylacetamide solution at 90°C under nitrogen for two hours. The appearance of a strong infrared absorbance at 1705 cm^{-1} was an indication of carbamate formation. The derivatized polymer was isolated as a white powder by precipitation of the reaction solution into a nonsolvent such as methanol. Alternatively thin films were cast directly from solution; the lithium salt could be removed by rinsing with acetone. Figure 1 illustrates the reaction of cellulose with phenyl isocyanate.

D. Ester Derivatives of Polysaccharides

Ester derivatives of cellulose, chitin, dextran, amylose, and amylopectin were prepared utilizing the acid chloride derivatives described in Part B of the Experimental Section.

In a typical example (Figure 2), the 2,2-dichloropropionate ester of chitin was prepared by reacting 1.0 g of chitin dissolved in 100 ml of a 5% LiCl/N,N-dimethylacetamide solution with 0.006 moles of 2,2-dichloropropionyl chloride at 140°C for three hours. The product was isolated by precipitation into methanol.

TABLE I. Physical Data of Derivatized Polysaccharides

Structure	Sample Wt. (mg)	D.S.[a]	IR Absorbance (cm^{-1})	$\eta_{sp/c}$[b] @0.25 dl/g
CHI-OR_1	2.1	0.65	1700	0.52
CHI-OR_2	2.5	0.68	1707	-
CHI-OR_3	2.6	0.70	1700	-
CHI-OR_4	1.8	0.69	1707	-
CHI-OR_5	1.4	0.19	1705	-
CHI-OR_6	1.5	0.73	1725	0.59
CHI-OR_7	1.7	0.41	1740	0.29
CEL-OR_1	1.9	0.98	1700	0.55
CEL-OR_2	1.4	0.87	1705	-
CEL-OR_3	1.3	0.81	1705	-
CEL-OR_4	2.2	0.70	1710	-
CEL-OR_5	1.0	0.12	1705	-
CEL-OR_7	2.1	0.50	1740	0.28
AM-OR_6	1.5	0.44	1725	0.37
AM-OR_7	1.7	0.22	1742	0.20
AMP-OR_1	1.2	0.65	1700	0.58
D-OR_1	1.5	0.82	1700	0.46
D-OR_2	1.3	0.90	1707	-
D-OR_3	1.5	0.88	1707	-
D-OR_4	1.4	0.50	1707	-

CHI - Chitin
CEL - Cellulose
AM - Amylose
AMP - Amylopectin
D - Dextran

R_1 = $(CH_3)_3$-C-C(=N-N=C(-SCH_3)-N(-NH-C(=O)-)-C(=O)-) (triazinone ring)

R_2 = C_6H_5-N(H)-C(=O)-

R_3 = CH_3-C_6H_4-N(H)-C(=O)-

R_4 = CH_3-C_6H_4-S(=O)(=O)-N(H)-C(=O)-

R_5 = Cl-C_6H_4-N(H)-C(=O)-

R_6 = Cl-C_6H_3(Cl)-O-CH_2-C(=O)-

R_7 = CH_3CCl_2C(=O)-

[a]Degree of substitution
[b]Reduced Viscosity

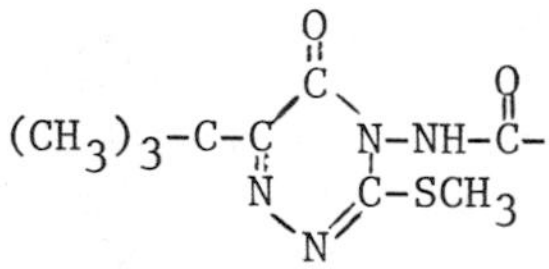

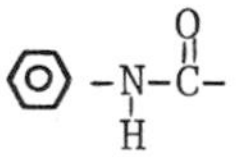

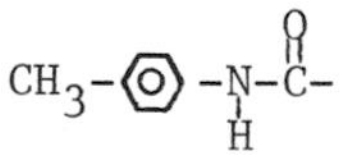

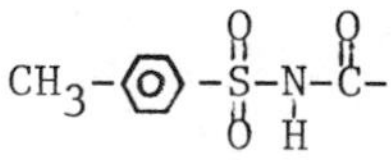

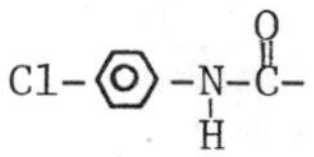

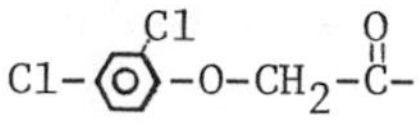

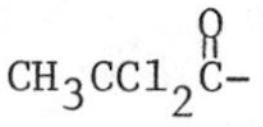

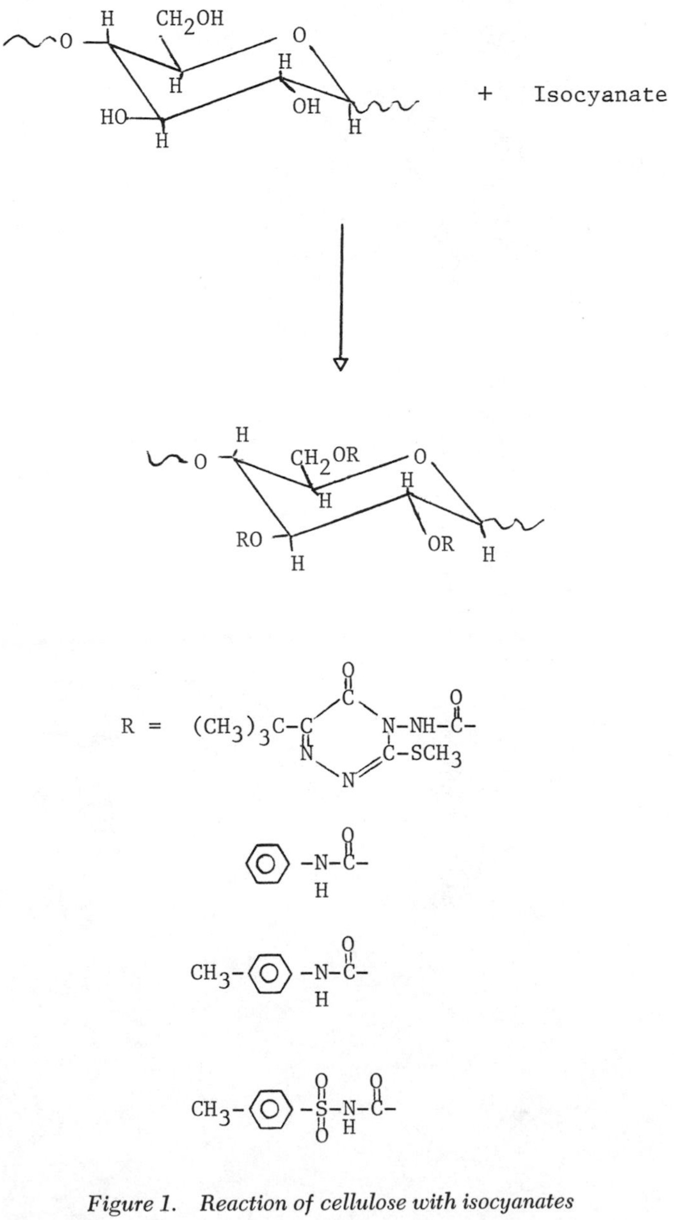

+ Isocyanate

R =

Figure 1. Reaction of cellulose with isocyanates

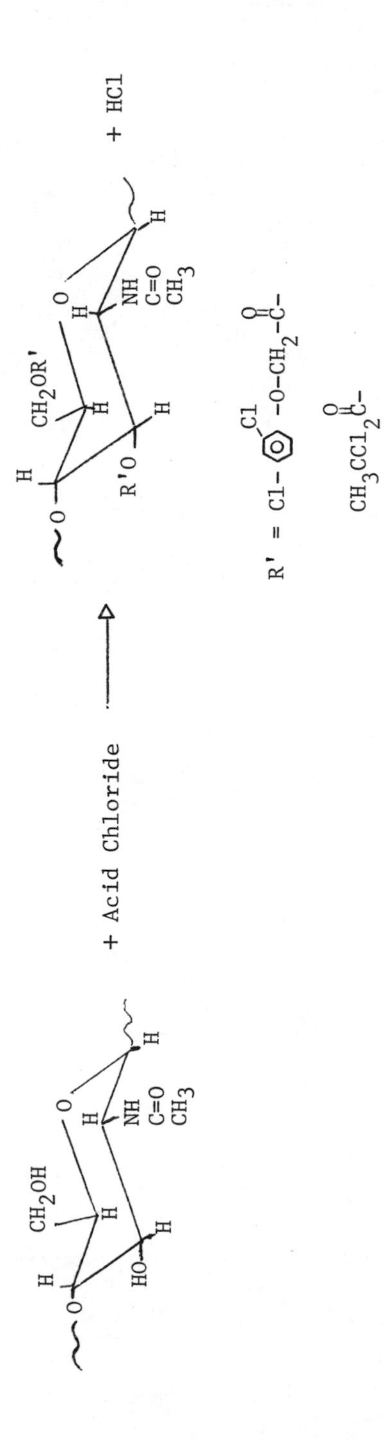

Figure 2. Reaction of chitin with acid chlorides

E. Determination of Degree of Substitution

The derivatized polysaccharides were extracted using a Soxhlet apparatus with tetrahydrofuran as the extracting solvent. The resulting polymer was weighed and then placed in an aqueous sodium hydroxide solution (0.1 N) for 24 hours at 60°C. From the weight of the polymer, the concentration of the amine or acid salt as observed by ultraviolet spectroscopy and the repeating unit weight, the degree of substitution (Table 1) was calculated. It should be noted that these values are probably lower than the actual values since complete hydrolysis for large particle sizes may not be possible in these unswollen polymers.

F. Hydrolysis of Carbamate Derivatives

Preweighed 1.0 mg samples of each polysaccharide carbamate derivative were submerged in aqueous solutions at three pH values: 3.1; 7.0; and 11.3. Three milliliter aliquots were withdrawn at periodic intervals and analyzed by ultraviolet spectroscopy. Typical results are shown in Figures 3 and 4 for pendant hydrolysis rates of carbamate derivatives of cellulose and chitin respectively as a function of pH.

G. Instrumental Analyses

Infrared spectra were recorded on a Perkin Elmer Model 567 Spectrophotometer. Ultraviolet spectra were obtained on a Cary 1756 Spectrophotometer. Gas chromatograms were recorded on a Tracor Model 220 with electron capture detector. High pressure liquid chromatography studies were conducted with a Waters Model ALC-200 with ultraviolet and refractive index detectors.

Results and Discussion

Table I lists physical data for a number of the carbamate and ester derivatives of cellulose, chitin, amylose, amylopectin, and dextran synthesized as described in the Experimental Section. The solubility of the polysaccharide starting materials as well as that of the produced derivatives allows for macromolecular characterization through techniques including UV, NMR, IR, high pressure liquid chromatography, etc.

The solubilization of polysaccharides such as chitin and cellulose apparently results from the disruption of strong intermolecular hydrogen bonding by the lithium ions in the N,N-dimethylacetamide. Interestingly under identical conditions, cations such as Na^+, K^+, Cs^+, Ca^+ , Ba^{++} showed no tendency to solvate the above polymers. Additionally, some specificity was shown for the anion type, i.e., Br^-, Cl^-, and NO_3^-. These trends are under further investigation.

It was anticipated that homogeneous reaction conditions

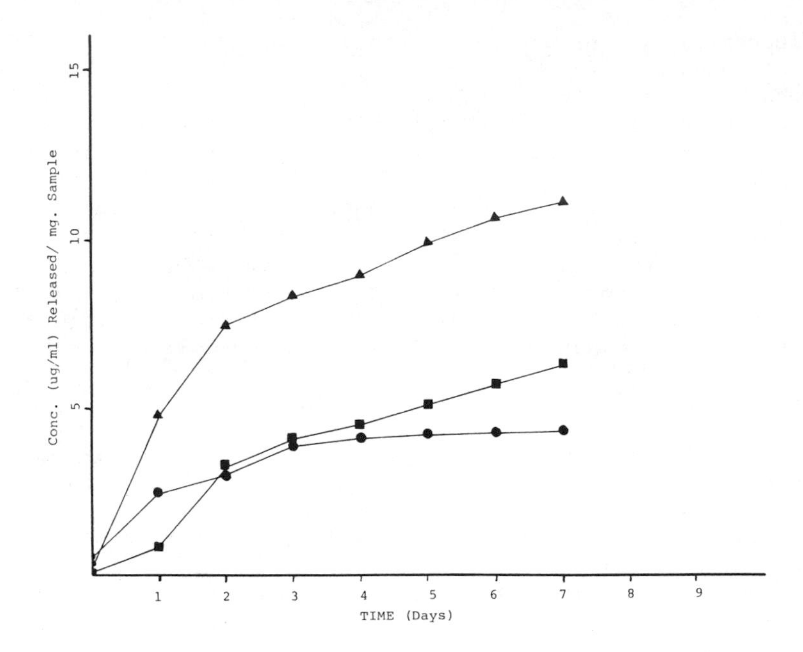

Figure 3. Hydrolysis of pendant urethane groups as a function of medium. Pendant: ⌬*—NHCOO—; substrate: cellulose; (▲) basic medium (pH = 11.3); (■) acidic medium (pH = 3.1); (●) deionized H_2O (pH = 7.0)*

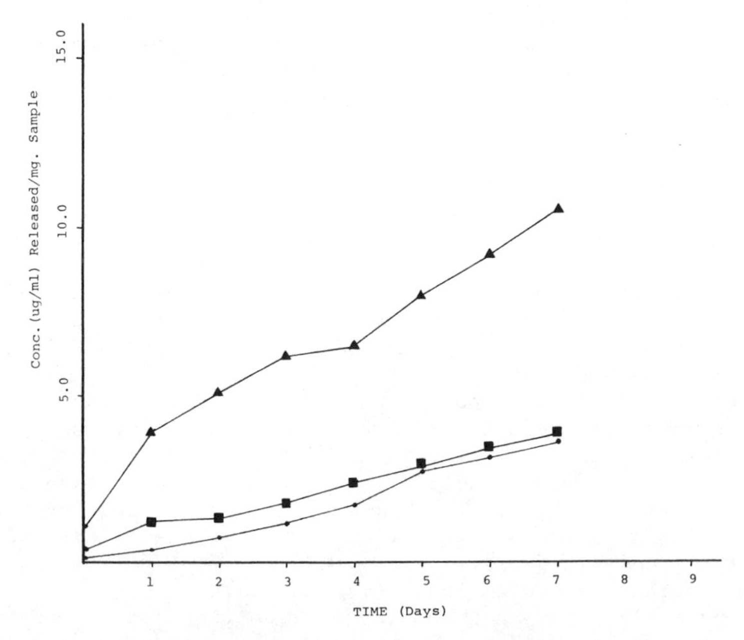

Figure 4. Hydrolysis of pendant urethane groups as a function of medium. Pendant: CH_3⌬*—NHCOO—; substrate: chitin; (▲) basic medium (pH = 11.3); (■) acidic medium (pH = 3.1); (●) deionized H_2O (pH = 7.0)*

should certainly lead to more favorable kinetics than heterogeneous conditions. The reactions reported were conducted without catalysts; no attempt was made to optimize yields. The presence of the carbamate or ester bonds in the derivatized samples was detected in the infrared in the region from 1695 to 1750 cm^{-1} (Table 1). The degrees of substitution (Table I) were determined by 24 hour saponification of the polysaccharide derivatives followed by ultraviolet spectroscopic analysis of the hydrolyzed products. Prior to hydrolysis, the polysaccharide derivatives were extracted with benzene and then methanol to remove any unattached reaction products. Particle size and the solubility parameter of the polysaccharide carbamate or ester derivatives undoubtedly have a great influence on the efficiency of an extraction procedure and upon the degree of hydrolysis during saponification. Therefore, a high pressure liquid chromatography method is being developed to more accurately determine the degree of substitution. In the N,N-dimethylacetamide/Li^+ salt solutions, low molecular weight (unattached) species can be separated from the derivatized polymers; however, strong solvent absorbance and high solution viscosities presently prevent quantitative analysis. At high degrees of carbamate substitution, other solvents may be used for dissolution and the method is more reliable.

Derivatives (Table I) formed by the reaction of the chosen polysaccharide with an isocyanate or acid chloride are carbamates and esters respectively. However, chitin or poly[(1→4)(N-acetyl-2-amino-2-deoxy-β-glucopyranose)] actually used in this experiment contains approximately 16% free amine groups which can form urea and amide derivatives with the above reagents.

Viscosity data are reported in Table I for a number of the polysaccharide derivatives in 5% LiCl/N,N-dimethylacetamide solutions. At low concentrations of polymers, an upward curvature in the $\eta_{sp/c}$ (reduced viscosity) vs c (concentration) plot was observed. Additionally, nonlinear increases in solvent viscosity were observed for increased lithium ion concentrations in the absence of polymer. Therefore, reduced viscosities at 0.25 dl/g are reported.

Hydrolysis studies have been conducted under acidic, basic, and neutral conditions for the polysaccharide derivatives. Figures 3 and 4 illustrate typical rates of hydrolysis for carbamate derivatives of cellulose and chitin. The rates of release at a pH value 11.3 were considerably higher in both systems than at pH values of 3.1 and 7.0. After seven days in the basic medium the cellulose derivative had delivered 27.3 percent of the available aniline. In the acidic medium and neutral medium 15.6 and 10.6 percent were delivered. After seven days the chitin derivative delivered 27.7, 10.0, and 9.5 percent of the available p-methylaniline in the basic, acidic, and neutral media, respectively.

Acknowledgements

Support for portions of this research have come from the NOAA Sea Grant Program and from Hopkins Chemical Company of Madison, Wisconsin.

Literature Cited

1. Seymour, R.B. and E.L. Johnson, Coatings and Plastics Preprints, 36(2), 668 (1976).
2. Seymour, R.B. and E.L. Johnson, Polymer Preprints, 17(2), 382 (1976).
3. Phillipp, B., H.S. Schleicher, and W. Wagennecht, "Non-Aqueous Solvents of Cellulose," Chem Tech, 702-709, November (1977).
4. McCormick, C.L. and M.M. Fooladi, "Synthesis, Characterization, and Release Mechanisms of Polymers Containing Pendant Herbicides," Controlled Release Pesticides, ACS Symposium Series 53, 112-125, H.B. Scher, ed., Washington, D.C., 1977.
5. McCormick, C.L. and K.E. Savage, "Development of Controlled Release Systems Containing Pendant Metribuzin," Proceedings of the 1977 Controlled Release Symposium, pp. 28-40, Corvallis, Oregon (1977).
6. McCormick, C.L., D.K. Lichatowich, and M.M. Fooladi, "Controlled Activity Pendant Herbicide Systems Utilizing Chitin and Other Biodegradable Polymers," Proceedings of the 5th International Symposium on Controlled Release of Bioactive Materials, pp. 3.6-3.17, Gaithersburg, MD (1978).
7. Savage, K.E., C.L. McCormick, and B.H. Hutchinson, "Biological Evaluation of Polymeric Controlled Activity Herbicide Systems Containing Pendant Metribuzin," Proceedings of the 5th International Symposium on Controlled Release of Bioactive Materials, pp. 3.18-3.28, Gaithersburg, MD (1978).
8. McCormick, C.L. and D.K. Lichatowich, "Homogeneous Solution Reactions of Cellulose, Chitin, and Other Polysaccharides to Produce Controlled Activity Pesticide Systems," submitted to the Journal of Polymer Science (1978).
9. deCindo, B. and C. Migliaresi, Polymer, 19, 526 (1978).
10. Morgan, P.W. and S.L. Kowler, Macromolecules, 8, 104 (1975).
11. Panar, M. and L.F. Beste, Polymer Preprints, American Chemical Society, p.   (1965).
12. Austin, P.R., German Offen 2,707,164 (1978); U.S. Patents 4,059,457 and 4,062,921 (1979).
13. Fooladi, M.M., University of Southern Mississippi, Ph.D. Dissertation in preparation (1979).
14. Hayes, G.M., "Chitin as a Chemical Raw Material," Encyclopedia of Chemical Technology, A. Standen, ed., Interscience, p. 222 (1960).

RECEIVED July 12, 1979.

25

Modification of Cotton with Tin Reactants

CHARLES E. CARRAHER, JR., JACK A. SCHROEDER,
CHRISTY McNEELY, and JEFFREY H. WORKMAN
Department of Chemistry, Wright State University, Dayton, OH 45435

DAVID J. GIRON
Microbiology and Immunology, Wright State University, Dayton, OH 45435

Cellulose is a naturally occurring polymeric carbohydrate, hydrolyzable to glucose, consisting of anhydroglucose units linked through a beta-glucosidic bonding. Natural cellulose exhibits usual chain lengths of 1000 to 3000 units long. It is a very common material making up about one-third of all vegetable matter. In actuality cellulose is quite complex and varying in exact composition. Cotton is a relatively pure natural cellulose, containing only 3-15% of noncellulosic material.

The modification of cotton has occurred for years being one of the earliest executed chemical processes in man's history. Even so, much still remains with many of the more recent studies catalyzed by the increasing need to utilize regenerable materials as feedstock in a widening variety of uses. Most of these modifications are topochemical in nature, occurring with cellulosic reactive groups which are available in the amorphorus regions and on the surfaces of crystalline areas. We chose to attempt more intimate, complete modification of cotton. There are few, if any techniques for completely solubilizing cellulose. Generally "solution" is effected through chain degradation where the cellulosic material actually forms a gell-like solution.

The use of bisethylenediamine copper (II) hydroxide solutions to effect solution of cotton has been practiced for many years and is still industrially practiced on a small scale as a method of regeneration of cotton. Copper-amine solutions were utilized for this study for a number of reasons including a. as noted above, an abundance of prior knowledge exists concerning the technique; b. it allows fairly good solution of the cotton; c. it was found, early in our work, to allow the execution of the types of modification desired; and d. it is easily handled and can be utilized on the gram as well as ton scale. Further purity of modified material, i.e. effectiveness of removal of unreacted, etc. material is easily followed through analysis of the copper present in the modified material.

0-8412-0540-X/80/47-121-381$05.00/0

Here we will concentrate on the modification of cotton through reaction with tin-containing reactants and thermal characterization of the products.

EXPERIMENTAL

Reaction apparatus was described elsewhere (1). Briefly it is a one pint Kimex emulsifying jar placed on a Waring Blendor (Model 1120). Dipropyltin dichloride, dibutyltin dichloride, diphenyltin dichloride, triphenyltin chloride, dimethyltin dichloride and dioctyltin dichloride (Alfa Inorganics, Inc., Beverly, Mass.) were used as received. A predetermined volume of bisethylenediaminecopper (II) hydroxide-cellulose solution formed by dissolving cotton (Padco Non-sterile Cotton manufactured by The Absorbant Cotton Co., Valley Park, Mo.) in bisethylenediaminecopper (II) hydroxide (Ecusta Paper Corp., Pisgah Forest, North Carolina; effected with two hours of mechanical stirring (2,3,4) was added to a stirred (ca 20,500 rpm, no load) solutions of organic solvent containing the organotin halide. The products were collected, after suction filtration, in a sintered glass funnel. Repeated washings with water were carried out until after the blue coloration ceased.

Dilute sulfuric acid was added to some reaction solutions to effect neutralization and consequently precipitation of unreacted (usual) and additional modified product. The precipitated material was tested for tin and where tin inclusion occurred, the samples were kept separate from the originally precipitated material.

Infrared spectra were obtained using a Perkin-Elmer 457 Grating Infrared Spectrometer utilizing KBr pellets. Thermal analysis was effected utilizing a duPont 951 Thermogravimetric Analyzer, duPont 990 Thermoanalyzer, duPont Differential Scanning Calorimeter and Fisher-Johns Melting Point Apparatus. Elemental analysis was effected for copper and tin utilizing a Varian Techtron AA6 Atomic Absorption Spectrophotometer utilizing Price's technique for sample preparation (4).

RESULTS

Synthesis

Reaction appears to be general with wide variation of product yield and tin-moiety inclusion ranging from low to high depending on the specific reaction conditions (for instance Table 1). Thus both amount (proportion) and frequency (extent) of cotton modification can be easily controlled lending itself to useful industrial utilization.

Previous work by us has led to the synthesis of silicon, germanium and tin polyethers utilizing interfacial systems (for instance 5-9). The tin-cotton products should possess an analogous structure from previously reported similarities in reactivity of soluble cellulosic hydroxyls with organic acid chlorides.

$$R_2\ MX_2 + HO\text{-}R\text{-}OH \longrightarrow \left(\begin{array}{c} R \\ | \\ M \\ | \\ R \end{array} \text{-O-R-O} \right)$$

I.

The structure of the modified products is probably a mixture of mono-, di- and trisubstituted materials with the amounts of di- and trisubstitution increasing as the proportion of tin-moiety increases even though for the products with dibutyltin dichloride tin moiety inclusion remained high throughout the tin:cotton range of 0.30:1 to 5:1 being about the value expected for trisubstitution.

Representative infrared spectra of modified cotton products appear in Figures 1A and 1B. Analysis of the infrared spectrum of the products were consistent with the tin-modified product being formed. For instance for the di-n-propyltin dichloride-cellulose product a large, broad band appeared from 3200-3500 cm^{-1} characteristic of Sn-OH and R-OH groupings. The aliphatic CH stretching vibrations for both the dipropyltin and cellulose are present between 2850 and 2860 cm^{-1}. Bands characteristic of aliphatic C-H out of place deformations characteristic of dipropyltin appear at 1425, 1415, 1380 and 1130 cm^{-1}. The Sn-O asymmetric stretch from Sn-O-R occurs at 670 cm^{-1} and the Sn-O-R symmetric stretch for tin ethers occurs at 610 cm^{-1} - both bands are present in the modified products (10,11). The Sn-Cl stretching band occurs between 320 to 350 cm^{-1} depending on tin substitution (12). Some products exhibit small bands in this region characteristic of unreacted Sn-Cl groups but spectra of most of the products are clear in this region. Further, negative $AgNO_3$-sodium fussion results were found for the products which were clear in the 320 to 350 cm^{-1} region. A broad band centering about 3300 to 3400 cm^{-1} is characteristic of cellulosic -OH groups. Bands characteristic of the Sn-OH moiety vary between 3400 to 3500 cm^{-1}, making identification of Sn-OH exhibiting bands in the lower (ca 3400 cm^{-1}) region difficult. Bands characteristic of the presence of Sn-OH are present and identifiable where the Sn-OH band occurs within the upper region (ca 3470 to 3500 cm^{-1}). Thus products from dibutyltin dichloride (even when employing a five fold excess of the tin monomer) exhibit a band centering about 3490 cm^{-1} characteristic of the Sn-OH grouping and no band in the 320 to 350 cm^{-1} region.

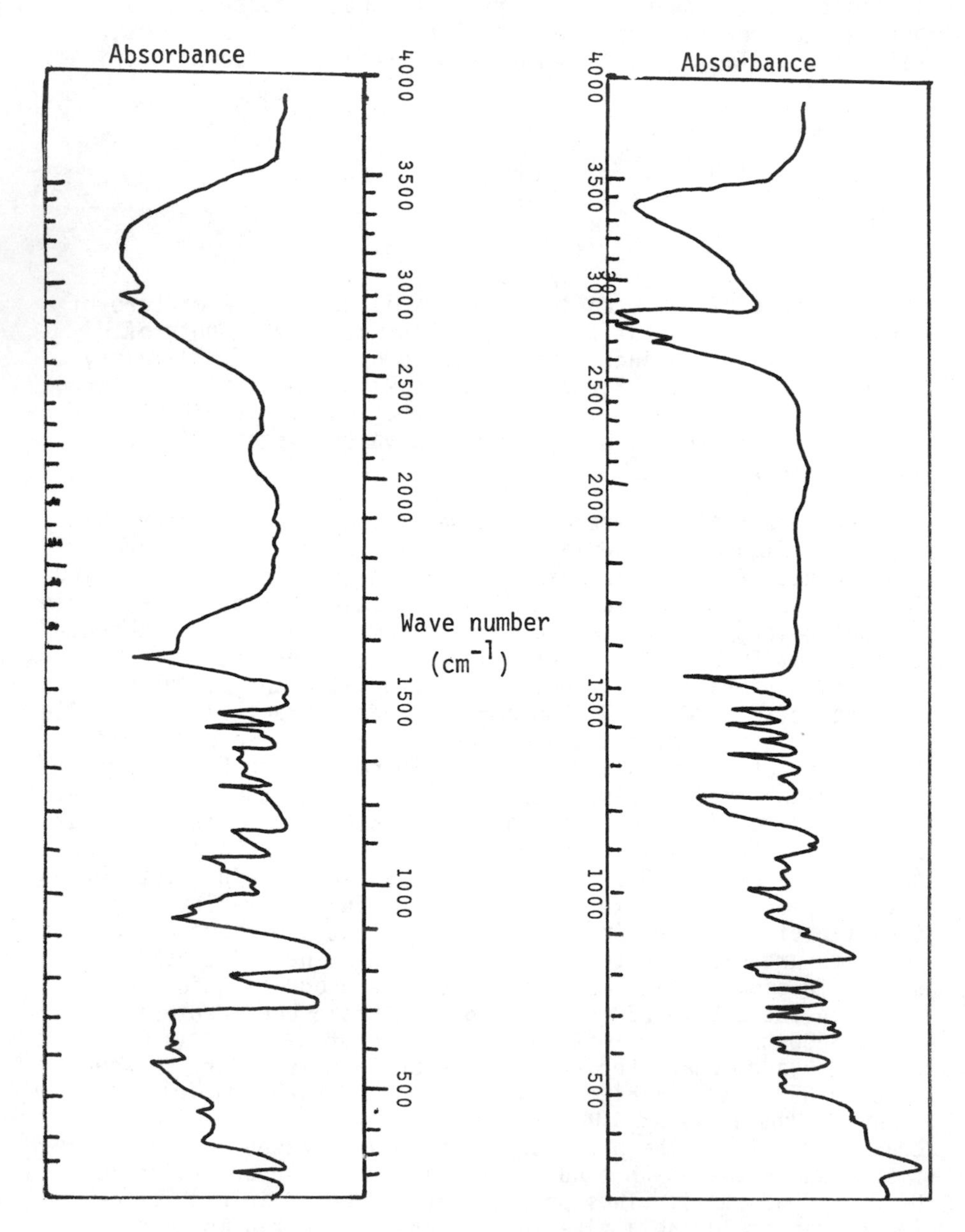

Figure 1. IR spectra of condensation product of cotton with (A) diphenyltin dichloride and (B) dibutyltin dichloride

Because of the shown presence of Sn-OH groups for products employing dihalo reactants and relatively high inclusion of tin-moiety calculations contained in Table 1 were based on Forms II and III.

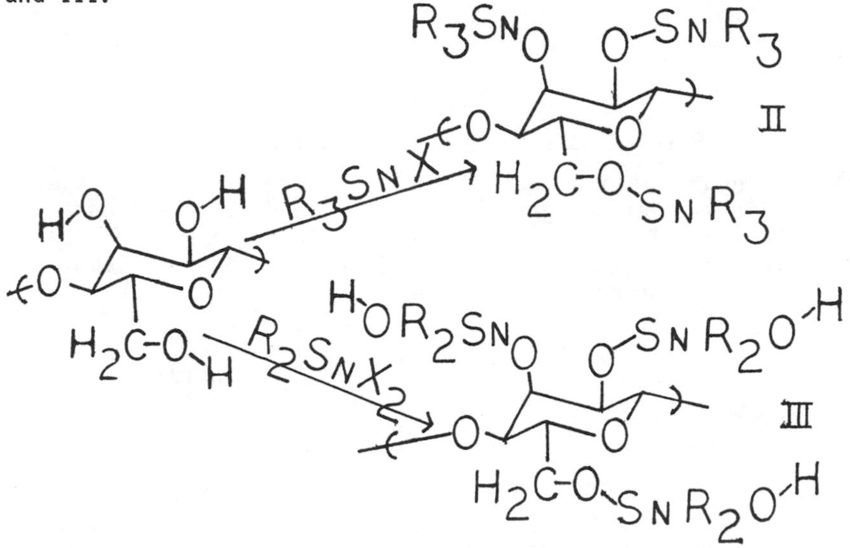

Product purification is easily followed by monitoring the decrease in blue-green coloration possible because of the high coloration of copper II complexes. The presence of copper in the modified product was followed using AA. Copper levels were typically less than 10^{-2}% for compounds reported in this paper. Thus simple, repeated water washings are sufficient to remove copper and presumably unreacted tin monomer. The latter may not always be true if quite excessive amounts of the tin reactant are employed such as in Table 1 where a 5:1 tin:cotton ratio was employed giving a yield in slight excess of 100%, presumably due to the inclusion of some unreacted tin monomer. While it may be desired that complete removal of the tin monomer be effected, it is probably not critical for a number of potential applications where the biological properties are experienced through control release since small amounts of unreacted tin monomer will probably only affect the initial rate of "controlled release" of the tin.

Physical Properties

All of the products are insoluble in all attempted solvents including DMSO, HMPA and DMF typical of crosslinked products. They are hydrophobic and resistant to hydrolysis because of their hydrophobic nature. This change to a hydrophobic nature is positive for applications requiring water stability, resistance and repellency and is typical of most water soluble polymers (such as poly(acrylic acid), polyethyleneimine and poly(vinyl alcohol)

Table I

Results as a function of tin monomer

Organotin Halide	Amount Tin Reactant (mmole)	Molar Ratio Tin: Cotton Reactive Groups	Yield (grams)	Yield Assuming Complete Inclusion (%)	Tin Found (%)	Tin Calculated Assuming Complete Inclusion	Initial Degradation Temp. Air (°C)[a]
Dipropyltin Dichloride	0.93	1:1	0.19	38	21	27	280
Dibutyltin Dichloride	0.28	0:30:1	0.0085	5	41	40	270
Dibutyltin Dichloride	0.47	0:50:1	0.047	15	44	40	
Dibutyltin Dichloride	0.69	0.75:1	0.087	21			
Dibutyltin Dichloride	0.93	1:1	0.077	14			
Dibutyltin Dichloride	1.86	2:1	0.29	52			
Dibutyltin Dichloride	2.33	2.5:1	0.43	77	41	40	

Dibutyl-tin Di-chloride	4.65	5:1	0.58	103	37	40	
Triphenyl-tin Chlor-ide	0.93	0.5:1	0.39	100			240
Triphenyl-tin Chlor-ide	1.86	1:1	0.50	65	19	29	
Dimethyl-tin Di-chloride	0.93	1:1	0.11	32			
Dimethyl-tin Di-chloride	2.79	3:1	0.14	41			
Dioctyl-tin Di-chloride	0.93	1:1	0.31	40			
Dioctyl-tin Di-chloride	2.79	3:1	0.66	86			210
Diphenyl-tin Di-chloride	0.93	1:1	0.024	4			280

Reaction conditions: Aqueous solutions of cotton (0.100 g; 0.62 moles) with bisethylenediamine copper (II) hydroxide to give 25 ml solution are added to rapidly stirred (about 20,500 rpm no load) carbon tetrachloride (25 ml) solutions containing the organotin halide at about 25°C, 30 secs stirring time.

a. For cotton = 310°C

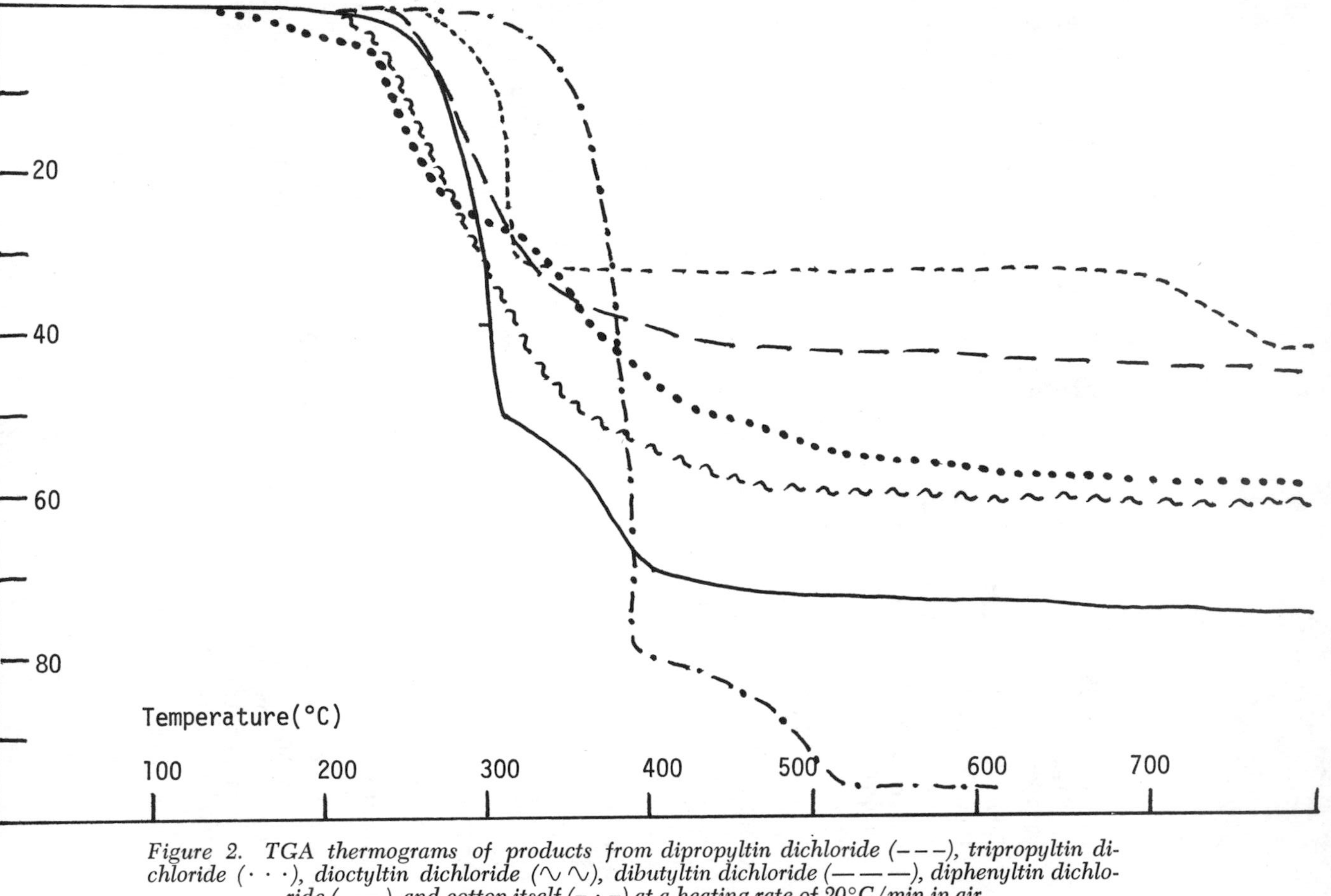

Figure 2. TGA thermograms of products from dipropyltin dichloride (– – –), tripropyltin dichloride (· · ·), dioctyltin dichloride (∼∼), dibutyltin dichloride (——–), diphenyltin dichloride (——), and cotton itself (– · –) at a heating rate of 20°C/min in air

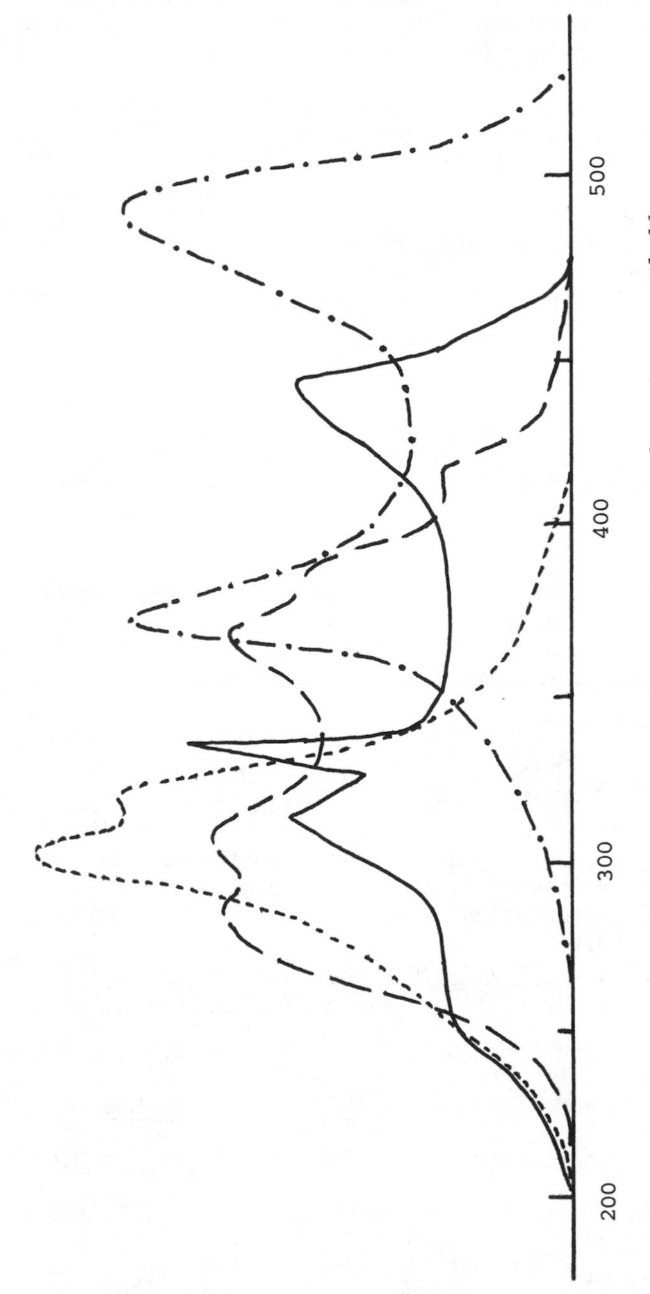

Figure 3. DSC thermograms of cotton (- · -) and condensation products of cotton with dibutyltin dichloride (——), diphenyltin dichloride (— —), and dipropyltin dichloride (- - -) in air at 20°C/min for 1.00 ± 0.05 mg samples. The X-axis is the ΔT = 0 line and the area above it is the exothermic heat change.

modified through condensation with organometallic halides. resulting from a replacement of the "hydrogen bonding proton" by a dipolar aprotic moiety (such as 13). (See Figures 2 and 3.)

The products are solid, generally exhibiting flexibility. They appear to degrade without softening with initial degradation near that of cotton itself (for instance Table I). The tin moiety typically remains as part of the residue to greater than 900°C in air and nitrogen. This is important for applications such as use in commercial and residential insulation where "safe" burning is necessary to meet building codes since many volatile organotin moieties appear to have some toxicity and are known to cause headaches.

Biological Activity

Insoluble cellulose and modified cellulose products were added to paper disks contained on plates seeded with approximately 1000 spores of tested organisms. Amount added was several granules (about 0.1 mg and less) per spot. The plates were then incubated at 25°C for 24 hours and the inhibition of confluent growth recorded. All but the products form dioctyltin dichloride showed good fungi inhibition. The fungi tested are typical and widespread and represent a good cross sectional test for the applicability of such modified cellulosic products as retarders of fungi related to rot and mildew.

Literature Cited

1. Carraher, C., J. Chemical Education (1969), 46, 314.
2. Whistler, R., "Methods in Carbohydrate Chemistry", Vol. III, (1963), 78-79.
3. McCaffery, E., "Laboratory Preparation for Macromolecular Chemistry", McGraw-Hill, New York, 1970, 140-144.
4. Price, W.J., "Analytical Atomic Absorption Spectroscopy", 1974, 151-153.
5. Carraher, C., J. Polymer Sci., A-1 (1969), 7, 2351 and 2357.
6. Carraher, C., and Klimiuk, G., J. Polymer Sci., A-1 (1970), 8, 973.
7. Carraher, C. and Scherubel, G., J. Polymer Sci., A-1 (1971), 9, 983.
8. Carraher, C. and Klimiuk, G., Makromolekulare Chemie (1970), 133, 211.
9. Carraher C. and Scherubel, G., Makromolekulare Chemie (1972), 152, 61.
10. Butcher, F., Gerrard, W., Mooney, E. and Rees, R. and Witlis, H., Spectrochim Acta, (1964), 20, 51.
11. Hester, R., J. Organometallic Chem., (1970), 23(1), 123.
12. Carraher, C., Angew. Makromolekulare Chemie, (1973), 31, 115.
13. Carraher, C., "Interfacial Synthesis", Vol. II, Edited by Millich, F. and Carraher, C., Chpt. 19, Dekker, New York, 1978.

RECEIVED September 28, 1979.

26

Surface Modifications of Cellulose and Polyvinyl Alcohol, and Determination of the Surface Density of the Hydroxyl Group

T. MATSUNAGA and Y. IKADA

Institute for Chemical Research, Kyoto University, Uji, Kyoto 611, Japan

The surface chemistry of polymeric substances is increasingly becoming important not only in the well-known phenomena such as adhesion, wettability, friction, and adsorption, but also in developing new materials such as composites, chromatographic gel beads, and biocompatible artificial organs. In studying the macroscopic interfacial properties of solid polymers, it is highly desired to have a detailed knowledge of the molecular surface properties such as the chemical constitution.

In spite of recent remarkable progress in surface analysis, there still remain many unresolved problems relating the surface characteristics. For instance, although X-ray photoelectron spectroscopies such as ESCA, Auger spectroscopy and Fourier transform attenuated total reflectance(ATR) IR spectroscopy are able to provide information regarding the chemical structure near the surface region, these are not particularly powerful in assaying functional groups present only at the surface in a quantitative level. This may be because the absolute amount of the groups directing to outside at the surface is too low to be quantitatively determined by the conventional analytical methods. However, in recent years it has been demonstrated that very small amounts of the functional groups newly created at or close to the surface of polyethylene films by oxidation(1) and poly(vinyl alcohol)(PVA) films by reaction with hexamethylene diisocyanate(2) can be determined with the use of fluorescence spectroscopy. Leclercq and his coworkers have also measured the surface density of functional groups generated on corona treated poly(ethylene terephthalate) by adsorption of radioactive calcium ions(3).

The objective of this work is to determine the surface concentration of the hydroxyl groups of cellulose and PVA films utilizing their chemical modification. We chose these polymers mainly because the hydroxyl group is their sole functional group. Recently we have reported that a cellulose film is more excellent in wettability towards water than PVA, though cellulose is insoluble in water, in contrast to PVA(4). Since only the chemical composition of the surface must be responsible for water

0-8412-0540-X/80/47-121-391$05.00/0

wettability provided that both films have a microscopically smooth surface, the density of the surface hydrophilic groups will be different between the two films. In this work we will attempt to react with isocyanates only the hydroxyl groups present at the film surface. This will be followed by hydrolysis to release the corresponding amines, identified by means of the fluorescence assay which has proved to be highly sensitive in determining extremely small amounts of amines(5, 6). In addition, we will describe surface reaction of the films with adipoyl chloride and subsequent reaction with 7-hydroxycoumarin.

Experimental

Materials. The cellulose films employed are a cellophane of 0.14 mm thickness, donated by Tokyo Cellophane Co., Inc., Tokyo, Japan(hereafter designated as cellophane) and a cellulose tubing for dialysis of 0.19 mm thickness produced by Visking Co., Inc.(hereafter designated as Visking). The PVA film with a thickness of 0.32 mm was supplied by Unitika Co., Inc., Osaka, Japan. The films were cut to 2 x 5 cm size and purified by the conventional extraction with water at room temperature, followed by extraction with a benzene-ethanol(1:2) mixture at 80°C. The cellulose films were further extracted with dimethyl sulfoxide(DMSO) at 45°C for 24 hrs, rinsed with water and then subjected to Soxhlet extraction with methanol for 24 hrs. Instead of boiling water and DMSO, 1 N NaOH was used for extraction of the PVA film at 45°C. After rinsing the purified films with plenty of water, they were dried under vacuum at 70°C for 48 hrs and stored in a desiccator containing $CaCl_2$. Just prior to use, the films were again dried under vacuum at 70°C for 12 hrs.

Reactions. n-Butyl-, phenyl-, α-naphthylisocyanate, adipoyl chloride, and tin octoate were extra pure grade and used without further purification. 7-Hydroxycoumarin was recrystallized from water. Toluene and dioxane were distilled after drying with anhydrous magnesium sulfate and a molecular sieve. Urethanations of films were carried out in toluene at 45°C without catalyst and/ or at 30°C with 0.67 $g.l^{-1}$ of tin octoate, the initial isocyanate concentration being 70 $g.l^{-1}$. After being allowed to proceed for a given period, the reaction was stopped by removing the urethanated film and then immersing in a methanol-toluene(1:3) mixture followed by Soxhlet extraction with methanol. Then the film was rinsed with water and dried under vacuum at 70°C for 24 hrs. Adipoylation of films was accomplished in toluene at 45°C and an adipoyl chloride concentration of 110 $g.l^{-1}$ without catalyst. The adipoylated film was washed several times with anhydrous toluene and dioxane, followed by reaction with 7-hydroxycoumarin in dioxane containing sodium hydride for 2 hrs at room temperature. After coupling the coumarin, the film was extracted with methanol in a Soxhlet apparatus, rinsed with water, and dried under vacuum at 70°C.

Hydrolysis and Fluorescence Assay. After measuring the exact surface area, the urethanated film was placed in 2 ml of 1 N NaOH in a glass tube, purged with nitrogen, and then sealed under a reduced pressure. After hydrolysis of urethanes at 50°C for a given period, we added 2 ml of 1 N HCl to neutralize the reaction product. Three ml of the neutralized solution was transferred to another glass tube. To this was added 7 ml of a borate pH 10 buffered solution containing 7 mg of o-phthalaldehyde and 7 μl of mercaptoethanol, except for the naphthylurethanated film. The fluorescence intensity for n-butylamine was read 5 mins later after addition of the o-phthalaldehyde solution at an excitation wavelength of 338 nm and an emission wavelength of 427 nm, while the fluorescence from aniline was read 60 mins later after addition of o-phthalaldehyde at an excitation wavelength of 350 nm and an emission wavelength of 430 nm. In the case of α-naphthylurethanated film, the fluorescence measurement was directly run on the neutralized solution without adding o-phthalaldehyde, since α-naphthylamine has strong fluorescence at pH 7. The excitation wavelength employed is 310 nm and the emission wavelength 440 nm.

The film reacted with adipoyl chloride followed by coupling with 7-hydroxycoumarin was subjected to methanolysis at 1 N HCl and 60°C. The regenerated coumarin was assayed at pH 10 by fluorescence spectroscopy at an excitation wavelength of 329 nm and an emission wavelength of 455 nm. A Hitachi MPF-4 Fluorescence Spectrophotometer was used for all fluorescence measurements.

Contact Angle Measurement. The contact angle towards water on films was measured at room temperature with the inverted bubble method to avoid drying of the film during the measurement. More than 5 readings on different strips from the same film were averaged. The deviation of each reading from the average was within ± 1°. The contact angle of the adipoylated film was measured prior to coupling of 7-hydroxycoumarin, but after hydrolyzing the acid chloride of half-ester to the carboxyl group by immersing in a water-acetone(1:3) mixture.

Results

The reaction scheme employed in the surface modification of films is depicted in Figure 1. Urethanation as well as adipoylation was in all cases performed in toluene to prevent the reactions from invading the interior of films. Since determination of the hydroxyl groups at the film surface requires complete removal of isocyanates and adipoyl chloride remaining unreacted at the surface and in the bulk polymer, the reacted films were subjected to rigorous extraction prior to further reactions. An example of the result of extraction carried out for the film which was naphthylurethanated at 30°C for 82 hrs is given in

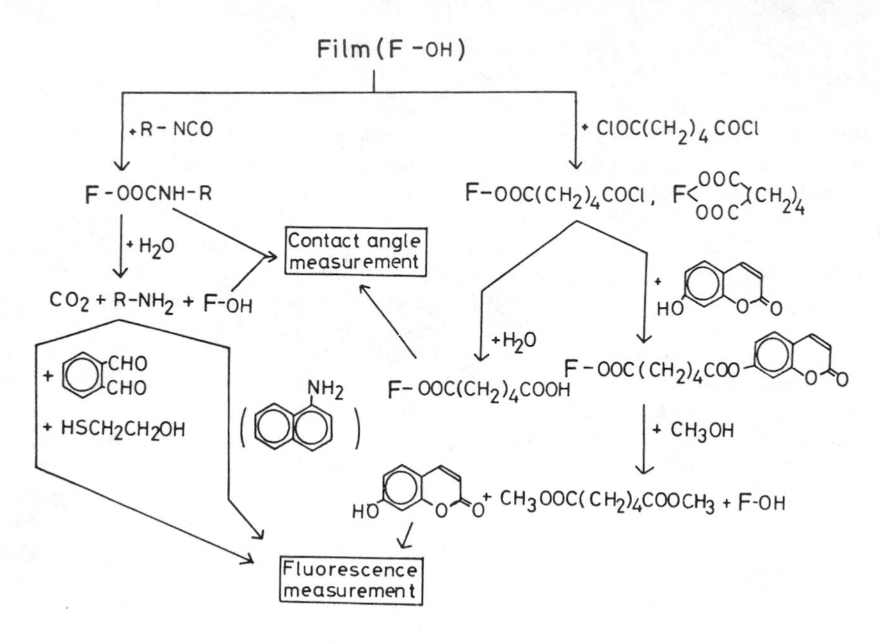

Figure 1. Outline of surface reactions and fluorescence assay

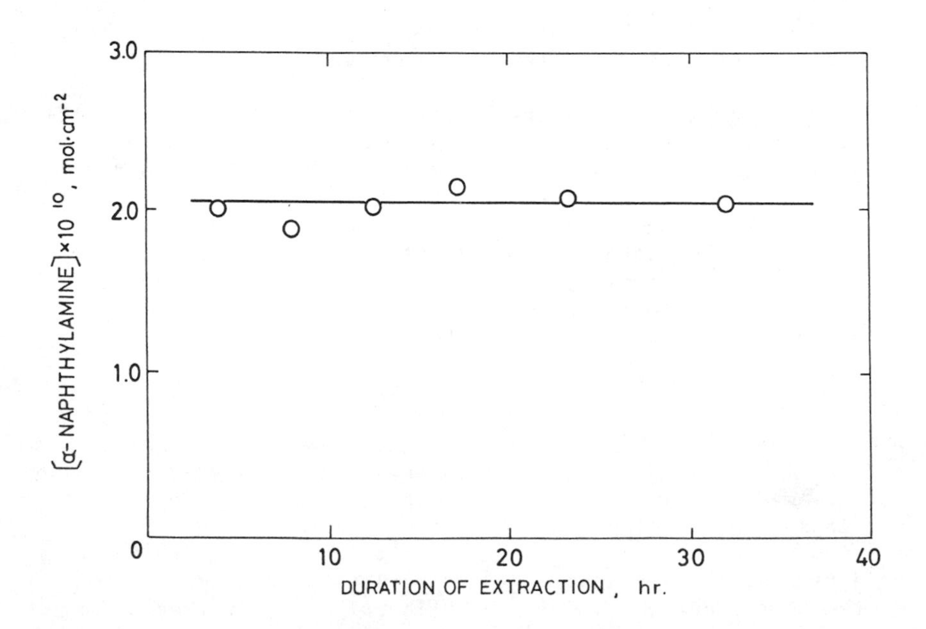

Figure 2. Effect of Soxhlet extraction with methanol on determination of α-naphthylamine liberated from cellophane reacted with α-naphthylisocyanate

Fig 2, where the observed amount of the naphthylamine generated from the reacted film is plotted against duration of extraction. If isocyanate remains unextracted, the extent of reaction would apparently decrease with the extraction time. However, the amount of naphthylamine is obviously constant in the range of extraction duration studied, supporting complete removal of the unreacted isocyanate from the film.

After extraction, the urethanated films were subjected to alkaline hydrolysis of urethanes to liberate the corresponding amines, while the adipoylated films were hydrolyzed after having reacted with 7-hydroxycoumarin. Amounts of the released amines and coumarin were determined by fluorescence spectroscopy as described in the Experimental section. Since aniline as well as butylamine has no appreciable fluorescence by themselves, their fluorescence assay was made after reacting with o-phthalaldehyde in the presence of mercaptoethanol. In Figure 3, where relative fluorescence intensities are plotted as a function of concentrations of amines and hydroxycoumarin, one can see that the fluorescence intensities vary linearly with their concentration to permit us the quantitative determination of extremely small amounts of amines and hydroxycoumarin.

In the following we will present some typical results of urethanation and adipoylation and, in addition, of contact angles of surface-modified films against water.

Phenylurethanation. As expected, the films of cellulose as well as PVA became less wettable towards water as the surface modification with phenylisocyanate proceeded. Change of advancing contact angles with duration of phenylurethanation is reproduced in Figure 4. Clearly, both of contact angles increase with the reaction time, except for the peculiar variation seen in the initial course of phenylurethanation of PVA. This is probably related to unusual solubilization behavior of PVA, which is commonly observed when some of the hydroxyl groups have been modified. As is shown in Figure 5, the increased contact angles fall again, upon hydrolysis, to those of films without phenylurethanation, indicating that the phenylurethanes can be hydrolyzed to completion within about 10 hrs under this reaction condition.

It is required to determine directly the yield of products in order to examine whether or not the surface-urethanated films really undergo complete hydrolysis. Figure 6 gives a plot of the concentration of aniline liberated on hydrolysis. Although some scatter is seen, variation of contact angle with time in Figure 5 seems to correspond to that shown in Figure 6. Based on the results in Figures 5 and 6, the phenylurethanated films were placed in 1 N NaOH for 24 hrs to hydrolyze all the urethanes. Figure 7 shows the result obtained for the cellulose films. It appears that the reaction comes to completion with a saturated surface concentration of about 1×10^{-9} mol per unit area of the

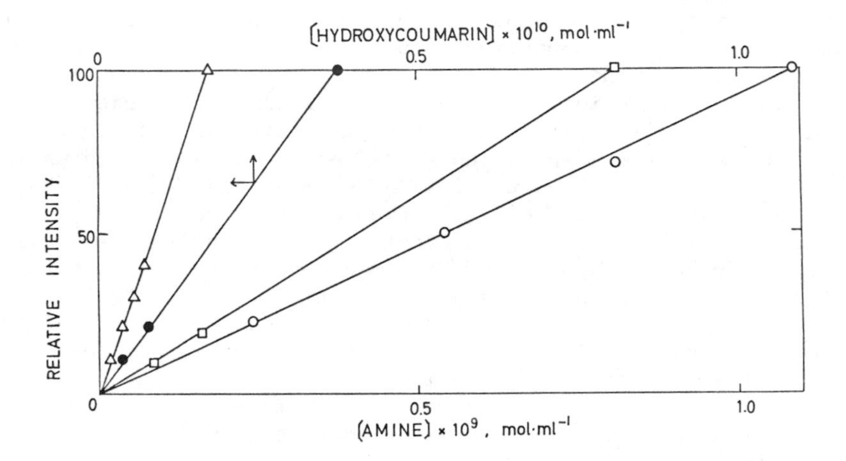

Figure 3. Relative fluorescence intensities of n-*butylamine and aniline after reaction with* o-*phthalaldehyde, α-naphthylamine, and 7-hydroxycoumarin: (○) aniline; (□)* n-*butylamine; (△) α-naphthylamine; (●) 7-hydroxycoumarin*

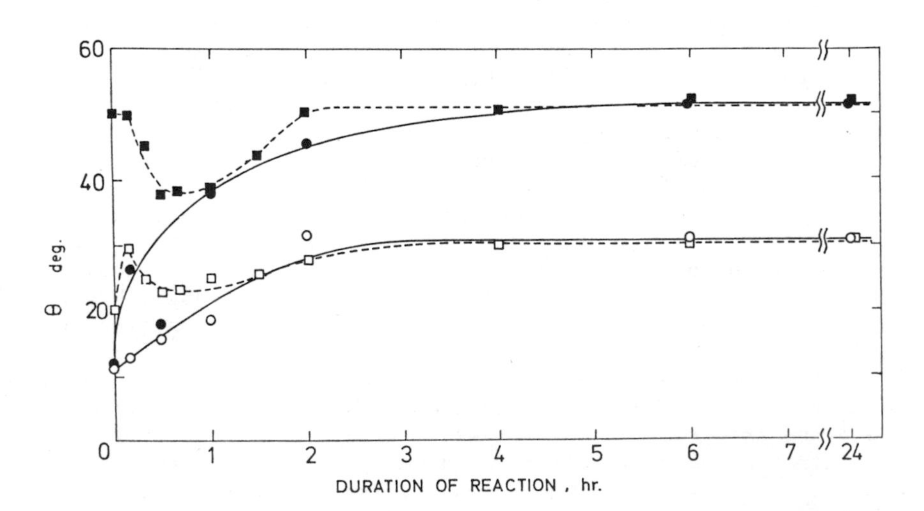

Figure 4. Contact angle θ of cellophane and PVA films as a function of phenylurethanation time: (——) cellophane; (– – –) PVA; (○, □) receding; (●, ■) advancing

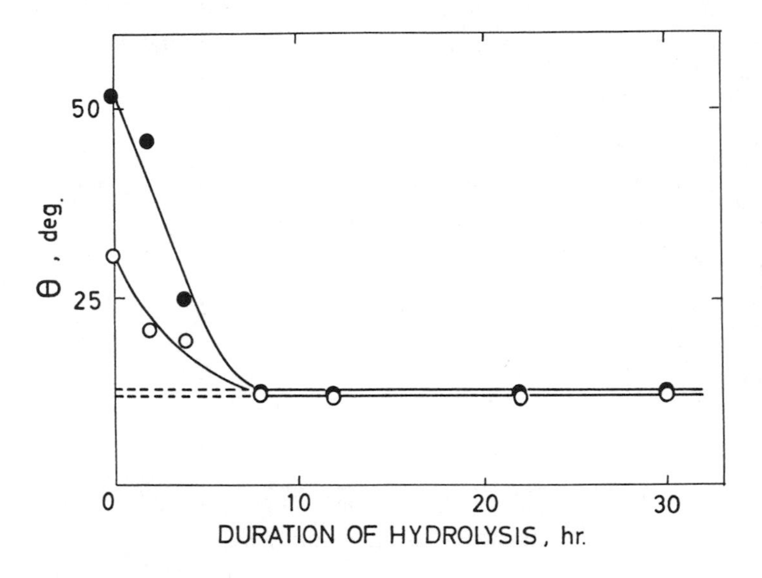

Figure 5. Contact angle θ of phenylurethanated Visking film as a function of hydrolysis time: (○) receding; (●) advancing; (– – –) unreacted film

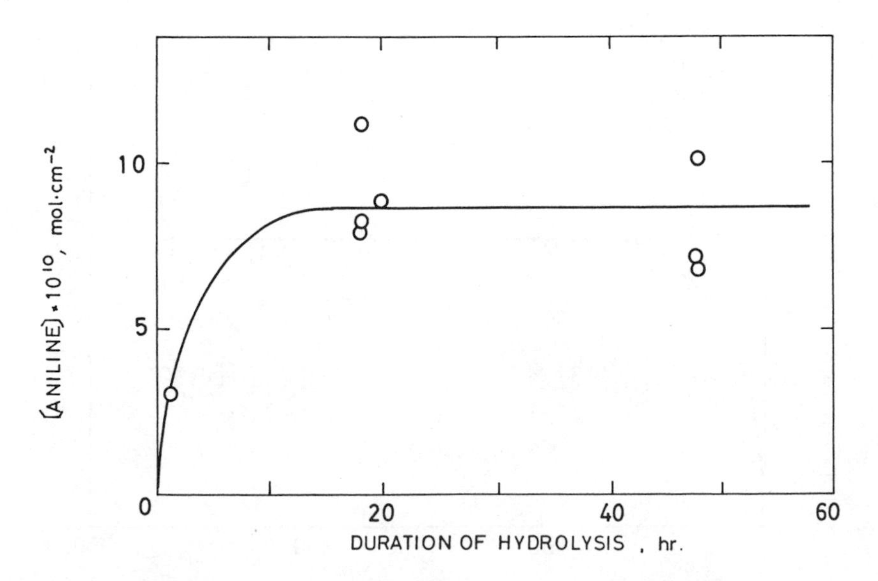

Figure 6. Amount of aniline liberated on hydrolysis of phenylurethanated Visking film

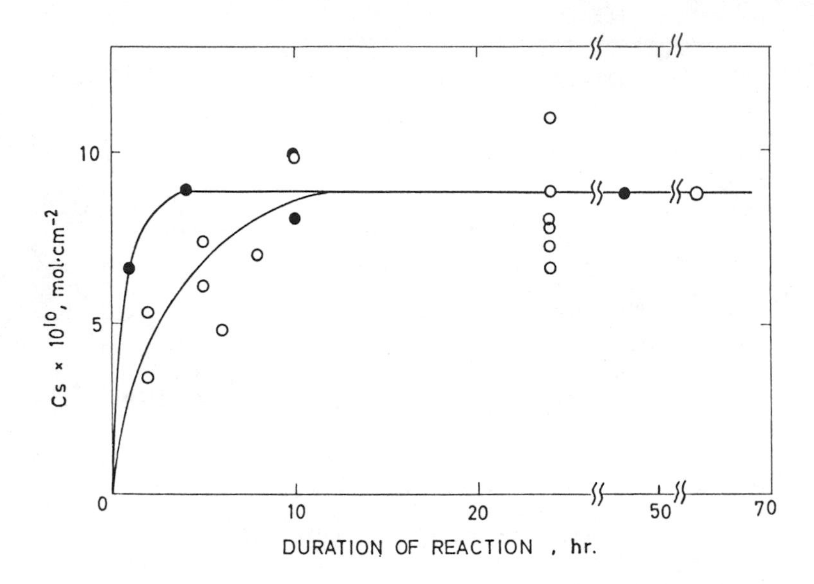

Figure 7. Surface concentration (Cs) of the hydroxyl groups phenylurethanated with and without tin octoate for cellulose films: (○) Visking without catalyst; (●) cellophane with tin octoate

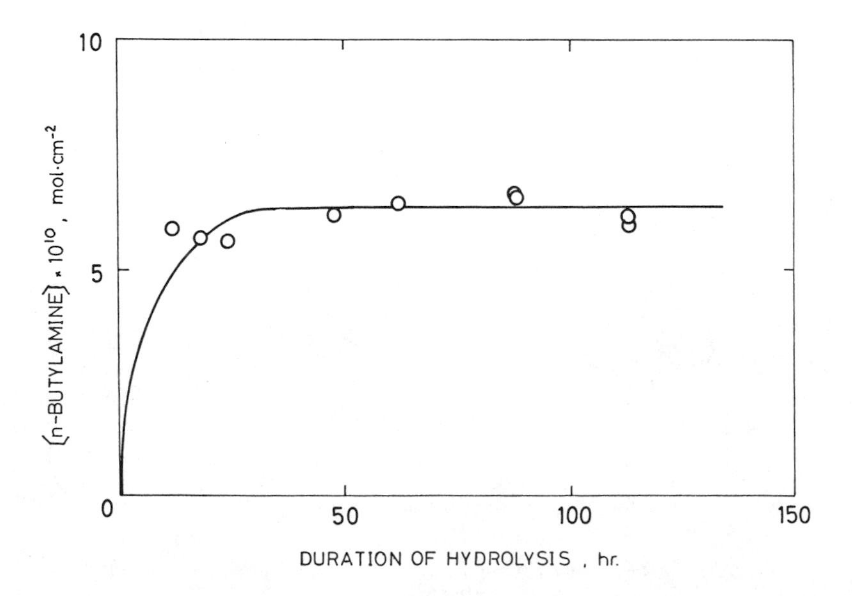

Figure 8. Amount of n-*butylamine liberated on hydrolysis of* n-*butylurethanated cellophane film*

film. Levelling-off of the reaction is an indication that the urethanation is expectedly restricted to the film surface or, at least, has an extremely low reaction rate in the bulk of film. It is also seen that when phenylurethanation is carried out at 30°C with the use of tin octoate as catalyst, the reaction proceeds with a higher rate than that without catalyst in spite of lower temperature, but the products show a similar saturated level of urethanation.

n-Butylurethanation. Figures 8 and 9 represent the results of n-butylurethanation of cellophane. Comparison of Figures 8 and 9 with Figures 6 and 7 reveals that there is no significant difference between phenyl- and butylurethanation, except that the butylurethanation has a lower reaction rate than the phenylurethanation.

α-Naphthylurethanation. As mentioned above, α-naphthylamine is intensely fluorescent and hence does not require the further reaction with o-phthalaldehyde for assay of the amine. Presumably because of omission of this final step of the assay sequence, accuracy in determining the naphthylurethane is much better than that of phenyl- and butylurethane. The amount of naphthylamine generated from a naphthylurethanated cellophane on hydrolysis is plotted as a function of the hydrolysis time in Figure 10. It is evident that the naphthylurethane bound to the film surface can be hydrolyzed in 1 N NaOH in about 30 hrs. Figure 11 represents surface concentrations of the hydroxyl groups reacted with α-naphthylisocyanate for three different films. Comparison of urethanation of the Visking cellulose in the presence of tin octoate with that in the absence of the catalyst clearly indicates an enhancement effect of the catalyst on the naphthylurethanation. It should be also pointed out that both of the cellulose films have the same urethane surface concentration of 2.0×10^{-10} mol·cm^{-2} at saturation, irrespective of the presence of catalyst, whereas the PVA film has half the surface concentration of cellulose.

Adipoylation. Since adipoyl chloride has two groups that are reactive with the hydroxyl group, adipoylation is expected to have some features different from urethanation with monoisocyanates. The results of adipoylation are given in Figure 12. In contrast to urethanations, the ordinate of Figure 12 does not represent the surface concentration of all the hydroxyl groups reacted with adipoyl chloride, but merely the amount of the hydroxyl groups that were half-esterified by adipoyl chloride, since we determined the coumarin that was coupled with the acid chloride group remaining at the other end of halfesterified adipoyl chloride. Therefore, adipoylation does not provide any means to evaluate the total surface concentration of the hydroxyl groups of films. As is seen in Figure 12, the amount of hydroxycoumarin

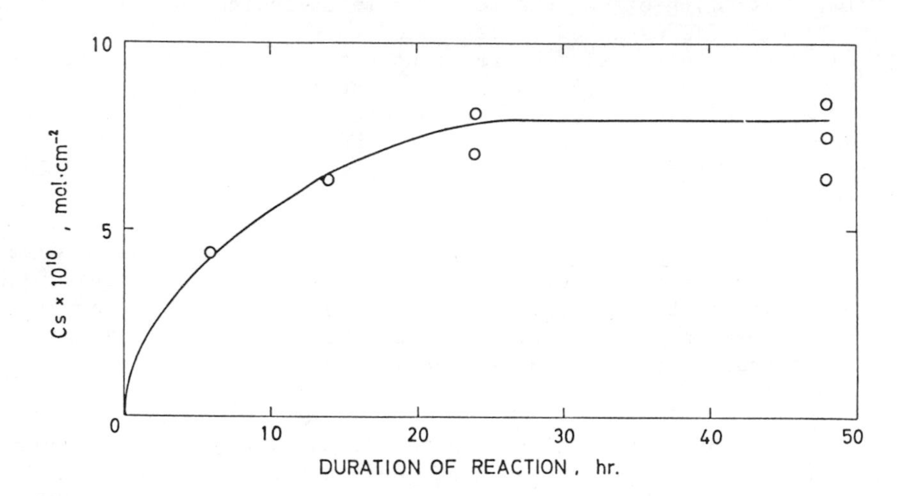

Figure 9. Surface concentration (Cs) of the hydroxyl groups n-*butylurethanated with tin octoate for cellophane film*

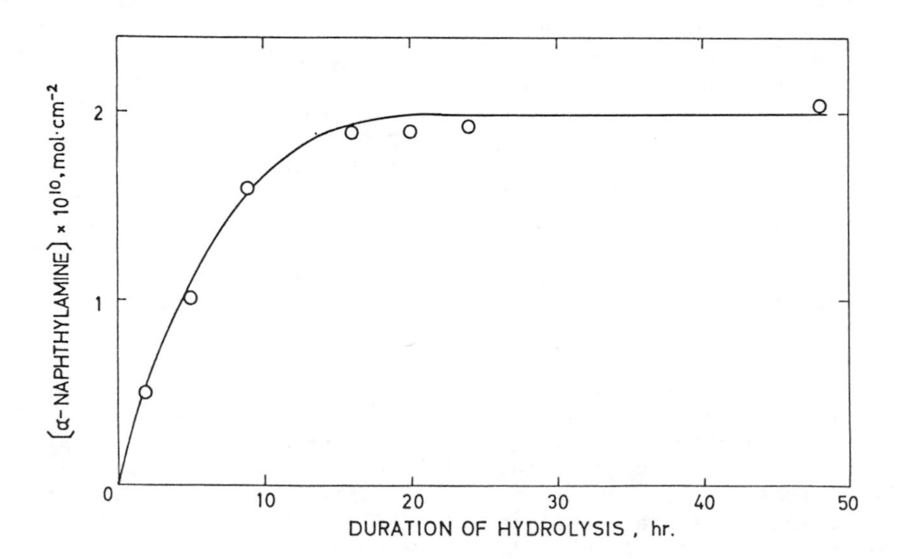

Figure 10. Amount of α-naphthylamine liberated on hydrolysis of α-naphthylurethanated cellophane film

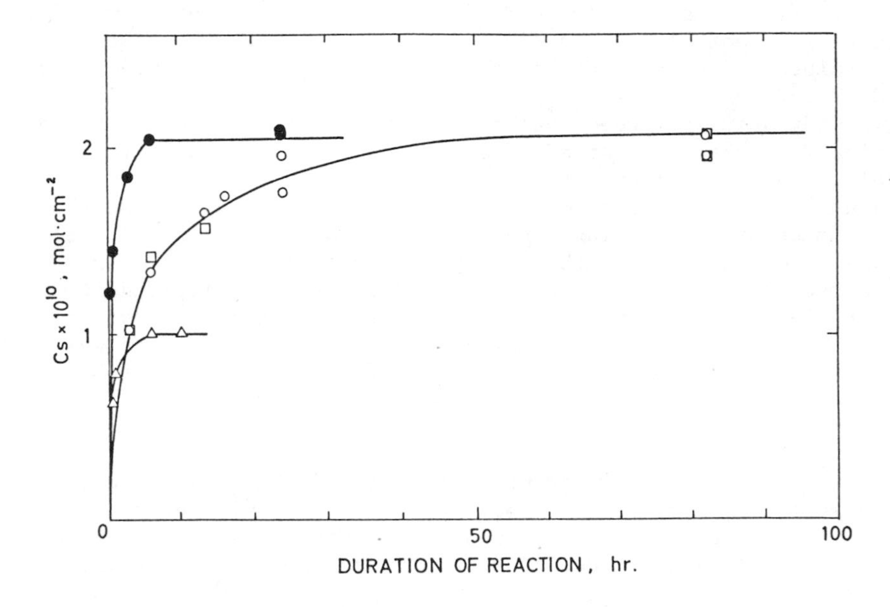

Figure 11. Surface concentration (Cs) of the hydroxyl groups α-naphthylurethanated with or without tin octoate for cellulose and PVA films: (○) Visking without catalyst; (●) Visking with tin octoate; (□) cellophane without catalyst; (△) PVA with tin octoate

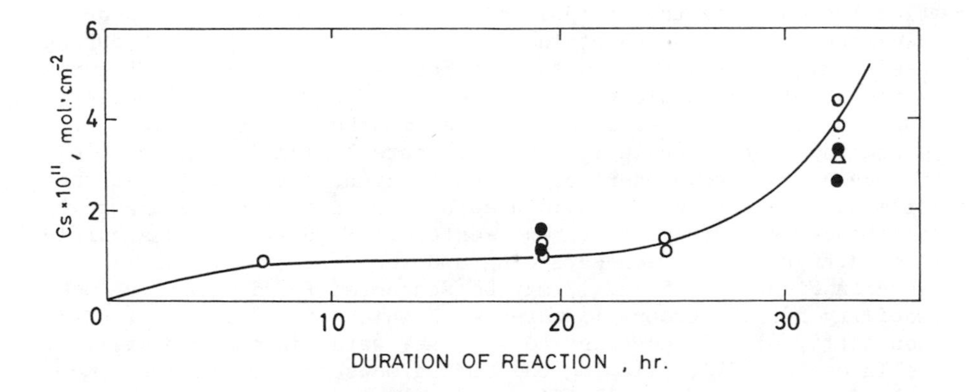

Figure 12. Surface concentration (Cs) of the hydroxyl groups half-esterified by adipoyl chloride for cellulose and PVA films: (○) Visking; (△) cellophane; (●) PVA

coupled is very low, but seems to increase rather rapidly with increasing reaction time. This observation is apparently different from that of urethanation. Another peculiar result was also obtained in the contact angle measurement. As an example, the data for PVA films are reproduced in Figure 13. Obviously, both the advancing and receding contact angles increase with the reaction time when determined after drying the adipoylated film in air. On the contrary, the contact angles of the reacted films are identical to those of the unreacted film, when the measurement was made without having exposed the films to air. A similar trend was also observed for the adipoylated cellulose films. The contact angle decreased if the dried films were immersed in water, but did not fall down to the value before immersing in water. Recently the effect of drying on the surface composition was also discussed by Ratner and his coworkers(7).

Discussion.

Surface Modification of Cellulose and PVA Films. Cellulose, as well as PVA, is known to be a typical non-ionic, hydrophilic polymer possessing hydroxyl groups. As this group has a high reactivity, chemical modification of these polymers is relatively easy and, in fact, has been the subject of extensive research. However, so far as we know, no work has been reported concerned with reactions occurring only at the surface of films or fibers from these polymers.

If the reaction is performed with reagents of large size in the medium which has no or extremely low affinity to these polymers, it is expected that the reaction proceeds only in the surface region. For this purpose we employed toluene as the non-swelling solvent and maintained the temperature below 45°C during reaction. Variation of contact angles with time shown in Figures 4 and 13 clearly indicates that the reactions with isocyanates and acyl chloride take place at least in the surface region, since the contact angle is a good measure reflecting the subtle change of surface properties. Occurrence of the reaction at the film surface can be also evidenced by the contact angle change accompanying hydrolysis as represented in Figure 5. The significant difference between advancing and receding contact angles seen in Figures 4, 5 and 13 may be accounted for in terms of mobility of side groups in water with which the films are in contact(4, 8). In contrast to toluene, water is able to swell cellulose and PVA, allowing the side groups to turn to a thermodynamically more favorable direction.

Although the contact angle provides valuable information about the surface characteristics, it has no relation with the change of chemical structure in the bulk of films. To examine whether or not the reaction is proceeding into the bulk films, the extent of reaction should be determined with good accuracy. In this respect, fluorescence spectroscopy gives us a good tool,

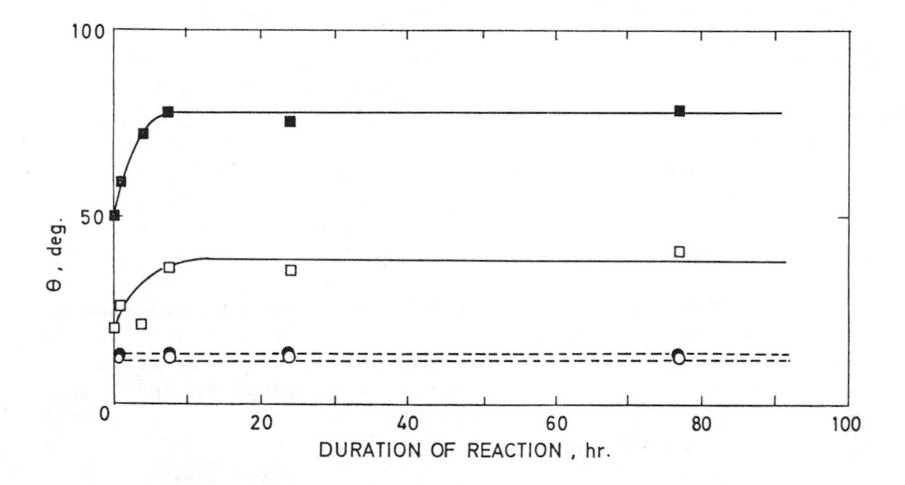

Figure 13. Contact angle θ of PVA film as a function of adipoylation time: (——) measured after drying; (– – –) measured without drying; (○, □) receding; (●, ■) advancing

because its sensitivity is exceedingly higher than usual spectroscopic methods such as UV and IR. It is interesting to point out that no difference was detected in ATR IR spectra between the unreacted and reacted films for all the samples obtained in this study. However, the released amines and hydroxycoumarin could be determined by fluorescence spectroscopy with fairly good accuracy, as Figure 3 demonstrates. In spite of some scatter in amine determination (see Figures 6-9), one can see a distinct trend in urethanation, that is, a saturation of reaction. This may be a strong evidence for restriction of the reactions only to the surface of films. Otherwise, the reaction yield would gradually increase without exhibiting a clear plateau.

As is apparent from Figure 12, adipoylation gives a somewhat peculiar yield-time curve which seems not to obey simple first-order kinetics. This may be related to the fact that two different types of reactions possibly occur in the adipoylation onto the films having many hydroxyl groups; the one is di-esterification and the other half-esterification. It appears that formation of the di-ester predominates over that of the half-ester in this surface reaction, particularly in the early stage of reaction, since only the half-ester is capable of being coupled with hydroxycoumarin (recall that the ordinate of Figure 12 represents the amount of coumarin coupled and hence the surface concentration of the hydroxyl groups half-esterified).

Surface Concentration of Hydroxyl Groups. Since it has been concluded that the urethanation with monoisocyanates in toluene is limited to the hydroxyl groups present at the film surface, we can evaluate the concentration of the hydroxyl groups that are present at the surface, more strictly, accessible to the reaction with isocyanates. The amounts of hydroxyl groups urethanated to saturation, obtained from Figures 7, 9 and 11 are summarized in Table I, which also contains the values obtained from experiments not shown here. These are expressed as surface concentration per cm^2 of films, which are assumed to have a perfectly flat

Table I
Surface Concentration, Cs, of the Hydroxyl Groups Urethanated for Cellulose and PVA Films and Surface Area, As, per Hydroxyl Group

	Cs x 10^{10}, mol·cm^{-2} Cellulose			As, $Å^2$ Cellulose		
	Cellophane	Visking	PVA	Cellophane	Visking	PVA
n-Butylisocyanate	8.0	10.6	8.7	20	16	19
Phenylisocyanate	8.2	10.3	4.6	21	16	37
α-Naphthylisocyanate	2.1	2.1	1.3	80	80	129

surface. It appears that this assumption does not lead to a serious errors, because the observed concentration difference between the cellophane and the Visking film is not significant, though their production methods are different. Even if there was an appreciable difference in the original films, it might have been greatly reduced during the exhaustive purification process.

Inspection of Table I reveals that the surface concentrations of the hydroxyl groups of cellulose reacted with α-naphthylisocyanate are 4 to 5 times lower than those reacted with butyl- and phenylisocyanate. This may be attributable to the high bulkiness of the naphthylisocyanate molecule, bringing about steric hindrance in the reaction. It is then concluded that the actual surface concentration of the hydroxyl groups of cellulose is identical or very close to those of the hydroxyl groups reacted with butyl- or phenylisocyanate. The relatively small size of the n-butylisocyanate molecule will preclude the possibility of an appreciable steric effect of this molecule. In this connection, the surface concentrations found for the PVA film may provide a useful suggestion. In this case, the concentrations of the hydroxyl groups urethanated are about half those of cellulose, except butylurethanation. This is, however, rather reasonable, if we remember that PVA has a higher contact angle than cellulose, suggesting the hydroxyl density at the surface of PVA to be lower than that of cellulose. The high concentration observed for butylurethanation of PVA may imply that butylisocyanate has been reacted also with the hydroxyl groups existing near the film surface.

Finally we will discuss the surface areas occupied by one hydroxyl group. Those calculated from the surface concentrations, are given in Table I. It is seen that the calculated area is 18 $Å^2$ on the average for cellulose and 37 $Å^2$ for PVA. If the hydroxyl groups are assumed to be distributed isotropically throughout in and on the cellulose film, we can estimate the surface area occupied by one hydroxyl group simply from $(dN_L/M_o)^{-2/3} \cdot 10^{-16}$ (in $Å)^2$, where d is the density of cellulose (1.52 $g \cdot cm^{-3}$), M_o is the molar mass per one hydroxyl group (54 $g \cdot mol^{-1}$), and N_L is Avogadro's number. Inserting the values characteristic to cellulose, we obtain 15 $Å^2$ as the surface area per hydroxyl group. Considering that this estimation is based on rough approximations, we may state that the observed area agrees well with this estimated one. Similar to cellulose, PVA is also found to have 15 $Å^2$ as the specific area when estimated from the density. This small area compared with 37 $Å^2$ (the observed one) is not surprising, because the main chain of PVA molecule must be much more flexible than that of cellulose consisting of rigid pyranose rings and consequently some of hydroxyl groups in the surface region of PVA may readily turn from the surface of the film to the bulk, where the free energy of hydroxyl groups should be much lower than in a hydrophobic region as in toluene(9). Again, the good agreement of the observed with the estimated area gives another strong evidence for the conclusion that the reactions with aromatic isocyanates in toluene are confined to the surface of cellulose and PVA films. It is noteworthy that quite similar restriction to the film surface was also observed for the reaction of cellophane with α-naphthylisocyanate in cyclohexane.

Abstract

The surface of cellulose and PVA films has been modified by urethanation with n-butyl-, phenyl-, and α-naphthylisocyanate and by esterification with adipoyl chloride. To restrict the reactions only to the film surface, toluene is employed as the reaction medium. The reactions accompany increase in contact angle against water, but no detectable change in ATR IR spectra as compared with the starting films. Hydrolysis of the urethanated films results in liberation of the corresponding amines. Urethanation can be followed by assaying the liberated amines with fluorescence spectroscopy. In the case of adipoylation, 7-hydroxycoumarin is further coupled to acid chloride present on the half-esterified films, regenerated by hydrolysis and then determined by fluorescence assay. Urethanation with aromatic isocyanates exhibits saturation to give strong evidence that the reaction is really limited to the film surface. The observed surface density of the hydroxyl groups reacted to saturation varies with the nature of reagents as well as films. Based on the results it is concluded that the cellulose and PVA films have the hydroxyl groups of approximately 1×10^{-9} and 5×10^{-10} mol·cm^{-2} at the surface, respectively.

Literature Cited

1. Rasmussen, F. R.; Stedronsky, E. R.; Whitesides, G. M. J. Amer. Chem. Soc., 1977, 99(14), 4736.
2. Caro, J_R., S. V.; Paik Sung, C. S.; Merril, E. W. J. Appl. Polym. Sci., 1976, 20, 3241.
3. Leclercq, B.; Sotton, M.; Baszkin, A.; Ter-Minassian-Saraga, L. Polymer, 1977, 18, 675.
4. Ikada, Y.; Mita, T. 14th symposium on polymers and water, 1976, reprint p.13 (Tokyo).
5. Yusem, M.; Delaney, W. E.; Lindberg, M. A.; Fashing, E. M. Anal. Chim. Acta, 1969, 44, 403.
6. Simons, J_R., S. S.; Johnson, D. F. Anal. Biochem., 1977, 82, 250.
7. Ratner, B. D.; Weathersby, P.K.; Hoffman, A. S. J. Appl. Polym. Sci., 1978, 22, 643.
8. Holly, F. J.; Refojo, M. F. ACS Symposium Series (Hydrogels for Medical and Related Applications, Andrade, J. D., ed.) 1976, 31, 252.
9. Matsunaga, T.; Ikada, Y.; Kitamaru, R. Polymer Preprints, Japan, 1978, 27(3), 475.

RECEIVED July 12, 1979.

27

Simultaneous Interpenetrating Networks Based on Castor Oil Elastomers and Polystyrene: A Review of an International Program

L. H. SPERLING, N. DEVIA, and J. A. MANSON

Materials Research Center, Coxe Laboratory #32, Lehigh University, Bethlehem, PA 18015

A. CONDE

Universidad Industrial de Santander, Ingenieria Quimica, Apartado Aereo 678 Bucaramanga, Colombia, S.A.

Among the renewable resources available in the world, plant products rank very high. Examples include cotton, which yields clothing; wood, for construction; and natural rubber, for automotive tires, etc. Many plants yield valuable oils, such as corn oil, linseed oil, and cotton seed oil (1). Besides food uses, these oils provide the basis for paints, adhesives and other industrial uses. The presence of multiple unsaturated sites allows for ready polymerization (2). Castor oil, which comes from the castor bean plant, is nearly unique among vegetable oils in containing hydroxyl groups in addition to points of unsaturation. Thus, there are two ways of polymerizing castor oil: through the use of sulfur or oxygen, which attacks the double bonds, or through the hydroxyl groups, to form polyurethanes, or polyesters, etc. (3,4,5).

This paper reviews a four-year international program between Colombia and the United States. Its objectives were two fold: (1) From a scientific point of view, this program provided basic information about the interrelationships among synthesis, morphology, and physical and mechanical behavior of interpenetrating polymer networks made from a step growth reaction and a chain growth mechanism. (2) Since there is thought in Colombia about developing a castor oil industry, the engineering information developed within the program will provide a basis for the development of a host of toughened plastics and reinforced elastomers. As a further point of interest, castor oil provides, in a modest way, an alternative source of useful chemicals in a petroleum starved world.

0-8412-0540-X/80/47-121-407$05.00/0

Interpenetrating Polymer Networks

Materials known as interpenetrating polymer networks, IPN's, contain two or more polymers, each in network form (6-9). A practical restriction requires that at least one of the polymer networks has been formed (i.e. polymerized or crosslinked) in the immediate presence of the other. Two major types of synthesis have been explored, both yielding distinguishable materials with different morphologies and physical properties.

The first type, termed sequential IPN's, involves the preparation of a crosslinked polymer I, a subsequent swelling of monomer II components and polymerization of the monomer II *in situ*. The second type of synthesis yields materials known as simultaneous interpenetrating networks (SIN's), involves the mixing of all components in an early stage, followed by the formation of both networks via independent reactions proceeding in the same container (10,11). One network can be formed by a chain growth mechanism and the other by a step growth mechanism.

The main path of the research employed both methods of synthesis in turn. At first, the graduate students Yenwo and Pulido explored the use of sequential IPN's based on castor oil urethanes and polystyrene (12-16). At the same time, the graduate student Devia, working in Colombia, explored an alternate synthetic route using latex technology (17). Since nothing was known about the behavior of such materials, their collective objective was to provide a map upon which further efforts could be intelligently based. This effort has now been reviewed (18).

Subsequently, Devia came to the U.S. to continue the research. A SIN type synthesis was selected because it offered two advantages: ease in processing and a lower glass transition temperature, T_g, of the rubber product. As shall be seen from the following, the IPN and SIN synthesis yield quite different materials, even if the chemistry is nominally the same. However, since the earlier work has already been described, the following will emphasize the SIN synthesis (19-22).

Castor Oil

Among all the vegetable oils, castor oil has very special characteristics that have made it one of the most important commercial oils. Extracted from the beans of the plant *Ricinus communis* that grows throughout much of the tropical world, it is one of the few naturally occurring triglycerides that approaches being a pure compound and is the only major oil composed essentially of the triglycerides of a hydroxy acid, ricinoleic acid.

As shown in structure (1), the number of double bonds and hydroxyl groups are identical at three each per oil molecule (90% pure).

$$
\begin{array}{l}
H_2C\text{-}O\text{-}\overset{\overset{O}{\|}}{C}\text{-}(CH_2)_7\text{-}CH{=}CH\text{-}CH_2\text{-}\overset{\overset{OH}{|}}{C}H\text{-}(CH_2)_5\text{-}CH_3 \\
\ \ | \\
HC\text{-}O\text{-}\overset{\overset{O}{\|}}{C}\text{-}(CH_2)_7\text{-}CH{=}CH\text{-}CH_2\text{-}\overset{\overset{OH}{|}}{C}H\text{-}(CH_2)_5\text{-}CH_3 \\
\ \ | \\
H_2C\text{-}O\text{-}\overset{\overset{O}{\|}}{C}\text{-}(CH_2)_7\text{-}CH{=}CH\text{-}CH_2\text{-}\overset{\overset{OH}{|}}{C}H\text{-}(CH_2)_5\text{-}CH_3
\end{array}
\qquad (1)
$$

Synthesis

In order to provide a better basis for comparison, the IPN synthesis will be described first.

Castor oil-urethane elastomers were prepared by reacting 2,4 tolylene diisocyanante, TDI, 80/20:2,4/2,6 TDI, or hexamethylene diisocyanate (HDI) with castor oil. The last reaction was rather slow and thus dibutyltin dilaurate, 0.001 gm per gm of HDI, was used as a catalyst. Since TDI hydrolyses significantly in the presence of trace amounts of water, DB grade castor oil from the Baker Castor Oil Company (NL Industries) was employed.

The reaction between TDI and castor oil is exothermic and bubbles are produced in the reaction mixture (castor oil contains a few tenths of a percent volatile material that will evaporate as the temperature of the reaction mixture goes up. Some of the bubbles produced are trapped in the mixture as the viscosity increases. Stirring with a teflon coated magnetic spin bar also produces some bubbles). In order to produce elastomers that are bubble free, the reaction is carried out in two stages.

Stage 1. A known weight of DB oil is mixed with excess TDI (the excess here refers to the ratio of NCO groups to OH groups being larger than 1.0) at room temperature to produce an isocyanate terminated prepolymer. The mixture is stirred vigorously for at least one hour. The bubbles present are removed by applying a vacuum to the prepolymer for about 15 minutes. This results in a clear bubble-free highly viscous liquid.

Stage II. In this stage the prepolymer is crosslinked with excess castor oil. The degassed prepolymer is mixed with enough DB castor oil to give a final predetermined NCO/OH ratio. The mixture is stirred vigorously for 20 minutes. It is then degassed again as in Stage I and poured into a mold and heated for two hours at 130°C to complete the reaction. The mold is allowed to cool and the product separated. The resulting elastomer is clear and tough, the modulus depending on the NCO/OH ratio.

In order to synthesize the IPN, the urethane elastomer was swelled with styrene containing 0.4% benzoin as initiator and 1% divinyl benzene (DVB) as crosslinker. Polymerization of the styrene was carried out by ultraviolet radiation at room temperature for 24 hours.

By comparison, the synthesis of SIN's involved a castor oil derived crosslinked elastomer and crosslinked polystyrene as constitutive networks synthesized simultaneously. Three different systems emerged from the use of the three different crosslinking agents for the castor oil: a) sebacic acid or derivatives to form a castor oil polyester network (COPEN), b) 2,4 tolylene diisocyanate (TDI) to form a castor oil polyurethane network (COPUN), and c) sebacic acid plus 2,4 TDI to form castor oil poly(ester-urethane) network (COPEUN). The synthesis procedure starts with the preparation of the corresponding prepolymers: a) castor oil polyester prepolymer, COPEP1, the resultant product of the reaction of one castor oil equivalent with one sebacic acid equivalent until the acid value fell to 33; b) COPEP4, an extended chain polyol obtained by completely reacting 0.6 acid equivalents with one castor oil equivalent; c) castor oil polyurethane prepolymer, COPUP1, an isocyanate terminated prepolymer formed from the reaction of 2.2 equivalents of 2,4 TDI with one equivalent of castor oil.

The styrene mixture was prepared by dissolving proportionally 0.4 gms of benzoyl peroxide and 1 ml. of commercial divinylbenzene solution (55%) in 99 mls of freshly distilled styrene monomer. This was polymerized to form a polystyrene network, PSN. SIN's containing 10 and 40% elastomer, as well as two corresponding homopolymer networks were studied, see Figure 1.

The general synthesis procedure involved the solution of the selected castor oil prepolymer in the styrene monomer mixture at 25°C, followed by the addition of the appropriate curing agent for the elastomer phase. Stirring for 5 to 10 mins under nitrogen atmosphere at 25°C yielded clear solutions. The polymerization of samples containing 40% elastomer was carried out without further stirring by pouring the clear solutions between teflon-lined glass plates and placing them into a constant temperature oven at 80°C for 24 hours, followed by reacting the elastomer at 180-200°C.

SIN's containing 10% elastomer were charged into a 500 ml. resin kettle provided with a nitrogen inlet, thermometer reflux condenser, and a high torque stirrer. Polymerization of the styrene took place with vigorous stirring at 80°C in a constant temperature oven for 24 hours. Subsequent heating to 180-200°C allowed the elastomer portion to fully react.

Morphology changes during the synthesis of SIN's. During the chemical process by which a solution of comonomers is transformed into a SIN, several morphological changes occur. The path of morphological changes depends on overall composition, solubility relationships, reaction rates, and the rate of stirring, if any. As an example, the morpholgical paths of the COPEN/PSN materials will be described below employing the synthetic scheme shown in Fig. 1.

Castor oil and sebacic acid are reacted at 180-200°C until the mixture approaches gelation, so a branched prepolymer having an equal number of both functional groups (COPEP1) is obtained (Fig. 1, upper left). Due to the high temperatures required for polyester formation, the reaction is readily stopped by cooling the prepolymer to 80°C. The styrene comonomer mixture is prepared at room temperature and charged to the reactor containing the polyester prepolymer, where mixing takes place (Fig. 1, lower left). This yields a mutual solution of all components required for the formation of both networks. The temperature is then raised to 80°C in order to initiate the styrene polymerization. (The polyester reaction rate is nil at this temperature.) In polymerizing the styrene component within the polyester prepolymer mixture, the first amounts of polystyrene produced early in the reaction remains dissolved until some critical concentration is reached, followed by phase separation, see Figs. 2 and 3.

The solution is transformed to an oil-in-oil emulsion in which a polystyrene solution forms the disperse phase and the elastomer polyester component solution the continuous phase. The point of phase separation is observed experimentally by the onset of turbidity, due to the Tyndall effect. The conversion required for phase separation to occur depends basically on the solubility of the polystyrene chains in the elastomer solution, which in turn is governed by the elastomer concentration and compatibility of the two polymers.

As polymerization proceeds, the total volume of polystyrene polymer particles increases rapidly at the expense of the styrene monomer from the solution. What happens next depends on several factors, mainly composition and stirring. It was found that for SIN formulations having an elastomer content greater than about 15%, no further changes occur and elastomer material will remain the continuous phase, regardless of the extent of agitation.

However, for SIN's having up to 10-15% elastomer content, it was found that stirring induces significant changes in the morphology of the mixture. If stirring is not provided, the polystyrene polymer particles will sink and coalesce giving rise to a two-layered system.

The upper layer possesses more elastomer which forms the continuous phase and the bottom layer has the PS as the continuous phase.

Samples prepared with stirring and poured into test tubes at different times (stopping the stirring) showed the sequence illustrated schematically in Fig. 2. The two layers were distinguishable because of dullness and hardness differences. At a reaction temperature of 80°C, the volume of the upper layer (elastomer continuous) decreases slowly and finally disappears at about 90 min. Samples of both top and bottom layers were studied by transmission electron microscopy techniques, and micrographs for a 10/90 COPE/PSN are shown in Fig. 3. Up to 90 min, samples exhibit elastomer continuous top and plastic continuous bottoms. At 90 min, coinciding with the disappearance of the upper layer (see Fig. 2), a phase inversion takes place. Micrographs T2A, T2B, and T2C in Fig. 3 were all taken from the top layer and illustrate the process of phase inversion. At T2A the castor oil

SYNTHESIS SCHEME FOR CASTOR OIL POLYESTER BASED SIN'S

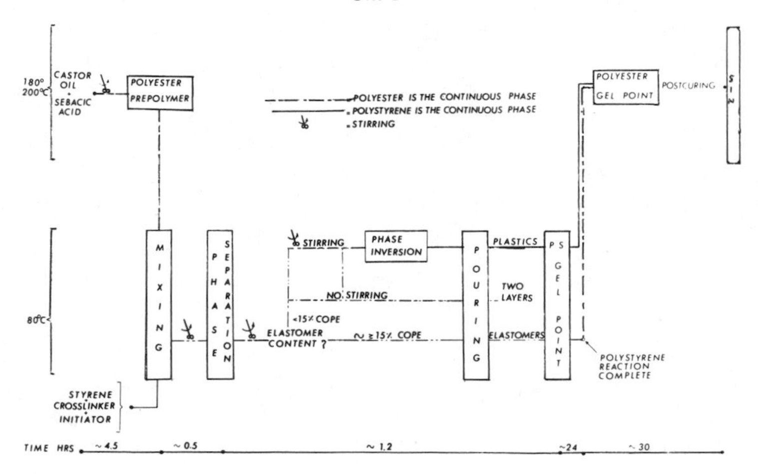

Macromolecules

Figure 1. Process scheme for the synthesis of castor oil polyester SIN's (20)

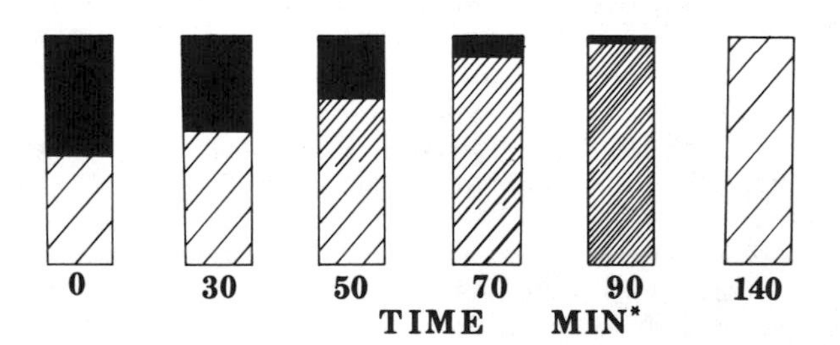

Macromolecules

Figure 2. Layering effect during the synthesis of a 10/90 COPEN/PSN SIN (20) (*) refers to reaction time with stirring prior to pouring. Shadowing proportional to softness

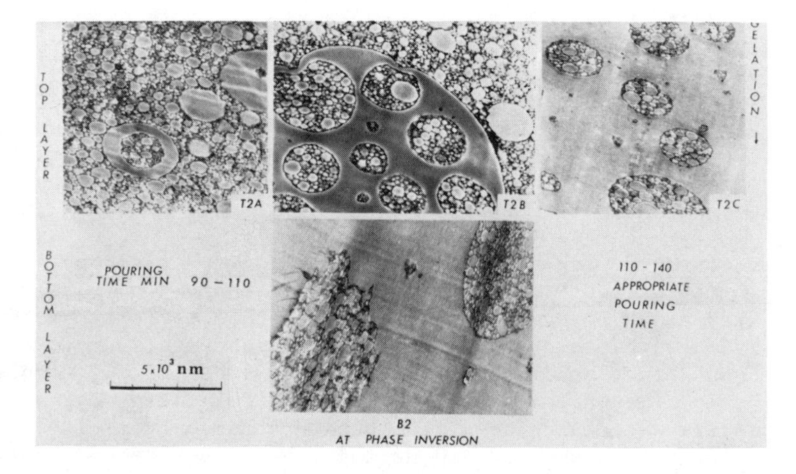

Macromolecules

Figure 3. Morphology changes induced by stirring during the synthesis of a 10/90 COPE/PSN SIN. Sample poured into the mold at the phase inversion point (20).

elastomer (stained dark by the OsO_4) forms the continuous phase. Micrograph T2B shows the polystyrene domain coalescence process by which elastomer domains are generated and a polystyrene continuous phase begins to form, thus illustrating the actual phase inversion point. Micrographs T2C and B2 show that at the end of the phase inversion, the top and bottom regions have identical morphologies and the sample attains macroscopic homogeneity.

In the time interval between phase inversion and gelation of the polystyrene continuous phase, the final morphological features such as size average and size distribution of elastomer domains become fixed. Since these morphological changes affect properties such as modulus and impact resistance, the characteristics of the system at and just after phase inversion and before gelation demand the closest scrutiny. The open time interval was found to decrease as the polyester prepolymer content increases, probably because higher polystyrene conversions are required for the system to reach suitable phase inversion conditions.

By way of comparison, the morphology of the sequential synthesis shows a much finer structure, the average domains being of the order of 300-600Å. Also, because of swelling restrictions, it was impossible to achieve materials having more than about 80% polystyrene (15).

Molecular mixing via dynamic mechanical spectroscopy. While electron microscopy yields the phase size, shape, etc., as delineated above, dynamic mechanical spectroscopy (DMS) yields the composition within each phase. The DMS measurements employed a Rheovibron direct reading viscoelastometer model DDV-II (manufactured by Toyo Measuring Instruments Co., Ltd., Tokyo, Japan). The measurements were taken over a temperature range from -120°C to 140°C using a frequency of 110 Hz and a heating rate of about 1°C/min. Sample dimensions were about 0.03 x 0.15 x 2 cms.

Each of the SIN's examined showed two glass transitions, one for each phase. In general, the transitions were shifted inward, suggesting small but significant extents of molecular mixing.

The systematic changes in the glass temperatures illustrated in Table 1 indicate quantitatively the changes in the composition within each phase. The random copolymer equation can be used to estimate the composition within each phase:

$$\frac{1}{T_g} = \frac{W_1}{T_{g_1}} + \frac{W_2}{T_{g_2}} \qquad (2)$$

where

$$W_1 + W_2 = 1 \qquad (3)$$

The quantity T_g represents the glass transition temperature of the phase in question, and W_1 and W_2 represent the weight fractions of elastomeric and plastic polymers, respectively. The quantities T_{g_1} and T_{g_2} stand for the homopolymer T_g's in eqn. (2).

Table 1. Glass Transition Temperature of SIN's Based in Castor Oil Elastomers and Crosslinked Polystyrene (21)

Composition	Lower T_g (°C)	Upper T_g (°C)
PSN	-----	117
10/90 COPEN/PSN	-60	117
40/60 COPEN/PSN	-58	114
COPEN	-66	-----
10/90 COPEUN/PSN	-30	114
40/60 COPEUN/PSN	-30	not observed
COPEUN	-50	-----
10/90 COPUN/PSN	6	105
40/60 COPUN/PSN	6	115
COPUN	-4	-----

Polymer Engineering and Science

The volume fraction of each phase was taken from the fractional area in the transmission electron micrographs. Combined with the values shown in Table 1, the compositions within each phase were calculated and are shown in Table 2. Overall, the results suggest variably 0-20% actual molecular mixing. Noting the probable errors in estimating the experimental T_g's, the W_1 and W_2 values are probably correct to within ± 0.05. Thus mixing plays an important role in interpenetration and influences the reinforcement within each phase.

Table 2. An Estimate of the Molecular Compositions Within Each Phase (21)

Composition	Elastomer Phase		Plastic Phase	
	W_1	W_2	W_1	W_2
10/90 COPEN/PSN	0.94	0.06	0.00	1.00
40/60 COPEN/PSN	0.92	0.08	0.01	0.99
10/90 COPEUN/PSN	0.81	0.19	0.01	0.99
40/60 COPEUN/PSN	0.81	0.19	----	----
10/90 COPUN/PSN	0.88	0.12	0.07	0.93
40/60 COPUN/PSN	0.88	0.12	0.01	0.99

Polymer Engineering and Science

Figures 4 and 5 show the stress-strain behavior of reinforced elastomers and toughened plastics, respectively. In each case, the corresponding homopolymers are included for comparison. In

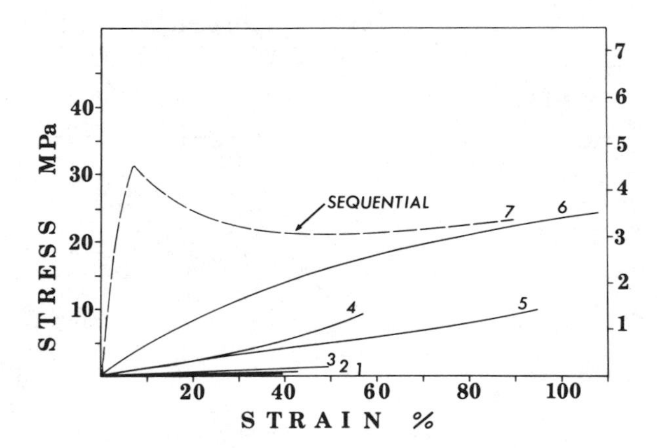

Polymer Engineering and Science

Figure 4. Stress–strain curves for SIN's containing 40% castor oil elastomer (21). Discontinuous curve adapted from sequential IPN synthesis: (1) COPEN; (2) COPEUN; (3) COPUN; (4) 40/60 COPEN/PSN; (5) 40/60 COPEUN/PSN; (6) 40/60 COPUN/PSN; (7) 40/60 COPUN/PSN

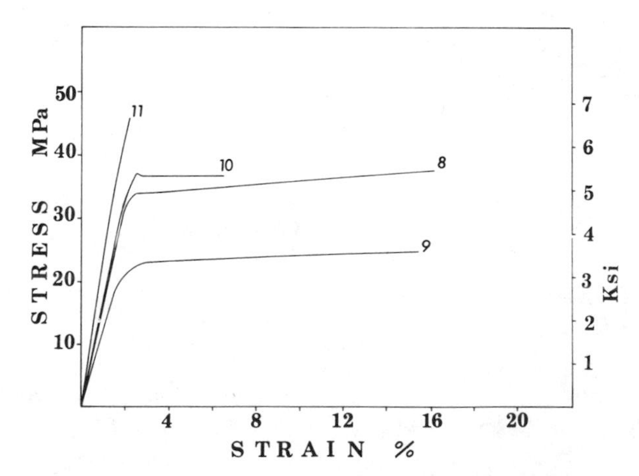

Polymer Engineering and Science

Figure 5. Stress–strain curves for SIN's exhibiting plastic behavior (ASTM D 1708): (8) 10/90 COPEN/PSN; (9) 10/90 COPEUN/PSN; (10) 10/90 COPUN/PSN; (11) PSN (21)

Figure 4, the SIN materials, curves 4, 5 and 6, show greater elongation and higher modulus than the homopolymer curves 1, 2 and 3. The sequential IPN curve, No. 7, shows a far stiffer initial behavior than its counterpart SIN. The corresponding morphologies show that, surprisingly, the sequential IPN of 40/60 composition probably has a greater degree of PS continuity.

Figure 5 shows a much greater strain to break for the SIN's, curves 8, 9 and 10, than homopolymer PS, curve 11. Tables 3 and 4 summarize the mechanical data. Also included in Table 4 are the impact strengths of the plastic materials. In each case, the fracture energy of the SIN is significantly higher than either homopolymer. Interestingly, while the COPEN/PSN made the best plastic material, the COPUN/PSN made the best elastomer.

Table 3. Tensile Properties of SIN's at Ambient Conditions Elastomeric Compositions (21)

Composition	Tensile Strength[1] (MPa)	% Strain At Break	Elastic Modulus (MPa)	Fracture Energy[2] (J/m^3)
40/60 COPEN/PSN	9.2	57	13.1	2.17
40/60 COPEUN/PSN	9.8	95	24.8	4.47
40/60 COPUN/PSN	24.1	108	65.5	16.85
COPEN	0.35	39	1.5	0.07
COPEUN	0.70	43	2.3	0.15
COPUN	0.97	58	2.8	0.29

(Reprinted from Polym. Eng. Sci.)

[1]Samples tested at 2.11×10^{-5} m/s (ASTM D 1708).
[2]Calculated based on the area under the stress-strain curves.

Polymer Engineering and Science

Table 4. Tensile Properties of SIN's Containing 10% Castor Oil Elastomer at Ambient Conditions (21)

Composition	Tensile Stress[1] At Yield MPa	Tensile Stress[1] At Break MPa	% Strain At Break	Elastic Modulus MPa	Fracture Energy[2] (J/m^3)	Izod Impact Strength J/m
PSN	---	46.1	2.2	2360	0.58	13.3
COPEN/PSN	31.1	37.02	16.0	1520	5.37	67.8
COPEUN/PSN	22.1	25.5	18.5	1090	3.46	44.4
COPUN/PSN	37.3	36.7	6.4	1680	2.01	24.6

(Reprinted from Poly. Eng. Sci.)

[1]Samples tested at 2.11×10^{-5}m/s (ASTM D 1708).
[2]Calculated based on the area under the stress-strain curves.

Polymer Engineering and Science

Discussion

Since the purpose of this paper is to review the synthesis and behavior of SIN's based on castor oil elastomers and crosslinked polystyrene, a summary of the main results and conclusions of this study is given below: (Refer to Ref. 22.)

1. A 100% natural polymer based entirely on agricultural products, the polyester elastomer obtained by reacting castor oil with a castor oil derivative, sebacic acid, was the basis for the synthesis of SIN's.

2. Based on castor oil derived elastomers and crosslinked polystyrene, a simultaneous mode of polymerization can be successfully employed to synthesize prototype engineering materials such as tough, impact resistant plastics and reinforced elastomers.

3. All of the SIN's studied formed two-phase systems with tensile properties that were similar to the continuous phase component but exhibiting a substantial improvement in toughness arising from the characteristics of the disperse phase. The SIN's required higher energies to fracture than either polymer component, indicating a synergistic mechanism.

4. At lower elastomer contents, the presence of stirring during the early stages of the polymerization induced morphological changes similar to that observed in the bulk synthesis of the commercial high impact polystyrenes, finally evoking a material bearing a close resemblance to the morphology of commercial HIPS. The polystyrene forms the continuous phase, while the elastomer forms domains ranging from 100 to 5000 nm in size and contain polystyrene subinclusions. The total energy imput and the shear conditions introduced by stirring during polymerization and the rate of elastomer network formation in the time interval between phase inversion and gelation seems to determine the elastomer domain size distribution of such materials.

The cellular structure within the elastomer domains seems to be controlled mainly by the rate of elastomer network formation and the solubility of the polystyrene in the elastomer. Pouring into the molds must be done in the time (or conversion) interval between phase inversion and gelation. Such time intervals decreased as the elastomer concentration increased, because the higher polystyrene conversions were required for phase inversion. Thus, a critical "window" exists for stopping the stirring and pouring into the molds. Pouring too early leads to poor products because of incomplete phase inversion. Pouring too late also yields inferior products because of the onset of gelation.

5. There is a limit in composition, at about 15% elastomer content, at which stirring alone can no longer induce the polystyrene phase to be continuous, and the quality of the mechanical properties of the materials change drastically.

6. The impact resistance of the SIN plastics increased with elastomer content and polyester content of the elastomer phase.

7. Dynamical mechanical spectroscopy and Izod impact results suggest that the glass transition temperature of the elastomer phase constitutes the most critical parameter in achieving impact resistance in these materials.

8. The modulus, elongation to break, ultimate strength, and toughness of the straight castor oil elastomers increase with increasing amounts of TDI employed as crosslinker. A similar trend was observed in the elastomeric SIN's (40% elastomer content) in which the elastomer phase was continuous.

9. While phase separation per se occurs when the free energy of mixing becomes positive, two different mechanisms of phase separation are postulated in the synthesis of SIN's. The first one originates through precipitation of polystyrene chains from the polymerizing solution and seems to occur periodically evolving a multi-modal domain size distribution. The second one originates in the reduction of the ability of the still reacting elastomer to hold the styrene monomer mixture. The process is referred to as microsyneresis and occurs mainly when the styrene monomer concentration exceeds that of equilibrium swelling conditions within the elastomer. Microsyneresis appears to be encouraged by the presence of network inhomogeneities such as local regions of different crosslink density.

10. The morphology and mechanical properties of SIN's evolve from a sequence of events, namely: separation of phase 1 from phase 2, separation of phase 2 from phase 1, gelation of polymer 1, gelation of polymer 2, and phase inversion. The crucial criteria center on which happens first and what morphological changes occur. Variables such as composition, compatibility, reaction rates and reaction conditions determine the morphological path followed by the reacting system.

11. Although the materials studied in this research program lack optimization, they already compare satisfactorily to commercial materials in many respects. With reasonable further research and development studies, high quality tough plastics and reinforced elastomers may be anticipated.

Conclusions

By way of conclusion, the SIN materials were quite different from the original sequential IPN compositions. While the polyester linkage could be used with the sequential mode of synthesis also, its presence in the SIN made for much better impact strengths, probably because of the lower elastomer T_g permitted. The SIN synthesis allowed a greater range of compositions to be made.

Processing, as shown in Figure 1, is far easier. The material can be extruded into a mold after phase inversion, Figure 3, but before gelation, allowing for a practical system of handling.

While value judgements on potential usefulness are fraught with uncertainties, probably the reinforced elastomers are more interesting than the toughened plastics. In significant measure, this conclusion is based on the premise of the original undertaking: to develop the potentials of a castor oil-based product. The elastomers use more of the natural material, percentage wise, than the plastic. Items ranging from shoe heels to gaskets might easily be made, based on the prototype compositions examined herein. However, the plastic compositions are exceedingly tough.

Acknowledgements

The authors wish to thank the National Science Foundation in the United States for support under Grant No. INT74-06791 AOI, and Colciencias in Colombia.

Literature Cited

1. Agricultural Marketing Service, U.S.D.A.
2. Chemical and Engineering News, 1974, 52(37), 8.
3. Naughton, F. C., J. Am. Oil Chemists Soc., 1974, 51, 65.
4. Conde-Cotes, A. and Wenzel, L. A., Revista Latinoamericana de Ingeniera Quimica y Quimica Aplicada, 1974, 4, 125.
5. Swern, D., Ed. "Bailey's Industrial Oil and Fat Products," 3rd Ed. Interscience, 1964.
6. Huelck, V.; Thomas, D. A. and Sperling, L. H. Macromolecules, 1972, 5, 340.
7. Frisch, H. L.; Klempner, D. and Frisch, K. C. J. Polym. Sci., 1969, B-7, 775.
8. Lipatov, Y. S. and Sergeeva, L. M. Russian Chem. Rev., 1976, 45(1), 63.
9. Meyer, G. C. and Mehrenberger, P. Y. European Polym. J., 1977, 13, 383.
10. Touhsaent, R. E.; Thomas, D. A. and Sperling, L. H. "Toughness and Brittleness of Plastics," R. D. Deanin and A. M. Crugnola, Eds., Advances in Chemistry Series 154, 1976, 205.
11. Kim, S. C.; Klempner, D.; Frisch, K. C.; Radigan, N. and Frisch, H. L. Macromolecules, 1976, 9, 258.
12. Yenwo, G. M.; Manson, J. A.; Pulido, J.; Sperling, L. H.; Conde, A. and Devia, N. J. Appl. Polym. Sci., 1977, 12, 1531.
13. Yenwo, G. M.; Sperling, L. H.; Pulido, J. and Manson, J. A. Polym. Eng. Sci., 1977, 17(4), 251.
14. Pulido, J. E.; Yenwo, G. M.; Sperling, L. H. and Manson, J. A. Rev. UIS, Colombia, 1977, 7(7), 35.

15. Sperling, L. H.; Manson, J. A.; Yenwo, G. M.; Devia, N.; Pulido, J. E. and Conde, A. "Polymer Alloys," D. Klempner and K. C. Frisch, Ed., Plenum, 1977, New York.
16. Devia, N.; Conde, A.; Sperling, L. H. and Manson, J. A. Rev. UIS, Colombia, 1977, 7(7), 19.
17. Devia-Manjarres, N.; Conde, A.; Yenwo, G.; Pulido, J.; Manson, J. A. and Sperling, L. H. Polym. Eng. Sci., 1977, 17(5), 294.
18. Yenwo, G. M.; Sperling, L. H.; Manson, J. A. and Conde, A. "Chemistry and Properties of Crosslinked Polymers," S. S. Labana, Ed., Academic Press, 1977, 257.
19. Devia, N.; Manson, J. A.; Sperling, L. H. and Conde, A. Polym. Eng. Sci., 1978, 18(3), 200.
20. Devia, N.; Sperling, L. H.; Manson, J. A. and Conde, A. Macromolecules, 1979, 12 (3), 360.
21. Devia, N.; Sperling, L. H.; Manson, J. A. and Conde, A. Polym. Eng. Sci., 1979, 19(12), 870, 878.
22. Devia, N.; Sperling, L. H.; Manson, J. A. and Conde, A. J. Appl. Polym. Sci., 1979, 24, 569.

RECEIVED July 12, 1979.

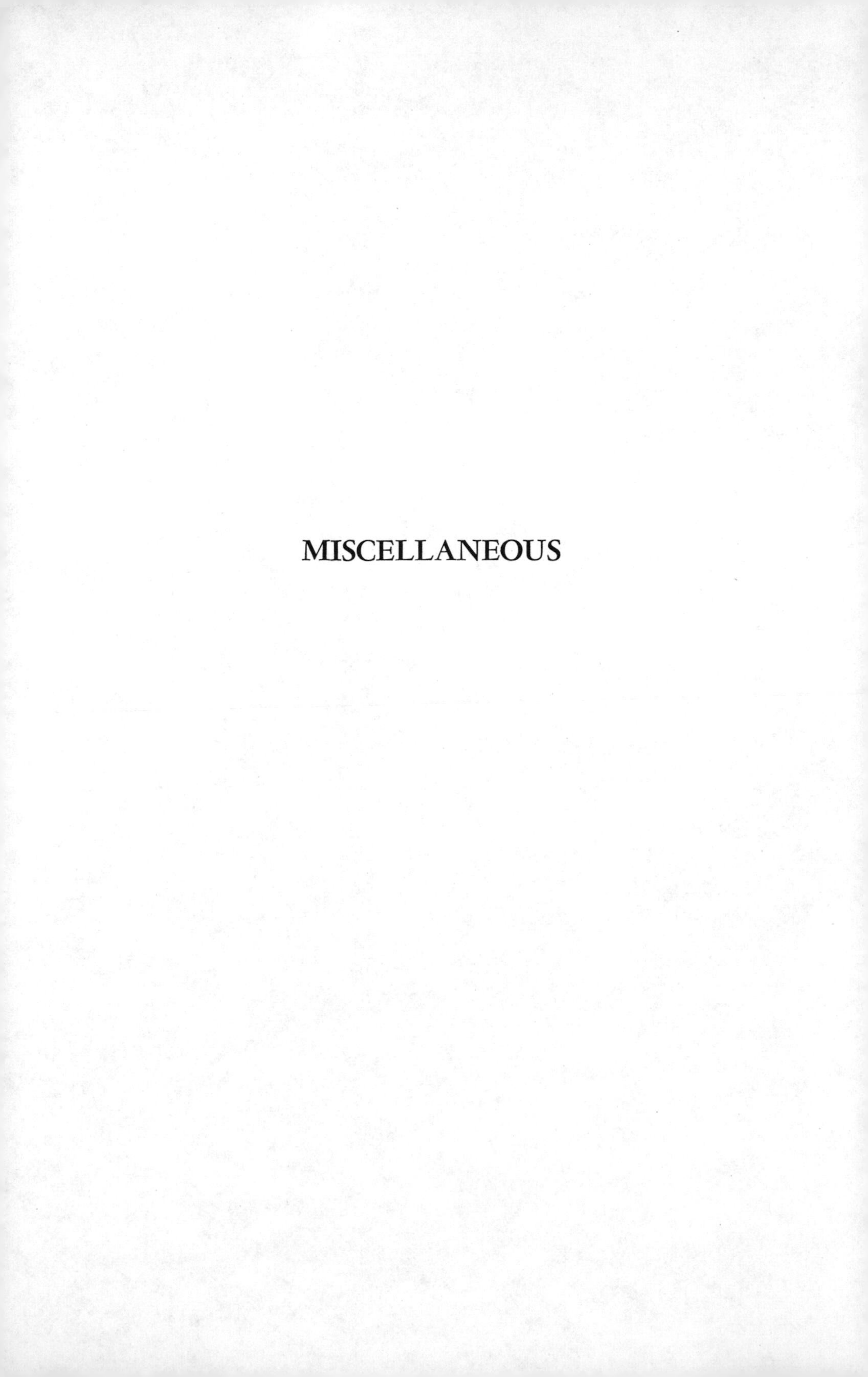

MISCELLANEOUS

Flammability of Phosphorus-Containing Aromatic Polyesters: A Comparison of Additives and Comonomer Flame Retardants

ROBERT W. STACKMAN

Celanese Research Company, Summit, NJ 07901

The use of phosphorus compounds as flame retardants has been reviewed by Lyons and others (1, 2, 3, 4, 5). The mechanism of the action of this element is generally accepted to involve decomposition to produce acids which function as char promoters. Phosphorus compounds are particularly effective flame retardants for polyesters where they function to increase the char yields.

While there are a large number of both phosphorus additive and comonomer compounds available, no direct comparisons have been reported between the effectiveness of the two methods of incorporation, aside from some references to the lack of permanency of many additive compositions. The use of additives, on the other hand, may provide a greater flexibility, allowing the production of polymeric compositions of varying degrees of flame retardance, from the same base resin. The purpose of this study was to determine whether any real differences in effectiveness are detectable due to the method of incorporation of phosphorus into a polymer system.

Aromatic polyphosphonates have been found to be especially effective flame retardant additives for polyester compositions (6, 7, 8), especially for polyethylene terephthalate. These additives are phosphorus esters of a di- phenol with the following structure:

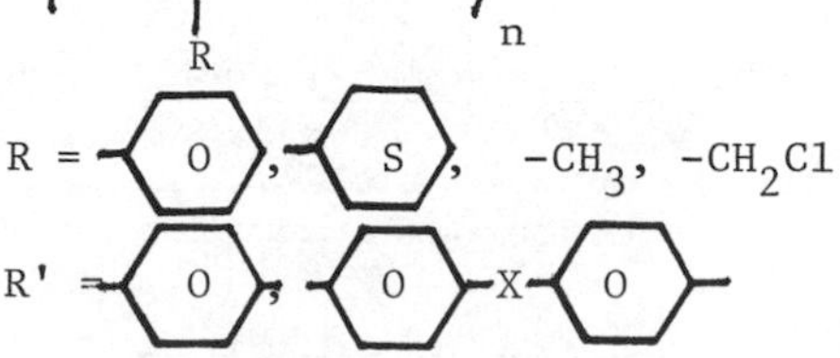

$$\left[O - \underset{R}{\overset{O}{\overset{\|}{P}}} - O - R' \right]_n$$

R = ⟨O⟩, ⟨S⟩, $-CH_3$, $-CH_2Cl$

R' = ⟨O⟩, ⟨O⟩-X-⟨O⟩

0-8412-0540-X/80/47-121-425$05.00/0

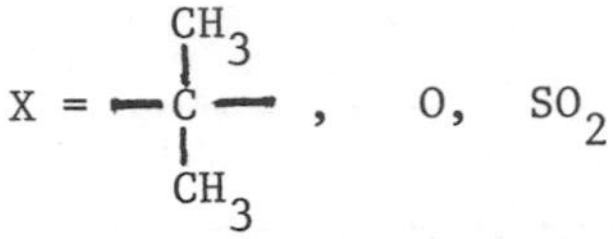

$$X = -\overset{CH_3}{\underset{CH_3}{C}}- , \quad O, \quad SO_2$$

Aromatic polyesters, commercially important molding resin materials, show a low degree of flammability and produce high percentages of char on exposure to a flame or on heating to pyrolysis conditions (9).

In view of the utility of the aromatic polyesters and the demonstrated effectiveness of the aromatic polyphosphonates as flame retardants, the combination of these two polymers was chosen for this study. In addition, this system provided a composition in which both copolymers and polymer blends could be prepared with identical chemical compositions. The polyesters were prepared from resorcinol with an 80/20 m/m ratio of iso-phthaloyl and terephaloyl chlorides while the polyphosphonates were resorcinol phenylphosphonate polymers. Copolymerized phosphorus was incorporated by replacement of a portion of the acid chloride mixture with phenylphosphonic dichloride.

The structures of the compositions are as shown below:

Copolymers.

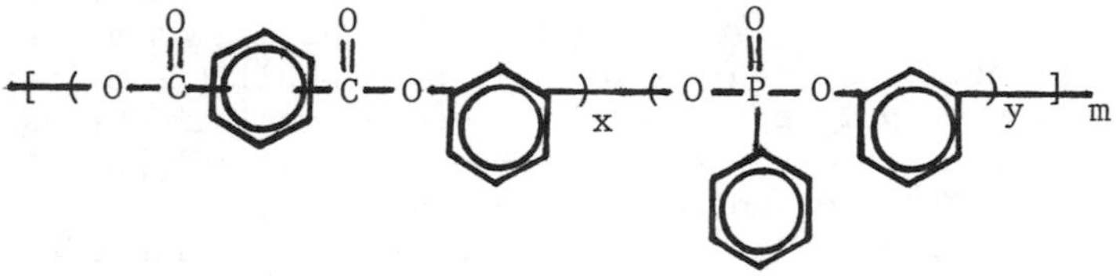

$$-[-(O-\overset{O}{\overset{\|}{C}}-C_6H_4-\overset{O}{\overset{\|}{C}}-O-C_6H_4-)_x-(O-\overset{O}{\overset{\|}{\underset{C_6H_5}{P}}}-O-C_6H_4-)_y-]_m-$$

Polymer Blends.

$$-(O-\overset{O}{\overset{\|}{C}}-C_6H_4-\overset{O}{\overset{\|}{C}}-O-C_6H_4-)_x- \qquad -[-O-\overset{O}{\overset{\|}{\underset{C_6H_5}{P}}}-O-C_6H_4-]_y-$$

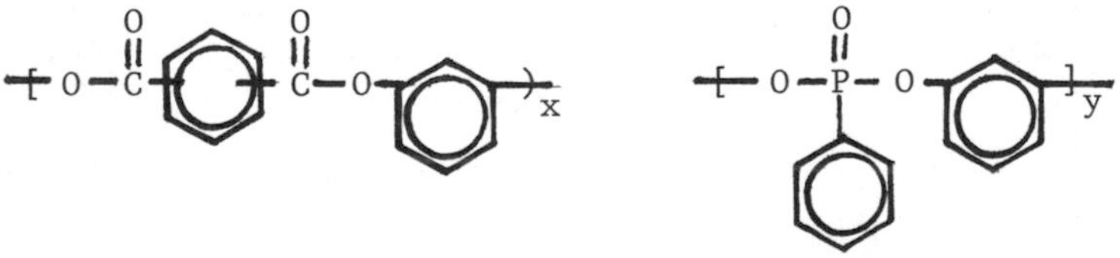

Experimental

Preparation of poly(m-phenylene)iso/terephthalate (80/20). The polymers were prepared by solution condensation of the acid chlorides with resorcinol in methylene chloride solution using triethyl amine as the acid acceptor as described by Korshak (10).

Preparation of Phosphorus Containing Terpolymers. Prepared by substitution of a portion of the acid chloride mixture by a calculated amount of phenyl phosphonic dichloride. The compositions of these polymers are summarized in Table 1.

Preparation of poly(m-phenylene)phenyl phosphonate. These polymers were prepared by the method of Toy using resorcinol and phenylphosphoric dichloride (11).

Preparation of Polymer Blends. A series of polymer blends was prepared by co-solutioning predetermined amounts of the poly(m-phenylene)isophthalate/terephthalate (80/20) with poly (m-phenylene)phenyl phosphonate, in methylene chloride. The polymer blends were recovered by evaporating the solution to dryness and ground to 40 mesh with a Wiley Mill. The composition of these blends and their analyses are summarized in Table I.

Table I
Composition and Self-Quenching Times for Phosphorus Containing CO Polymers

Sample	Inherent Viscosity dl/g*	%P	Oxygen Level for Burn Time of:			
			1 sec.	3 sec.	5 sec.	9 sec.
A	0.39	0	18.7	20.1	20.9	21.2
B	0.35	1.7	20.2	22.5	23.1	23.8
C	0.35	3.4	22.5	24.5	25.3	26.5
D	0.38	4.2	21.6	24.5	26.0	27.0
E	0.32	6.7	23.0	26.2	27.2	-
F	0.31	7.8	23.5	25.5	26.3	27.5

*Inherent viscosity measured at 0.1% concentration in 90% phenol/10% tetrachloro ethane solvent.

Preparation of Samples for Flammability Testing. Samples of the phosphorus containing terepolymers and of the polymer blends were converted to film by compression molding on a Carver Laboratory Press with electrically heated platens. The films were prepared at 250°C and 20,000 lb. pressure, using a 10 mil thick frame mold. Samples (2" x 1/4") were cut from this film for flammability testing.

Flammability Testing of Polymer Films. The flammability characteristics of the polymers and blends were determined using a specially designed oxygen index apparatus (Figure 1). Self-quenching time (SQT) analyses were performed as described by Stuetz (12). The time for film samples to self-extinguish, at varying oxygen concentrations, were determined. The results of these timed burnings are summarized in Table II and in Figures 2 and 3.

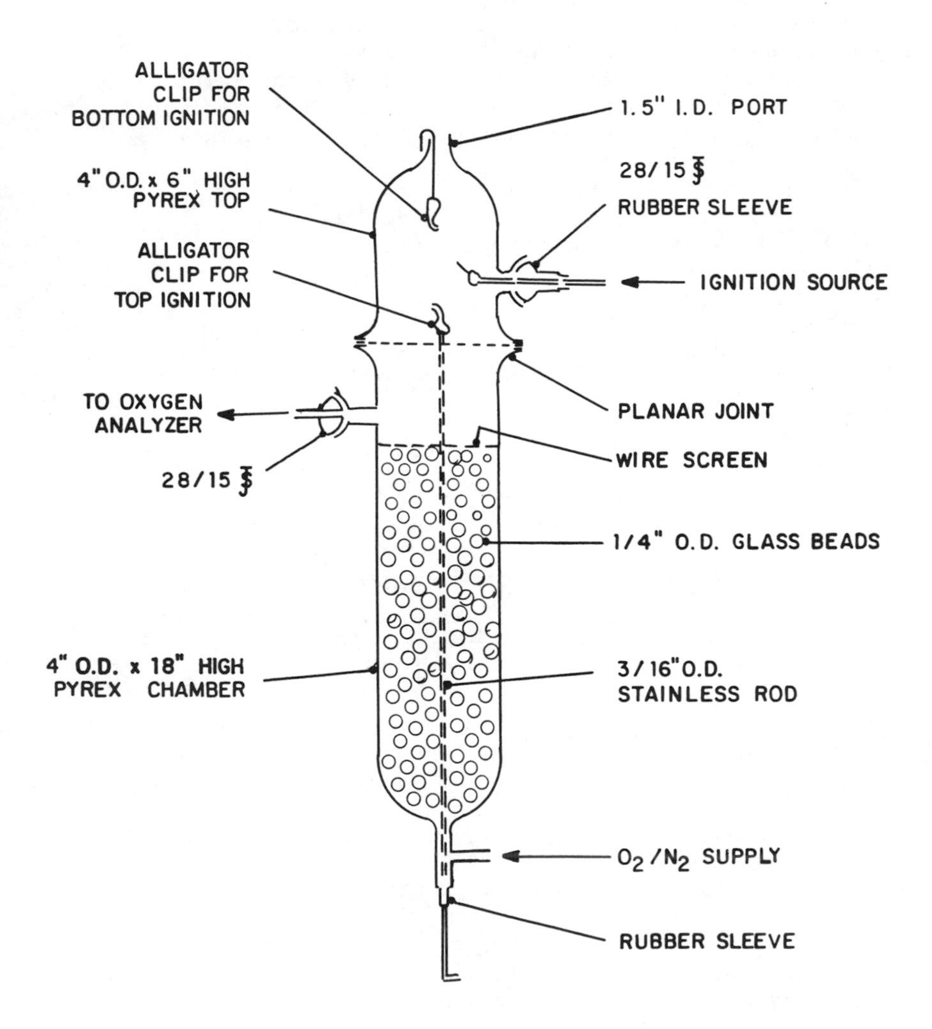

Figure 1. Oxygen index apparatus

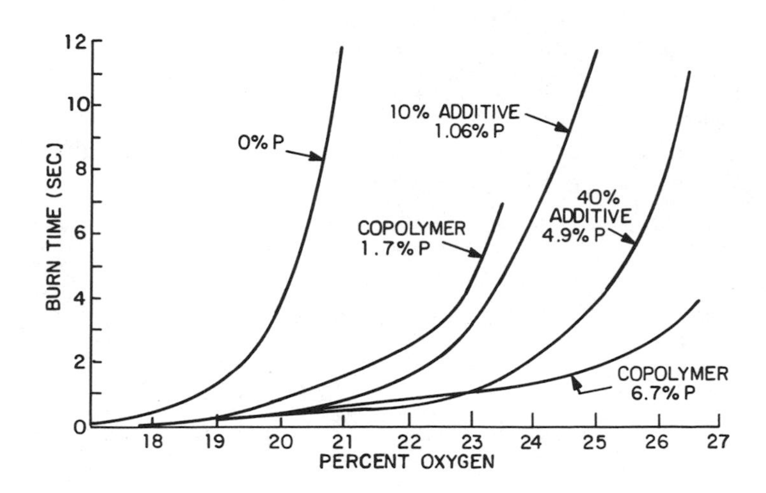

Figure 2. Self-quenching time vs. atmospheric oxygen content for terpolymer and blend compositions

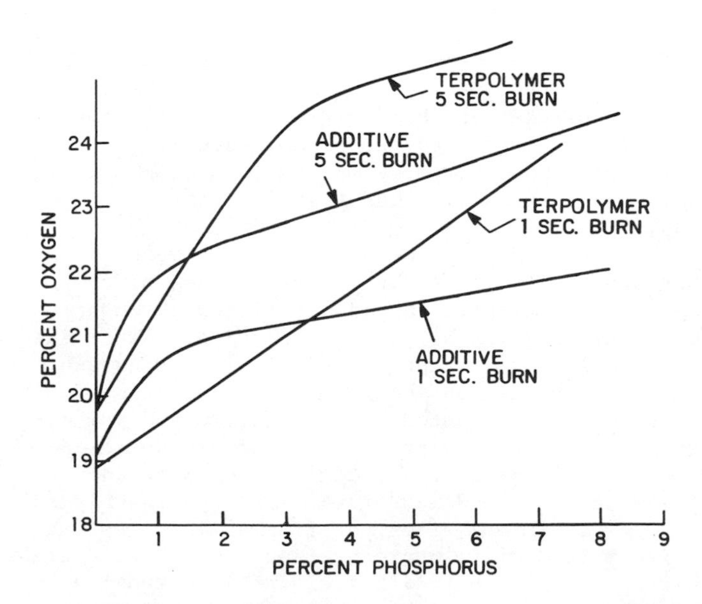

Figure 3. Oxygen content required to maintain 1 and 5 sec burn time as a function of phosphorus content

Table II
Composition and Self Quenching Times
for Phosphorus Containing
Aromatic Polyester Blends

Sample	%P	Oxygen Level for Burn Time Of:			
		1 sec.	3 sec.	5 sec.	9 sec.
A-1	0.56	19.7	21.1	21.7	22.2
A-2	1.06	21.0	22.8	23.5	24.4
A-3	1.36	20.7	22.6	23.2	24.0
A-4	2.20	21.2	23.2	24.0	24.8
A-5	2.86	22.5	25.0	25.5	25.9
A-6	3.69	21.7	23.0	22.8	24.5
A-7	4.52	20.9	22.8	24.0	24.7
A-8	5.96	22.5	24.6	25.5	26.2
A-9	6.22	21.8	24.6	25.3	25.7
A-10	8.04	22.4	24.4	25.3	26.3

Discussion

Sample Uniformity. It is apparent that the phosphorus atoms in both the copolymer and blend systems are all in the same environment, i.e. bonded to resorcinol moieties. In addition, since all of the phosphorus is in the form of polymeric species, differences in flammability due to volatility can be eliminated from consideration.

The only real sources of difference between the additive and comonomer systems would be the potential lack of homogeneity of the polymer blends, which may result either through incomplete blending or segregation by phase separation. Since such nonhomogeneity could lead to irreproducibility in the burning trials, this would be detected by the flammability testing.

Ester interchange between the additive and polymer would be expected if the mixture were held for an excessive time in the melt. In order to avoid the potential for interchange, the blendings were conducted in solution and heated only for a time necessary to melt press films for flammability testing. An alternative method of film preparation by solvent casting of films was discarded due to difficulties in preparing uniform, sufficiently thick, solvent free films for evaluation.

Flammability Analysis. The ease of self-extinguishment has been proposed as a criterion for flammability by Stuetz (12) and Miller (13). The SQT analyses of Stuetz is particularly appropriate to this system since aromatic polyesters such as poly(m-phenylene)isophthalate/terephthalate (80/20) are non dripping compositions. They are also only marginally flammable as can be seen in the results of Table 1. This unmodified polyester self-extinguishes in about 5 seconds in an oxygen concentration equal to that of air (20.9), even though it ignites and burns at a much lower oxygen level. The addition of phosphorus to this composition, even at low levels, gives a significant increase in the flame retardancy as measured by the SQT bottom ignition oxygen index techniques.

Figure 2 illustrates the effect of phosphorus upon the SQT for copolymers and blends at various phosphorus levels. It is obvious from this curve that the effect of the phosphorus is to make the polyester more self-extinguishing and has little effect upon ignition.

The differences in the effect of phosphorus as an additive or as a comonomer can be seen more clearly in Figure 3. In these curves, the percent phosphorus in the sample is plotted against the oxygen level required for a given time of burning. In Figure 3 the data show that the oxygen level required to maintain burning for one second (SQT = 1 sec.) is practically linear with increasing phosphorus content for the phosphorus containing terpolymers.

Lower levels of phosphorus are required for the additive compositions to maintain the one-second burn. At some higher level (∿4% phosphorus) the two curves cross and at still higher phosphorus contents, the copolymers appear to be more flame retardant than are the compositions containing the phosphonate additive.

The same trend holds even when longer burn times (SQT = 5 sec.) are selected. While the point of equivalence of oxygen level shifts to lower phosphorus contents, the compositions containing additive phosphonate are always superior to the terpolymers, at low phosphorus composition.

The mechanism of the action of the phosphonate as a flame retardant is generally believed to be decomposition into acid fragments which contribute to char formation. These acidic species catalyze decomposition of the polyester, and give rise to species which on reaction with the phosphorus moiety cause char formation. TGA curves of the copolymers confirm that the incorporation of phosphorus into the polymer increases the char residue (Figure 4). These curves, however, show little evidence that the presence of phosphorus has any effect upon the temperature or rate of decomposition of the polyester. The curves are all fairly similar up to about 450°C. After that point, the amount of residue is proportional to the amount of phosphorus in the terpolymer.

This would be expected since any change in the onset of decomposition or initial rate of decomposition would affect the oxygen concentration for ignition of the sample. As mentioned

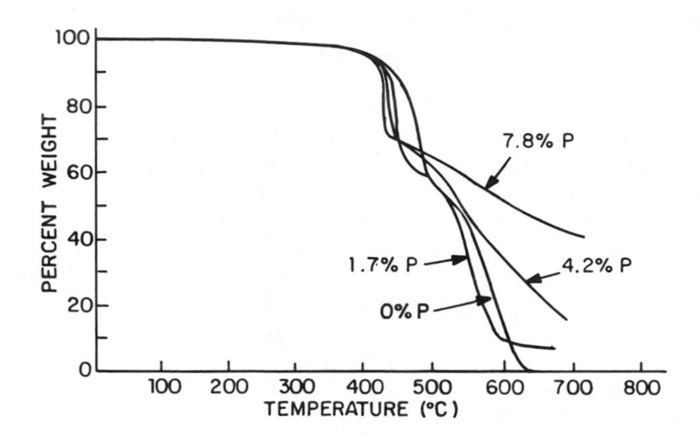

Figure 4. Thermogravimetric analysis curves of polyester terpolymers

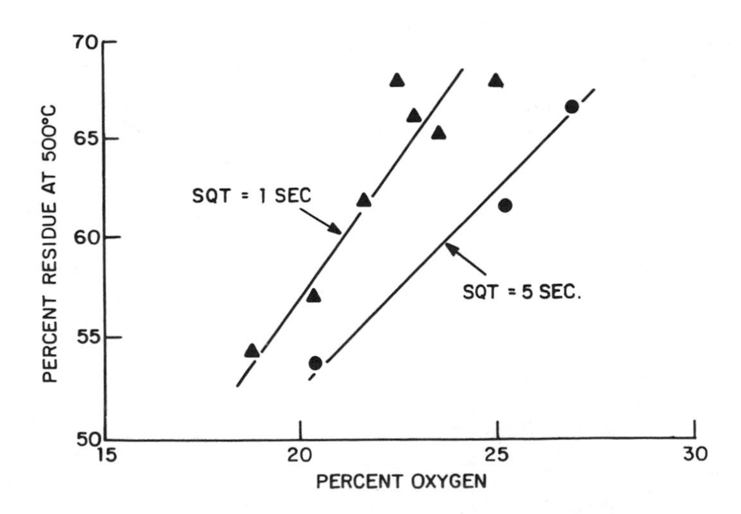

Figure 5. Percent oxygen in atmosphere for 1- and 5-sec self-quenching time as a function of TGA residue at 500°C

above, no change was noted for the oxygen concentration required for ignition of any of the samples tested.

The increases in char residue for the samples is readily correlatable with the oxygen level required for SQT's of 1 and 5 seconds for the samples (Figure 5). This lends support to a mechanism in which the surface of the burning polymer is screened from oxygen by a phosphorus rich char barrier which snuffs out the flame (14). Since the ignition of the substrate is a necessary condition for the generation of this barrier, no effect is found on the ignition by the presence of the phosphorus compound. Higher phosphorus contents result in faster char build-up; therefore, a faster, more complete screening of the surface and a more rapid extinguishment of the flame (Figure 2).

While it is felt that the mechanism of the extinguishment in both the copolymer and blend systems is the same, the differences in efficiency seen in the data are probably due to transitory differences in the nature of the char barrier. In the case of the additive compositions, which appear to be more efficient flame quenching agents, it is possible that the barrier is richer in phosphate (-POPO-) linkages which contain less flammable components than would a barrier from the copolymer. The phosphate rich barrier may alternatively merely be a more efficient oxygen screen either due to trapping of oxygen or being a more fluid, cohesive film as opposed to a foamy porous barrier from a higher melting char former such as the copolymer. The SQT's might be expected to converge and equalize at high phosphorus levels (Figure 3). Since the greatest deviation occurs at high phosphorus levels, it is possible that the additive segregates out and flows away from the flame first before it can decompose and form a barrier at high levels. The low melting point and low melt viscosity would favor this behavior. The net effect would then appear to be a lower effective additive concentration. The leveling off of the SQT-oxygen level curve for the additive would indicate this type of behavior.

The conclusion of the study would, therefore, be that: (a) phosphorus based additives and comonomer are effective flame quenching agents for aromatic polyesters; (b) the additive compositions are more effective at low phosphorus contents and (c) the mechanism of quenching involves an increase in char residue which is related to the phosphorus content of the composition.

Acknowledgments

The author would like to express his appreciation to the Management of the Celanese Research Company for permission to publish this work, to Edward Kuczynski for technical assistance, Andrew DiEdwardo for the TGA and other analyses and to Arnold Rosenthal for support and encouragement.

Abstract

Two series of aromatic polyester compositions, containing varying levels of phosphorus, were prepared by different methods; (a) blending with an aromatic polyphosphonate and (b) copolymerization by incorporation of a phosphonate unit into the polyester backbone.

Flammability of the blends and terpolymers was determined by bottom ignition oxygen index measurements performed on compression molded films. The results indicate that the use of phosphorus based additives is the more effective method for improving the flammability of this polyester up to a level of about 20% additive. At higher levels, the copolymers appear to be less flammable. Ignition takes place with all the samples at fairly low oxygen levels; however, in the presence of either the phosphorus additives or comonomers the samples self extinguish rapidly. The higher the phosphorus content, the more rapid is the self extinguishment.

Thermal gravimetric analysis shows that the increase in phosphorus content results in char yield increases which are correlatable with the self-extinguishment time decreases. This led to the conclusion of phosphorus "rich" barrier shielding the surfaces as a mechanism of the flammability decrease.

Literature Cited

1. Lyons, J. W., "The Chemistry and Uses of Fire Retardants," Interscience Publishers, a division of John Wiley & Sons, Inc., New York, 1970.
2. Lyons, John W., J. Fire Flammability 1, 302 (1970).
3. Delman, Alvin D., J. Macromol. Sci. - Revs. Macromol. Chem., c3(2), 281 (1969).
4. Papa, Anthony J., Ind. Eng. Chem. Prod. Res. Develop., 9, 478 (1970).
5. Grunzow, Albrecht, Accts. of Chem. Res., 11, 1979 (1978).
6. Cohen, Stuart L. and Stackman, Robert W., U. S. 3,829,405 to Celanese Corp., August 13, 1974.
7. Cohen, Stuart L. and Stackman, Robert W., U. S. 3,830,771 to Celanese Corp., August 20, 1974.
8. Bostic, James E., Jr., Yeh, Kwan-Nam and Barker, Robert H., J. Appl. Poly. Sci., 17, 471 (1973).
9. Steuben, Kenneth C. and Imhof, Lawrence G., J. Fire and Flammability 4, 8 (1973).
10. Korshak, Von V. and Vinogradova, S. V., "Polyesters," Pergamom Press, Oxford 1965.
11. Toy, Arthur Dok Fon, U. S. 2,435,252 to Victor Chemical Works, February 3, 1948.
12. Stuetz, D. E., DiEdwardo, A. H., Zitomer, F., and Barnes, B. P., J. Polymer Science, (A); accepted for publication.
13. Metler, B., Goswani, B. C. and Turner, R., Textile Research J., 61, 1973.
14. Learworth, G. S. and Thevarti, D. G., Brit. Polymer J. 2(5), 249 (1970).

RECEIVED September 24, 1979.

29

Electronic Cooperativity in New π-Donor Polymers Prepared by Modification Reactions of Poly(vinylbenzyl chloride)

FRANK B. KAUFMAN, EDWARD M. ENGLER, and DENNIS C. GREEN

IBM T. J. Watson Research Center, Yorktown Heights, NY 10598

I. Introduction

This paper discusses the synthesis and physico-chemical consequences of binding low ionization π-donors, such as TTF monocarboxylic acid (I) and 1,3-di-(p-methoxyphenyl)-5-(p-hydroxyphenyl)-Δ^2-pyrazoline (II) to a poly(vinylbenzylchloride) backbone.

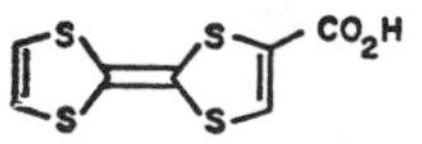

CO_2H

I

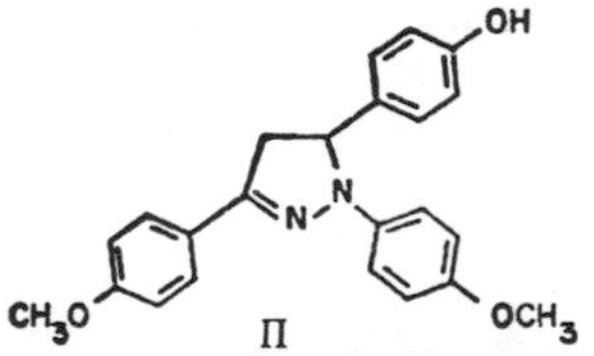

OH

N N

CH_3O II OCH_3

These particular molecules were chosen since they exemplify the interesting and unique properties(1) of an entire class of π-donor species. These properties include the formation of stable, colored radical cations and the known ability to interact strongly in the crystalline solid state.(2)

Poly(vinylbenzylchloride) was used as the substrate polymer because both the linear and the lightly cross-linked material were readily obtainable and because they had both been previously shown to be capable of undergoing a variety of chemical transformations(3,4). A second advantage to the use of this polymer was the possibility of determining how the presence or absence of crosslinking, without changing the nature of the backbone, affected the properties of the resulting polymer. Finally, earlier studies(4) of reactive groups bound to these cross-linked polymers had established the ability of these groups to interact chemically with each other thus suggesting the presence of site-site intermoleculer interactions. It seemed reasonable to us that the presence of these chemical interactions was a necessary but not a sufficient condition for observing nontrivial electronic interactions between the polymer bound donors.

Previous work on the synthesis of TTF (tetrathiafulvalene) containing polymers(5-11) has been reported by at least seven groups of researchers. Most of this work concerns condensation(5,6,7,8,9) polymers or polymers made from vinyl substituted TTF molecules(11-12). Without exception, the polymers produced by these methods have been largely unacceptable for subsequent physical study because of their brittle,intractable, highly insoluble nature. Only by reaction of a suitably monofunctionalized TTF derivative with the preformed polymer poly(vinylbenzylchloride) has it been found possible(10) to prepare soluble TTF homopolymers with more manageable physical properties.

0-8412-0540-X/80/47-121-435$05.00/0

II. Experimental

A. Donor-Bound Polymer Preparation

The polymers and copolymers discussed here were all prepared by reaction of the homogeneous (linear) or heterogeneous (cross-linked) poly(vinylbenzylchloride) substrate polymer with the potassium or cesium salts of the suitably monofunctionalized donors (Reaction 1).

$$(CH_2-\underset{\displaystyle CH_2Cl}{CH})_n + Donor-\bar{Y}M^+ \rightarrow (CH_2-\underset{\displaystyle CH_2YDonor}{CH})_n + MCl \qquad (1)$$

For the homogeneous reactions, pyrazoline and tetrathiafulvalene donors were attached to the polymer when Y = 0 and M = K by refluxing for several days in dry tetrahydrofuran. In the case of the heterogeneous polymers, the solvent employed was found to be crucial in achieving high donor coverage onto the polymer as shown in Table I. TTF was also attached to the polymer backbone via a carboxylate linkage (i.e. Y= $C0_2$, M=Cs) and this reaction was conveniently carried out by heating, either the linear or cross-linked polymers, at 50-55°C in purified dimethyl formamide. Higher temperatures led to donor decomposition.

TABLE I

Monomer	Polymer[a] %Cross-linking[b]	Polymer[a] % CH_2Cl[c]	Experimental Conditions: Equation 1 Y	Equation 1 M	Solvent[d]	Reflux Time (days)	% Reaction[e]	% Coverage[f]
Pyrazoline I	1	84	0	K	1:1 D-E	4	72	60
Pyrazoline I	1	84	0	K	4:1 D-E	4	88	74
Pyrazoline I	0.5	76	0	K	4:1 D-E	4	90	68
Pyrazoline I	2	70	0	K	4:1 D-E	4	71	50
Pyrazoline I	4	76	0	K	4:1 D-E	4	64	49
Pyrazoline I	8	74	0	K	4:1 D-E	4	48	35
Pyrazoline I	0[g]	100	0	K	THF	4	74	74
TTF	1	84	CO_2	K	4:1 D-E	4	25	21
TTF	1	84	CO_2	Cs	DMF	4[h]	90	76
$TTFCH_2$	1	84	0	K	4:1 D-E	4	17	14
$TTFCH_2$	1	84	0	K	THF	4	87	73
$TTFC_6H_4$	1	84	0	K	4:1 D-E	4	20	17
$TTFC_6H_4$	1	84	0	K	THF	4	90	76
$TTFC_6H_4$	0[g]	100	0	K	THF	4	95	95

[a]Poly(vinylbenzylchloride). [b]Cross-linked using divinylbenzene. [c]Chloromethylated, cross-linked polystyrene resins were obtained commercially from Bio-Rad Laboratories. Percent chloromethylation is based on the available phenyl groups in the polymer; that is minus the percent cross-linking. [d]D=dioxane; E = ethanol. [e]Percent of available chloromethyl groups reacted with donor. [f]Percent reaction x percent chloromethylation. [g]Polymer prepared by free-radical polymerization of 60.40 para-meta chloromethylated sytrene (Dow Chemical). [h]Reaction heated at 50-55°C.

The degree of donor coverage could be easily controlled by varying the initial donor-polymer repeat-unit mole ratio. No difficulty was encountered in achieving rather high coverages of donors. For example, reaction of ~90% of the chloromethyl groups with TTF was possible (see Table I). For cross-linked chloromethylated polystyrene resins, the degree of coverage was dependent on solvent and the percent cross-linking as seen in Table I. Increasing the "swelling" ability of the solvent (i.e. 1:1 Dioxane-EtOH to 4:1 Dioxane-EtOH) or decreasing the percent cross-linking (from 8% through 0.5%) both lead to increased coverage of the donor on the polymer.

Characterization of the donor bound polymers follows from their spectroscopic (ir and uv-vis: KBr) properties in comparison with the starting donor monomers, and from elemental analyses. That the donors are covalently bound to the polymer and not present as unreacted monomers can be seen by the absence of the characteristic monomer functional group absorption (i.e. -OH, CO_2H) in the donor bound polymer. For example in Figure 1, the comparative ir spectra of p-hydroxyphenyl-TTF monomer and this donor covalently bound to linear and to cross-linked polysytrene are given. Except for the presence of the hydroxyl absorption in the monomer, all three spectra are essentially identical, indicating a rather clean polymer attachment reaction.

The percent reaction was typically determined by elemental analysis; however, in the case of cross-linked polymers, the percent reaction was also evaluated by titrating the amount of chloride liberated and by measuring the weight gain of the polymer resin. All three methods gave comparable results (within 2-3%).

B. Monofunctionalized Donor Preparation

The key to the attachment of the donor to the polymer lies in the preparation of suitably monofunctionalized donors. A variety of synthetic procedures have been employed and are outlined in Table II. Conversion of the functionalized alkali salt derivatives were accomplished by reaction with the alkali hydroxide (CsOH or KOH in alcohol) or by reaction with an alkali hydride (i.e. KH in tetrahydrofuran). The latter reaction was found to proceed more cleanly and conveniently.

TABLE II

Monofunctionalized Donors

Monomer Donor	Salt Preparation Condition (Equation 1)	Monomer Preparation
1, 3-di-(p-methoxyphenyl)-5-(p-hydroryphenyl)-[2]-pyrazoline	KOH in dioxane-EtOH KH in THF	a
tetrathiafulvalene carboxylic acid	CsOH in DMF	b
hydroxymethyl tetrathiafulvalene	KOH in dioxane-EtOH	c
p-hydroxymethyl tetrathiafulvalene	KH in THF	d

[a]Reference 20 [b]D. C. Green, J.C.S. Chem. Comm., 1978. [c]D. C. Green, J. Org. Chem., in press. [d] E. M. Engler, V. V. Patel and F. B. Kaufman, to be published.

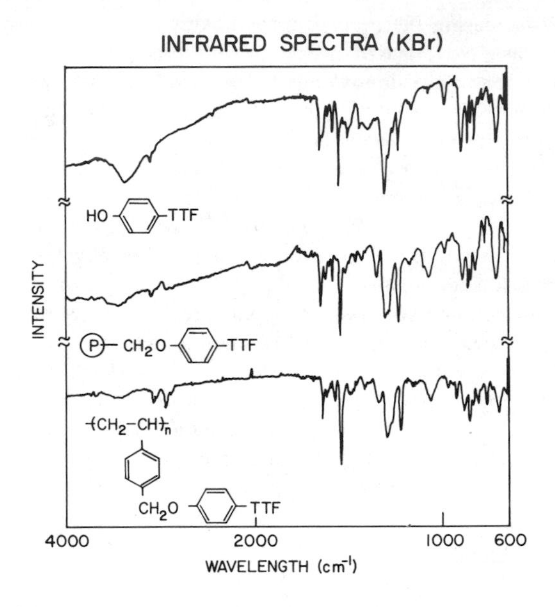

Figure 1. Comparative IR spectra in KBr of monomer phenoxyphenyl TTF, linear polymer-bound phenoxy TTF, and 2% cross-linked polymer-bound phenoxy TTF

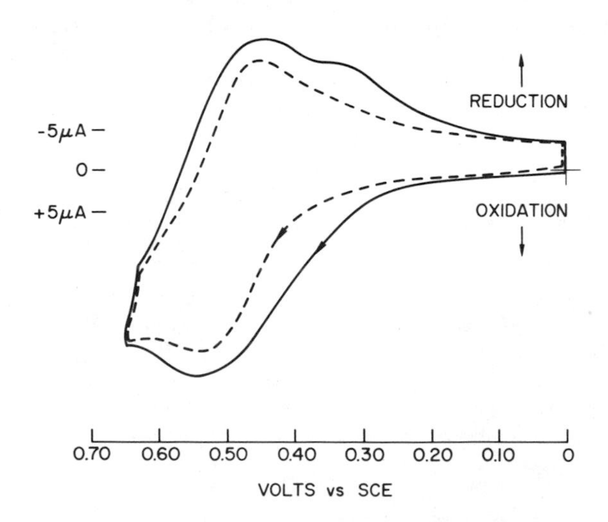

*Figure 2. Cycle voltammogram of high (——) vs. low coverage (– – –) carboxy TTF linear polymer in DMF (0.1*N *TEAP vs. SCE, Pt electrode)*

C. Polymer Film Preparation

The cross-linked heterogeneous polymer films for electrochemical and optical measurements were prepared by first suspending the polymer resin in benzene solvent (1:30;w:w) and stirring via a magnetic stirrer for ~15 minutes. Then, a drop of the suspension was applied to the substrate (~80 Å evaporated Pt, ultrasonically cleaned with DMF, acetone, CH_2Cl_2) from a pipette. The film was allowed to air dry, followed by 30 minutes of evacuation in vacuum. Linear polymer films could be prepared using spin-coating techniques. Typically a concentrated solution of the polymer in tetrahydrofuran was prepared; and a drop placed onto a rapidly spinning substrate. After drying by air while spinning, the substrate was ready for use.

D. Electrochemical and Optical Measurements

Electrochemical measurements were performed in an electrochemical cell equipped with quartz windows which fit into the sample compartment of a Cary 14 spectrometer. The cell (CH_3CN, 0.1N TEAP vs S.C.E.) employed three electrode (Pt auxiliary electrode) potentiostatic control. A Tacussell PRT Potentiostat and PAR model 175 signal generator were used for the measurements.

III. Results and Discussion

A. Linear TTF Polymers: Oxidation-Reduction Properties

Cyclic voltammograms of high coverage (90%) and low coverage (30%) poly(vinylbenzylcarboxytetrathiafulvalene) linear polymers dissolved in DMF solution are shown in Figure 2. Two points of interest will be mentioned here. First, the magnitude of the first oxidation potentials observed for both polymers (0.50-0.70V) are in the range of typical chemical oxidants such as Br_3^-, Br_2, etc. Thus it should be possible to use these reagents to selectively oxidize the polymers (see below). Secondly, the solution electrochemistry indicates that in going to the high coverage material there are slight shifts in the voltammetry peaks and an additional reduction peak appears at low voltages. These differences appear to be related to the high TTF site densities, and shorter coverage interdonor distances along the high coverage chain (see below).

Partial oxidation of the high coverage and low coverage polyers using the chemical oxidant Br_3^- in DMF was found to give solutions with the spectra shown in Figure 3. For comparison, the monomeric model compound[14] cation radical spectrum is shown in Figure 4. In obtaining these spectra, it was observed that the results were independent of the oxidation (chemical vs. electrochemical) procedure.

The data shows that the high coverage partially oxidized polymer exhibits new absorptions at 2μ and at 0.8μ, while the low coverage polymer only has a new absorption at 0.8μ, with no electronic transitions observed at lower energy. In contrast to this the monomeric species, oxidized under the same conditions, shows no electronic absorptions in this region throughout the complete range (0-100%) of donor oxidation.

From work on crystalline TTF salts in the solid state[15,16] these two new absorptions can be assigned as the dication dimer, D_2^{2+}, absorbing at 0.8μ,

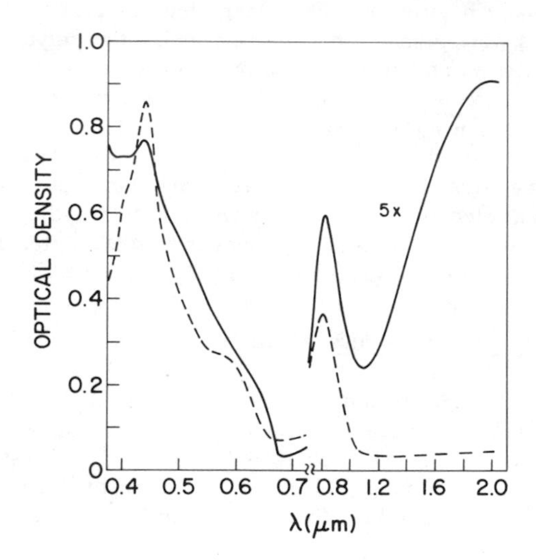

Figure 3. Visible–near IR spectra of high coverage (——) and low coverage (– – –) carboxy TTF polymer (2.5 × 10^{-3}M in DMF solution to which 2 mL of a 5 × 10^{-3}M Br_3^- solution were added)

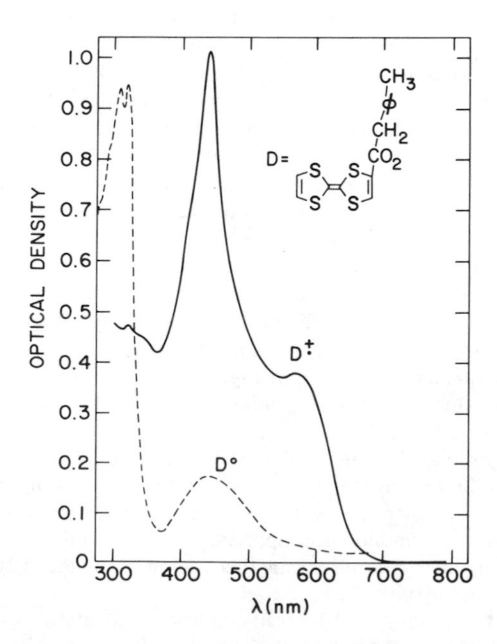

Figure 4. Monomeric visible–near IR spectra in DMF (neutral and oxidized species)

and a mixed valence aggregate $[D_2^+]_n$ absorbing at 2μ. No spectral evidence for aggregation is seen in the monomer spectra (Figure 4). Instead, the monomer solution at all stages of oxidation merely displays spectra typical of neutral and monomer TTF cation radical species. Thus, a prime prerequisite for the observation of these aggregate spectra in solution is the binding of TTF molecules along the polymer chain at some critical intrasite density.

This is the first time, to our knowledge, that ionized TTF clusters or aggregates have been observed in a fluid medium. From previous work, it can be inferred that an ordered crystalline matrix was necessary to stabilize a mixed-valence monomer array[17], while a low temperature solvent matrix[16] gave dimeric spectra. We interpret these results in the following way. Ion-radicals of the TTF class undoubtedly have large negative enthalpies for dimerization typical[18] of planar ion-radicals of this type. Due to the prohibitive positive entropy contribution to the free energy, dimeric (D_2^{2+}) aggregation is only observed under conditions where this term is reduced, i.e. at low temperatures or at some "enforced" critical site density.

Mixed valence aggregation, on the other hand, has never been observed[19] in low temperature matrices under conditions where dimeric aggregates do exist. Apparently, some minimum number of TTF molecules (greater than 2) appropriately oriented with respect to each other, are required for the stabilization of the mixed valence state. Both a crystalline matrix or a suitably constructed polymer could and seemingly does provide this environment.

With regard to the latter point, the absence of a mixed valence transition in the oxidized low coverage polymer case is an important point. No mixed valence transtiion was observed over the whole range of oxidation (0-100%) studied. This indicates that whereas the D_2^{2+} aggregate is stable under these conditions, the mixed valent dimer analog, D_2^+, is not. At least in these polymeric matrices, therefore, the stoichiometric requirement for observation of the mixed valence state appears to involve $(D_2^+)_n$ where $n \geq 2$.

These spectroscopically observed aggregates could be important to the mechanical, structural and electronic properties of the partially oxidized solid polymers. In order to begin to explore this possibility, we compared the use of two different oxidation procedures in preparing solid high coverage TTF polymer films. In the first method, quantitative chemical titration using Br_3^- in DMF solution was followed by the addition of the nonsolvent, ether, to precipitate out the partially oxidized polymer. The electronic spectrum of the polymer film produced (Figure 5) is similar to that found in solution with aggregate absorptions of 0.8μ and 2μ. In the second method, a solid film of the high coverage TTF polymer was deposited onto a transparent substrate from THF solution, and then oxidized using small amounts of Br_2 vapor. In all cases, the electronic spectrum obtained (Figure 6) showed only a dimer transition at 0.8μ.

We suggest that these results can be explained if the aggregation process in these solid TTF polymers proceeds by means of a two-step mechanism (Figure 7) in which the fast oxidative electron transfer step is followed by a slow process of ion clustering/reorganization which is favored by a low viscosity environment. This mechanism is consistent with the fact that the starting neutral homopolymer shows no spectroscopic evidence for site-site interaction between the pendant donors. The absorption spectrum of the polymer is

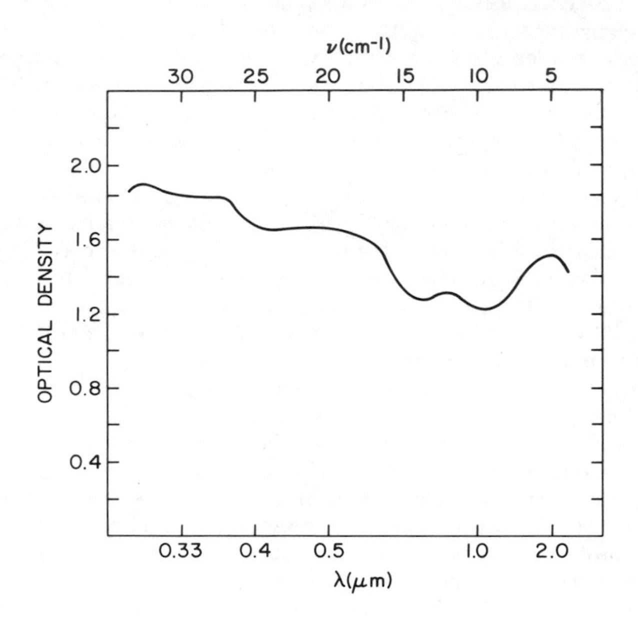

*Figure 5. Partially oxidized polymer film spectrum (*see *text for oxidation procedure)*

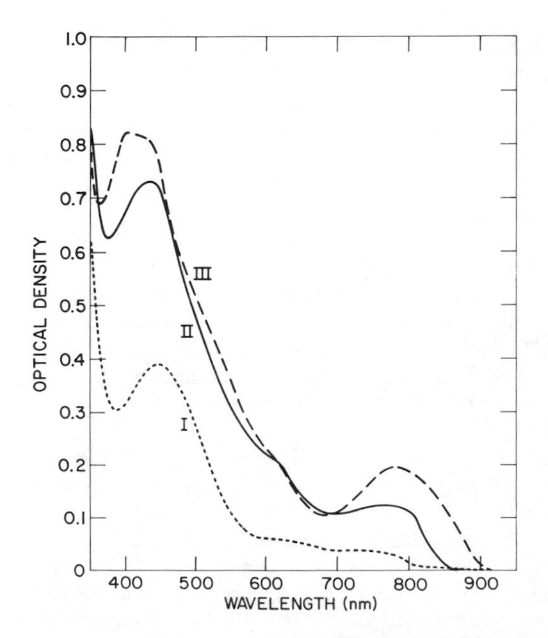

Figure 6. Changes in absorption spectrum of neural polymer film on exposure to Br_2 vapor: (I) no Br_2; (II, III) after increasing exposure times

virtually superimposable with that of the neutral monomer (Figure 4, dashed line). Thus, upon oxidation it would be expected that if this pictue were valid, spectroscopically observable polymer aggregates would form readily in low viscosity environments (such as solution), where structural rearrangements could easily take place to give specific donor-donor overlaps. In higher viscosity situations such as for solid films, on the other hand, only those aggregates with the smallest structural reorganization requirements would form. In fact, this is what has been experimentally observed since, in the high viscosity solid film case, only dimeric aggregates are formed. The mixed valence clusters which involve $n \geq 2$ molecules present too large a barrier to structural reorganization under these conditions and thus cannot form in the film environment.

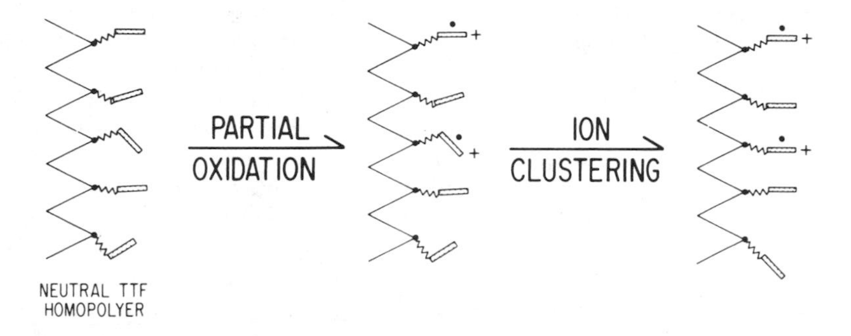

Figure 7. Proposed mechanism for polymer oxidation

B. Pyrazoline and TTF Cross-Linked Polymers

Although the cross-linked pyrazoline polymer is insoluble in common organic solvents by virtue of the crosslinks present, it was discovered that films of this material could be cast onto a variety of substrates from a suspension in a swelling solvent such as benzene. The ability to cast films is presumably related to the substantial change in polymer resin morphology after the coupling reaction has proceeded. The initial resin is hard, spherical and white in color, while the reacted bead is powdery, irregular in appearance and yellow, the color of the pyrazoline monomer.

Films of this polymer were coated onto metallic electrodes and examined for electrochemical response using the standard equipment and procedures typically applied to solution electrochemical investigations. The films were observed to be strongly electroactive: cyclic voltammograms were recorded (CH_3CN 0.1 N TEAP vs S. C. E.) which looked similar to the unbound monomeric pyrazoline in solution. In addition, it was observed during the course of the voltammetric experiment that the films changed color uniformly from yellow to green as the film was being oxidized. Exhaustive electrolysis at potentials from 0 to 1V., or open circuiting the cell in this potential range gave a film whose absorption spectrum (Figure 8), is typical of isolated pyrazoline cation, and which was constant with time. Reversal of the potential caused the film to go back to its original yellow state.

This phenomenon is interesting because it provides a new way of synthetically modifying the properties of an electrode surface. In this regard, it is important to understand how the electrochemical coloration process proceeds. From the optical changes observed, the number of electrochemical equivalents passed in these experiments, and from the measured weights of the films deposited on the electrode surfaces, we have determined that ~50% of the bound donors in these polymer films are oxidized under the conditions of the experiment. Since these are relatively thick films, this means that the electrochemical coloration is occurring in the film bulk and not in a few monolayers at the electrode surface.

In order to determine the generality of π-donor polymer films to undergo electrochemical oxidation, the film forming ability and the presence of electrochemical activity for TTF bound cross-linked polymers was studied. The data (Table III) shows that only a single case provided reasonably adherent, electrochemically active films. This suggests that some subtle, but important factors related to donor identity and chemical reaction conditions (i.e. to give a particular polymer morphology) are operating to give these desirable film forming properties. S.E.M. (scanning electron microscopy, 1μ resolution) studies of these cross-linked polymers showed important differences in morphology between the electroactive and inactive polymer materials. For the pyrazoline materials, the microscopy indicated a smooth, regular, high surface area polymeric material with little if any indication of structure or discontinuities. For all the inactive polymer systems, however, the films were grossly discontinuous and characterized by an appearance of inherent granularity. Even the electroactive TTF bead films appeared somewhat discontinuous in the SEM photographs.

TABLE III

Electrochemical Properties of Functionalized Polymer Bead Films[a]

Donor Polymer[b]	Film Formation	Electrochemical Activity
$-OC(=O)TTF$ (KOH-D/E)	-	-
$-OC(=O)TTF$ (CsOH-DMF)	-	-
$-OCH_2TTF$ (KOH-D/E)	+	-
$-OCH_2TTF$ (KH-THF)	+	+
$-O\phi TTF$ (KOH/D/E)	-	-
$-O\phi TTF$ (KH (THF)	++	++

a. Relative activity (+) or inactivity (-) observed

b. Functionalized Bead polymer (SX-1, 5.4 meq/1 gm starting chloromethyl coverage) with preparation conditions; solvents, D = Dioxane, E = Ethanol.

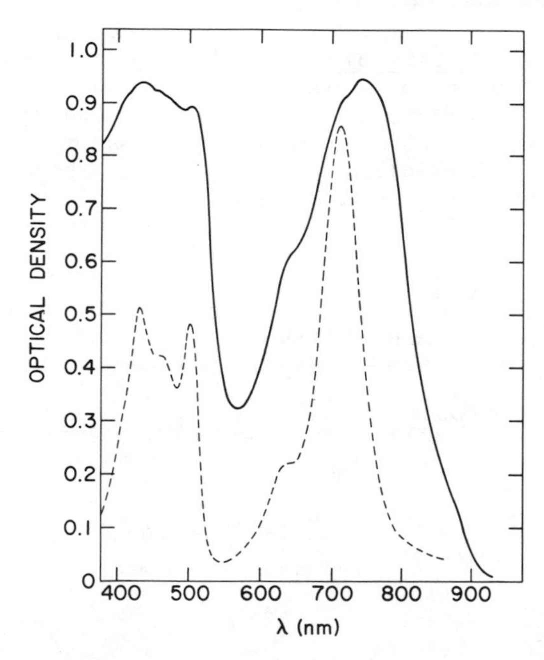

Figure 8. Absorption spectra of oxidized cross-linked polymer film (——) and pyrazoline monomer cation in CH_3CN *(– – –)*

These differences in film morphology were also reflected as differences in film formation conditions, film adhesion, and in electrochemical properties. The pyrazoline beads readily formed films from solvents such as benzene. For the phenoxy TTF system, however, only CH_2Cl_2 was effective in forming films. In general, the TTF cross-linked polymers were found to be less adherent to the metallized substrates than the pyrazoline cross-linked polymers. Electrochemically, it was found that the pyrazoline films showed complete activity after one potential sweep. The TTF polymer films, on the other hand, required from 5 to 20 cycles to reach full electrochemical activity as evidenced by a constant voltammogram with cycling. Furthermore, it was observed that the TTF polymer films were much less electroactive than the pyrazoline materials as shown by optical densities and total coulombs passed which were several times less for the TTF systems.

C. Mechanism of Charge Transport in Cross-Linked Donor Polymer Films

The electrochemistry of the solid donor polymer films involves ejection of an electron to form pyrazoline or TTF cations. This means some kind of electron transport mechanism[20] must exist to move electrons in the film bulk between the film extremities and the electrode surface (see schematic, Figure 9). We suggest that the important requirement for charge transport in these materials is the mobility of electroactive groups attached to the polymer chain. This suggestion is consistent with previous work using chemical[4] and physical[21] probes which established that in lightly cross-linked polystyrene beads, in contact with appropriate solvents, these polymers have bound molecular groups which possess considerable flexibility.

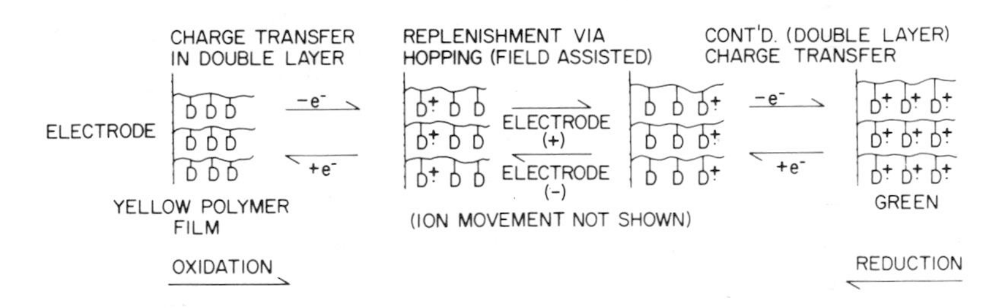

Figure 9. Schematic of electron transport at metal–pyrazoline cross-linked polymer interfaces

An important point is that the electrochemically driven charge transport in these polymeric materials is not dependent on the presence of mixed valence interactions which are well known to give rise to electronic conductivity[17] in a number of cation radical crystalline salts. This is clearly seen from the absorption spectrum of the electrochemically oxidized pyrazoline films (Figure 8) which show no evidence for the mixed valence states that are the structural electronic prerequisites for electrical conductivity in the crystalline salts. A more definitive confirmation of this point is provided by the absorption spectrum (Figure 10) of electrochemically oxidized TTF polymer films which shows

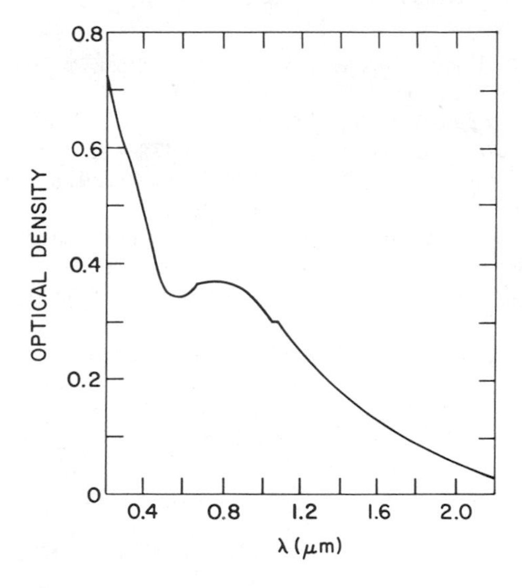

Figure 10. Spectrum of electrochemically oxidized cross-linked TTF polymer films

only absorptions due to TTF^+ and $(TTF^+)_2$. In the TTF system where mixed valence absorption have been identified in the solution studies on the polymeric systems (Part IIIA) and are known to lead to high electrical conductivities in the crystalline salts[15], the absence of these absorptions in the solid film polymer electrochemistry experiments suggests that stable mixed valence clusters are not essential to the charge transport process.

The electrochemical and spectroscopic data indicates that sites on these polymers can communicate with each other, in the electron transfer sense, on a relatively short time scale and without the formation of stable mixed valence clusters. Electronic tranport via hopping or tunnelling and modulated by means of neighboring molecular group collisions would be consistent with these requirements. The relative molecular nonspecificity of this mechanism suggests that other polymeric materials would show similar effects and this has been seen for thin films of poly[22](vinylferrocene) and poly[23](nitrostyrene).

IV. Conclusions

Electroactive donors, such as TTF or triarylpyrazoline, can be bound in high yield to polymeric matrices. The TTF linear polymers show interesting cooperative properties (i.e., ion-radical cluster formation) that is not observed for the isolated monomers in solution or the low coverage polymers. Furthermore, thin solid films of these donors bound to cross-linked polymer backbones display remarkably facile charge transport through the film bulk which is accompanied by dramatic and reversible optical changes.

Literature Cited

1. Engler, E. M., Chem. Tech. 1976, 6, 274. Pragst, F. and Bock, C., J. Electroanal, Chem. 1975, 61, 47.
2. Miller J. S., Epstein A. J., Eds. Annals N. Y. Acad. Sci., "Synthesis and Properties of Low Dimensional Materials" 1978 Vol. 313
3. Gibson H. W., and Bailey F. C., Macromolecules 1976 9, 688. Cohen H. L., J. Polym. Sci. Polym. Chem Ed. 1976 14, 863.
4. See Crowley J. I., and Rappoport H., Acc. Chem. Res. 1976, 9, 135 for a recent review.
5. Welcome J. R., Theses M. S., Department of Chemistry, SUNY at Buffalo, 1973.
6. Ueno Y., Masuyama Y., and Okawara M., Chem. Lett., 1975, 603.
7. Pittman C. U. Jr., Narita M., Liang Y. F., Macromolecules, 1976 9, 360.
8. Hertler W. R., J. Org. Chem., 1976 41, 1412.
9. Kossmehl G., Rohde M., Makromol. Chem., 1977 178, 715.
10. Green D. C. and Kaufman F. B., IBM Tech. Discl. Bull. 1977 20, 2865.
11. Kaplan M. L., Haddon R. C., Wudl F., Feit E. D., J. Org. Chem., 1978 43, 4642.
12. Green D. C. and Allen R. W., Chem. Commun., 1978, 832.
13. 80Å Pt Electron Beam Evaporated onto 30Å Nb.
14. Prepared in similar manner as polymer; cf ref. 20.
15. Scott B. A., LaPlaca S. J, Torrance J. B., Silverman B. D. and Welber B., J. Amer. Chem. Soc. 1977 99, 6631.
16. Torrance J. B., Scott B. A. , Welber B., Kaufman F. B. and Seiden P. E., Phys. Rev. (B), 1979 19, 730.
17. Torrance J. B., Acc. Chem. Res. 1979 12, 79.
18. Kaufman F. B., J. Amer. Chem. Soc., 1976 98, 5339.
19. Kaufman F. B., unpublished results.
20. Kaufman F. B. and Engler E. M., J. Amer. Chem. Soc., 1979 101, 547.
21. Veksli Z., Miller W. G. and Thomas E. L., J. Polymer Sci. Symp. No. 54, 1976, 299.
22. Merz A. and Bard A. J., J. Amer. Chem. 1978 100, 3222.
23. VanDeMark M. R. and Miller L. L., ibid, 1978 100, 3223.

RECEIVED July 12, 1979.

30

Structure and Property Modification of *m*-Carborane Siloxanes

EDWARD N. PETERS

Union Carbide Corporation, One River Road, Bound Brook, NJ 08805

Carborane-siloxanes are a family of polymers which have the linear structure, I,

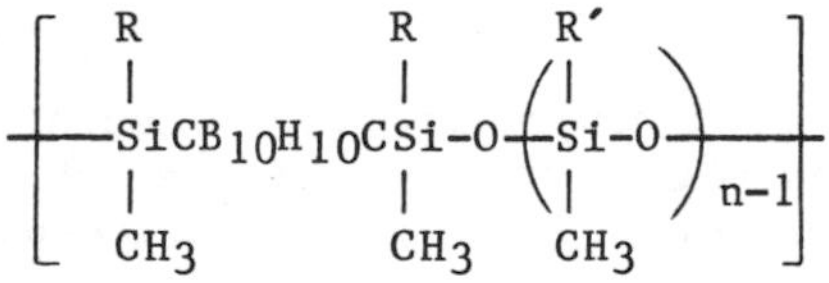

$$\left[-\underset{CH_3}{\overset{R}{\underset{|}{\overset{|}{Si}}}}CB_{10}H_{10}C\underset{CH_3}{\overset{R}{\underset{|}{\overset{|}{Si}}}}-O\left(-\underset{CH_3}{\overset{R'}{\underset{|}{\overset{|}{Si}}}}-O-\right)_{n-1} \right]$$

I

where R and R′ can be methyl or fluoroalkyl, and in addition R′ can be phenyl. The interest in carborane-siloxanes centers around the need for new polymers having enhanced flame resistance, and greater thermal and oxidative stability (1). Indeed the incorporation of the m-carborane moiety into the siloxane backbone has resulted in significant enhancement of properties.

Several approaches are available for modifying polymer properties in order to maximize the performance of a polymeric system for a particular end use. These methods include polymer backbone modification, polymer blends, grafting, and the use of additives such as reinforcements, cross-linking agents, stabilizers, plasticizers, etc. The properties of carborane-siloxane polymers can be modified and optimized through a combination of structural changes and the use of supplementary agents. The work from several laboratories including Olin Corporation, Princeton University, and Union Carbide Corporation will be reviewed in this chapter.

Backbone Modification

Different structures of carbone-siloxane polymers are obtainable via several synthetic routes which have been developed.

0-8412-0540-X/80/47-121-449$05.00/0

Polymers (I) where n is 3 and 5 can be prepared by hydrolysis-condensation of a carborane-based siloxane as shown in equation 1 (2).

$$\mathrm{Cl}{-}\left(\underset{\mathrm{CH_3}}{\overset{\mathrm{R'}}{\mathrm{Si}}}{-}\mathrm{O}\right)_{\frac{n-1}{2}}\underset{\mathrm{CH_3}}{\overset{\mathrm{R}}{\mathrm{Si}}}\mathrm{CB_{10}H_{10}C}\underset{\mathrm{CH_3}}{\overset{\mathrm{R}}{\mathrm{Si}}}\left(\mathrm{O}{-}\underset{\mathrm{CH_3}}{\overset{\mathrm{R'}}{\mathrm{Si}}}\right)_{\frac{n-1}{2}}{-}\mathrm{Cl} \xrightarrow{H_2O} \underline{\mathrm{I}} \qquad (1)$$

Polymers (I) where n is 4, 5, and 6 are prepared by the cohydrolysis-condensation of a bis(chlorodisiloxyl)carborane and a dichlorosilicon compound which appear in equation 2 (2).

$$\mathrm{Cl}{-}\underset{\mathrm{CH_3}}{\overset{\mathrm{R'}}{\mathrm{Si}}}{-}\mathrm{O}{-}\underset{\mathrm{CH_3}}{\overset{\mathrm{R}}{\mathrm{Si}}}\mathrm{CB_{10}H_{10}C}\underset{\mathrm{CH_3}}{\overset{\mathrm{R}}{\mathrm{Si}}}{-}\mathrm{O}{-}\underset{\mathrm{CH_3}}{\overset{\mathrm{R'}}{\mathrm{Si}}}{-}\mathrm{Cl} + \mathrm{Cl}{-}\underset{\mathrm{CH_3}}{\overset{\mathrm{R'}}{\mathrm{Si}}}\left(\mathrm{O}{-}\underset{\mathrm{CH_3}}{\overset{\mathrm{R'}}{\mathrm{Si}}}\right)_{n-4}{-}\mathrm{Cl} \xrightarrow{H_2O} \underline{\mathrm{I}} \qquad (2)$$

The polymer where n is 2 can be prepared according to equation 3 in which a bis(hydroxysilyl)carborane is reacted with a bisureido silane where X is an N-phenyl-N′-tetramethyleneureido group (3).

$$\mathrm{HO}{-}\underset{\mathrm{CH_3}}{\overset{\mathrm{R}}{\mathrm{Si}}}\mathrm{CB_{10}H_{10}C}\underset{\mathrm{CH_3}}{\overset{\mathrm{R}}{\mathrm{Si}}}{-}\mathrm{OH} + \mathrm{X}{-}\underset{\mathrm{CH_3}}{\overset{\mathrm{R'}}{\mathrm{Si}}}{-}\mathrm{X} \longrightarrow \underline{\mathrm{I}} \qquad (3)$$

Equation 4 shows the ferric chloride catalyzed synthesis of polymer where n is 1 (4).

$$\mathrm{Cl}{-}\underset{\mathrm{CH_3}}{\overset{\mathrm{R}}{\mathrm{Si}}}\mathrm{CB_{10}H_{10}C}\underset{\mathrm{CH_3}}{\overset{\mathrm{R}}{\mathrm{Si}}}{-}\mathrm{Cl} + \mathrm{CH_3O}{-}\underset{\mathrm{CH_3}}{\overset{\mathrm{R}}{\mathrm{Si}}}\mathrm{CB_{10}H_{10}C}\underset{\mathrm{CH_3}}{\overset{\mathrm{R}}{\mathrm{Si}}}{-}\mathrm{OCH_3} \xrightarrow[\Delta]{FeCl_3} \underline{\mathrm{I}} \qquad (4)$$

Transition Temperatures

The two main transitions in polymers are the glass-rubber transition (Tg) and the crystalline melting point (Tm). The Tg is the most important basic parameter of an amorphous polymer because it determines whether the material will be a hard solid or an elastomer at specific use temperature ranges and at what temperature its behavior pattern changes.

Physical transitions for a series of linear carborane-siloxane polymers, I, have been examined and the data appear in Table I.

TABLE I

Transitions and Stability of Carborane-Siloxanes

n	R	R′	Tg, °C	Tm, °C	Wt. Loss by 700°C
1	CH_3	--	25	260	20%
2	CH_3	CH_3	-42	90	32%
2	CH_3	C_6H_5	-12	A	~5%
2	CH_3	CH_3 (33) C_6H_5 (67)	-22	A	~5%
2	CH_3	CH_3 (67) C_6H_5 (33)	-27	A	~5%
3	CH_3	CH_3	-68	A	45%
4	CH_3	CH_3	-75	A	48%
5	CH_3	CH_3	-88	A	60%
1	$CH_2CH_2CF_3$	--	28	A	--
2	CH_3	$CH_2CH_2CF_3$	-29	A	25%
2	$CH_2CH_2CF_3$	CH_3	-12	A	59%
2	$CH_2CH_2CF_3$	$CH_2CH_2CF_3$	-3	A	69%
3	$CH_2CH_2CF_3$	$CH_2CH_2CF_3$	-15	A	78%

A = Absent

For the homologous series of carborane-dimethylsiloxanes (I, R = R′= CH_3, n = 1 through 5), the glass transition temperature decreases with increasing siloxane content (i.e., greater n) (5). When n is 1 or 2 the polymers have a crystalline melting point. This crystalline phase adversely affects elastomeric properties. In the case where n = 2, this crystallinity can be eliminated by the incorporation of phenyl groups on the polymer backbone (R′ = C_6H_5). Thus, phenyl modification results in a completely amorphous system; however, the Tg has risen from -42° to -12°C. Partial phenyl modification (R′ = CH_3/C_6H_5 [67/33]) produces an amorphous polymer with a low Tg (-37°C) (6). Polymers where n = 3,4, and 5 are amorphous with decreasing Tg′s of -68°, -75°, and -88°C, respectively. The trifluoropropyl modified polymers are amorphous and exibit decreasing Tg with increasing n and increasing Tg with increasing $CH_2CH_2CF_3$ content (7, 8).

A linear relationship was obtained for the homologous series of carborane-dimethylsiloxanes and carborane-methyltrifluoropropylsiloxanes by considering their structures to be an alternating copolymer of $\leftarrow O-SiR_2 \rightarrow_n$ and $\leftarrow CB_{10}H_{10}CSiR_2 \rightarrow$ linkages and by plotting weight fraction of $\leftarrow CB_{10}H_{10}CSiR_2 \rightarrow$ versus 1/Tg as shown in Figure 1 (1, 5).

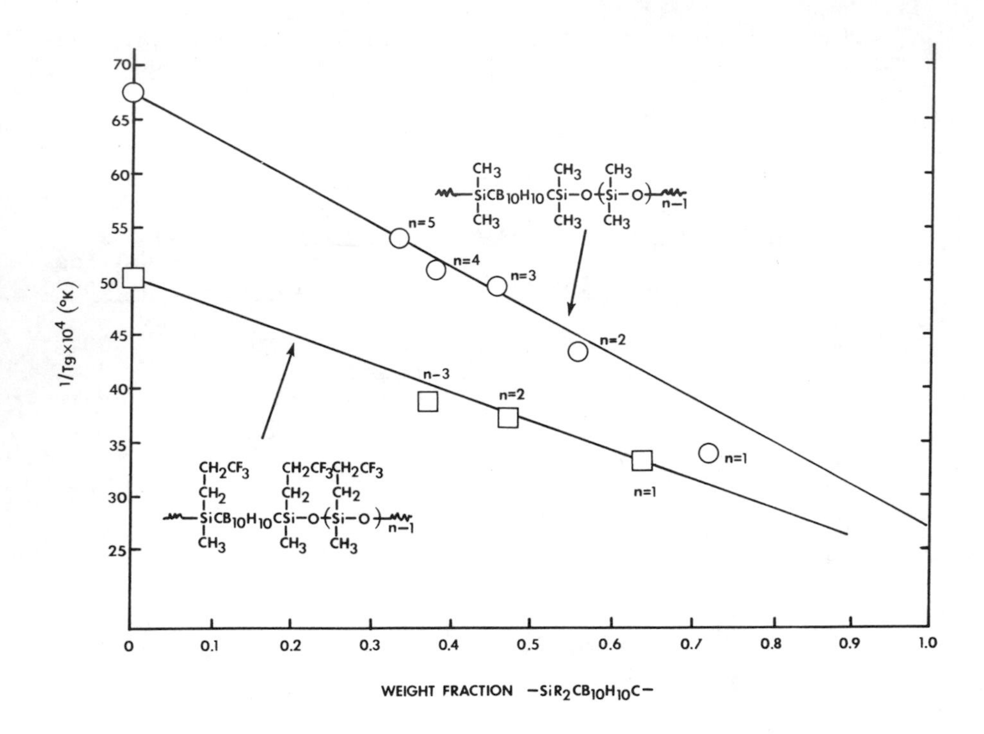

Figure 1. Correlation of glass transition temperatures

Thermal Stability

One of the unique properties of carborane-siloxane polymers is their outstanding thermal stability. Thermogravimetric analyses in an inert atmosphere reveal that rapid weight loss does not occur until above 400°C. For the homologous series of carborane-dimethylsiloxanes the stability of the polymers increases with decreasing siloxane content per repeat unit (5, 9). The total weight lost by 700°C appear in Table I. A correlation of weight loss as a function of siloxane linkages per repeat unit is shown in Figure 2.

The phenyl modified polymers show a significant decrease in weight loss compared to their all methyl analog (6). In a study of the thermal breakdown of phenyl substituted carborane-siloxane polymers, it has been reported that the presence of phenyl groups in carborane-siloxanes leads to cross-linking and less loss of weight (9).

The presence of trifluoropropyl groups on the polymer backbone leads to an increase in weight loss (7, 8).

Thus, by modifying the polymer backbone the stability and lower use temperature can be varied.

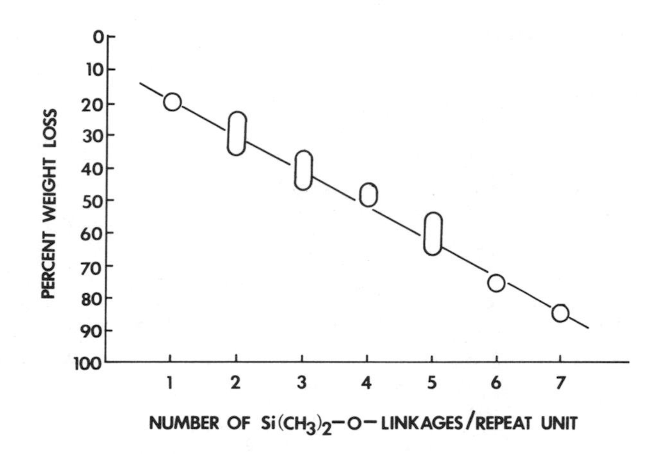

Figure 2. Correlation of weight loss at 700°C

Formulation Parameters

The phenyl modified polymers possess the optimum combination of high temperature and elastomeric properties and were used in the study of formulation parameters. These variables can have an important effect on the thermal stability and property profile of vulcanized systems. For example, the use of reinforcing silicas, peroxide content, and oxidative stabilizers have been shown to be important (3, 10, 11). However, polymer-silica interactions had the most pronounced effect on retaining properties during high temperature aging studies.

As shown in Figure 3, heat aging at 315°C in air of a sample reinforced with hydrophilic silica (amorphous) leads to a rapid loss of elongation after 150 hours. However, modifying the silica by trimethylsiloxation (hydrophobic silica) results in substantial improvements--even after 1000 hours at 315°C in air, elastomeric properties were retained.

The adverse effect of the hydrophilic silica was attributed to the condensation reaction of surface silanol groups on the silica and phenylsilane moieties on the polymer backbone. This results in increased cross-linking via formation of siloxane bonds between the polymer and silica.

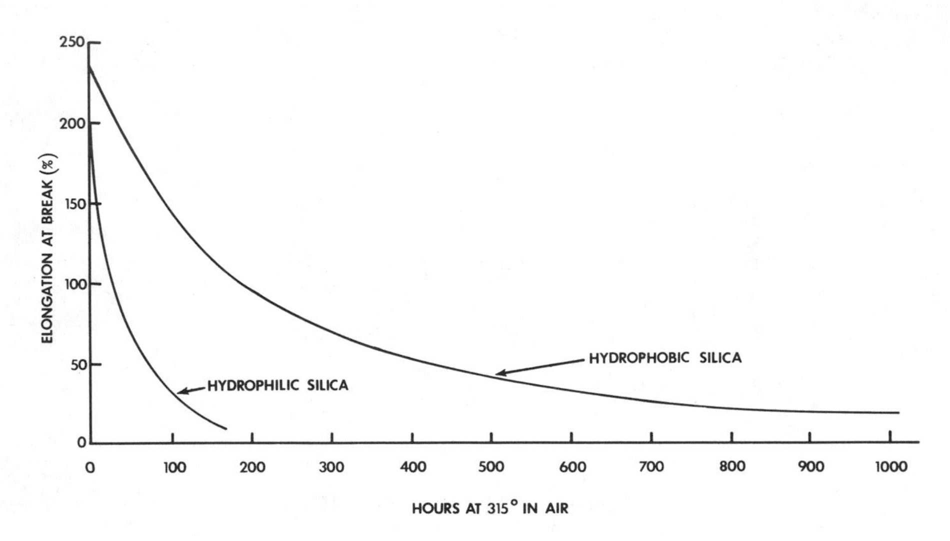

Figure 3. Elongation after heat aging as a function of type of reinforcement

Solvent Resistance

Resistance to solvents is important in several end use situations. The degree of swelling in various solvents can be modified by changes in the polymer backbone and the type of reinforcement. The effect of hydrophobic and hydrophilic reinforcing silicas on the swelling of vulcanized trifluoropropyl-modified carborane-siloxane polymer (I, n = 2, R = CH_3, R′ = $CH_2CH_2CF_3$) appears in Table II. In acetone the swelling is significantly less when hydrophilic silica is used in reinforcement. A similar trend was noted in toluene.

TABLE II

Effect of Type of Reinforcing Agent on Swelling

Solvent	Swelling, %* Hydrophobic Silica	Hydrophilic Silica
Acetone	71	2
Toluene	130	89

*ASTM D-471; Vulcanizates with 30 phr silica.

Modification of the polymer backbone by the incorporation of trifluoropropyl groups leads to substantial decreases in swelling. In vulcanized systems reinforced with hydrophilic silica (30 phr) the swelling decreased with increasing $CH_2CH_2CF_3$ content as shown in Table III.

TABLE III

Swelling as a Function of Polymer Modification

n	R	R′	Swelling, %	
			Toluene	Reference Fuel B
2	CH_3	CH_3 (67) C_6H_5 (33)	160	113
2	CH_3	$CH_2CH_2CF_3$	98	80
2	$CH_2CH_2CF_3$	CH_3	63	43
2	$CH_2CH_2CF_3$	$CH_2CH_2CF_3$	23	20

Conclusions

Properties of carborane-siloxanes can be readily modified through changes in the polymer backbone and by the use of supplementary agents in order to optimize the property-performance profile for different use situations.

Thus a strong technology base now exists for carborane-siloxane polymers which adds measureably to the arsenal of design engineers for advanced systems. The visionary support of the Office of Naval Research has been largely responsible for the important advancements in this area.

Literature Cited

1. Peters, E. N., J. Macromol. Sci., Rev. Macromol. Chem., in press.
2. Knollmueller, K. O.; Scott, R. N.; Kwasnik, H.; Sieckhaus, J. F.; J. Polym. Sci., A-1, 1971, 9, 1071.
3. Peters, E. N.; Hedaya, E; Kawakami, J. H.; Kwiatkowski, G. T.; McNeil, D. W.; Tulis, R. W.; Rubber Chem. Technol., 1975, 48, 14.
4. Papetti, S; Schaeffer, B. B.; Gray, A. P.; Heying, T. L.; J. Polym. Sci., A-1, 1966, 4, 1623.
5. Roller, M. B. and Gillham, J. K., Polym. Eng. Sci., 1974, 8, 567.

6. Peters, E. N.; Kawakami, J. H.; Kwiatkowski, G. T.; McNeil, D. W.; Hedaya, E; J. Polym. Sci., Polym. Phys. Ed., 1977, 15, 723.
7. Scott, R. N.; Knollmueller, K. O.; Hooks, Jr., H.; Sieckhaus, J. F.; J. Polym. Sci., A-1, 1972, 10, 2303.
8. Peters, E. N.; Stewart, D. D.; Bohan, J. J. Moffitt, R.; Beard, C. D.; J. Polym. Sci., Polym. Chem. Ed., 1977, 15, 973.
9. Andrianov, K. A.; Pavlova, S.-S. A.; Zhuravleva, I. V.; Tolchinskii, Yu. I.; Astapov, B. A.; Polym. Sci. U.S.S.R., 1977, 19, 1037.
10. Peters, E. N.; Stewart, D. D.; Bohan, J. J.; Kwiatkowski, G. T.; Beard, C. D.; Moffitt, R.; Hedaya, E.; J. Elastomers Plast., 1977, 9, 177.
11. Peters, E. N.; Stewart, D. D.; Bohan, J. J.; McNeil, D. W.; J. Elastomers Plast., 1978, 10, 29.

RECEIVED July 12, 1979.

Siloxane-Modified Polyarylene Carbonates

HAROLD ROSENBERG, TSU-TZU TSAI, BORIS D. NAHLOVSKY, and CSABA A. KOVACS[1]

Nonmetallic Materials Division, Air Force Materials Laboratory, Wright-Patterson Air Force Base, OH 45433

Polycarbonates and, in particular, the polycarbonate derived from bisphenol A, are useful as transparent and impact-resistant structural plastics (1). However, further modification of the arylene carbonate backbone by incorporation of siloxane segments has been of interest and has been investigated by a number of workers during the last decade. The main objective of such modification has been to decrease the glass-transition temperature of the polycarbonate and thereby obtain transparent elastomeric polymers. Such macromolecules are hybrids between arylene carbonate and siloxane polymers. Two main types of hybrid arylene carbonate-siloxane polymers were obtained in these investigations: (a) block copolymers with backbones comprised of arylene carbonate and siloxane blocks, and (b) alternating homopolymers, I, the backbones of which consisted of regularly alternating siloxanylene and arylene carbonate segments. While the length of the

$$\left[-Ar-O-\overset{\overset{O}{\|}}{C}-O-Ar-\underset{R^2}{\overset{R^1}{Si}}\left(-O-\underset{R^4}{\overset{R^3}{Si}}- \right)_x O-\underset{R^2}{\overset{R^1}{Si}}- \right]_n \qquad \text{I}$$

R^1, R^2 = CH_3 or C_6H_5

R^3, R^4 = same or different than R^1 or R^2

x = 0 - 3

Ar = $\underline{p}$-C_6H_4, $\underline{m}$-C_6H_4, $\underline{p}$-C_6H_4-$C(CH_3)_2$-$\underline{p}$-C_6H_4

blocks in the block copolymers investigated varied within the given copolymer and usually contained several tens or hundreds

[1] Current address: Research Laboratory, Kodak Park Division, Eastman Kodak Company, Rochester, NY 14650.

of Si and O atoms (i.e., as polysiloxane moieties), the length of the siloxanylene segments in the alternating homopolymers of interest, I, is much shorter (x = 0 - 3). Due to the different structure of the backbone in the arylene carbonate-siloxane block copolymers and alternating homopolymers, I, both types of polymers would be expected to exhibit somewhat different physical and chemical properties.

The majority of those involved in research on siloxane-modified poly(arylene carbonates) have chosen to synthesize copolymers of arylene carbonates and polysiloxanes (2-17). Various properties (3,4,5,15,16,18-32) as well as the morphology (3,18,19,23,33) of such copolymers were investigated in great detail. Synthesis of these block copolymers , while complex, is potentially not costly since materials and methods used in the manufacturing of polysiloxanes and polycarbonates can be used. The properties of the arylene carbonate-siloxane block copolymers synthesized vary from those of elastomers to tough plastics, depending on the length and relative concentration of both blocks in such copolymers. The improved mechanical strength of these copolymers was attributed to formation of microdomains containing arylene carbonate blocks which act as virtual crosslinks and filler particles (19). Arylene carbonate-siloxane block copolymers are thus considered potentially useful as tough elastomeric sheets and membranes (15,34,35). However, most copolymers of this type contain hydrolytically-unstable Si-O-C linkages. Moreover, thermal stability of these copolymers does not exceed the thermal stability of polysiloxanes since thermal degradation of the polymer backbone can occur by chain scission of the siloxane blocks leading to formation of low-molecular weight cyclic decomposition products.

To-date, only scant reports have been published on attempted synthesis of alternating homopolymers of the type I (36,37,38,39, 40), and no well-characterized high molecular weight polymers of the type I are reported in the literature. Preparation of such polymers requires initially the synthesis of new and more complex monomers. It appeared that a number of obstacles had to be overcome to permit synthesis of the monomers and their polymerization. Our efforts to overcome such obstacles and modify aromatic polycarbonates by incorporation of siloxanylene segments are described herein. It was the objective of this investigation to modify the backbone of aromatic polycarbonates in this manner so as to lower the glass-transition temperature (T_g) and obtain polymers with elastomeric properties while retaining desired thermal stability and transparency. To achieve this goal, it was considered essential to synthesize a number of polymers of the type I in order to derive structure-properties correlations for the tailoring of such polymer structures.

In this research three synthetic approaches leading to siloxane-modified poly(arylene carbonates) were investigated: (1) the polycondensation of siloxanylene-linked bisphenols, II, with phosgene (37,38,39,40); (2) the homopolycondensation of bis-silanols containing phenylene carbonate groups, III, (36); and (3) the heteropolycondensation of bis-silanols, III, with diacetoxysilanes, diacetoxydisiloxanes, or diacetoxytrisiloxanes.

Results and Discussion

1. Polycondensation of Bisphenols, II, with Phosgene. Polycondensation of siloxane-linked bisphenols, II, with phosgene is the most obvious synthetic approach leading to siloxane-modified poly(arylene carbonates) since the phosgene-bisphenol polycondensation is used in the synthesis of aromatic polycarbonates (1). This method was used initially to prepare polymer 1 (as indicated in reaction 1) as well as for the attempted synthesis of polymers 2 and 5:

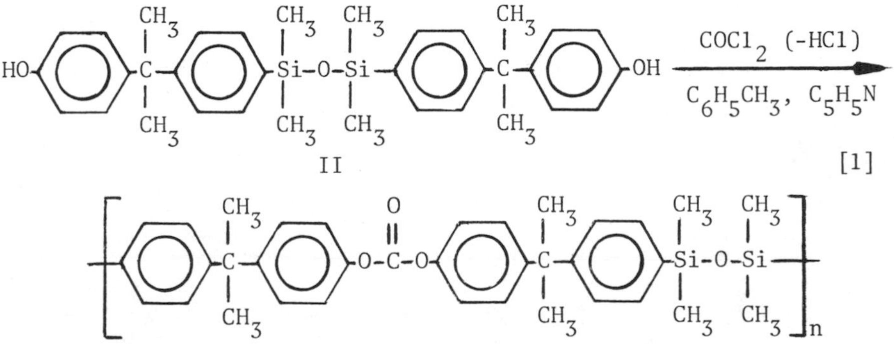

$$HO-C_6H_4-C(CH_3)_2-C_6H_4-Si(CH_3)_2-O-Si(CH_3)_2-C_6H_4-C(CH_3)_2-C_6H_4-OH \xrightarrow[C_6H_5CH_3,\ C_5H_5N]{COCl_2\ (-HCl)} \quad [1]$$

II

$$\left[-C_6H_4-C(CH_3)_2-C_6H_4-O-\overset{O}{\overset{\|}{C}}-O-C_6H_4-C(CH_3)_2-C_6H_4-Si(CH_3)_2-O-Si(CH_3)_2-\right]_n$$

1

However, the glass-transition temperature of 80°C obtained for polymer 1 was determined to be too high for the bis(phenylene)-isopropylidene system, even if modified with longer siloxanylene segments, to be useful as an elastomer. Moreover, the synthetic route based on the bisphenol-phosgene polycondensation (reaction 1) gave polymers of relatively low molecular weight. For these reasons it was decided to concentrate on the synthesis of a series of polymers 2-12 (Table I) containing p- and m-phenylene groups and siloxanylene segments of varying length while investigating alternate synthetic approaches (reactions 2 and 3).

2. Homopolycondensation of Bis-silanols, III. Homopolycondensation of bis-silanols, III, was a second choice as a synthetic approach leading to siloxane-modified poly(arylene carbonates) (reaction 2). Phosgene in pyridine was found to be the most effective medium of a number of related reagents (oxalyl chloride, acetyl chloride, HCl) having potential for promoting the homopolycondensation of bis-silanols, III. Polymers obtained through its use exhibited higher molecular weights

TABLE I
PROPERTIES OF SYNTHESIZED POLY(SILOXANYLENEARYLENE CARBONATES)

$$\left[-\text{Ar}-\text{O}-\overset{\overset{\text{O}}{\parallel}}{\text{C}}-\text{O}-\text{Ar}-\underset{\underset{R^2}{|}}{\overset{\overset{R^1}{|}}{\text{Si}}}\left(-\text{O}-\underset{\underset{R^4}{|}}{\overset{\overset{R^3}{|}}{\text{Si}}}-\right)_x-\text{O}-\underset{\underset{R^2}{|}}{\overset{\overset{R^1}{|}}{\text{Si}}}- \right]_n$$

Polym. No.	Ar	x	R^1	R^2	R^3	R^4	Synth. Meth.[a]	$\overline{M}_w$[b]	$[\eta]$[c]	T_g[d] (°C)	T_d[e] (°C)
1	$-C_6H_4-C(CH_3)_2-C_6H_4-$	0	CH_3	CH_3	---	---	A	---	0.43	80	---
2		0	CH_3	CH_3	---	---	B	630,000	2.07	49	421
3		1	CH_3	CH_3	CH_3	CH_3	C	60,000	0.42	-1	433
4		2	CH_3	CH_3	CH_3	CH_3	B	---	1.40[f]	-24	456
							C	64,000	0.31	--	---
5		0	CH_3	CH_3	---	---	B	---	1.30[f]	14	455
6		1	CH_3	CH_3	CH_3	CH_3	C	38,000	0.24[f]	-19	401

TABLE I (CONTINUED)

Polym. No.	Ar	x	R^1	R^2	R^3	R^4	Synth. Meth.[a]	$\overline{M}_w$ [b]	$[\eta]$ [c]	T_g [d] (°C)	T_d [e] (°C)
7	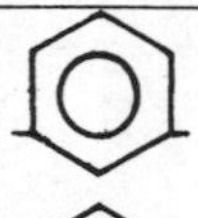	2	CH_3	CH_3	CH_3	CH_3	B	---	0.87[f]	-35	402
							C	50,000	0.40	--	---
8		2	CH_3	CH_3	C_6H_5	CH_3	C	121,000	0.43	-5	422
9		2	C_6H_5	CH_3	CH_3	CH_3	C	22,000	0.25	-1	421
10		2	C_6H_5	CH_3	C_6H_5	CH_3	C	20,000	0.21	18	445
11		3	CH_3	CH_3	C_6H_5	CH_3	C	81,000	0.31	-7	440
12		3	C_6H_5	CH_3	C_6H_5	CH_3	C	20,000	0.20	11	385

(a) A - Polycondensation of bisphenols, II, with phosgene; B - phosgene-induced homopolycondensation of bis-silanols, III; C - heteropolycondensation of bis-silanols, III, with diacetoxysilanes or diacetoxydi-(or tri-)siloxanes; (b) corresponds to the peak on the GPC curve; (c) THF, 25°C; (d) DSC; (e) "decomposition temperature" determined as the intercept of the tangent of the steepest part of the TGA curve (obtained in a helium atmosphere at a heating rate of 20°C/min.), representing the first major weight loss; (f) toluene, 30°C.

$$HO\text{-}Si(CH_3)_2\text{-}\left(O\text{-}Si(CH_3)_2\right)_y\text{-}C_6H_4\text{-}O\text{-}\overset{O}{\overset{\|}{C}}\text{-}O\text{-}C_6H_4\text{-}\left(Si(CH_3)_2\text{-}O\right)_y\text{-}Si(CH_3)_2\text{-}OH \xrightarrow[C_5H_5N]{COCl_2(-H_2O)}$$

III [2]

$$\left[\text{-}C_6H_4\text{-}O\text{-}\overset{O}{\overset{\|}{C}}\text{-}O\text{-}C_6H_4\text{-}Si(CH_3)_2\text{-}O\text{-}\left(Si(CH_3)_2\text{-}O\right)_{2y}\text{-}Si(CH_3)_2\text{-}\right]_n$$

(Table I) than those obtained by use of other reagents. The main limitation of this approach is that only polymers containing an even number of silicon atoms in the siloxanylene segment can be synthesized; polymers 2, 4, 5, and 7 (Table I) were prepared by this method (reaction 2, y = 0 or 1). Another disadvantage of this approach is that the synthesis of polymers containing a tetrasiloxanylene segment (polymers 4 and 7), which are of particular interest due to their low value of glass-transition temperature, requires a complex synthesis of bis-silanols, III (y=1). At present, this method leads to siloxane-modified poly(arylene carbonates) with higher molecular weights than are obtained by all other methods investigated in this laboratory.

3. Heteropolycondensation of Bis-silanols, III, with Diacetoxysilanes. Heteropolycondensation of bis-silanols, III, was investigated since this approach appeared to offer a relatively rapid preparative route to a number of various siloxane-modified poly(arylene carbonates) which were required in order to obtain structure-properties correlations for such polymers. In addition, polymers containing an odd number of silicon atoms in the siloxanylene segment can be prepared by this approach. However, reaction conditions employed in condensation of bis-silanols, III, with diaminosilanes (41) or siloxazanes (42) led to the cleavage of the carbonate group. Consequently, the recently reported silanol-acetoxysilane polycondensation reaction (43) was investigated for the polycondensation of bis-silanols, III, with diacetoxysilanes or diacetoxydisiloxanes (reaction 3).

It was found that the reaction conditions which were optimized for the synthesis of poly(arylene siloxanylenes) (43) could be employed for the synthesis of siloxane-modified poly-(arylene carbonates). 2,4,6-Trimethylpyridine (collidine) was selected as the most suitable of all catalysts investigated (43) for the synthesis of the siloxane modified poly(arylene carbonates). Properties of polymers prepared by this method are given in Table I. In comparision to the phosgene-catalyzed homopolycondensation of bis-silanols, III, the inherent viscosities

$$HO-\underset{CH_3}{\overset{CH_3}{|}}{Si}-C_6H_4-O-\overset{O}{\overset{\|}{C}}-O-C_6H_4-\underset{CH_3}{\overset{CH_3}{|}}{Si}-OH + CH_3CO\left(O-\underset{CH_3}{\overset{CH_3}{|}}{Si}\right)_x OOCCH_3 \xrightarrow[-CH_3COOH]{base}$$

III

$$\left[C_6H_4-O-\overset{O}{\overset{\|}{C}}-O-C_6H_4-\underset{CH_3}{\overset{CH_3}{|}}{Si}-O\left(\underset{CH_3}{\overset{CH_3}{|}}{Si}-O\right)_x\underset{CH_3}{\overset{CH_3}{|}}{Si}\right]_n \quad [3]$$

of polymers obtained by the silanol-acetoxysilane condensation were found to be significantly lower.

4. Properties of Polymers. Since the siloxane-modified poly(arylene carbonates) are of potential interest as specialty elastomers for certain aerospace applications, it is important to achieve as low a value as possible for the glass-transition temperature in order to achieve flexibility at low temperatures in such systems. From Table I it is readily apparent that, of the various dimethyl-substituted polymers investigated, polymer 7 best meets this requirement. As would be expected, the value of the glass-transition temperature for the polymers listed in the table decreases with the increasing length of the siloxanylene segments and with the decreasing length of the arylene carbonate segment. Furthermore, polymers containing *m*-phenylene groups in the backbone have a considerably lower value for the glass-transition temperature than the analogous polymers containing *p*-phenylene groups. For this reason, in further investigation of the effect of varying pendant groups (synthesis of polymers 8-12) efforts were concentrated on polymers containing *m*-phenylene groups. Introduction of pendant phenyl groups on the polymer backbone increases the glass-transition temperature (T_g) considerably, particularly when the phenyl groups are introduced on silicon atoms attached to the phenylene moiety. This increase of the T_g is apparently caused by limited flexibility of the siloxanylene segments containing pendant phenyl groups, as is also evidenced from nmr spectra.

At the other end of the temperature spectrum, with high thermal stability of siloxane-modified poly(arylene carbonates) also a desired property, the onset of thermal decomposition (40) for polymers 1-12 was found to be in the range of 385-456°C (as determined from TGA curves obtained by heating polymer samples in nitrogen at a heating rate of 20°C/min.). There does not appear to be any pronounced trend in regard to variation of the thermal stability with structure in polymers 1-12. The small differences in the values of T_d for these polymers can be due to differences in molecular weights or to the possible presence of trace impurities.

Data on the analytical characterization of polymers (i.e., the elemental analyses and nmr spectral characteristics) are summarized in Table II.

Experimental

1. General. Intrinsic viscosities, [η], were measured on a Cannon-Ubbelohde dilution viscometer at 30°C in THF. The GPC analyses were carried out using a Waters Model 6000A liquid chromatograph, equipped with a UV detector and a set of micro-styragel 30-cm columns (10^5, 10^4, 10^3,500, 500 and 100 Å). Polymers were analyzed in THF solutions and the values of $\bar{M}_w$ were determined by the universal method, using polystyrene standards. Glass-transition temperatures of polymers were determined by differential scanning calorimetry (DSC) in an atmosphere of nitrogen at a heating rate of 20°C/min. The thermogravimetric analyses were carried out in a duPont Model 950 thermogravimetric analyser at a heating rate of 20°C/min. and at a helium flow-rate of 60 ml/min. Synthesis of monomers will be described in forthcoming publications.

2. Polycondensation of Bisphenols, II, with Phosgene. A freshly-prepared solution of phosgene (8 mmole) in 10 ml of methylene chloride was added dropwise to a stirred solution of 5 mmole of type II bisphenol in 2 ml of dry pyridine and 30 ml of methylene chloride. After the mixture was stirred for seven hours, the resulting solution was extracted with 50 ml of 10% HCl and then several times with distilled water until neutral. The solution was dried over anhydrous calcium sulfate, concentrated under reduced pressure, and the polymer precipitated by methanol. The polymer was redissolved in methylene chloride, reprecipitated with methanol, and dried overnight at 100°C (0.01 mm Hg).

3. Homopolycondensation of Bis-silanols, III. Into a flask equipped with a magnetic stirring bar and a gas inlet adapter tube extending just above the surface of the reaction mixture was placed 10 mmole of a type III bis-silanol dissolved in 30 ml of freshly-distilled pyridine. The flask was cooled by a cold water bath and phosgene introduced slowly while the reaction mixture was stirred vigorously. A precipitate was formed initially and, after 15 minutes of continued phosgene addition, the mixture became quite viscous indicating that the reaction was essentially complete. Excess phosgene was removed by sweeping the system with dry nitrogen. The reaction mixture was diluted with 20 ml of methylene chloride, filtered and the resulting polymer product was isolated by precipitation with methanol as described above.

TABLE II

ANALYTICAL DATA ON POLY(SILOXANYLENEARYLENE CARBONATES)

Poly. No.	C, % Calcd.	C, % Found	H, % Calcd.	H, % Found	Si, % Calcd.	Si, % Found	^1H NMR Chemical Shifts* δ_a	δ_b	arom.
1	72.37		6.94		9.67		s 0.31(12)	---	m 7.0-7.6(16)
2	59.30		5.81		16.28		s 0.33(12)	---	m 7.2-7.5(8)
3	54.51	54.12	6.62	6.47	20.13	20.26	s 0.33(12)	s 0.06(6)	m 7.3-7.5(8)
4	51.18	51.32	6.55	6.42	22.79	22.92	s 0.33(12)	s 0.07(12)	m 7.3-7.6(8)
5	59.30		5.81		16.28		s 0.33(12)	---	m 7.2-7.5(8)
6	54.51	54.58	6.26	6.52	20.13	20.17	s 0.35(12)	s 0.07(6)	m 7.2-7.5(8)
7	51.18	51.07	6.55	6.46	22.79	22.79	s 0.37(12)	s 0.07(12)	m 7.2-7.5(8)
8	60.35	60.71	5.88	5.99	18.21	18.47	m 0.34-0.23(18)		m 7.2-7.6(18)
9	60.35	60.57	5.88	5.89	18.21	18.26	s 0.59(6)	s 0.01(12)	m 7.1-7.6(18)
10	66.45	66.37	5.44	5.29	15.16	15.44	m 0.55(6)	m 0.13(6)	m 7.0-7.6(28)
11	60.60	60.62	5.89	6.08	18.65	18.66	m 0.34-0.22(21)		m 7.2-7.6(23)
12	65.71	65.88	5.51	5.41	16.01	16.19	m 0.55(6)	m 0.1-0.2(9)	m 7.0-7.6(33)

*Reported as ppm downfield from TMS internal standard with a and b representing assignment of protons in $SiCH_3$ (a: R^1, R^2, = CH_3; b: R^3, R^4 = CH_3) and arom the protons in m-phenylene and phenyl groups.

4. Polycondensation of Bis-silanols, III, with Diacetoxysilanes, Diacetoxydisiloxanes or Diacetoxytrisiloxanes. A mixture of 12.00 mmole of a bis-silanol, III, 12.00 mmole of a diacetoxysilane, diacetoxydisiloxane or diacetoxytrisiloxane, 4.4 ml of collidine, and 10 ml of toluene was magnetically stirred at room temperature in a sealed flask in an atmosphere of dry nitrogen for two days. The solution was then either allowed to stand for an additional four days or was refluxed for four hours under a positive pressure of dry nitrogen. The resulting polymer product was isolated by precipitation with methanol as described above.

Acknowledgment

The financial support of two of us (B.D.N. and C.A.K.) by the Air Force Systems Command through a National Research Council Postdoctoral Research Associateship is acknowledged. We also wish to thank Mr. D. Crawshaw, Mr. W. Price and Mrs. N.K. Ngo for assistance with experiments.

Literature Cited

1. Bottenbruch, L., "Polycarbonates", in "Encyclopedia of Polymer Science and Technology", Mark, H.F. and Gaylord, N.G., Eds. John Wiley: New York, N.Y., 1969; Vol. 10, p. 710.
2. Kambour, R.P. Polymer Preprints,1969; 10 [No.2], 885.
3. Vaughn, H.A.,Jr. J. Polym. Chem., Polym. Letters Ed., 1969, 7, 569.
4. Vaughn, H.A., Jr. Amer Chem. Soc., Div. Org. Coatings Plast. Chem., Papers, 1969, 29, 133; Chem. Abstr., 1971, 74, 4096.
5. Matzner, M.; Noshay, A. U.S. Patent 3,579,607 (1971).
6. Union Carbide Co. French Patent 2,076,507 (1971); Chem. Abstr., 1972, 77, 35981.
7. Noshay, A.; Matzner, M.; Williams, T.C. Ind. Eng. Chem., Prod. Res. Develop., 1973, 12, 268.
8. Buechner, W.; Noll. W.; Bressel, B. German Patent 2,162,418 (1973); Chem. Abstr., 1973, 79, 92984.
9. Bayer A.-G. French Patent 2,163,700 (1973); Chem. Abstr., 1974, 80, 60397.
10. Merritt, W.D., Jr. U.S.Patent 3,821,325 (1974); Chem. Abstr., 1974, 81, 136962.
11. Merritt, W.D., Jr. U.S. Patent 3,832,419 (1974); Chem. Abstr., 1974, 81, 153605.
12. Raigorodskiy, I.M.; Bakhayeva, G.P.; Makarova, L.I.; Savin, V.A.; Andrianov, K.A. Vysokomol. Soedin., 1975, A17, 84.
13. Holdschmidt, N.; Bressel, B.; Buechner, W.; De Montigy, A. German Patent 2,343,275 (1975); Chem. Abstr., 1975, 83, 11389.
14. Kambour, R.P.; Corn, J.E.; Miller, S.; Niznik, G.E. J. Appl. Polym. Sci., 1976, 20, 3275.

15. Kambour, R.P.; Corn, J.E.; Klopfer, H.; Miller, S.; Orlando, C.M. U.S. Gov. Report AD-A041-087 (1977); Chem. Abstr., 1978, 88, 7723.
16. Friedrich, T.K.; Maass, G.; Beck, M. German Patent 2,555,746 (1977); Chem. Abstr., 1977, 87, 54061.
17. Kambour, R.P. J. Polym. Chem., Polym. Letters Ed., 1969, 7, 573.
18. LeGrand, D.G. J. Polym. Sci., Polym. Letters Ed., 1969, 7, 579.
19. LeGrand, D.G. and Gaines, G.L., Jr. Polym. Preprints, 1970, 11 [No.2], 442.
20. Magilla, T.L.; LeGrand, D.G. Polym. Eng. Sci., 1970, 10, 349.
21. Kambour, R.P. in "Block Copolymers", S.L. Aggarwal, Ed. Plenum: New York, N.Y., 1970; p.263.
22. Narkis, M.; Tobolsky, A.V. J. Macromol. Sci., Phys. Ed., 1970, 4B, 877.
23. Gaines, G.L., Jr. J. Polym. Sci., Part C, 1971, 34, 115.
24. LeGrand, D.G. Trans. Soc. of Rheology, 1971, 15, 541; Chem. Abstr., 1971, 75, 141290.
25. LeGrand, D.C. J. Polym. Sci., Polym. Letters Ed., 1971, 9, 145.
26. Kaniskin, V.; Kaya, A.; Ling, A.; Shen, M. J. Appl. Polym. Sci., 1973, 17, 2605.
27. Smirnova, O.V.; Klenova, T.S.; Khatuntsev, G.D; Shedulyakov, V.D.; Mironova, N.V.; Sebernikova, A.I. Vysokomol. Soedin., 1975, 16A, 1940.
28. Stefan, D.; Williams, H.L. J. Appl. Polym. Sci., 1974, 18, 1451.
29. Raigorodskiy, I.M.; Lebedev, V.P.; Savin, V.A.; Bakhayeva, G.P. Vysokomol. Soedin., 1975, 17A, 1267.
30. Kambour, R.P.; Ligon, W.V.; Russell, R.R. J. Polym. Sci., Polym. Let. Ed., 1978, 16, 327.
31. Niznik, G.E.; LeGrand, D.G. J. Polym. Sci., Polym. Symp., 1978, 60, 97.
32. Allport, D.C. in "Block Copolymers", Allport, D.C. and Janes, W.H., Eds. John Wiley: New York, N.Y., 1973; p.532.
33. General Electric Co. Netherlands Patent Appl. 7,407,871 (1974); Chem. Abstr., 1975, 83, 132868.
34. Stark, L.; Auslander, D.M.; Mandell, R.B.; Marg, E. French Patent 2,185,653 (1974); Chem. Abstr., 1974, 81, 50730.
35. Lloyd, N.C.; Pearle, C.A.; Pattison, J. U.S. Patent 3,595,974 (1971); Chem. Abstr., 1971, 75, 152338.
36. Greber, G.; Lohman, D. Angew. Chem., Internat. Ed., 1967, 6, 462.
37. Greber, G. Angew. Makromol. Chem., 1969, 4/5, 212.
38. Greber, G. J. Prakt. Chem., 1971, 313, 461.
39. Sheludyakov, V.D.; Mkhitaryan, S.S.; Gorlov, E.G.; Zhinkin, D.Ya. USSR Patent 550,407 (1977); Chem. Abstr., 1977, 86, 172154.

40. Ehlers, G.F.L. Air Force Materials Laboratory Technical Report No. AFML-TR-75-202, Part I, AD-A019-957 (1975).
41. Burks, R.E.; Covington, E.R.; Jackson, M.V. and Curry, J.E. J. Polym. Sci., Polym. Chem. Ed., 1973, 11, 319.
42. Breed, L.W.; Elliott, R.L. and Whitehead, M.E. J. Polym. Sci., Part A-1, 1967, 5, 2745.
43. Rosenberg, H. and Nahlovsky, B.D. Polym. Preprints, 1978, 19, [No.2], 625.
44. Ehlers, G.F.L. Air Force Materials Laboratory Technical Report No. 74-177, Part I , AD-A008-187 (1974).

RECEIVED July 16, 1979.

Polymer–Filler Composites thru In Situ Graft Copolymerization: Polyethylene–Clay Composites

NORMAN G. GAYLORD, HANS ENDER, LINWOOD DAVIS, JR., and AKIO TAKAHASHI

Gaylord Research Institute, New Providence, NJ 07974

The use of inorganic additives as extenders in thermoplastic polymers is a long established practice. In recent years, the role of such additives has changed from that of cost-reducing fillers to property enhancing reinforcing agents. This conversion has come about as a result of the compatibilization of the additive with the thermoplastic polymer, by interaction at the polymer-filler interface.

The interaction of the polymer with the filler is promoted by the presence of reactive functionality in the polymer, capable of chemical reaction or hydrogen bonding with the functionality, generally hydroxyl, on the surface of the filler. Thus, carboxyl-containing polymers, e.g. ethylene-acrylic acid copolymers and maleic anhydride- and acrylic acid-grafted polyethylene and polypropylene interact readily with fillers.

Another route to such interaction is graft copolymerization of an unsaturated trialkoxysilane, e.g. methacrylatopropyl- or vinyltrialkoxysilane, or alkoxytitanate with the polyolefin, prior to or concurrent with compounding with the filler. An alternative route is pretreatment of the filler with the silane or titanate so that the pendant unsaturation reacts with the radical sites generated on the polymer during compounding.

In the present paper, we report on the compatibilization of clay with polyolefins, specifically low and high density polyethylenes (LDPE, HDPE), through the radical-induced polymerization of maleic anhydride (MAH) in the presence of the polymer and clay, so that the MAH is grafted on the PE and the anhydride groups concurrently react with the filler surface (1, 2).

Experimental

Two procedures were used to prepare compatibilized polymer-filler composites:

One-step process, wherein the filler was compounded into the polymer with the desired degree of loading, in the presence of MAH and a peroxide;

0-8412-0540-X/80/47-121-469$05.00/0

Two-step process, wherein the filler was compounded into the polymer at high levels of loading, e.g. 50/50-90/10 clay/PE, in the presence of MAH and a peroxide, and the resultant concentrate then compounded with additional PE to the desired final degree of loading.

The peroxide catalyzed polymerization of MAH in the presence of PE and clay was carried out in a Brabender Plasticorder (Model PL-V151, Roller No. 6/RH/EH mixer) at 75 rpm at a temperature of 130-165°C. The total charge was 40-60 g and one-half of the PE was charged initially and allowed to flux. The MAH and peroxide were premixed. After one-fifth of the clay was added, one-fifth of the MAH-peroxide mixture was added, followed by one-fifth of the remaining PE. This sequence was repeated four times (total charging time 3-4 min) and, after the torque indicated complete fluxing (about 1 min after charging), mixing was continued for 20 min.

In the two-step process, the polymer-treated clay, i.e. 70/30-90/10 clay/PE, prepared as described above, was added alternately with PE in five increments to one-half of the PE initially charged to the mixing chamber.

Tensile properties and flexural modulus were determined on 20 mil pressed films in accordance with ASTM D882 and D790, respectively. Izod (notched) impact strength was determined on compression molded plaques in accordance with ASTM D256. Water vapor transmission was determined on 5 mil pressed films in accordance with ASTM E96.

Results and Discussion

The coupling of PE and clay by means of MAH is accomplished by simultaneously mixing the reactants in the presence of a free radical catalyst. The compatibilization presumably involves the reaction of MAH with PE to form an MAH-grafted PE, followed by reaction of the pendant anhydride groups with the hydroxyl groups on the clay. The bonding between the MAH moieties and clay hydroxyls takes the form of covalent ester linkages and/or hydrogen bonds.

~$CH_2CH_2CH_2$~   HC=CH   O=C–O–C=O   H–O (clay surface)   —R·→   ~CH_2CHCH_2~ / HC—CH- / O=C C=O / O OH (clay surface)   +   ~CH_2CHCH_2~ / HC—CH- / O=C C=O / OH OH / H–O (clay surface)

In the one-step process, the reactants are mixed in the proportions desired in the final composite. In the two-step process, the polymer-coated clay prepared in the first step is compounded with additional PE. The two-step process yields composites with better mechanical properties than the one-step process.

The temperature for the PE-MAH-clay reaction is generally 150°C. However, in the case of LDPE, the second step in the two-step process may be conducted at 130°C.

Solomon (3, 4, 5) reported that various clays inhibited or retarded free radical reactions such as thermal and peroxide-initiated polymerization of methyl methacrylate and styrene, peroxide-initiated styrene-unsaturated polyester copolymerization, as well as sulfur vulcanization of styrene-butadiene copolymer rubber. The proposed mechanism for inhibition involved deactivation of free radicals by a one-electron transfer to octahedral aluminum sites on the clay, resulting in a conversion of the free radical, i.e. catalyst radical or chain radical, to a cation which is inactive in these radical initiated and/or propagated reactions.

An evaluation of numerous clays, including kaolins and bentonites of comparable particle size and distribution, revealed a wide variation in the properties of LDPE-clay composites prepared under identical conditions with similar loadings. This may be attributed to the interference with the radical reactions involved in the coupling sequence. This is confirmed, in part, by the finding that the most effective clays were those which were reported to have been treated with sodium polyphosphate to improve their dispersibility in water during papermaking processes. Solomon reported (4) that treatment of the clays which inhibited radical reactions with sodium polyphosphate reduced the inhibition.

The "ene" reaction of MAH with a polymer containing unsaturation occurs at elevated temperatures in the absence of a catalyst. The presence of a radical catalyst is necessary to promote the reaction between MAH and a saturated polymer. However, the mere generation of radical sites on the polymer, e.g. by shear or attack by catalyst radicals, is insufficient to promote reaction to a significant extent. Thus, the appendage of MAH onto PE requires the use of either high concentrations of catalyst or the use of a catalyst which has a short half-life at the reaction temperature.

These conditions are identical to those required for the homopolymerization of MAH, i.e. the catalyst has a half-life of up to about 30 min at the reaction temperature (6). Thus, t-butyl perbenzoate (tBPB) is effective in the preparation of LDPE-clay composites at 130° and 150°C and HDPE-clay composites at 150°C while dicumyl peroxide is less effective in LDPE-clay composites at 130°C than at 150°C.

The attempted polymerization of MAH in mineral oil with various catalysts demonstrated that those catalysts which were most effective in promoting MAH polymerization were most effective in the preparation of PE-clay composites at the same temperature. Further, those clays which gave composites suggesting the absence of compatibilization were found to interfere with the homopolymerization of MAH when present in the MAH-mineral oil mixture.

The optimum concentration of MAH in the preparation of a clay/PE concentrate was generally 10-20 weight-% based on the PE, while the catalyst concentration was 2-5 weight-% based on the PE. The necessity to use a catalyst which initiates the homopolymerization of MAH at the mixing temperature, suggests that the grafted MAH is present as a polymeric or oligomeric sequence rather than as a single unit. It has already been suggested (7, 8, 9) that in the grafting of MAH on PE in the presence of a free radical catalyst, poly(maleic anhydride) rather than monomeric MAH is the species which is attacked by free radicals to generate thereon radical sites which couple with radical sites generated on the PE by radical attack. However, it is our belief that propagating poly(maleic anhydride) couples with radical sites on the PE, i.e. grafting occurs as a result of termination of propagating chains onto PE.

Since MAH has a low melting point and readily sublimes, it is sometimes advantageous to use maleic acid. The latter is a high melting solid with little tendency to sublime and is converted to MAH above 140°C, i.e. during the processing. Poly(maleic anhydride) may also be used in lieu of MAH, albeit not as effectively. In this case, the mechanism involving radical attack on both the former and PE is probably operative.

The reaction between MAH, PE, clay and the catalyst must take place essentially simultaneously. If the clay is first reacted with MAH at elevated temperatures, analogous to the treatment of clay with silane coupling agents, a clay maleate half ester or diester is probably formed. The latter does not readily react with PE even in the presence of a free radical catalyst. If the catalyst is allowed to react prior to the addition of MAH, crosslinking of the PE occurs, yielding a brittle composite with poor flow properties. In the preferred procedure, incremental and sequential additions of an MAH-peroxide mixture and clay are made to the molten PE.

The effect of the components and conditions of preparation on the properties of a 70/30 LDPE/clay composite is shown in Table I. The 10/90 mixture of LDPE Bakelite Polyethylene Resin DYNH-1 (Union Carbide Corp.) and Hydrite 10 clay (Georgia Kaolin Co.) was compounded at 150°C in the Brabender Plasticorder in the presence of MAH and/or t-butyl perbenzoate (tBPB). The PE-coated clay was then mixed with additional DYNH-1 LDPE at 130°C to yield a 70/30 PE/clay composite. A 30/70 PE/clay concentrate was prepared in a similar manner at 150°C and converted to a 70/30 PE/clay composite at 130°C. The 10/90 PE/clay concentrate is an easily handled, clay-like product while the 30/70 concentrate is a tougher lumpy material.

Table I
Compatibilized 70/30 LDPE/Clay (DYNH-1/Hydrite 10) Composites

Concentrate				Mechanical Properties			
Clay pts	LDPE pts	MAH pts	tBPB pts	Yield Strength psi	Break Strength psi	Elongation %	Flexural Modulus psi
45	5	1	0.25	1700	1700	97	42,800
45	5	5	1.00	1840	1840	36	51,600
45	5	0	1.00	1590	1590	44	41,600
45	5	5	0	1760	1760	57	46,900
35	15	0.93	0.23	1810	1807	95	47,800

The 30/70 PE/clay concentrate yields 70/30 composites having slightly better properties at a lower MAH-tBPB concentration, based on PE, than does the 10/90 PE/clay concentrate.

The influence of the various components of a HDPE/clay composite on the properties is shown in Table II. A 30/70 PE/clay concentrate was prepared from 15 parts HDPE Fortiflex A60-70R (Allied Chemical Corp.) and 35 parts of Hydrite 10 by mixing at 150°C in the presence of 3 parts of maleic acid (MA) and 0.75 parts of tBPB. The concentrate was then blended at 150°C with additional HDPE to yield a 50/50 HDPE/clay composite.

Table II
Compatibilized 50/50 HDPE/Clay (Fortiflex A60-70R/Hydrite 10) Composites

Components				Mechanical Properties			
HDPE	Clay	MA	tBPB	Break Strength psi	Elongation %	Flexural Modulus psi	Izod Impact (Notched) ft-lb/in
+	-	-	-	4070	750	201,000	2.1
+	+	-	-	4040	2	447,000	0.3
+	+	-	+	4430	2	432,000	0.4
+	+	+	-	4780	3	434,000	0.5
+	+	+	+	5160	3	499,000	2.8

The PE-MAH-clay composite prepared by the in situ polymerization of MAH in the presence of PE and clay actually contains three phases:

(1) clay particles
(2) crosslinked and MAH-grafted PE bonded to the clay particles
(3) uncrosslinked PE

The modulus increases on going from the uncrosslinked matrix PE through the crosslinked bonded PE to the clay core. The effect of such a gradient of modulus on impact strength and other properties has previously been noted in kaolin-reinforced HDPE(10).

The level of clay loading dramatically influences the mechanical properties of HDPE Fortiflex A60-70R/Hydrite 10 clay composites prepared from MAH-tBPB coupled 30/70 PE/clay concentrates. The 30/70 HDPE/clay mixture was prepared either in the absence of MAH and tBPB or in the presence of 20% MAH and 5%tBPB, based on PE, at 150°C and then compounded with additional HDPE to yield the final PE/clay composites shown in Table III.

Table III
Effect of Clay Loading in HDPE/Clay
(Fortiflex A60-70R/Hydrite 10) Composites

Composite PE/Clay wt. ratio	MAH-tBPB	Mechanical Properties: Yield Strength psi	Break Strength psi	Elongation %	Flexural Modulus psi
100/0[a]	-	4270	2840	222	198,000
80/20	-	-	4130	5	239,000
80/20	+	4465	4380	9	252,000
70/30	-	-	4140	7	316,000
70/30	+	4870	4800	7	292,000
50/50	-	-	4040	2	447,000
50/50	+	5500	5540	6	413,000
30/70[b]	-	-	-	-	-
30/70	+	-	4200	1	633,000

[a]Mixed in Plasticorder at 150°C for 20 min in absence of clay, MAH and tBPB

[b]Poor flow properties and poor film formation; no properties determined

The maximum yield and break strengths were attained in 50/50 HDPE/clay compositions, e.g. the break strength increased 105% and the yield strength increased 29% over the values for unfilled HDPE. In the absence of the coupling components, untreated clay gave compositions with no yield strength, and the break strengths were independent of loading from 30/70 to 80/20 PE/clay and were 45% greater than the value for unfilled HDPE. The flexural modulus of both treated and untreated clay filled compositions increased with extent of loading and was only slightly changed by the coupling process.

The water vapor transmission of compatibilized PE/clay composites are lower than those of the unfilled PE and PE/clay mixtures and do not increase until the clay content is greater than 40%. The values are similar to those of crosslinked PE, in the absence of clay, and suggest that the clay acts as a crosslinking agent for the PE.

Summary

The compatibilization of clay with LDPE and HDPE is accomplished by the in situ polymerization of MAH or its precursor maleic acid, in the presence of a radical catalyst. The latter must be capable of initiating the homopolymerization of MAH, i.e. it must be present in high concentration and/or have a half-life of less than 30 min at the reaction temperature, e.g. t-butyl perbenzoate (tBPB) at 150°C. In a one-step process, the clay and PE are mixed with MAH-tBPB in the desired PE/clay ratio. In the preferred two-step process, a 70/30-90/10 clay/PE concentrate is prepared initially in the presence of MAH-tBPB and then blended with additional PE to the desired clay loading. The compatibilized or coupled PE-MAH-clay composites have better physical properties, including higher impact strengths, than unfilled PE or PE-clay mixtures prepared in the absence of MAH-tBPB.

Literature Cited

1. Gaylord, N.G. (to Champion International Corp.), U.S. Patent 3,956,230 (May 11, 1976).
2. Gaylord, N.G. (to Champion International Corp.), U.S. Patent 4,071,494 (January 31, 1978).
3. Solomon, D.H., and Rosser, M.J., J. Appl. Polym. Sci. (1965), 9, 1261.
4. Solomon, D.H., and Swift, J.D., J. Appl. Polym. Sci. (1967), 11, 2567.
5. Solomon, D.H., and Loft, B.C., J. Appl. Polym. Sci. (1968), 12, 1253.
6. Gaylord, N.G., and Maiti, S., J. Polym. Sci., Polym. Lett. Ed. (1973) 11, 253.
7. Gabara, W., and Porejko, S., J. Polym. Sci. A-1 (1967), 5, 1539.
8. Gabara, W., and Porejko, S., J. Polym. Sci. A-1 (1967), 5, 1547.
9. Porejko, S., Gabara, W., and Kulesza, J., J. Polym. Sci. A-1 (1967), 5, 1563.
10. Fallick, G.J., Bixler, H.J., Marsella, R.A., Garner, F.R., and Fettes, E.M., Mod. Plastics (1968), 45 (January, No. 5), 143.

RECEIVED July 12, 1979.

INDEX

A

B

D

F

G

H

I

K

N

U

V

W

X

Z